Environmental History

Volume 5

Series editor

Mauro Agnoletti, Florence, Italy

More information about this series at http://www.springer.com/series/10168

Mauro Agnoletti · Francesca Emanueli
Editors

Biocultural Diversity in Europe

Springer

Editors
Mauro Agnoletti
GESAAF
University of Florence
Florence
Italy

Francesca Emanueli
GESAAF
University of Florence
Florence
Italy

ISSN 2211-9019 ISSN 2211-9027 (electronic)
Environmental History
ISBN 978-3-319-26313-7 ISBN 978-3-319-26315-1 (eBook)
DOI 10.1007/978-3-319-26315-1

Library of Congress Control Number: 2015956375

Printed on acid-free paper

This Springer imprint is published by SpringerNature
The registered company is Springer International Publishing AG Switzerland

Contents

Part II Management and Conservation Approaches

Contributors

Mauro Agnoletti CULTLAB—Laboratory for Landscape and Cultural Heritage, University of Florence, Florence, Italy

Eduardo Aguilera Agro-ecosystems History Laboratory, University Pablo de Olavide, Seville, Spain

Miguel A. Altieri Department of Environmental Science, Policy, and Management, University of California, Berkeley, CA, USA

Martin Arnold Institut für Sächsische Geschichte und Volkskunde, Dresden, Germany

Giuseppe Barbera Dipartimento Scienze Agrarie e Forestali, Università di Palermo, Palermo, Italy

Boris Beláček Landscape Protected Area Poľana, Zvolen, Slovakia

Barbara Bożętka Institute of Geography, Gdańsk University, Gdańsk, Poland

Robert Gerald Henry Bunce Institute of Agricultural and Environmental Sciences, Estonian University of Life Sciences, Tartu, Estonia

Francesca Camilli Institute of Biometeorology, National Research Council, IBIMET-CNR, Florence, Italy

Laura Cassi Department of SAGAS, University of Florence, Florence, Italy

Aurora Cavallo DiBT University of Molise, Campobasso, Italy

Csaba Centeri Faculty of Environmental and Agricultural Sciences, Department of Nature Conservation and Landscape Ecology, Institute of Nature Conservation and Landscape Management, Szent István University, Gödöllő, Hungary

Antonio Cid Agro-ecosystems History Laboratory, University Pablo de Olavide, Seville, Spain

Tim Collins Collins and Goto Studio, Glasgow, Scotland, UK

Sebastiano Cullotta Dipartimento Scienze Agrarie e Forestali, Università di Palermo, Palermo, Italy

Barbara Czesak Department of Land Management and Landscape Architecture, University of Agriculture in Kraków, Kraków, Poland

Stanislav Dávid Institute of Landscape Ecology, SAS, Nitra, Slovak Republic

Catherine Dezio University La Sapienza, Rome, Italy

Sara Di Lonardo Institute of Biometeorology-National Research Council (IBIMET-CNR), Florence, Italy

Robert Dixon-Gough Department of Land Management and Landscape Architecture, University of Agriculture in Kraków, Kraków, Poland

Marta Dobrovodská Institute of Landscape Ecology, Slovak Academy of Sciences, Bratislava, Slovak Republic

Pablo Domínguez Laboratory of Socio-Ecological Systems in the Globalization (LASEG), Institut de Ciencia i Tecnologia Ambientals (ICTA), Universitat Autònoma de Barcelona, Bellaterra (Cerdanyola del Vallès), Spain; Departament de Antropologia Cultural i Història d'Amèrica i Àfrica, Universitat de Barcelona, Barcelona, Spain

Rebekka Dossche Laboratory of Archaeology and Environmental History (LASA), Department of Antiquities, Philosophy and History (DAFIST), University of Genoa, Genoa, Italy; Department of Agriculture and Society, Institute for Agricultural and Fisheries Research (ILVO), Merelbeke, Department of Geography, Belgium and Ghent University, Ghent, Belgium

Johannes Dreer Hof und Leben GmbH, Unternehmensberatung für Landwirte, Johannes Dreer, Kirchdorf, Germany

David Edwards Social and Economic Research Group, Forest Research, Roslin, Scotland, UK

Sebastian Eiter NIBIO—Norwegian Institute of Bioeconomy Research, Ås, Norway

Francesca Emanueli CULTLAB—Laboratory for Landscape and Cultural Heritage, University of Florence, Florence, Italy

Vladimíra Fabriciusová Landscape Protected Area Pol'ana, Zvolen, Slovakia

Bruno Foggi Department of BIO, University of Florence, Florence, Italy

Fabrizio Frascaroli Religion and Nature Group, Department of Religion, University of Florida, Gainesville, FL, USA; Institute of Evolutionary Biology and Environmental Studies, University of Zurich, Zurich, Switzerland

Federica Frondizi Rignano sull'Arno, Florence, Italy

Peter Gajdoš Institute of Landscape Ecology, SAS, Nitra, Slovak Republic

Jean Gault Ingénieur général des ponts des eaux et des forêts, ministère français de l'agriculture, de l'alimentation, et de la forêt, CGAAER, Paris, Cedex 15, France

Manuel González de Molina Agro-ecosystems History Laboratory, University Pablo de Olavide, Seville, Spain

Reiko Goto Collins and Goto Studio, Glasgow, Scotland, UK

Andrew Paul Gutierrez Center for the Analysis of Sustainable Agricultural Systems, Kensington, CA, USA; College of Natural Resources, University of California, Berkeley, CA, USA

Gloria Guzmán Agro-ecosystems History Laboratory, University Pablo de Olavide, Seville, Spain

Ľuboš Halada Institute of Landscape Ecology, SAS, Nitra, Slovak Republic

John M. Halley Department of Biological Applications and Technology, University of Ioannina, Ioannina, Greece

Robert Hearn Laboratory of Archaeology and Environmental History (LASA), Department of Antiquities, Philosophy and History (DAFIST), University of Genoa, Genoa, Italy

Józef Hernik Department of Land Management and Landscape Architecture, University of Agriculture in Kraków, Kraków, Poland

Juan Infante-Amate Agro-ecosystems History Laboratory, University Pablo de Olavide, Seville, Spain

Peter Jančura Technical University in Zvolen, Zvolen, Slovakia

Jadran Kale Department of Ethnology and Anthropology, University of Zadar, Zadar, Croatia

Jutta Kapfer NIBIO—Norwegian Institute of Bioeconomy Research, Tromsø, Norway

Tony Knight School of Anthropology and Conservation, University of Kent, Canterbury, UK

Jana Krčmářová Faculty of Humanities, Department of Common Anthropology, Charles University, Prague, Czech Republic

Alexandra Kruse Institute for Research on European Agricultural Landscapes e. V, Overath, Germany

Zdeněk Kučera Faculty of Science, Department of Social Geography and Regional Development, The Research Centre for Cultural and Historical Geography, Charles University in Prague, Praha 2, Czech Republic

Davide Marino DiBT University of Molise, Campobasso, Italy

Vasilis Nitsiakos Department of History and Archeology, University of Ioannina, Ioannina, Greece

Luca Ongaro IAO (Istituto Agronomico per l'Oltremare), Florence, Florence, Italy

Maria Pazdan Department of Land Management and Landscape Architecture, University of Agriculture in Kraków, Kraków, Poland

Sandro Pieroni Unione dei comuni della Garfagnana, Castelnuovo di Garfagnana, Lucca, Italy

Luigi Ponti Laboratorio Gestione Sostenibile degli Agro-ecosistemi, Unità Tecnica Agri, ENEA Centro Ricerche Casaccia, Rome, Italy; Center for the Analysis of Sustainable Agricultural Systems, Kensington, CA, USA

Zlatica Prídavková Technical University in Zvolen, Zvolen, Slovakia

Hans Renes Faculty of Geosciences, Utrecht University, Utrecht, The Netherlands; VU University Amsterdam, Amsterdam, The Netherlands

Michael Roth School of Landscape Architecture, Urban and Environmental Planning, Nürtingen-Geislingen University, Nürtingen, Germany

Seddik Saïdi Chef de Départernent Amélioration et Conservation des Ressources Génétiques, Division Scientifique, INRA Maroc, BP 6356 Rabat Institut, Rabat, Morocco

Dénes Saláta Faculty of Environmental and Agricultural Sciences, Department of Nature Conservation and Landscape Ecology, Institute of Nature Conservation and Landscape Management, Szent István University, Gödöllő, Hungary

Emma Salizzoni Interuniveristy Department of Regional and Urban Studies and Planning, Politecnico and Università di Torino, Turin, Italy

Antonio Santoro Laboratory for Landscape and Cultural Heritage, University of Florence, Florence, Italy

Markéta Šantrůčková Silva Tarouca Research Institute for Landscape and Ornamental Gardening, Průhonice, Czech Republic

Bianca Maria Seardo DIST, Dipartimento di Scienze, Progetto e Politiche del Territorio, Politecnico e Università di Torino, Turin, Italy

Kalev Sepp Institute of Agricultural and Environmental Sciences, Estonian University of Life Sciences, Tartu, Estonia

Maria Adele Signorini Department of BIO, University of Florence, Florence, Italy

Maurizia Sigura Department of Agriculture and Environmental Sciences, University of Udine, Udine, Italy

Martina Slámová Department of Landscape Planning and Design, Technical University in Zvolen, Zvolen, Slovakia

Mateja Šmid Hribar Scientific Research Centre of the Slovenian Academy of Sciences and Arts, Anton Melik Geographical Institute, Ljubljana, Slovenia

Jana Špulerová Institute of Landscape Ecology, SAS, Bratislava, Slovak Republic

Kalliopi Stara Department of Biological Applications and Technology, University of Ioannina, Ioannina, Greece

Dagmar Štefunková Institute of Landscape Ecology, Slovak Academy of Sciences, Bratislava, Slovak Republic

Giacomo Tagliaferri Institute of Biometeorology-National Research Council (IBIMET-CNR), Florence, Italy

Pille Tomson Institute of Agricultural and Environmental Sciences, Estonian University of Life Sciences, Tartu, Estonia

Eva Torremocha Agro-ecosystems History Laboratory, University Pablo de Olavide, Seville, Spain

Rigas Tsiakiris Forestry Service of Ioannina, Ioannina, Greece

Mimi Urbanc Scientific Research Centre of the Slovenian Academy of Sciences and Arts, Anton Melik Geographical Institute, Ljubljana, Slovenia

Anna Varga Centre for Ecological Research, Hungarian Academy of Sciences, Vácrátót, Hungary

Bas Verschuuren Sociology of Development and Change Group, Sub-department of Sociology and Anthropology of Development, Wageningen University, Wageningen, The Netherlands

Inmaculada Villa Agro-ecosystems History Laboratory, University Pablo de Olavide, Seville, Spain

Salvador Villacreces Technical University of Madrid, Madrid, Spain

Martin Weber Silva Tarouca Research Institute for Landscape and Ornamental Gardening, Průhonice, Czech Republic

Katarína Zrníková Technical University in Zvolen, Zvolen, Slovakia

Chapter 1
Biocultural Diversity and Landscape in Europe: Framing the Issue

Mauro Agnoletti and Francesca Emanueli

Abstract The International Conference on Biological and Cultural Diversity held in Montreal on June 2010, produced the Declaration on Biocultural Diversity and the UNESCO-SCBD Joint Programme on the linkages between cultural and biological diversity. The first meeting for the implementation of the Joint Programme was held in Florence (Italy) in April 2014. The scientific and policy dimensions of the linkages between cultural and biological diversity are of utmost importance in Europe where policies are devoted to the conservation of biodiversity and cultural heritage, but rarely focused on the result of interactions between nature and culture expressed by the rural landscape. The Florence Conference gathered scientists from different disciplines considering biocultural diversity as a good example of a topic requiring a transdisciplinary approach not always supported by university and research. This not only for an effective understanding of the biodiversity associated with landscapes shaped by the man, but also for the further development of the Joint Programme in terms of research and political implementation. The meeting was organized into a scientific part and a workshop for the drafting of a declaration on biocultural diversity. The declaration states that the European rural landscape (about 80 % of the European Union territory) is predominantly a biocultural multifunctional landscape, while the current state of biological and cultural diversity in Europe results from the combination of historical and ongoing environmental and land-use processes and cultural heritage. This book shows the existence and the importance of biocultural diversity associated to European landscape. This heritage should be studied, preserved and valorized by public policies.

Keywords Biocultural diversity · Biodiversity · Transdisciplinarity · Biodiversification

---

M. Agnoletti (✉) · F. Emanueli
Laboratory for Landscape and Cultural Heritage, School of Agriculture, University of Florence, Via San Bonaventura 13, Florence, Italy
e-mail: mauro.agnoletti@unifi.it

F. Emanueli
e-mail: francesca.emanueli@unifi.it

M. Agnoletti and F. Emanueli (eds.), *Biocultural Diversity in Europe*,
Environmental History 5, DOI 10.1007/978-3-319-26315-1_1

1.1 The UNESCO—SCBD Joint Programme

The discussion about the linkages between biological and cultural diversity in research started about thirty years ago (see Fig. 1.1), but only recently the topic was brought to the attention of relevant international organizations. The International Conference on Biological and Cultural Diversity was held in Montreal on June 2010 and produced two main documents, the Declaration on Biocultural Diversity and the Joint Programme on the linkages between cultural and biological diversity, developed in collaboration with the UNESCO and the Secretariat of the Convention on Biological Diversity of the United Nations (JP-BiCuD). The joint programme promoted a vision of the world in which the global community sustains biological and cultural diversity, with the Convention on Biological Diversity acting as global focal point for biodiversity and UNESCO acting as global focal point for cultural diversity. The programme strengthens the linkages between biological and cultural diversity initiatives, and enhances synergies between interlinked provisions of conventions and programmes dealing with biological and cultural diversity at relevant scales. The general principle for the implementation of the joint programme concerns a full and effective participation of all relevant actors, and in particular indigenous and local communities in the establishment and implementation of the JP, as also a collaborative engagement of policy and decision-makers, in both rural and urban contexts. Among the specific objectives of the JP, an important one is to build bridges between

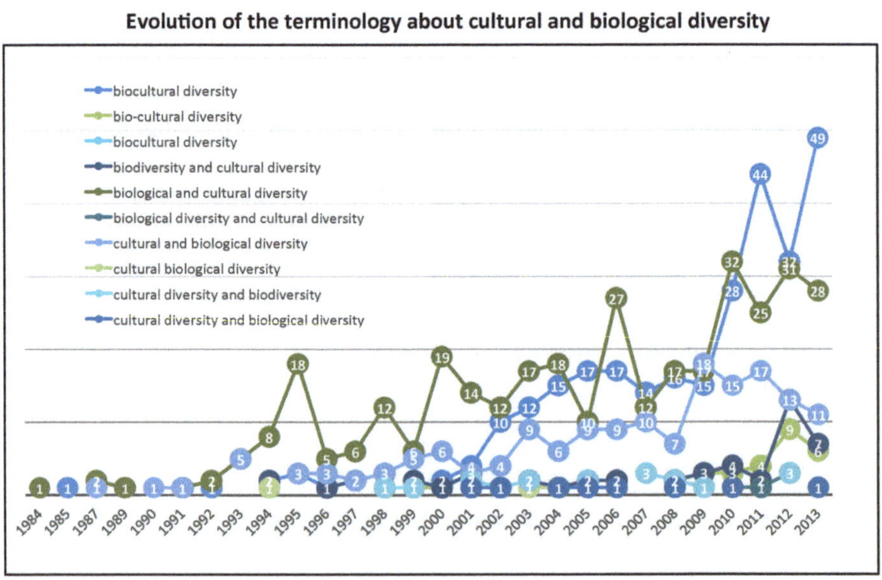

Fig. 1.1 The terminology used in scientific articles in the last decades about cultural and biological diversity put more and more into evidence the term "biocultural diversity". *Source* Science Direct

ongoing work on biodiversity and cultural diversity, the promotion of synergies and information sharing among already-existing programmes, and projects and activities that focus on links between biological and cultural diversity.

The JP also encourages to further explore conceptual and methodological issues related to the links between biological and cultural diversity, supporting and fostering learning networks on biocultural approaches, raising awareness about the importance of interdependent biological and cultural diversity in resource management and decision-making processes. It also emphases the need to develop interdisciplinary conceptual and methodological frameworks, taking into consideration the many ways in which cultures have shaped and continue to shape biodiversity in sustainable way (e.g. cultural landscapes, traditional agricultural systems, sacred sites, culturally significant species and urban biodiversity). The programme also intended launch pilot projects in appropriate sites, to apply the knowledge on the links between biological and cultural diversity in equitable management and governance practices.

1.2 The Florence Meeting and the Need for an Interdisciplinary Approach

The first meeting for the implementation of the Joint Programme was held in Florence (Italy) in April 2014. This is supposed to be the first of a number of meetings that the JP-BiCuD intended to organize in various continents, but there were good reasons for organizing this first meeting in Europe. The scientific and policy dimensions of the linkages between cultural and biological diversity are of utmost importance in the European context where cultural, environmental and rural policies are devoted to the conservation of biodiversity and cultural heritage, but rarely focused on the interactions between nature and culture. Yet, throughout European history, the outcomes of such interactions, including through traditional farming and forestry practices, have been critical for creating resilient landscape patterns, diversifying biological and cultural resources and shaping cultural identity of different European regions. In the last two decades, landscape has also become an important element in European Policies, especially considering the European Landscape Convention established in the year 2000, signed by more than 35 countries, as well as the Common Agricultural Policies, that put the conservation of the landscape asset of Europe among the objectives of rural policies together with the conservation of biodiversity.

Several international scientific organizations and the United Nations have designated *landscape* as one of the primary concerns of upcoming sustainability policies, and the last two summits of the United Nation organized on Climate Change in Lima in 2014 and Paris in 2015 included a “Global Landscape Forum”. From the point of view of research, it is also evident the growth of the importance of the term “landscape” compared to term like “ecosystem” in the published books, as also the importance of “cultural landscape” compared to “natural landscape”

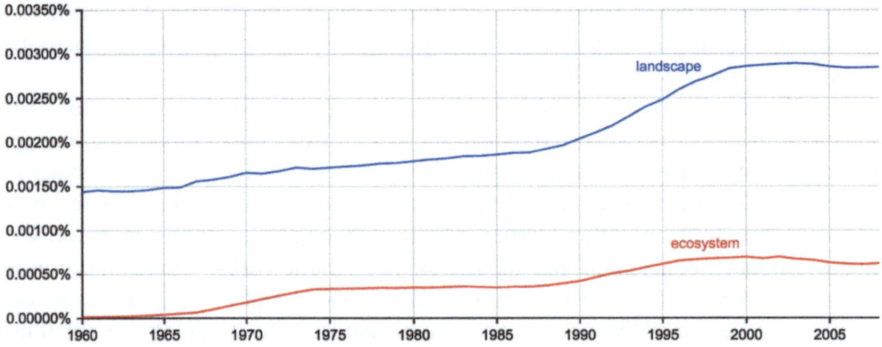

Fig. 1.2 The graph compares the usage frequency of the terms landscape and ecosystem as found in publications by Google Books. *Source* Google Ngram Viewer

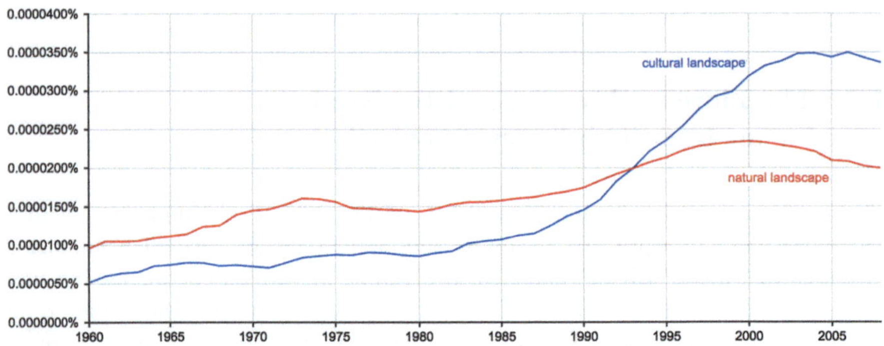

Fig. 1.3 The graph compares the usage frequency of the terms cultural landscape and natural landscape as found in publications by Google Books. *Source* Google Ngram Viewer

(see Figs. 1.2 and 1.3). This reflects the general growth of the landscape perspective in the scientific literature, but it has some interesting relations also with the governance of the European territory. The quality of European landscapes and the associated quality of life is determined by the tight interplay between their economic, social and cultural aspects, through time and space, and is often grounded in specific landscapes features. The preservation of such features contributes towards higher quality of life for local populations through material and immaterial means. These features improve people's lives and contribute to local economy while fulfilling their recreational, emotional and spiritual needs, and their sense of identity.

However, the conservation of complex landscape patterns that reflect the identity of different regions, their historical management practices and related biological and cultural diversity needs to be ensured through integrated planning and management strategies, particularly in view of the need to adapt to global climate change, but requires an interdisciplinary and transdisciplinary approach in scientific research. For these reasons, the conference wanted also to address the need of more research focusing on the links between cultural and biological diversity. This is an important task as these links have been rarely formalized from a scientific perspective, partly as a consequence of a separation between nature and culture not only in policies, but also between science and humanities. An important project, named RESCUE (Responses to environmental and societal challenges for our unstable earth), was developed in order to address this problem by the European Science Foundation and COST, in 2011. The project explored the matter of "collaboration between the natural, social and human sciences in global change studies", stressing that effective research collaboration can take place only if research programmes and projects are developed in proactive collaboration of scientists of different disciplines from their conception and this interaction exists to their final evaluation. Definition of the problem often sets the terms of engagement, expected outcomes and who is involved. This helps ensure that a joint research or conceptual framework is developed with all the necessary commonalities, such as semantics and a common agenda (Holm et al. 2012). Human and social sciences benefit from embedding anthropogenic research questions in an understanding of environmental forces.

The report developed during the RESCUE programme wanted to propose a strategic vision to break down the individual and institutional barriers that hamper collaboration across Europe between natural and social sciences and humanities in global change studies. Although the report showed that good examples of interdisciplinary research exist, the present situation is not fit for dealing with global change issues. The main reason is that interdisciplinary and transdisciplinary research is not yet evident. At most universities and other (academic) research institutions, the monodisciplinary approach has the upper hand. The traditional view is that real interdisciplinary research too often only occurs occasionally, not in a structured way and that scientists usually become interdisciplinary due to market forces after achieving expertise in their core fields. There is a very real risk that interdisciplinary researchers are downgraded by monodisciplinary colleagues in many major institutions and organizations. University powerbases traditionally rest with monodisciplinary colleagues. This problem has been noted also at political level in relation to the development of the European research programmes. A meeting organized by the European Presidency took place in Vilnius in 2013, producing the Vilnius Declaration on Social Science and Humanities, calling for a better integration of social science and humanities in the HORIZONS 2020.

Taking into consideration the problems of the scientific approach, the Scientific Committee of the Florence Conference encouraged the attendance of scientists from different disciplines considering biocultural diversity as a perfect example of a topic requiring an interdisciplinary approach. This not only for an effective understanding of the biodiversity associated with landscapes shaped by the man, but also for the further development of the joint programme in terms of research and political implementation. In this respect, the committee of the conference organized the meeting into a scientific part, presenting studies addressing the linkages between cultural and biological diversity, followed by a workshop dedicated to elaborate a political document for the implementation of a biocultural diversity approach in the European Continent. The committee received 165 paper proposals, and the selection procedure accepted 63 papers considered highly relevant for the topic of the conference and also 11 posters, from 25 countries. The scientific committee selected some of the papers presented that have been published in a special issue of the journal Biodiversity and Conservation and in this book published in the Environmental History Series of Springer-Verlag.

The expert meeting for the drafting of the UNESCO-SCBD declaration on biocultural diversity was attended by 42 experts from 14 countries and about 33 organizations, including FAO, ICOMOS, IUCN, and IUFRO. The Florence Declaration (see Annex) takes into account the results of the conference works and has not only produced political indications for the implementation on the Joint programme, but addressed important issues concerning the research and the political activities needed for the promotion of the concept of biocultural diversity:

1. The current state of biological and cultural diversity in Europe results from the combination of historical and ongoing environmental and land-use processes and cultural heritage.
2. Since it assimilates economic, social, cultural and environmental processes in time and space, the European landscape is predominantly a biocultural, multi-functional landscape. As such, it provides a crucial and effective space for the integration of biological and cultural diversity for human well-being, including in the context of rural territories.
3. Landscapes rich in biocultural diversity are often those managed by small-scale or peasant farmers, traditional livestock keepers/pastoralists, and small-scale/ artisanal fishermen.

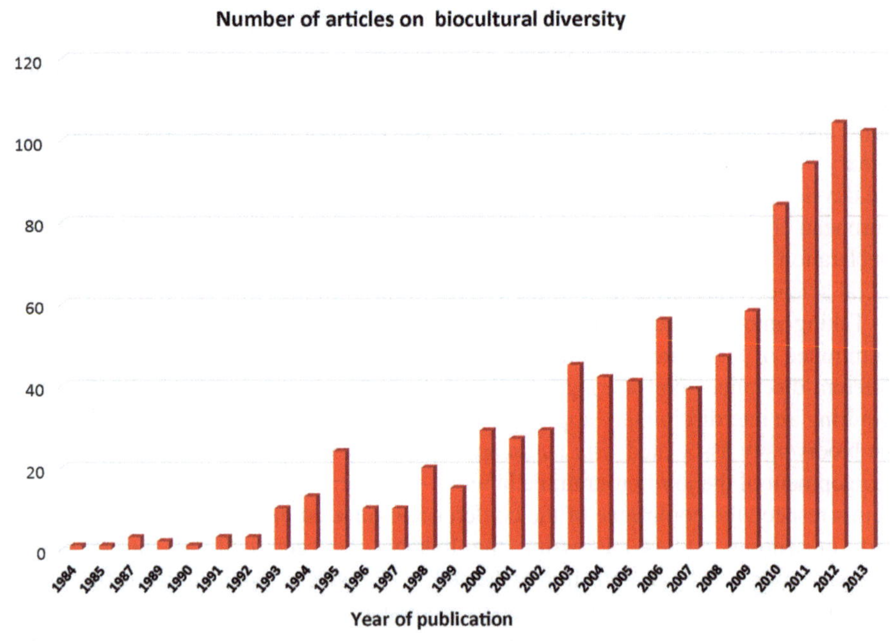

Fig. 1.4 There is an increasing number of articles dedicated to biocultural diversity in recent years although most of them are focused to linguistic diversity. *Source* Science Direct

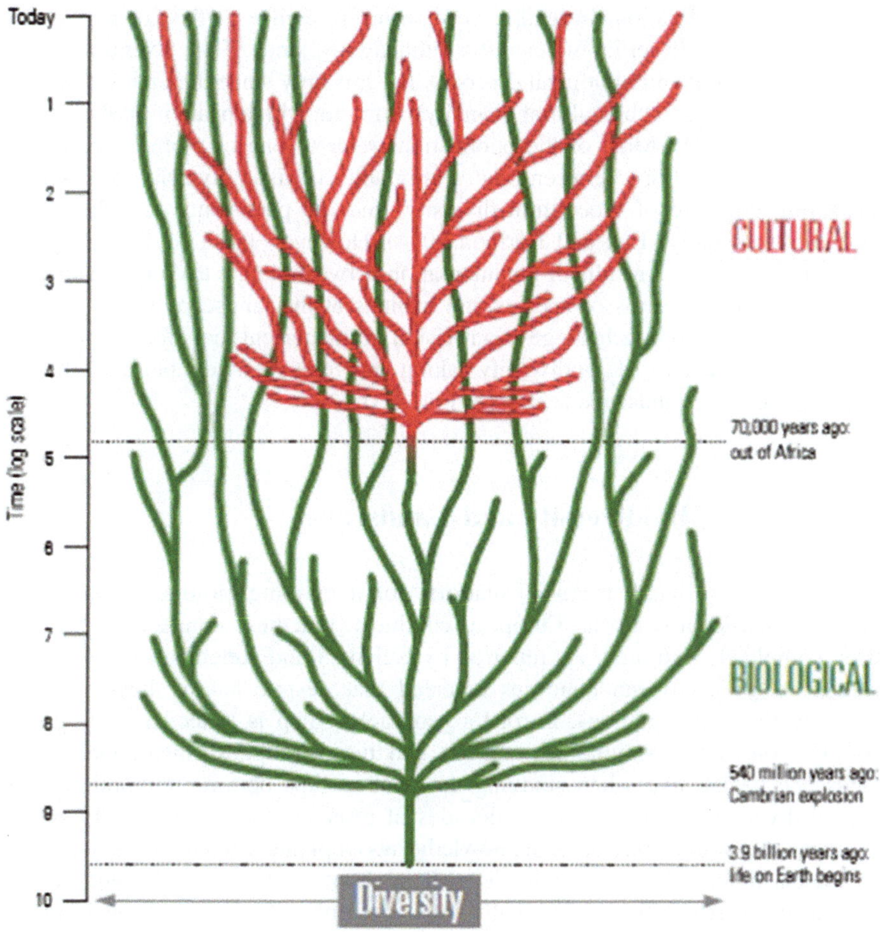

Fig. 1.5 The biological tree of species diversity began to diversity with the Cambrian explosion, about 540 millions years ago. The cultural tree of linguistic diversity began to diversify 70–80,000 years ago (Loh and Harmon 2014). The development of culture also affected animal and plant species, starting a widespread biodiversification process

4. To better understand the dynamic interplay between biological and cultural diversity at the landscape level and its implications for livelihoods and well-being, there is a need for enhanced interdisciplinary and transdisciplinary research of the links between biological and cultural diversity at the national and sub-national levels, including their historical background.

The Declaration was presented to the Conference of the Parties of the CBD in Pyeongchang, Republic of Korea, in October 2014, as well as during the ICOMOS world general assembly, in Florence, in November 2014. Considering the text of the joint programme on the linkages between biological and cultural diversity of 2010, the Florence Declaration presents a very strong reference to the concept of biocultural diversity. In terms of scientific publications, since 1984, scientific articles have cited cultural and biological diversity, but it is only since the end of the 1990s that clear reference to "biocultural diversity" is found, with a peak of publications in 2012 (see Fig. 1.4). Most of the publications refer to one or more of three main themes: the correlations between biodiversity and linguistic diversity, the tools for measuring the state of biocultural diversity, and the persistence and the loss of biocultural diversity (Loh and Harmon 2014). However, these studies rarely take into consideration the results of the integration between nature and culture affecting the biodiversity in terms of species and habitats at the landscape scale. The biocultural diversity approach suggests a revision of the current tools for the protection and the management of biodiversity taking into consideration biodiversification processes and the landscape scale (see Fig. 1.5).

1.3 History, Biodiversity and Landscape

The landscape provides a crucial platform for integrating biological and cultural diversity, particularly in the European continent, but these connections have not been completely explored. As reported by Kelemen and Kelemen (2015), genetic studies on 36 European countries revealed that despite having lower genomic diversity than other regions, Europe's geographic map is reflected in its genetic composition with an astounding precision, and this also suggests the importance of a landscape perspective when studying the relationships between cultural and biological diversity in Europe. Some decades of research had accumulated evidence about the depth and detail of local knowledge about plants and animals, habitats, and ecological functions and relations, as well as about the low environmental impact and indeed sustainability—historically and at present—of many traditional forms of natural resource use (Agnoletti and Rotherham 2015). This is also the research ground of disciplines such as Historical Ecology or Forest History. The research evidence also pointed to a variety of ways, in which humans have maintained, enhanced and even created biodiversity through culturally diverse practices of management of "wild" resources and the raising of domesticated. However, countering the image of "pristine" environments, unaffected by humans, that could be brought back to their "original" state by fencing them off to protect them from human activity, has not really changed conservation strategies, nor has helped for an effective consideration of biodiversification processes (Agnoletti 2014).

The biodiversity of cultural landscapes, especial rural ones, is revealed especially when their individual historical dimension is directly explored in the field. A historical perspective can recognize the environmental systems and processes that shape each rural landscape as true functional nodes in a more general historical process of "environmental biodiversification" (Ingold et al. 2003). Under adequate conditions, landscape can play a direct role in the preservation of "agricultural biodiversity". It is less well-known, however, that by reconstructing the historical role of agricultural, silvicultural and husbandry practices in local processes of biodiversification, we can better define the features of animal and species in combination with the biodiversity at landscape level. The problem of traditional practices is one of the principal nodes in the perspective of the redefinition of biodiversity, especially at forest level (Parrotta and Trosper 2012). Biodiversification processes today are occurring not only in conjunction with the transformation and abandonment of agricultural spaces, the two general historical phenomena involving the dynamics of contemporary agriculture. In several well-known cases—for example the ones of "ancient woods"—, they appear to be specifically connected with the local environmental-historical legacy, that is, the earlier manipulations that activated their ecological resources in the first place. Scholars have said more than once that, in the perspective of historical ecology, the vegetable component of a rural landscape, its present biodiversity, can be regarded as an artefact which, not unlike a Medieval cathedral, displays series—or, at least, traces—of historical stratifications. There are studies indicating an impact of management regimes on the heterogeneity of within-population genetic structure in forest landscapes (Piotti et al. 2013), although the range of research results shows that the response of forest trees to management techniques can be highly species-specific. However, independently by the negative or positive effects of the human impact on forest landscapes, density, structure and species composition have been affected by culture, as happened to the genetic structure of the species.

1.4 History and Biology

Historical rural landscapes, especially those presenting a high heterogeneity (Agnoletti 2013), are often related to traditional practices and represent good examples for understanding the biocultural diversity. It is also known that traditional rural landscapes maintained complex land-use mosaics, especially like those existing in Europe in the nineteenth century, before the agricultural industrialization turned them into homogeneous land covers polarized between intensive monocultures and afforestation of abandoned lands from the 1960s onwards (Agnoletti 2013; Marull et al. 2015). The spatial component of biological diversity (Tilman 1994) is of crucial importance for biodiversity, but is also one of the most interesting scientific perspectives to understand biocultural diversity. This approach often focuses on the interaction between disturbances and land cover diversity, in relation to ecosystem services (Tscharntke et al. 2005). The role played by the spatial heterogeneity of the land matrix in respect to ecosystem complexity and resilience in cultural landscapes is a central element (Bengtsson et al. 2003; Benton et al. 2003). However, these approaches together with putting into evidence the interplay between patch disturbance and land-cover diversity as a key mechanism in biodiversity, also stress the role of rural landscape mosaics able to offer habitats to different inner species, increasing the number of ecotones and species as well (Harper et al. 2005). Much of this biological diversity is located at scales higher than plot or farm level and depends on keeping a landscape-wide variety of land covers. When high species richness is kept at landscape level thanks to land cover heterogeneity, the decrease of biological diversity typical of large monocultures can be counterbalanced. This way, a disturbance-complexity interplay leads to divergent and compensatory trends followed by α-diversity at plot scale (within-patch or within each community), β-diversity at landscape level (between-patch or between-communities) and γ-diversity of the species pool hosted at regional scale (Loreau 2000; Roxburgh et al. 2004; Gabriel et al. 2006; Loreau et al. 2010). The predominance of β-diversity kept by the spatial heterogeneity of a variety of intermingled land covers is key mechanism of biocultural diversity maintenance in cultural landscapes. The effectiveness of farm management in increasing biodiversity reaches a peak at intermediate levels of landscape heterogeneity, given that simple landscapes tend to behave as a single monoculture poorly endowed of biological diversity, whereas highly complex ones retain great biodiversity anyway. Hence, a complex landscape matrix may enhance the overall biological diversity except when it comes to species that require specific habitats and other conservation policies. The role of landscape in the relationships between cultural and biological diversity becomes more evident when we consider the classic ecosystem theories and the trophic levels. In the ideal state of an undisturbed environment, the producers allowing the consumers to survive are identified in the "natural systems", but in the present period of the history of the earth almost all the previous natural systems are affected by the man and they can be rather described as cultural landscapes. These landscapes are characterized by β-diversity mostly depending from agricultural, grazing and forest practices.

1.5 The Operational Level

The concept of biocultural diversity has important effects in terms of management, because not only landscape management, but also ecosystem management, is now primarily about biocultural values. It has been common in the past to show ecosystem management as the point where two areas, the ones of socioeconomic system and natural systems overlap, with the commonly shared space being where ecosystem management occurs (Vogt et al. 1997). This type of representation leads one to assume that most of the natural system and most of the social system not contained in the overlapping zones, function independently from one another. According to the developments of man–nature relationships of the last decades, this should be redrawn into a conceptual figure where no part of the natural system is isolated from the social system, not only because there are many legacies of human activities in landscapes that appear to be "virgin", or unaffected by humans, but also because climate change induced by the man affects also ecosystems physically isolated from human societies. From the scientific point of view, according to this perspective, papers and studies describing biodiversity and landscapes, even though not directly addressing the term, are basically about biocultural diversity, especially in Europe where the man has strongly affected the natural basis at least since the development of agriculture.

At the world level, besides CBD and IUCN, important conservation programmes where the concept of biocultural diversity could be introduced are the UNESCO World Heritage List (WHL) and FAO's Globally Important Agricultural Heritage Systems (GIAHS) programme (Koohafkan and Altieri 2011). Presently, these programmes do not specifically address biocultural diversity, although most of the biodiversity included in the protected areas clearly has a cultural origin (Agnoletti 2014). In the dossier prepared to nominate sites in these conservation programmes, biodiversity is still mostly assessed according to the traditional wildlife and natural habitats approaches. Even in the cultural landscapes of the UNESCO WHL, what is not farmed land is usually considered natural or semi-natural, but not as a biocultural entity. In the FAO GIAHS, the focus is more on the traditional practices related to farming, but the assessment of biodiversity in the applications is also focused on the same approach, as also the actions listed in the management plans. However, in the recommendations indicated by UNESCO for the candidature of the Cultural Landscape of Bourgogne in the World Heritage List, defined during the 39th Ordinary Session of the World Heritage Committee in Bonn (Germany), the inclusion of biocultural diversity was required for the description of the biodiversity related to the forests included into the buffer zone of the dossier.

The biocultural diversity concept presents an opportunity to revise some of the current approaches to biodiversity, recognizing the wider meaning of this term and the need for a revision of the current conservation strategies (Agnoletti and Rotherham 2015). While UNESCO, FAO, and the CBD might adapt their guidelines in the near future, the 28 member states of the European Union can surely be considered as the best ground for a political implementation for the existing conservation tools in Europe, especially considering the Habitat Directive. Together with the European Landscape Convention, the Habitat Directive makes the European Continent the part of the world where the most important tools for biodiversity and landscape conservation have been developed to date. The Habitat Directive (together with the Birds Directive) forms the cornerstone of the Europe's nature conservation policy. It is built around two pillars: the Natura 2000 network of protected sites and the strict system of species protection. The articles of the Directive are clear about the aims, which are to contribute towards ensuring biodiversity maintenance through the conservation of natural habitats and of wildlife in the European territory of the Member States to which the treaty applies. However, it does not consider biocultural diversity and landscape mosaics as a goal of conservation. A similar problem can be found in the European Criteria for Sustainable Forest Management, where values associated with landscape and biodiversity play a very minor role, as well as the cultural features of forests and woodlands (Agnoletti and Santoro 2015). Nevertheless, the role of traditional forest-related knowledge in shaping the forests and their biodiversity in Europe and across the world has already been stressed in several publications. Europe is basically a cultural landscape, and the cultural origin of the European Union territory has been recognized by the European Commission at least since 1999, when only 5 % of the territory was classified as natural. In this respect, the fact that the Natura 2000 network covers 20 % of the European territory, poses questions about the naturalness of the areas included, as well as on the features and the origin of biodiversity, not only in the protected Sites of Community Importance (SCI), but also in the entire EU territory, as in other parts of the world.

Many of the world's "primary forests" and biodiversity "hot spots" are located in regions with a high diversity of local populations who manage their natural resources based on distinctive cultures and their associated traditional knowledge and wisdom. In other rural environments, a long history of integration of forestry and agricultural activities has created land-use forms and biological diversity that is closely connected to complex landscape patterns. Landscapes shaped by culture often show a high level of habitat diversity tighten into a versatile mosaic produced by the application of different management regimes, and the introduction of a great mixture of species over the years, that came to meet specific economic, social or environmental roles. Failure to effectively and coherently address culture and history may very well be an emerging weakness that needs to be reconciled. This is necessary to give the public and local communities confidence in the protocols designed to recognize well-managed landscapes and in moving towards the goals of sustainable management. The practical implementation of nature conservation strategies in rural territories affected by centuries of human influence, without a redefinition of biodiversity targets, taking into consideration the historical relationships between agricultural practices and animal and plant species may result in conflicts with local populations. On the other hand, considering biocultural diversity and recognizing the need to conserve key habitats resulting from the reciprocating influences between people and nature, might help solve some contradictions between landscape and nature conservation. This would also help to counterbalance the widespread belief that abandonment of cultural landscapes to supposedly more "natural" successions is inherently and intrinsically good for "conservation" and the environment. Increasingly, this misinformation is promoted though some both popular and scientific texts. As a final consideration, we should add that in consideration of the current growth of the world's population, the food challenge and the climate challenge, different goals and evaluating tools might be required, also for biodiversity. In this respect, biocultural diversity seems to be a better tool at planning and management level, involving a different view in terms of species and habitats, considering a hierarchy taking into account the landscape level. At the same time, the landscape, considered as the result of the integration of social, economic and environmental processes in time and space, may offer a better perspective, in comparison with the ecosystem approach, to harmonize these processes in planning and management.

Annex: UNESCO-SCBD Florence Declaration on the Links Between Biological and Cultural Diversity

Florence (Italy), 11 April 2014

We, the participants of the First European Conference for the Implementation of the UNESCO-SCBD Joint Programme on Biological and Cultural Diversity, held from 8 to 11 April 2014 in Florence, Italy:

Recognizing the vital importance of cultural and biological diversity for present and future generations and the well-being of contemporary societies in urban and rural areas;
Recognizing further the importance of the links between cultural and biological diversity, and in this context noting the concept of **Biocultural Diversity** and the relevance of cultural services provided by ecosystems;
Acknowledging the important progress made in building the knowledge base on the links between biological and cultural diversity and its implications for policy and decision-making in the area of sustainable development;
Taking into account the Convention on Biological Diversity's relevant provisions in the text of the convention and the many decisions which have flowed from them, the UNESCO Culture-related Conventions and other relevant instruments, programmes and initiatives, and related International decisions on Human Rights and the Rights of Indigenous Peoples, and pertinent regional arrangements and agreements in the European context (Annex);
Highlighting the need to further strengthen the cooperation between the relevant international agreements and bodies in the field of bio-cultural diversity, and, in this context noting the importance of the joint programme between UNESCO and the Secretariat of the Convention on Biological Diversity on the links between biological and cultural diversity;

1. Agree that the following conclusions are of special relevance in the European context:

   (a) Rural and urban livelihoods and well-being are closely connected to the status and trends in biological and cultural diversity;
   (b) The current state of biological and cultural diversity in Europe results from the combination of historical and ongoing environmental and land-use processes and cultural heritage;
   (c) As it assimilates economic, social, cultural and environmental processes in time and space, the European landscape is predominantly a biocultural multifunctional landscape. As such, it provides a crucial and effective space for integration of biological and cultural diversity for human well-being, including in the context of rural territories;
   (d) Temporary, semi-permanent and permanent migratory human movements and associated exchange of skills, knowledge and goods between town and countryside have in many cases shaped the local biological and cultural diversity and still provide ample opportunities for their enhancement;
   (e) Landscapes rich in biocultural diversity are often those managed by small-scale or peasant farmers, traditional livestock keepers/pastoralists, and small-scale/artisanal fishermen;
   (f) The involvement of local communities, and recognition of and respect for their cultural heritage, traditional knowledge, innovations and practices can assist in more effective management and governance of multifunctional biocultural landscapes and contribute to their resilience and adaptability;

(g) To better understand the dynamic interplay between biological and cultural diversity at the landscape level and its implications for livelihoods and well-being, there is a need for enhanced interdisciplinary and transdisciplinary research of the links between biological and cultural diversity at the national and sub-national levels, including their historical background;
(h) Public awareness of the links between biological and cultural diversity and political action that considers these links in policy and decision-making processes are needed to effectively implement international and national commitments dealing with environmental, social and economic sustainability and human well-being at different scales.

2. Invite Governments and relevant bodies, including the European Union and the Council of Europe to initiate and undertake the following actions at the appropriate levels:

Regional level:
Noting the Common Agricultural Policy and the European Environmental Policies and legislation(s) and their relevance for strengthening the links between biological and cultural diversity

(a) Promote action at the regional level, especially in the context of the European Union, to enhance the implementation of international and regional agreements mentioned in Annex;
(b) Take specific actions for the valorisation and promotion of both cultural and biological diversity by, inter alia:

- Promoting the inclusion of cultural and biological diversity into national and local planning for nature conservation and landscape management, including protected areas, agricultural and forest landscapes;
- Designing and implementing procedures for the participation of different stakeholders, including local communities, in the implementation of policies relevant for linking biological and cultural diversity;
- Promoting the incorporation of the links between biological and cultural diversity in sustainable forest management, including through the development of appropriate criteria and indicators;
- Promoting the recognition of positive interactions between biological and cultural diversity through appropriate certification processes and effective product labelling;
- Promoting the use of the European Innovation Partnership (EIP) for the promotion and valorization of the links between biological and cultural diversity for scientists, policy-makers and other stakeholders.

National level:

(a) Adopt an integrated and transdisciplinary approach when implementing relevant international and regional agreements at the national level;

(b) Take action in the framework of current rural and environmental policies to incorporate biological and cultural diversity and the links between them;
(c) Make available funding such as research grants for transdisciplinary and multidisciplinary research on biological and cultural diversity and the links between them;
(d) Establish cross-sectorial and/or cross-departmental approaches to biological and cultural diversity;
(e) Recognize the existence of collective commons, traditional tenure and resource rights, customary laws and management useful for promoting and conserving biological and cultural diversity, including agricultural diversity in traditional crops and livestock breeds;
(f) Use bottom-up legislative development processes, inclusive of local communities, including both women and men;
(g) Recognize and ensure that legal regulatory arrangements do not run counter to sustainable local livelihoods dependent on biological and cultural diversity;
(h) Recognize and protect sacred natural sites noting they play an important role in local ecosystem regeneration and culture;
(i) Provide small grants and incentives for local biological and cultural diversity initiatives, including community knowledge exchanges;
(j) Provide mechanism for the review of relevant European and national policies and to take into account the links between biological and cultural diversity;
(k) Recognize and promote good practices for the conservation and promotion of interlinked biological and cultural diversity and provide resources for up-scaling and replication, where appropriate;
(l) Promote, through marketing techniques, biological and cultural diversity as underlying the competitiveness of rural landscapes which generate local produce and tourism;
(m) Support cultural, educational and recreational initiatives for the promotion of the links between biological and cultural diversity;
(n) Ensure subsidies are available to small-scale and diverse agricultural farmers, and farmers promoting and preserving agricultural biodiversity in crops, livestock breeds and landscapes, on an equal basis to large scale farmers.

Local level:

(a) Encourage partnerships such as participatory research opportunities between researchers and local communities;
(b) Encourage bottom-up inclusive development of policy and legislation relevant to biological and cultural diversity;
(c) Recognize, promote and where appropriate scale up and replicate good local practices for maintaining biological and cultural diversity;
(d) Encourage and empower local communities and their organizations to continue, transmit and innovate traditional practices compatible with the sustainable use of biocultural diversity;

(e) Support local community organisations to develop initiatives for the conservation of biocultural diversity as the basis for resilient farming systems and local economies;
(f) Promote urban biological and cultural diversity through community grants, creation, preservation and maintenance of public green spaces, community gardens and greening of cities.

3. Request that this Declaration is brought to the attention of Member States of the United Nations and presented for consideration at the relevant intergovernmental bodies of the Convention on Biological Diversity and UNESCO, as well as the governing bodies of the European Union, the Council of Europe, and in meetings and deliberations of expert communities.

The relevant instruments referred to above include but are not limited to:
The Convention on Biological Diversity's:

- Strategic Plan 2011–2020 and the 20 Aichi Targets as a framework for international efforts for biodiversity conservation, sustainable use and equitable sharing of benefits arising from the utilization of genetic resources;
- Articles 8(j) (traditional knowledge) and 10(c) (customary use) and related provisions of the Convention on Biological Diversity, and taking into account the related initiatives such as the Biodiversity Indicators Partnership;
- (with UNESCO) the Montreal Declaration on Bio-cultural Diversity (2010).

UNESCO's:

- Universal Declaration on Cultural Diversity (2001);
- World Heritage Convention (1972) and its guidelines concerning the inclusion of Cultural Landscapes in the World Heritage List (1992);
- Convention for the Safeguarding of the Intangible Cultural Heritage (2003);
- Convention on the Protection and Promotion of Diversity of Cultural Expressions (2005);
- Seville Strategy for Biosphere Reserves designated under UNESCO's Man and Biosphere (MAB) programme (1995).

Other relevant declarations, programmes and initiatives:

- the Universal Declaration of Human Rights(1948),
- the United Nations Declaration on the Rights of Indigenous Peoples (2007),
- the Universal Declaration on Bioethics and Human Rights (2005),
- the Earth Charter (2000),
- the Bali Declaration on the FAO International Treaty on Plant Genetic Resources for Food and Agriculture (2001),
- the Declaration of Belém of the International Society of Ethnobiology (1988),
- the Yamato Declaration on Integrated Approaches for Safeguarding Tangible and Intangible Cultural Heritage (2004),
- the Tokyo Declaration on the Role of Sacred Natural Sites and Cultural Landscapes in the Conservation of Biological and Cultural Diversity (2005),

- the Bonn Declaration on Education for Sustainable Development (2009),
- the United Nations Development Group Guidelines on Indigenous Issues,
- the UNESCO Local and Indigenous Knowledge Systems Programme,
- the FAO Globally Important Agricultural Heritage Systems Initiative.

Relevant regional arrangements and agreements in the European context, inter alia:

- European Council Directive 92/43/EEC of 21 May 1992 on the conservation of natural habitats and of wild fauna and flora (Habitat Directive);
- The European Landscape Convention (2000);
- The Framework Convention on the Protection and Sustainable Development of the Carpathians (Carpathian Convention) (2003);
- The Alpine Convention (Framework Convention) (1991).

References

Agnoletti M (ed) (2013) Italian historical rural landscapes. Cultural values for the environment and rural development. Springer, Dordrecht

Agnoletti M (2014) Rural landscape, nature conservation and culture: some notes on research trends and management approaches from a (southern) European perspective. Landscape Urban Plann 126:66–73

Agnoletti M, Rotherham ID (2015) Landscape and biocultural diversity: an introduction. Biodivers Conserv. doi:10.1007/s10531-015-1003-8

Agnoletti M, Santoro A (2015) Cultural values and sustainable forest management: the case of Europe. J Res. doi:10.1007/s10310-015-0500-7

Bengtsson J, Angelstam P, Elmqvist T, Emanuelsson U, Folke C, Ihse M, Moberg F, Nyström M (2003) Reserves, resilience and dynamic landscapes. AMBIO 6:389–396

Benton TG, Vickery JA, Wilson JD (2003) Farmland biodiversity: is habitat heterogeneity the key? Trends Ecol Evol 18:182–188

Gabriel D, Roschewitz I, Tscharntke T, Thies C (2006) Beta diversity at different spatial scales: plant communities in organic and conventional agriculture. Ecol Appl 16:2011–2021

Harper KA, MacDonald SE, Burton PJ, Chen J, Brosofske KD, Saunders SC, Euskirchen ES, Roberts D, Jaiteh MS, Esseen PA (2005) Edge influence on forest structure and composition in fragmented landscapes. Conserv Biol 19:768–782

Holm P, Goodsite ME, Cloetingh S, Agnoletti M, Bedrich M, Lang DJ, Leemans R, Moeller JO, Buendía MP, Pohl W, Scholz RW, Sors A, Vanheusden B, Yusoff K, Zondervan R (2012) Collaboration between the natural, social and human sciences in Global Change research. Environ Sci Policy 28:25–35

Ingold T, Sanga G, Ortalli G (eds) (2003) Nature knowledge: ethnoscience, cognition and utility. Berghahn Books, New York

Kelemen Z, Kelemen R (2015) Are cultural values linked to genetics in Europe? Biodiv Conserv 24. doi:10.1007/s10531-0150965-x

Koohafkan P, Altieri MA (2011) Globally important agricultural heritage systems: a legacy for the future. FAO, Rome

Loh J, Harmon D (2014) Biocultural diversity: threatened species, endangered languages. WWF Netherlands, Zeist, The Netherlands

Loreau M (2000) Are communities saturated? On the relationship between α, β and γ diversity. Ecol Lett 3:73–76

Loreau M, Mouquet N, Gonzalez A (2010) Biodiversity as spatial insurance in heterogeneous landscapes. Nat Acad Sci 100(22):12765–12770

Marull J, Tello E, Fullana N, Murray I, Jover G, Font C, Coll F, Domene E, Le V, Decolli T (2015) Long-term socio-ecological transition at different spatial scales: exploring the intermediate disturbance hypothesis in cultural landscapes (Mallorca, 1856-2012). Biodiv Conserv 24. doi:10.1007/s10531-015-0969-6

Parrotta J, Trosper RL (2012) Traditional forest-related knowledge: sustaining communities, ecosystems and biocultural diversity. Springer, Dordrecht

Piotti A, Leonardi S, Heuertz M, Buiteveld J, Geburek T, Gerber S, Kramer K, Vettori C, Giuseppe Vendramin G (2013) Within-population genetic structure in beech (*Fagus sylvatica* L.) stands characterized by different disturbance histories: does forest management simplify population substructure? PloS ONE 8(9). doi:10.1371/journal.pone.0073391

Roxburgh SH, Shea K, Wilson JB (2004) The intermediate disturbance hypothesis, patch dynamics and mechanisms of species coexistence. Ecol 85:359–371

Tilman D (1994) Competition and biodiversity in spatially structured habitats. Ecol 1:2–16

Tscharntke T, Schmidt MH, Roschewitz I, Thies C (2005) Differential effects of landscape and management on diversity and density of ground-dwelling farmland spiders. J Appl Ecol 42:281–287

UNESCO, SCBD (2010) UNESCO–CBD Joint program between biological and cultural diversity. UNESCO, Paris. http://www.unesco.org/new/en/natural-sciences/special-themes/biodiversity-initiative/biodiversity-culture/unesco-cbd-joint-programme/. Accessed 19 Nov 2015

UNESCO, SCBD (2014) Florence declaration on the links between biological and cultural diversity. UNESCO, Florence. http://www.landscapeunifi.it/images/pdf/UNESCO-CBD_JP_Florence_Declaration.pdf. Accessed 19 Nov 2015

Vogt K, Gordon J, Wargo J (1997) Ecosystem balancing science with management. Springer, New York

Part I
Landscape Characters and Biocultural Diversity

Chapter 2
The Traditional Mediterranean Polycultural Landscape as Cultural Heritage: Its Origin and Historical Importance, Its Agro-Silvo-Pastoral Complexity and the Necessity for Its Identification and Inventory

Giuseppe Barbera and Sebastiano Cullotta

Abstract Today, the Mediterranean is characterized by landscape patterns whose compositions result from countless, long and complex cultural and historical processes. However, the pressure on these landscapes and their rapid transformation into more modern forms call out for a better knowledge of the more complex forms of traditional land use and relative landscapes. In this context, an identification and clarification of the role of such mixed and complex forms of agro-forestry systems and landscapes, named "*giardino Mediterraneo*" ("Mediterranean garden") is necessary. This term is often applied to and associated with numerous different agricultural and agro-forestry systems as well as to numerous different kinds of rural landscapes, due to the complex and intricate historical process that has led to their identification and cultural evolution over time. This study identifies the characteristics of the polycultural and polyspecific Mediterranean garden's landscape, characterized by the presence of trees (both wild and cultivated), starting from a historical overview of Sicily. The analysed *Halaesa* landscape (Sicily) case study, as one of the first historical detailed description of a complex Mediterranean cultural landscape, is the result of a polycultural agro-silvo-pastoral system which guarantees complexity and richness (in terms of structural and biological diversity), as well as with reference to others environmental, cultural and economic multi-functionality. However, a comparison with the typological systems currently used for the cataloguing and mapping of traditional Mediterranean landscapes at different

G. Barbera (✉) · S. Cullotta
Dipartimento Scienze Agrarie e Forestali, Università di Palermo,
Viale delle Scienze, Edif.4—Ingr.H, 90128 Palermo, Italy
e-mail: giuseppe.barbera@unipa.it

S. Cullotta
e-mail: sebastiano.cullotta@unipa.it

M. Agnoletti and F. Emanueli (eds.), *Biocultural Diversity in Europe*,
Environmental History 5, DOI 10.1007/978-3-319-26315-1_2

scales shows the transversal importance of polycultural Mediterranean garden landscapes, and demonstrates how much a better definition and characterization of them is needed.

Keywords Biocultural diversity · Rural landscape · Landscape history · Landscape type · Landscape pattern · Landscape character · *Mediterranean garden* · *Coltura promiscua* · Material and non-material heritage · Multifunctionality · Mediterranean basin

2.1 Introduction

The different regions of the Mediterranean Basin show a high degree of both physical and climatic unity, despite the landscape's diversification by mountains, plateaus and plains, especially the coast. This environmental and ecological diversity (e.g. Braudel 1986) is also the consequence of the intersection of three different continents, and hence of their genetically different flora and fauna as well as their different civilizations (e.g. Braudel 1986; Grove and Rackham 2002; Mazzoleni et al. 2004; Blondel 2006). A complex "co-evolution" has shaped the interactions between natural ecosystems and the constantly evolving human land use practices (Di Castri et al. 1981), resulting in a mosaic of traditional rural landscapes which conserve many of the biological and cultural characteristics of those from the past. The first significant impact made by humans on forests and other natural ecosystems in the Mediterranean took place before the Neolithic revolution (Terral 2000), when permanent settlements were established. Forest management through wood cutting and coppicing, controlled burning, plant domestication, livestock husbandry, grazing and browsing, as well as through water management and terracing, has been the main tool for producing intermediate disturbance regimes for millennia (Zohary and Hopf 1993; Blondel 2006). These practices gradually led to complex and heterogeneous agro-silvo-pastoral rural patterns typified by a fine-grained mosaic pattern of land use (Sirami et al. 2010; Cullotta and Barbera 2011). This mosaic was made up of relatively small patches and corridors, and had a great species and interspecies diversity as a consequence of the cyclical disturbances introduced by rotational grazing, cutting and coppice regimes, fire management, as well as of cultivation and other human land use practices (Naveh 1995). Thus, these rural activities have played a major role in shaping the traditional Mediterranean landscape composition. The cultivation of arable land, olive groves, vineyards, mixed crops and fruit orchards and other multifunctional agricultural and agro-forestry systems are among the most important examples of traditional farming in the Mediterranean. Moreover, the conventional subdivision of properties into small units, due to the long intervals in ownership succession and property transfers, has further augmented this structural heterogeneity, thus influencing the contemporary cultural practices regarding rural landscapes (Horden and Purcell 2000). The physical expression of land division and property ownership is visually reinforced by the presence of stone walls and other artefacts or features such as terraces, hedgerows, canals, stone heaps, etc. (Grove and Rackham 2002; Brown et al. 2007; Petanidou et al. 2008; Barbera and Cullotta 2012). Today, the entire Mediterranean is characterized by landscape patterns whose compositions result from countless long and complex cultural and historical processes that developed in an equally complex and varied environment. However, the pressure on these landscapes and their rapid transformation by the current practices of agricultural intensification, crop abandonment and urbanization (Farina 2000; MacDonald et al. 2000; Stoate et al. 2009) into more modern landscapes, calls out for a better definition of some of the more complex traditional ways of using land and its relative rural landscapes. Traditional agro-silvo-pastoral Mediterranean landscapes, particularly those characterized by the presence of trees (both as strictly agricultural cultivations as well as in wild woods or as isolated trees) have maintained some defining characteristics regarding their composition, structure and function during the course of their slow evolution (Bignal et al. 1995; Vos and Meekes 1999; Vicente and Alés 2006). Traditional landscapes refer to these landscapes with a long history, which evolved slowly and where it took centuries to form a characteristic structure reflecting harmonious integration of abiotic, biotic and cultural elements (Antrop 1997). Thus, they must be considered dynamic landscapes which main historical identity was left unchanged until the mid-twentieth century (Bignal et al. 1995; Vos and Meekes 1999; Antrop 2005). Trees on farmland have a long tradition in Europe as they offer multiple functions (Biasi et al. 2012; Nerlich et al. 2013). These features are particularly expressed in those landscapes that are characterized by complex agricultural forms, mixed agro-forestry systems and landscapes that are capable of generating and guaranteeing an articulated environmental as well as economic and social multifunctionality (Vos and Klijn 2000; Pinto-Correia and Vos 2004; Eichhorn et al. 2006; Jose 2009; Jones-Walters 2008; Sánchez et al. 2010; Pinto-Correia et al. 2011). Considering these mixed and complex forms of agro-forestry systems and landscapes, in which the tree holds a central role, a better definition and clarification of the role of the "*Mediterranean garden*" in the traditional rural landscape certainly seems necessary. This definition is so often generically associated with the countless different agricultural and agro-forestry systems and landscapes that have been produced by the complex and intricate historic processes that have lead to their cultural definition and evolution. In this paper we use the term "Mediterranean garden" (*giardino Mediterraneo*) starting with the use and definition of it given by Emilio Sereni in its famous "*History of the Italian Agricultural Landscape*" (1961, trans. Litchfield 1997). This study traces the major historic, agricultural and rural processes that have characterized the Mediterranean basin in order to better define:

- the concept of a traditional rural Mediterranean landscape, with particular reference to that including fruit and non-fruit (wild) trees;
- the concept and importance of the polycultural systems and landscapes (*coltura promiscua*, *giardino Mediterraneo*), both because it is a historic landscape and because of its environmental complexity and value (multifunctionality);

- the transversal importance of the polycultural systems and landscape in the recent typological inventorying and mapping efforts of traditional rural landscapes;
- a proposal of a set of guiding structural elements and characters, configuration and elements, at the landscape and stand-system level that define these systems and landscapes and allow it to be recognized.

This paper works to identify the defining features of the polycultural systems and landscapes by following the historic development of this landscape, beginning with a historic description of the territory of Sicily (Central Mediterranean) as case study, in which all of the defining features of this complex landscape come together.

2.2 Origin and Historical Path of the Mediterranean Garden and Polycultural Landscape

During its slow evolution over the millennia, the Mediterranean agricultural landscape characterized by the presence of trees maintained some of it's initial properties from the foundation of its unique tradition: its peri-urban (just outside a village, quarter, or city) closed and protected location, its polycultural and polyspecific make-up, its irrigated and multifunctional character, its close relationship to culture and its continuous use as a source of artistic inspiration. We know from Shay et al. (1992, cit. in Blondel and Aronson 1999) that a peri-urban landscape made up of cultivated fields and fruit orchards alternating with wooded areas, such as those that can still be seen today in the various parts of the Mediterranean from the Iberian peninsula and Southern France in the West (Firmino 1999; Pinto-Correia and Vos 2004; Sirami et al. 2010) all the way to the far eastern regions (Braudel 1986; Kaldjian 2004; Kizos and Koulouri 2006), was already present in the ancient polycultural landscape located in the peri-urban areas of the island of Crete during the 4–3 millennium B.C. This fragmented landscape is both temporally and spatially heterogeneous because of its environmental, climatic and topographic variability and its interactions with various historic events and cultures (Naveh and Liebermann 1994; Cowling et al. 1996; Pinto-Correia and Vos 2004; Blondel 2006). These interwoven semi-natural and cultivated landscapes have been the cradle of man's relationship with nature for thousands of years and are real biodiversity hotspots due to their exceptional number of endemic and cultivated species. The three main crops —grapes, olives and grains (e.g. Olmo 1995; Terral 2000), are those which Braudel (1986) defined as the "trinity born from the union between climate and history"—however, in general, there is a great diversity of crops which has determined the area's food security and ecological stability over the centuries (Loumou and Giourga 2003). This diversity was created by history (with the main contributions being, in brief, the pre-Classical and Classical introduction of Asiatic species, the Islamic "agricultural revolution", the introduction of American species and finally the introduction of the species that arrived via the activities of plant collectors and

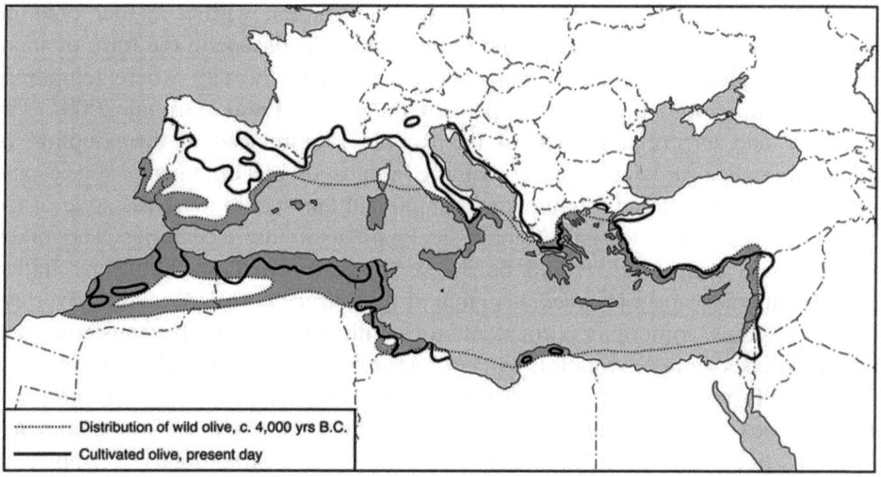

Fig. 2.1 Physical and biological delimitation of the Mediterranean region: the olive tree (wild and cultivated) is probably the most representative indicator of the mediterraneity and its related rural landscape (*source* Zohary 1995)

European scientific institutions) and by the Mediterranean's heterogeneous environmental characteristics. However, the Mediterranean landscape was unified by a climate so perfectly adapted to wild and cultivated tree and shrub species (Fig. 2.1) that their cultivation represents "a sort of natural monopoly" (Bevilacqua 1996). The most famous polycultural garden of the Mediterranean rural landscape tradition from the classical age is that of Alcino, in the Odyssey:

"*Outside the gate of the outer court there is a large garden of about four acres with a wall all round it. It is full of beautiful trees—pears, pomegranates, and the most delicious apples. There are luscious figs also, and olives in full growth …. Pear grows on pear, apple on apple, and fig on fig, and so also with the grapes, for there is an excellent vineyard …. In the furthest part of the ground there are beautifully arranged beds of flowers that are in bloom all the year round. Two streams go through it, the one turned in ducts throughout the whole garden, while the other is carried under the ground of the outer court to the house itself, and the town's people draw water from it*". (Homer, trans. Butler 1900). Another Homeric garden is that of Laertes, Odysseus's father—*a* "*great orchard*" (Ibid.) surrounded by a dry stone wall. For example, in the Italian landscape (which is very representative of the entire Mediterranean area's geography, environmental variability and history), the presence of fruit trees is generally celebrated with authoritative testimonies by the geoponic Latin writers: Columella, Pliny the Elder and Terrenzius Varro. In *Rerum Rusticarum*, Varro symbolically asks: "*Is not Italy so covered with fruit trees that it seems one vast orchard?*" (Varro, trans. Fairfax 1918). Fruit trees can be effectively identified as the most distinctive trait showing environmental diversity as well as the complex course of human history so clearly legible in the traditional Mediterranean landscapes.

The Greek model—an enclosed polyspecific garden, with a regular planting distance between trees—is also found in the Roman landscape in the form of small fruit and vegetable gardens (*hortuli*), which surround the city where temporary dwellings (*tabernae*) house worshipers of the sacred Lares (Grimal 2000). The economic and territorial growth of the Roman Empire lead to the adoption of Oriental paradise garden models, containing both useful and ornamental species and a strong architectural and monumental component that would mark the style of the Renaissance as well as the later Neoclassical gardens in future centuries. The mixed fruit crops of the *hortuli* would be the hallmark of what would be the Italian agricultural landscape *par excellence*, that of the *coltura promiscua* which overlays trees with grains, combining permanent crops with temporary herbaceous crops and natural patches (forest, woodlot, *macchia Mediterranea*, etc.), and occurring in a variety of forms. One of the most complex, for example, being the *alberata* (Desplanques 1959) in Central Italy, a mixed cultural system that trains grapevines on living trees (Agnoletti 2013). In Southern Italy, exstraordinary examples of *coltura promiscua* wich still survive are the terraced almond orchards of Gargano (Apulia) and the carob orchards of the province of Ragusa (South-eastern Sicily) (Barbera et al. 2010). Further development, both in terms of biodiversity and cultivation techniques, would come to the Mediterranean polycultural and polyspecific garden and its relative landscape, from the Arabic agricultural revolution between the ninth and twelfth centuries. In Sicily, the Caliphates and the Norman kings adopted Islamic landscape styles and agricultural systems in their gardens and parks where "business is mixed with pleasure, with science and with the arts" (Watson 1983). They were the privileged sites of the introduction of new species and techniques. This innovation does not regard a specific species or technology, but a systemic change in the vision of the agricultural system and its landscape—defined as a space in which different technologies (machines and hydraulic systems, mills, land improvement, rotations, inter-cropping, crops) contribute to a better use of water resources, a temporal and spatial differentiation of production and the connection of the various irrigation, energy, micro-climate and aesthetic functions into a system (Barbera 2005; El Faïz 2005; Cressier 2006). Citruses would begin to spread throughout the gardens and parks of the Mediterranean. Their presence would most define the characteristics of the Mediterranean landscape and affirm its place in the European imagination—not because of the vast number of trees, bath for the utilitarian or beauty products made from them—as a place of scenic beauty and eternal spring (Barbera 2000). In recent classifications of European landscapes of historical interest, this type of landscape is generally defined as the *huerta* or as one of its components (e.g. Courtot 1989; Meeus 1995; Zimmermann 2006).

The Mediterranean landscapes would increase their biodiversity again when new species were brought in by the conquest of the American continent: the most important for its impact on the natural and rural landscape would be the cactus pear (Barbera et al. 1992). Gardens would once again be the sites where unknown plants were introduced and would subsequentially affect European culture, some whose botanical characteristics were totally unknown, and others who were previously considered poisonous. In reference to the Italian territory in particular, in the next century fruit orchards, in various land mosaic pattern, would find a privileged place in the South. Apart from the small plots belonging to rich peasants or the middle-class bourgeois, "there was also the more massive initiative of feudal and ecclesiastical lords, whose possessions almost always contained gardens and preserves (*starze*), closed and well-defended plantations destined for the pleasure of the lords and to increase their revenues". (Sereni, trans. Litchfield 1997). The oldest records date back to the sixteenth century when Leandro Alberti (quoted. Bevilacqua 2000, trans. authors) writes of Ostuni's specialized orchards in Apulia (South-eastern Italy): "one sees many trees and lots of almonds planted in this way, how marvellous to consider that it was possible for so many trees to be planted by those men". In the 1500s, what is believed to be the first treatise specialized in fruit cultivation in the world was written in Southern Italy (Janson 1996): in *Agriculture Opusculum*, the Sicilian Antonino Venuto (1516, trans. authors) focuses on the typical species of southern fruit as well as a newcomer, "The orange, king, prince and master of all trees". Specialized fruit cultivation would be further established in the southern regions of Italy over the next centuries, especially after the crisis of the *latifundium* (large estates). Fruit trees (almonds, olives) and grapes would become the protagonists of the transformation of the South and its agricultural landscape (with terracing, hydraulic management, stone-clearing and stone-made artefacts), creating a more patched land mosaic. The South would participate in the agricultural revolution of the nineteenth century that transformed the European countryside with its fruit trees. While in Northern Italy, the same revolution would regard the spread of crop rotation with foraging, especially on arable land (Bevilacqua 1996). Finally, Southern Italy's countryside landscapes were finally made accessible by the spread of steamship and railroad connections and have been an integral part of European culture since the end of the eighteenth century. At the end of their *Grand Tour*, young German, French and British intellectuals spread the myth of the Mediterranean landscape. The polycultural and polyspecific *Mediterranean garden* landscape would keep these same distinctive characteristics until the crisis of *agricoltura promiscua*, and the phenomena of urbanization in the 1960s/1970s. The spread of industrial models destined to fail and devastating construction speculation along the coastal plains and in the peri-urban areas would devastate the landscape's beauty and production. The effects of this cataclysmic change can be seen today in degraded agricultural landscapes and systems that are no longer efficient in terms of their production or in terms of their environmental sustainability and cultural impact. However, in recent inventories and catalogues of European landscapes, historical and complex landscapes (coltura promiscua, polyspecific gardens) are important still today in various parts of the Mediterranean (e.g. Firmino 1999; Kizos and Koulouri 2006; Pinto-Correia and Vos 2004; Sirami et al. 2010). Once again, looking for example at national level in Italy, the recent Historical Rural Landscape Catalogue (Agnoletti 2013) highlights many extraordinary examples of still surviving complex and polyspecific landscapes. According to the main statistics of this catalogue, about the 40 % of the national historical landscapes of Italy are characterized by mixed crops and inter-crops, as an expression of different gradation of *coltura promiscua* land mosaic pattern. So that, today many of these landscapes are distributed along the entire Italian peninsula. To cite some of most important examples that express their polycultural character and complexity, already in their denomination (Agnoletti 2013): the "Terraced and irrigated chestnut groves and vegetable gardens in Alta Valle Sturia" (Liguria, Northern Italy); the "Landscape mosaic of Montalbano" (Tuscany, Central Italy); the "Polyculture of Loretello" (Marche, Central Italy); the "Mixed hill cultures of the lower Irpinia", the "Terraced orchard-gardens o f the hills of Naples" (Campania, Southern Italy); the "Mixed orchards of the Temples Valley", the "Polyculture on the slopes of Mt. Etna" (Sicily, Southern Italy). The general distribution shows and increasing importance of these complex systems of land use/cover types, and related patched landscapes, moving from Northern Italy to the South (in the Central Mediterranean Basin).

2.2.1 The Mediterranean Garden and Landscape Definition Requires Clarification

The word "garden" ("*giardino*" in Italian, "*jardin*" in French) with which fruit orchards are frequently defined in the Mediterranean area, goes back to the ancient Indo-Germanic *ghordo*, which means an enclosure, and the Greek *chortos*, Latinized as *hortus* meaning "small cultivated enclosure" (Venturi Ferriolo 1989). The confined space allows for the trees and the factors affecting their production to be kept under control; it is where man reaffirms and refines his supremacy over a friendly and collaborative nature, who, in return, rewards him with the very best she has to offer, her fruits (Barbera 2007). According to Sereni (1961), in typically Mediterranean regions, such as Sicily and Southern Italy, the origins of the Mediterranean garden can be found in the Classical period, with the colonization of Magna Graecia, as much as in the subsequent Arabic colonization (between the ninth and twelfth centuries). This shows how multi-temporal and complex the evolutionary process that shaped the Mediterranean garden landscape was.

In the example of an in-depth study of Sicily, but also in many other Mediterranean regions, the term "garden" has, at least since the age of the Normans (tenth to twelfth century), referred to an orchard or garden orchard. As shown by the term's development through history, this has determined its closed nature, the presence of evergreen species and irrigation, and non-production-related uses. It can have the extremely simplified characteristics of the island of Pantelleria's (Central Mediterranean Basin) "*jardino*" (a single tree—usually a citrus—protected by a dry stone wall that creates a micro-climate that allows the tree to grow without irrigation; this is indispensable in such a profoundly arid Mediterranean edaphic-climatic context (Barbera 2000) (Fig. 2.2a)—thus connecting itself directly to the primitive idea expressed by Venturi Ferriolo (1989) starting from the Sumerian gardens represented by a proto-Elamite pictogram showing a single tree enclosed by a fence (Fig. 2.3). Conversely, if we retrace the historical-cultural events mentioned above, the concept of the Mediterranean garden can be expanded

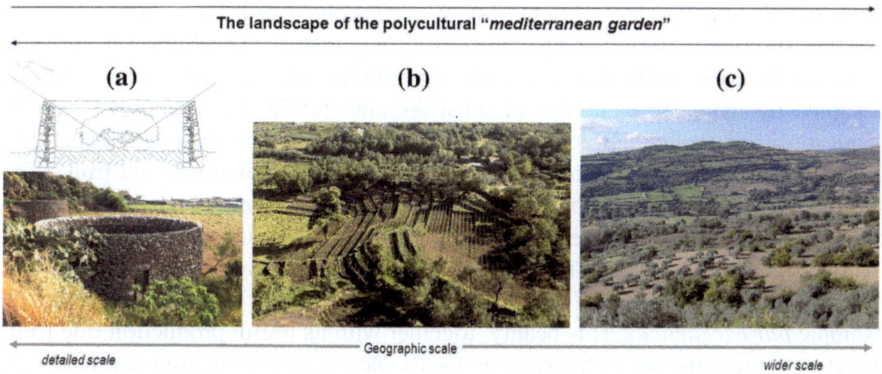

Fig. 2.2 The landscape of the polycultural "Mediterranean garden": from the protection of a single tree to a complex land use mosaic. Examples: **a** Mediterranean garden of Pantelleria (Sicilian channel), with e single citrus tree enclosed and protected by dry stone artefacts; **b** landscape of the Mediterranean *coltura promiscua* at fine-grained land use mosaic pattern (Noeth-eastern Sicily); **c** landscape of the Mediterranean *coltura promiscua* at corse-grained land use mosaic pattern at inter-plot level (SE-Sicily)

Fig. 2.3 The tree of life enclosed by a Sumerian pictogram from the III millennium B.C. Probably the first representation of a garden design that is still repeated today, for example in the "*jardini*" of Pantelleria (Southern Italy) or in Folegandros (Southern Greece) (*source* Venturi Ferriolo 1989)

and complicated, in terms of its spatial and structural complexity, its organizational forms and in the plants present in it. As Venturi Ferriolo says "planting fruit trees is the proper meaning of the garden" (Ibid.). In fact, for the Mediterranean gardens of Biblical Palestine, to those of Homeric Greece, to the Roman *horti*, to the *noharia*, *sanya*, *xirba* and *bahira* of Islamic Sicily, crop specialization (specialized or mixed) is defined differently according to time and place. This crop (or these crops), tend to be mixed, relatively small and located near the house and settled areas maintaining a polyspecific inter-cropping mosaic pattern of land use, in the large coastal Sicilian citrus groves and, in general, throughout the Mediterranean coastal regions.

For example, in Syracuse (South-eastern Sicily) the word "*giardino*" or garden (Sestini 1963) is substituted by "*paradiso*" or paradise (Trischitta 1983). These citrus landscapes, with that harmonic equilibrium characteristic of a "nature" designed by man, are examples of what Assunto (1973, trans. authors) called a "widespread aesthetic" not contained in a limited space. The same author believed them to be the perfect landscape due to their "simultaneity of flower and fruit". This simultaneous presence of the "two moments in the plant which represent the playfulness of beauty destined for contemplation and the delectable result of cultivation" is possible due to the coexistence of different trees with different phenologies or the presence of one re-flowering citrus tree (the lemon, being the example *par excellence*). This beauty, together with its useful production (the fruit), therefore represents one of the reasons for its success, its production and landscape are emblematic of the Mediterranean garden. In testimony to this, the historian Lupo (1990) emphasizes the attention that the world of arts and letters has given to the citrus garden. However, the citrus garden, because of its size and shape, does not always maintain the complexity of the mosaic land use pattern that is typical of the larger and more widespread polycultural and polyspecific *Mediterranean garden* sensu (Sereni 1961), and was reduced to one of the many patches that compose it. In this respect, Horden and Purcell (2000) state that the term *Mediterranean garden* is not to be understood as a particular crop type, but, rather as "... a location, a type of use of labour and resources ...". The Mediterranean garden/orchard is often associated with irrigation (Bolens 1989; Horden and Purcell 2000) and interacts with all aspects of the generally complex and articulated agro- and forest-ecosystem that surrounds it.

This analysis, with its two extremes, can give rise to a semantic confusion that isn't useful for the recognition and better definition of the complex *Mediterranean garden* landscape (Fig. 2.2a–c). Therefore a reflection on today's idea of these landscape, which contains various landscape types and different definitions (from Pantelleria's cell-like garden to the larger citrus plantations, from small mixed orchards to the varyingly large and complex mosaics of Mediterranean polyculture—*coltura promiscua*, from both principally ornamental gardens to productive ones) is therefore considered useful (Fig. 2.2).

2.3 The Mediterranean Garden and Polycultural Landscape as Cross-Reference Type in Current Landscape Inventories and Typologies

In order to better understand the complexity of the landscape and the large environmental value inherent in the historical definition of the polycultural systems and landscapes (in terms of complexity, structure and function), it may be useful to refer, where possible, the typology of these landscapes to the different recent inventories of the Mediterranean landscape that have been defined at the national and international level. For example, studies have been carried out over the last few decades, both in order to provide national landscape inventories (Blasi et al. 2000; Bunce et al. 1996; Pinto-Correia et al. 2002; Agnoletti 2013) as well as to define what connects them in typological terms and terminology (Meeus 1995; Zimmermann 2006; Mücher et al. 2010). However, only a few examples and experiences of landscape typologies have exclusively concerned the Mediterranean area (Vogiatzakis 2011), and existing inventory are not fully able to identify traditional cultural landscapes (Solymosi 2011). One of the most important typologies at the European level is the typological prospect proposed by Meeus (1995), which identifies the following major landscapes for the Mediterranean area (Table 2.1): *Mediterranean open field*, *Coltura promiscua*, *Mediterranean semi-bocage*, *Montado/Dehesa*, *Delta* and *Huerta*, *Terraces* and *Mountains*. Again, Pinto-Correia and Vos (2004) describe the following main traditional multifunctional landscapes for the Mediterranean area: the *Pastoral and forestry landscapes where transhumance takes place* (in a Mediterranean-montane climate), the *Small-scale mixed farming landscapes with marshes* in North-eastern Portugal (Atlantic–Mediterranean mountain climate), the *Coltura promiscua with métayage* (*sharecropping*) in Central Italy (Mediterranean climate Submediterranean hills), the *Chestnut grove landscapes* in France and Italy (Submediterranean hills and mountains climate), and *the Landscapes with montados and dehesas* in Southern Portugal and Spain (Mediterranean-continental climate plains). Many of these take on a strong European identity at the regional level and are consequentially considered *Regional Landscapes* (e.g. the *Coltura promiscua*, the *Montado/Dehesa*, the *Huerta*, etc.) (Meeus 1995). The complexity and importance of the *Mediterranean garden* described above regards many of these landscape types in practice (Table 2.1).

If we go to the national scale, only a few heterogeneous (in their methodological approach) published inventories are available. For example, looking at the Italian nation (Barbera and Cullotta 2009) and more specifically at Sicily (which is quite representative of the rest of the Central Mediterranean Basin in terms of its geography, morphology, environment and according to historic and cultural development described above), it is possible to highlight the presence of a wide range of landscapes. For example, in an analysis that is fundamentally based on physical aspects, Blasi (2007) identifies 34 "Landscape Systems" on the island (27 from the Mediterranean climate region and 7 from the temperate one) out of a national total of 67, making Sicily one of the Italian regions with the highest landscape variability. This wealth of landscapes is also confirmed by the results of the "Geographic Landscapes of Italy" inventory (Amatucci et al. 2001), proving once again that Sicily is one of the regions with the highest number of landscapes (13), as well as by the "Type and Unit Map of the Physiographical Landscapes of Italy" (*Carta della Natura* Project, APAT 2003) which identifies 37 types of landscapes grouped into 5 major physiographic units. These landscape inventories are primarily based on physical elements (geomorphology, lithology and hydrography) although they also consider anthropic elements (i.e. main soil covers and land use). Recently a first survey ever done on the traditional rural landscapes of

Table 2.1 Types of traditional Mediterranean rural landscapes at different scales (i.e. Meeus 1995; Barbera and Cullotta 2012), and their relative most representative characters (physiography, crop/cover types, rural architectures), compared to the transversality inherent in the definition and historical path of the polycultural Mediterranean garden

Main traditional landscapes of Europe (Mediterranean Europe) (after Meeus 1995) *(at European level)*	**Main traditional landscapes of Sicily** (after Barbera & Cullotta 2012) *(at sub-national / regional)*	**Most representative characters:** *land morphology*	*crop/vegetation cover*	*rural architecture*	**Transversal importance of Mediterranean garden characteristics**
Mediterranean open land	***Mediterranean open land***	plains, rolling hills	wheat, natural pasture	isolated human settlements	
Coltura promiscua	***Coltura promiscua***	various (marginal lands are frequent)	mixed crop and orchard, olive, almond, citrus, orchards, vineyard, hazelnut, pistachio, ash tree, patches of natural vegetation	diffused human settlements and considerable presence of rural architectures (stone-made terraces, enclosures, pathways, etc.)	
	Circum-Sicilian islands landscape	various	small mixed crop and orchard	diffused small human settlements and considerable presence of rural architectures (terraces, enclosures, pathways, etc.)	
Terraces	***Terraced landscape***	various (marginal lands are frequent)	Various mixed and specialized crops	considerable presence of terraces and other stone-made artefacts	
Huerta Delta	***Coastal and sub-coastal landscape with intensive crops***	alluvial plains	orange, lemon, mandarin, herbaceous crops	medium presence of rural artefacts	
Mediterranean semi-bocage	***Mediterranean semi-bocage/semi-open land***	plateaus and slopes	olive, carob tree, almond, patches of natural vegetation	medium presence of rural artefacts (enclosures, pathways, constructions of varying size)	
Montados/Dehesa	***Rainfed specialized fruit tree landscape***	plains, rolling hills and lowlands	olive, almond, vineyard	isolated human settlements	
Mountains	***Forest mountain and sub-mountain landscape***	Mountains	forest, other wooded lands	Rare/isolated rural constructions (to shelter people and animals)	
	High-mountain natural landscape	Mountains and peaks	open land, high-mountain shrubs	None rural architecture	

Italy (Agnoletti 2013) based on a more diversified environmental approach, including also cultural, historical and social aspects; it presents more than a hundred case studies distributed through the country as a preliminary study conducted in view of the compilation of a national register of historical rural landscapes and traditional agro-forestry practices. An initial overview of the main Sicilian macro-landscapes has recently been proposed (Cullotta and Barbera 2011; Barbera and Cullotta 2012) as part of a research project directed towards making the inventory and classification systems for landscapes at different scales homogenous and interconnectable in order to use a more detailed scale at the regional level (i.e. NUTS of II-level—Nomenclature of Territorial Units for Statistics by regional level) for the Sicilian landscapes. Based on a multidisciplinary approach, this typology lends itself to parallels with the landscapes defined by Meeus (1995) and other Authors (e.g. Zimmermann 2006) (Table 2.1) because of its shared nomenclature. In total, nine macro-landscapes (Fig. 2.4) (Barbera and Cullotta 2012) expressive of traditional landscapes have been described for Sicily, that is, land uses and covers that have a long history and that have been subject to slow changes over time (sensu Antrop 1997). In Table 2.1, these most representative landscape characteristics (according to: size, physiography, crop/vegetation cover types, rural architectures and artefacts) are reported for Sicily, and their typological link with the Meeus (1995) landscape types, for an easy upscaling at European level, is given. The aforesaid inventories, and most important characters, show the presence of *Mediterranean gardens* and polycultural systems in various types of traditional rural landscapes, and at various geographic levels (Table 2.1): in the *Coltura promiscua*, in the *Coastal and sub-coastal landscape with intensive crops*, in the

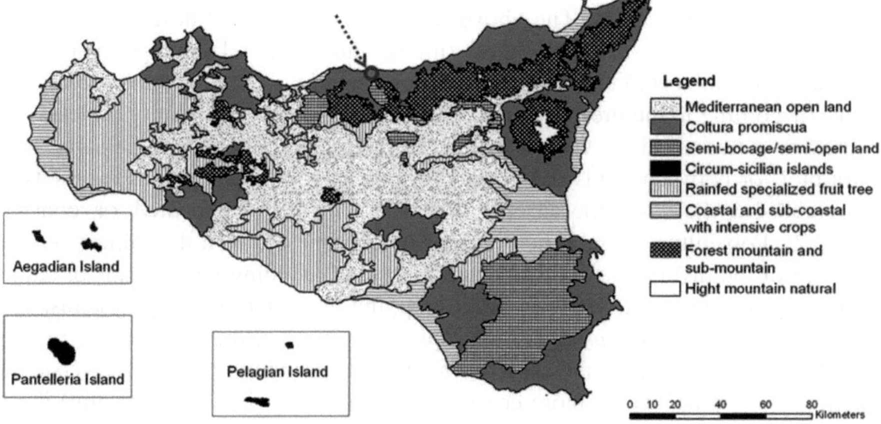

Fig. 2.4 Map of the main traditional landscapes of Sicily (after Barbera and Cullotta 2012); the location of the *Halaesa* site in the northern coast is indicated by the arrow

Mediterranean *Semi-bocage* and, in part, in other types of landscapes. The historical evolution of the definition of the *Mediterranean garden* and polycultural systems has given it, in practice, a multitude of elements that characterize its macroscopic landscape pattern. These classification systems, with their different details and scales, finally make sense of the complexity of the Mediterranean area at any level.

2.4 Variability of the Complex Mediterranean Polycultural Land Mosaic Pattern Characterized by the Presence of Trees

The definitions of the Mediterranean, specifically Italian, rural landscape derived from the historic development described above are attributable to a landscape that is so complex and diverse that it includes several of the major Mediterranean agricultural and agro-forestry landscapes that have been catalogued and described to date. In it, the coexistence of spaces dedicated to agriculture, forest and pasture refer back to the Latin categories of *ager*, *saltus* and *silva.* Geographers such as Vidal de la Blanche (1922 cit. in Claval 2007) or, more recently, historians like Aymard (1992) and landscape ecologists such as Pinto-Correia and Vos (2004) still consider these categories relevant, especially considering that the *ager* not only includes crops but also shrub and fruit orchards, the *saltus* mainly concerns aspects of maquis and garrigue scrublands affected by grazing, while the *silva* is made up of woodlands (where different traditional silvo-pastoral uses are practiced). These are all polycultural systems, composed of agricultural and agro-forestry patches (Cullotta et al. 1999; Barbera and Biasi 2011) that, through wood cutting, felling and the use of branches as animal feed, controlled fires, crop cultivation, livestock and their transhumance, water control and terracing, define the prerequisites—or "the golden rules" (Blondel 2006)—of the ancient Mediterranean agro-silvo-pastoral systems. The above-reported concepts and descriptions of the traditional Mediterranean polycultural (mixed cultivation) landscape in practice generate different spatial combination (i.e. configuration) of the *coltura promiscua.* In fact, according to the presence and location of the plant promiscuity (at least one of them is a tree species), we have different landscape patterns: in the same field (intra-plot) or between fields (inter-plot) (Fig. 2.2). Morover, this heterogeneity is reinforced by the land mosaic patchiness (i.e. more fragmented or coarser) (Fig. 2.2). Polyculture is an agriculture using multiple crops in the same space, in imitation of the diversity of natural ecosystems and avoiding large stands of single crops, or monoculture. It includes multi-cropping and inter-cropping systems. The "*coltura promiscua intra-plot*" is a multiple cropping, i.e. the practice of growing two or more crops in the same space during a single growing season. Vice versa, the practice of growing two or more crops in proximity (i.e. the inter-cropping) is the expression of the "*coltura promiscua inter-plot*" (Fig. 2.2c).

2.5 Approaching a Polycultural Mediterranean Garden Definition

*2.5.1 A First Description of the Complex Mediterranean Rural Landscape: The "*Tavola Di Halaesa*"*

Emilio Sereni's "*History of the Italian Agricultural Landscape*" (1961, trans. Litchfield 1997) is quite helpful in the search to identify a more comprehensive definition of the polycultural *Mediterranean garden* landscape that includes all of the different historical interpretations that it has gone through. This author attributes the characteristics of the traditional Mediterranean rural tree landscape to the landscape of the "*Mediterranean garden*", and shows how those characteristics are maintained all the way through time—from the Greek colonization to the post-war agriculture following WWII—remaining well-defined and long-lasting with regard to their formal structure and functional biotic and abiotic elements, their ecological and geographical context, their plurality of functions, their social uses and economic determinants (Sereni 1961). It is a landscape with irregular closed lots … defined by "*the polygonal irregularity of contours*" (Sereni, trans. Litchfield 1997), that is fragmented, twisted, squished, and formed by a "*tangle of little wooded plots divided by walls or hedges*" (Ibid.). Until the eighteenth century, this landscape was mainly "*restricted (to) suburban or coastal zones. It was thus still isolated amid vast extents of uncultivated land or open fields*" (Ibid.). It was "*an agricultural landscape of closed fields, vineyards, gardens, and fruit trees*" (Ibid.), and often even of arable and pasture land "*imprinted with a suburban physiognomy by dividing walls and the contiguous placement of houses and rustic storage sheds*" (Ibid.). "*Terraced arrangements became the chosen location for the most valuable crops, and particularly for trees and shrubs*" (Ibid.).

For the first time in landscape literature, Sereni (1961) retraces the elements of such mixed and complex forms of agro-forestry systems and landscapes, named *giardino Mediterraneo* (*Mediterranean garden*) starting with the forms and functional characteristics found in the *Tavola di Halaesa* (literally: *Stele of Halaesa*) (Fig. 2.5), a Greek colony founded in 403 B.C. (Northern Sicily), and the plan based on it drawn by Sicca (1924) which provides schematic information on the land use and landscape structure of the city of *Halaesa* (Fig. 2.6) (near the present Tusa) on the Tyrrhenian coast of Sicily (Fig. 2.4), between the second half of III and I B.C., during the establishment of Roman rule and following a probable redistribution of land (Barbera and Cullotta 2014). A reconstruction similar to that one, although more lean but with the same elements, can be seen in the two drawings made by Arangio Ruiz and Olivieri in 1925 (Fig. 2.7a), which graphically portray the information found on both of the columns of the marble epigraph (in total three marbles discovered in different moments: 1558, 1885 and 1958; the first citation by Fazello 1558) published by Torremuzza in 1753 (Burgio 2008). The text and drawings allow a first reading of the area (Figs. 2.6 and 2.7), but to arrive at its more precise definition, in as much as it is a cultural landscape derived from an

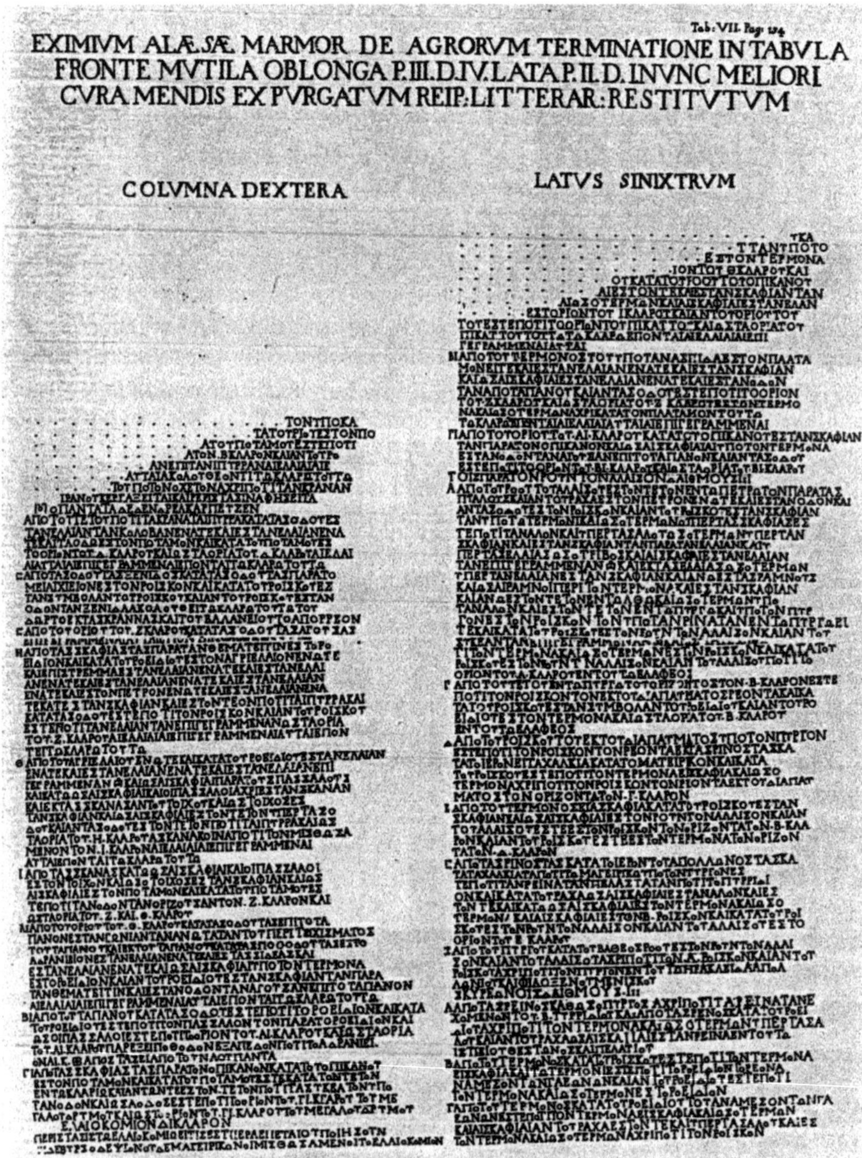

Alesa, rilievo della *tabula* pubblicato dal Torremuzza nel 1753.

Fig. 2.5 The "*Tavola di Halaesa*" (Stele of *Halaesa*) that describes the land use and landscape composition of the Greek colony founded in 403 B.C., Northern Sicily

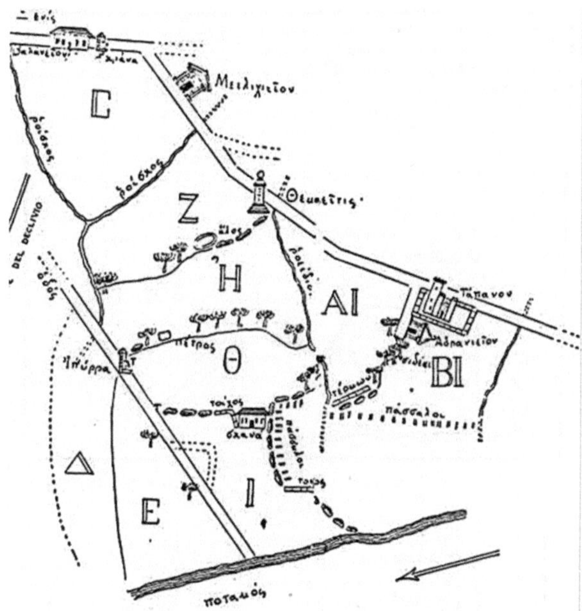

Fig. 2.6 The elements that define the traditional Mediterranean complex tree landscape through the plan based on it drawn by Sicca (1924) which provides schematic information on the land use and landscape structure of the city of *Halaesa* (Greek colony founded in 403 B.C., Northern Sicily)

encounter between natural characteristics, human history and perception, the reading by Belvedere (2008) of the epigraph based on the historical and archaeological findings, and literary references to the area is fundamental. From this multidisciplinary reading and analysing in details the information reported in the two columns of the marble epigraph (see Fig. 2.5), the analysed territory is defined by what appears to be a highly fragmented and heterogeneous agro-silvo-pastoral land use mosaic pattern, despite the fact that the exact size and shape of the patches which compose it cannot be specified. It is located in a suburban area, as is demonstrated by its proximity to the city walls, and is situated on the slopes interrupted by plateaus next to the river and stream beds (Fig. 2.7b). These, along with the springs present, constitute a dense hydrographical network which also involves aqueducts formed by clay channels and pipes as well as drains which come out from the walls. The fragmented nature of the mosaic is also determined by the network of various sized roads (Figs. 2.6 and 2.7): at least two connecting routes between the coast, the city and the interior as well as side roads and private unpaved pathways. These access ways link the sacred areas and important buildings together with some rural constructions presumably used for housing agricultural products or gear. *Aie* (farmyard for threshing too) are also present. The main crop grown is the olive, as shown by the frequent presence of cultivated specimens, along with ones growing wild in the maquis, those that have sprouted from the coppicing, and those that were abandoned after being planted. Some trees are located on boundaries and have the double function of a marker. The importance of olive trees and oil production is shown, not only by the widespread presence of the species throughout all of the areas described, but also by the existence of a nursery or, according to a

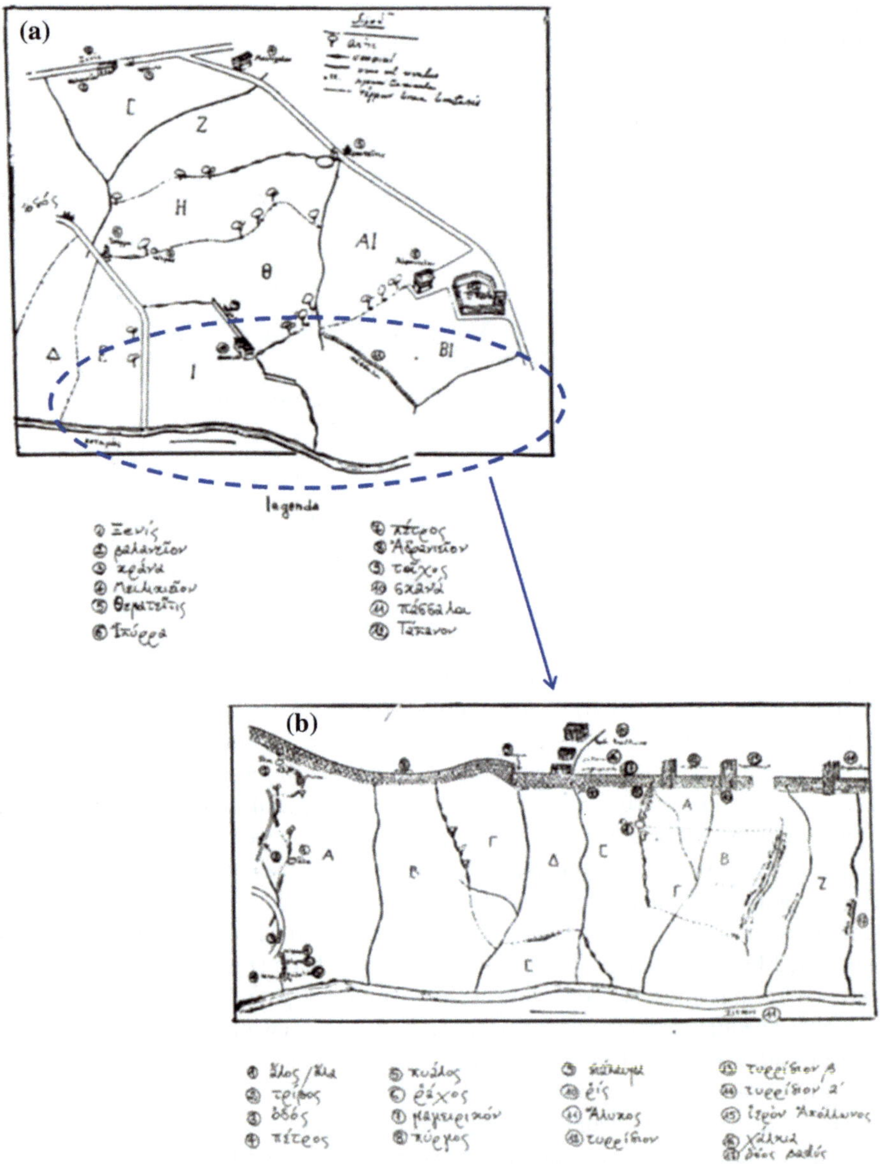

Fig. 2.7 Plan based on the *Halaesa* marble epigraph (**a**); and (**b**) landscape characters of the suburban slopes located between the city walls (*top*) and the today "Tusa" river (*lower*) (Greek colony founded in 403 B.C., Northern Sicily). By Arangio Ruiz and Olivieri (1925)

different hypothesis, an oil press. The olive tree is probably also present as a sacred tree. The crop system in areas with a lower slope is that of arable land and wooded pastures following a simple system of fallow fields. The presence of hydraulic structures, found during archaeological investigations, has made it possible to imagine the presence of irrigated orchards. Some of the other tree species present are: figs, pomegranates, pears and plums. They are certainly grown in *coltura promiscua* but are also used as hedges or grow wild, especially plums and pears. The latter is also used to mark the plot boundaries. Even briars and palisades, stone walls and boulders contribute to score and mark the boundaries by integrating the subdivisions marked by the hydrographical network and roadways. The grapevine is also cultivated. The soil, due to the slope's gradient, had to be at least partially terraced. The livestock present are sheep, goats and pigs. These were kept free range and also fed cork oak acorns, which were also used to tan leather. The forest and maquis saw grazing, hunting, as well as traditional wooden and non-wooden productions. These latter natural patches contributed to the landscape mosaic of the area as core and buffer areas, as well as corridors of the ecological networks.

The long-lasting historical landscape of *Halaesa* shows its polycultural characters still today. Through the centuries and the following historical and cultural processes, this landscape was able to conserve the characters of its foundations, later on enriched by additional and traditional crops and land uses. The landscape is still complex and irregularly patched; with crops, fields and natural cover types expressive of a traditional agro-silvo-pastoral Mediterranean landscape (Fig. 2.8). Currently, the olive grove is the most important cultivation yet. Most important

Fig. 2.8 A today view of the *Halaesa* area with an example of intra-plot mixed cultivation (*giardino Mediterraneo—coltura promiscua*) at stand level with various cultivated and wild plants (olive, fig, almond, pomegranate, citrus, cactus pear, apricot, peach, walnut, cherry, vegetables, oaks, woodlots of Mediterranean maquis-forests, elements related to the management and the use of water for irrigation)

veteran trees, especially olives, are still today used as historical landmarks, for instance as boundary trees between land properties and fields. The historical rural architecture and artefacts are currently marked by the archaeological site and evidences of the *Alesa* city, small settlements, rural buildings, stone-made boundaries, terraces, enclosures, pathways, aqueducts, etc.

2.5.2 Landscape Ecology, Multifunctionality and Cultural Heritage of the Complex Mediterranean Agro-Silvo-Pastoral Landscape

The historical and cultural importance of the *Halaesa* landscape is not the only thing that justifies the preciousness of these landscapes or the traditional farming practices that allow for their perpetuation. In fact, their strategic role in the preservation of a multifunctional agro-ecosystem ensured by the sustainability of the agricultural practices congenital to them is just as recognizable.

First, the complexity of the *Halaesa Tavola's* landscape is caused by the wealth of its physiographic context, common to many areas from West to East, from North to South bordering the Mediterranean (e.g. Braudel 1986). The city of *Alesa* is located on one of the foothills of the Tyrrhenian Nebrodi Mountains in a position close to the coast (see Fig. 2.4). The heterogeneous physiographic and environmental characteristics, following a hypothetical NS transect, clearly show that the territorial complexity results from the contact between very different environments, ranging from coastal to mountain ones. A great morphological diversity is expressed by plateaus, hills, summits, coastal and sub-coastal areas, rivers, etc. and results in an equally great biological and soil diversity contributing to the richness of each individual ecosystem as well as to that which comes from the ecotone relationship between the natural and cultivated patches.

The city's central location within the wider Mediterranean geographical area and human history have both contributed to further increasing the complexity and diversity that has made *Halaesa* a frontier town, and therefore a place of exchange between Sicilians, Greeks, Carthaginians and Romans marked by a multiculturalism that Cusumano (2006) applies to the sacred cults but that can also be applied to a plurality of agro-ecological knowledge (different species and varieties, agricultural techniques, etc.). The fact that *Halaesa* was an active centre of exchange and biological and cultural hybridization is also due to its location near the sea and along the NS coast of Sicily, an important route for the transportation of agricultural products from the hinterland to the coast (mainly wheat: during the Roman period, Sicily was considered the granary of Italy) as well as its location on the E–W coastal connection (see Fig. 2.4).

From this physical, geographical and historical diversity, a large, complex and heterogeneous mosaic follows, consisting of irregular patches of different shapes and sizes, with different morphological and typological traits in relation to their position in the area (plains, hilly areas, etc.), soil use (forest, scrub, grassland and cereal crops, tree crops, shrubs and presumably mixed and specialized vegetable gardens and orchards), the spatial and temporal succession (rotations, alternations, transhumance). The patches were enclosed by hedges and groups of trees and shrubs, or by non-living barriers (walls, palisades) or separated from natural or semi-natural areas by *buffers*. Linear structures in the form of ecological corridors—rivers, ditches, aqueducts, living and non-living barriers and terraces—are additional elements that diversify the ecosystem by functioning as ecotones. Further biodiversity comes from large boulders, piles of stones, isolated trees, which not only provide agricultural products but also ensure shade and shelter from the rain for humans and animals alike, and act as micro-sites with the function of *stepping zones* for a countless number of plants and animals.

The *Halaesa* landscape is the result of a polycultural Agro-silvo-pastoral system which, observed under the lens of *landscape ecology*, highlights very positively in terms of complexity and richness with reference to specific and intraspecific biodiversity, and the ecosystem's structure.

In Table 2.2 all the detailed information reported in the two columns of the marble epigraph are given for this case study, and grouped according to the following main aspects of the landscape ecology (e.g. see Forman and Godron 1986; Farina 1998): shape of patches, patchiness, crop types, core areas and buffer areas, corridors and ecological network, linear and point elements, rural architecture and other stone-made artefacts. It is possible to locate these main landscape elements to the different lots (about twelve) in which the *Halaesa* landscape is divided. Thus, while the left column of the marble describes the city's lots located on the north-eastern side of the area, the western and south-western part is described on the right column (see Table 2.2). The information reported in this last column describes a more complex and structured land mosaic comparing to the other part of the city and surroundings. Each lot is in detail described by both natural and artificial boundary lines of fields and patches (small streams, creeks, holes, trenchs, rocks, trees, woodlots, stone pathways and vias, sacred areas, urban walls, etc.) (Table 2.2), although no details are given on its dimension and shape. The temporal and spatial complexity of the *Halaesa* land mosaic manifests itself in the economic productivity and environmental services that participate in that multifunctionality belonging to complex Mediterranean agricultural and agro-forestry systems (Firmino 1999; Barbera et al. 2004; Pinto-Correia and Vos 2004; Blondel 2006; Eichhorn et al. 2006; Kizos and Koulouri 2006; Brown et al. 2007; Biasi et al. 2012; Otero et al. 2013; Agnoletti 2013). Agricultural production and forestry, such as animal breeding, are carried out in a context of environmental protection that comes from the high biodiversity and the complex structure of the fields and their relationship in a network of connectivity (Fig. 2.7) with the woodland and scrub areas. The water cycle is assured by the territorial hydrographical and hydraulic structures that contribute, with the terraces, to protect the soil and slope stabilization. The organic matter cycle is guaranteed by the presence of mixed crops and the integration of crops/livestock/woodlots. These landscape material elements and characters are visually (at least, but many other functional aspects are involved)

Table 2.2 Landscape patches and elements, grouped in main landscape ecology aspects, reported in the *columna dextera* (Dx) and *latus sinixtrum* (Sx) of the *Tavola di Halaesa* (Greek colony founded in 403 B.C., Northern Sicily), that describes the landscape Between The Second Half of III and I B.C. (after Barbera and Cullotta 2014)

Landscape elements reported in the column Dx	Landscape elements reported in the column Sx
Shape of patches	
/	Irregular and twisted division of field plots
Patchiness and crops	
• Olive groves all around the main sectors of the city and presence of olive nursery • Vineyard, fruit trees (pear, pomegranates, figs) • Sheep–caprine livestocks arable land and mixed arable–olive systems • Grazing lands	• Patched land mosaic and closed fields • Olive groves all around the main sectors of the city and presence of olive nursery • Grapes, olive groves, pomegranates, figs, • Fruit orchards, vegetable gardens, closed fields • Grain growth • Arable land and mixed arable–olive systems • Sheep–caprine farm (livestock) • Pastuelands
Core areas and buffer areas	
/	• Cork oak (patches of different size) • Other woodlots and shrublands • Mediterranean maquis
Corridors and ecological network	
• *River* (*the today Tusa river*) • Mediterranean riparian vegetation	• Streams, rivers and riparian medit. vegetation
Linear and point elements	
• Small stream • Hedgerows, plums & pears (wild and cultivated) • Wild pears and thorn hedgerows (*rhamnoi*) (as field boundary) • Holes (natural) and trenchs (artificial) • Boundary rocks	• Small streams, creeks, springs, holes (natural) and trenchs (artificial) • Aqueduct lines • Wild pears and hedgerows (*rhamnoi*) (as field boundary) • Other wild trees and fruit trees as boundary (big olive, pomegranates, figs) • sacred trees
Rural architecture and other stone-made artefacts	
• Rural buildings • Fortified walls (*pyrgos*) • Public fountains, landry, drinking trough • Viability network • Terraces • Trench (artificial)	• Rural buildings • Fortified walls (*pyrgos*) • Sheds and storerooms for agronomic cultivation tools • Tower • Canals (of "U"-shape of cut stone or *terracotta*; *terracotta* tubes) • Public fountains, landry, drinking trough • Sacred areas • Stone pathways and vias • Stone-made terraces • Stone-paved farmyards
other	
	• Poles for grapes

reinforced by the presence of stone walls and other artefacts or features such as: countless types of rural buildings (country residences, storeroom for fruit conservation, wine cellar, storeroom for agronomic cultivation tools, terraces, hedgerows, canals, stone heaps, etc. (Grove and Rackham 2002; Brown et al. 2007; Petanidou et al. 2008; Barbera and Cullotta 2012). Along with the multifunctional and environmental production, these spaces also have a cultural production shown by indirect literary references (Theocritus, cf. Belvedere 2008), and more generally contribute to the complex and rich *non-material heritage* (dialects, music, narratives, toponyms, etc.) (Scazzosi 2004; Moreira et al. 2006; Cullotta and Barbera 2011; Otero et al. 2013) and the appeal that is characteristic of polycultural Mediterranean agricultural systems.

2.6 Conclusions

The long-lasting, complex- and mixed forms of agro-forestry systems and landscapes, such as the polyculture (*coltura promiscua*), is ascribable and can be associated with different and innumerable agricultural land uses and different kinds of landscape configurations. These landscapes have developed from the intricate historic processes that have produced their cultural identification and evolution over the passage of time. The pre-Classical and Classical introduction of Asiatic species, the Islamic "agricultural revolution", the introduction of American species, the introduction of the species that arrived via the activities of plant collectors and European scientific institutions were most important historic milestones in this long-lasting process. The historic development traced above and the approaches currently used to classify, map and characterize the principal agrarian and agro-forestry landscapes in the Mediterranean context requires some reflection, so that even these latest useful and indispensable landscape inventory tools can better help define and unquestionably individuate those historic landscapes typical to the Mediterranean context, which are now, more than ever, subject to processes of abandonment and transformation.

The analysed *Halaesa* landscape, located in the geographic centre of the Mediterranean Basin, as one of the first historical detailed description of a complex Mediterranean rural landscape, is the result of a polycultural agro-silvo-pastoral system which guarantees complexity and richness (in terms of structural and biological diversity), as well as with reference to others environmental, cultural and economic aspects.

The presence of historical features, of traditional crops and land uses, of traditional land management and the conservation of architecture and other material cultural heritage related to the agricultural activity (i.e. historical rural monuments, rural country houses and settlements, terraces, stonewalls and related rural artefacts, agricultural and forestry tools and machines, manuscripts, poems, paintings and pictures), as well as non-material cultural heritage (e.g. dialects, music, narratives, etc.), are particularly important aspects considered by international and European organization toward their valorisation and conservation. This knowledge of traditional landscapes and of the polycultural *Mediterranean garden* landscapes in particular, is particularly urgent. A nomenclatural and characterizational definition that is preferably as holistic (multidisciplinary) as possible needs to be developed with the aim of individuating and planning strategies that conserve the landscape's characteristics, functions and identity. This theme is quite present in the PAC, as demonstrated by the recent indications of the European Council relative to its strategic orientation concerning rural development that should reflect the multi-functional role that the agricultural activity carries out in terms of the richness and diversity of the landscapes, of food and cultural (material and immaterial) products and natural heritage. Their constitutive complexity and multifunctionality should be dealt with from different points of view so as to reflect their over all inter-cultural value.

Acknowledgements The research was supported by the Italian Ministry of Research and University Education's PRIN 2010–2011 Project (University Programs of Scientific Relevance)—Project title '*I paesaggi tradizionali dell'agricoltura italiana: definizione di un modello interpretativo multidisciplinare e multiscala finalizzato alla pianificazione e alla gestione*' ('Traditional agricultural landscapes in Italy: a multidisciplinary and multi-scale assessment for the development of an integrated model for landscape planning and management') (prot. 2010LE4NBM).

References

Agnoletti M (2013) Italian historical rural landscapes: dynamics, data analysis and research findings. Springer, Dordrecht

Amatucci M, Buscaroli A, Degli Esposti AV, Ghepardi M, Rosetti-Vianello G (2001) Paesaggi geografici—Territorio senza confini nel sistema informativo scolastico. Ministero Ambiente, Ministero Pubblica Istruzione, ANPA, Ministero Difesa, Roma

Antrop M (1997) The concept of traditional landscapes as a base for landscape evaluation and planning. The example of Flanders region. Landscape Urban Plan 38:105–117

Antrop M (2005) Why landscapes of the past are important for the future. Landscape Urban Plan 70:21–34

APAT (2003) Il Progetto Carta della Natura alla scala 1:250.000: Metodologia di realizzazione. APAT, Manuali e Linee Guida 17, Roma

Arangio Ruiz V, Olivieri A (1925) Inscriptiones Graecae Siciliae et infimae Italiae ad jus pertinentes. Milano

Assunto R (1973) Il Paesaggio e l'Estetica. Giannini, Napoli

Aymard M (1992) Spazi. In: Braudel F (ed) Il Mediterraneo. Lo spazio, la Storia, gli uomini, le tradizioni. Bompiani, Milano, pp 123–144

Barbera G (2000) L'Orto di Pomona. Sistemi tradizionali dell'arboricoltura da frutto in Sicilia. L'Epos, Palermo

Barbera G (2005) Agricoltura e paesaggio nella Sicilia arabo-normanna. I Georgofili, Serie VIII 1:597–608

Barbera G (2007) L'albero da frutto nel paesaggio agrario del giardino mediterraneo. In: Sansavini S (ed) Nuove frontiere dell'arboricoltura italiana. Perdisa Editore, Bologna, pp 83–96

Barbera G, Biasi R (2011) I paesaggi agrari tradizionali dell'albero: il significato moderno di forme d'uso del suolo del passato. Italus Hortus 18:23–40

Barbera G, Carimi F, Inglese P (1992) Past and present role of the Indian-fig prickly-pear (Opuntia *ficus-indica* (L.) Miller, Cactaceae) in the agriculture of sicily. Econ Bot 46:10–20

Barbera G, Cullotta S (2009) Classificare i paesaggi culturali tradizionali: criteri metodologici e applicazione (Inventoring traditional cultural landscapes: methodology and application). In: AISF, Congresso Nazionale Selvicoltura, vols II–III, Florence, pp 960–967

Barbera G, Cullotta S (2012) An inventory approach to the assessment of main traditional landscapes in Sicily (Central Mediterranean Basin). Landscape Res 37:539–569

Barbera G, Cullotta S (2014) The Halaesa landscape (III B.C.) as ancient example of the complex and bio-diverse traditional Mediterranean polycultural landscape. Landscape Hist 35:53–66

Barbera G, Cullotta S, Pizzurro G (2004) Agroforestry systems of Mt Etna, Italy: biodiversity Analysis at landscape, stand and specific level. In: Marchetti M (ed) EFI Proceedings of monitoring and indicators of forest biodiversity in Europe—From Ideas to Operationality, vol 51, Joensuu, pp 481–492

Barbera G, Cullotta S, Rossi-Doria I, Ruhl J, Rossi-Doria B (2010) I paesaggi a terrazze in Sicilia: metodologie per l'analisi, la tutela e la valorizzazione. In: ARPA Sicilia, Collana Studi e Ricerche, vol 7, Palermo

Belvedere O (2008) Paesaggio catastale, paesaggio letterario e archeologia del paesaggio. Tre percezioni a confronto. In: Burgio A (ed) Il paesaggio agrario nella Sicilia ellenistico-romana. Alesa e il suo territorio, L'Erma di Bretschneider, Roma, pp 1–10

Bevilacqua P (1996) Tra natura e storia. Ambiente, economie, risorse in Italia, Roma

Bevilacqua P (2000) Il linguaggio degli alberi nel paesaggio agrario meridionale. Italus Hortus 7:7–9

Biasi R, Botti F, Barbera G, Cullotta S (2012) The role of mediterranean fruit tree orchards and vineyards in maintaining the traditional agricultural landscape. Acta Horticolturae 940:79–88

Bignal EM, McCracken DI, Corrie H (1995) Defining European low-intensity farming systems: the nature of farming. In: McCracken DI, Bignal EM, Wenlock SE (eds) Farming on the edge: the nature of traditional farmland in Europe. Joint Nature Conservation Committee, Peterborough, pp 29–37

Blasi C (2007) Biodiversity and landscape. In: Blasi C, Boitani L, La Posta S, Manes F, Marchetti M (eds) Biodiversity in Italy—contribution to the national biodiversity strategy. Palombi, Roma, pp 97–103

Blasi C, Carranza ML, Frondoni R, Rosati L (2000) Ecosystem classification and mapping: a proposal for Italian landscapes. Appl Veg Sci 2:233–242

Blondel J (2006) The 'design' of Mediterranean landscapes: A millennial story of humans and ecological systems during the historic period. Hum Ecol 34:713–729

Blondel J, Aronson J (1999) Biology and wildlife of the Mediterranean region. Oxford University Press, Oxford

Bolens L (1989) L'irrigation en al-Andalus: une société en mutation, analyse des sources juridiques (les "nawâzil" d'al-Wansharîsî). El agua en zonas áridas: arqueologia e historia. Istituto de Estudios Almerienses, Almeria, pp 71–87

Braudel F (1986) Civiltà e imperi del Mediterraneo nell'età di Filippo II. Einaudi, Torino

Brown RD, Lafortezza R, Corry RC, Leal DB, Sanesi G (2007) Cultural patterns as a component of environmental planning and design. In: Hong SK, Nakagoshi N, Fu BJ, Morimoto Y (eds) Landscape ecological applications in man-influenced areas: linking man and nature systems. Springer, Dordrecht, pp 395–415

Bunce RGH, Barr CJ, Gillespie MK, Howard DC (1996) The ITE land classification: providing an environmental stratification of Great Britain. Environ Monit Assess 39:39–46

Burgio A (2008) Il paesaggio agrario nella Sicilia ellenistico-romana. Alesa e il suo territorio. L'Erma di Bretschneider, Roma

Butler S (1900) The Odyssey. http://www.gutenberg.org/ebooks/1727, Accessed 14 Mar 2013

Claval P (2007) About rural landscapes: the invention of the Mediterranean and the French school of geography. Die Erde, Berlin, pp 7–23

Courtot R (1989) Campagnes et villes dans les huertas valenciennes, mémoires et documents de géographie. Editions du CNRS, Paris

Cowling RM, Rundel PW, Lamont BB, Arroyo MK, Arianoutsou M (1996) Plant diversity in Mediterranean-climate regions. Trends Ecol Evol 11:352–360
Cressier P (2006) La maîtrise de l'eau en al-Andalus. Paysages, pratiques et techniques. Casa de Velazquez, n° 93, Madrid
Cullotta S, Barbera G (2011) Mapping traditional cultural landscapes in the Mediterranean area using a combined multidisciplinary approach: method and application to mount Etna (Sicily; Italy). Landscape Urban Plan 100:98–108
Cullotta S, La Mantia T, Barbera G (1999) Descrizione e ruolo dei sistemi agroforestali in Sicilia. In: Authors Various (ed) Secondo Congresso Nazionale di Selvicoltura, Venezia, Italy. EdAs, Italy, pp 429–438
Cusumano N (2006) Culti nelle Tabulae Halaesinae: continuità e interculturalità. In: Various Authors (eds) La Sicilia romana tra Repubblica e Alto Impero. Atti del Convegno di Studi, Caltanissetta, Italy, pp 72–90
Desplanques H (1959) Il paesaggio rurale della coltura promiscua in Italia. Rivista Geografica Italiana, pp 29–61
Di Castri F, Goodall DW, Spechi RL (eds) (1981) Mediterranean-type shrublands. Elsevier, Amsterdam
Eichhorn MP, Paris P, Herzog F, Incoll LD, Liagre F, Mantzanas K, Mayus M, Moreno G, Papanastasis VP, Pilbeam DJ, Pisanelli A, Dupraz C (2006) Silvoarable systems in Europe—past, present and future prospects. Agrofor Syst 67:29–50
El Faïz M (2005) Les maîtres de l'eau. Histoire de la'hydraulique arabe. Actes Sud, Arles (France)
Fairfax H (1918) Roman farm management, the treateses of Cato and Varro done into English with notes of modern instances by a Virginia farmer. The Macmillan Company, New York
Farina A (1998) Principles and methods in landscape ecology. Chapman & Hall, London
Farina A (2000) The cultural landscape as a model for the integration of ecology and economics. Bioscience 50:313–320
Fazello T (1558) De rebus Siculus decades duae. Maida, Palermo
Firmino A (1999) Agriculture and landscape in Portugal. Landscape Urban Plan 46:83–91
Forman R, Godron M (1986) Landscape ecology. John Wiley & Sons, New York
Grimal P (2000) I giardini di Roma antica. Garzanti, Milano
Grove AT, Rackham O (2002) The nature of Mediterranean Europe: an ecological history. Yale University Press, New Haven
Horden P, Purcell N (2000) The corrupting sea. A study of the Mediterranean history. Blackwell Publishing, Singapore
Janson HF (1996) Pomona's harvest. An illustrated chronicle of antiquarian fruit literature. Timber Press, Portland
Jones-Walters L (2008) Biodiversity in multifunctional landscapes. J Nat Conserv 16:117–119
Jose S (2009) Agroforestry for ecosystem services and environmental benefits: an overview. Agrofor Syst 76:1–10
Kaldjian PJ (2004) Istanbul's bostans: a millennium of market gardens. Geogr Rev 94:284–304
Kizos T, Koulouri M (2006) Agricultural landscape dynamics in the Mediterranean: Lesvos (Greece) case study using evidence from the last three centuries. Environ Sci Policy 9:330–342
Litchfield B (1997) History of the Italian agricultural landscape. Princeton University Press, Princeton
Loumou A, Giourga C (2003) Olive groves—the life and identity of the Mediterranean. Agric Hum Values 20:87–95
Lupo S (1990) l giardino degli aranci. Il mondo degli agrumi nella storia del Mezzogiorno. Marsilio, Venezia
MacDonald D, Crabtree JR, Wiesinger G et al (2000) Agricultural abandonment in mountain areas of Europe: environmental consequences and policy response. J Environ Manage 59:47–69
Mazzoleni S, Di Pasquale G, Mulligan M, Di Martino P, Rego F (2004) Recent dynamics of the Mediterranean vegetation and landscapes. John Wiley & Sons Ltd, Chichester
Meeus JHA (1995) Pan-European landscapes. Landscape Urban Plan 31:57–79

Moreira F, Queiroz AI, Aronson J (2006) Restoration principles applied to cultural landscapes. J Nat Conserv 14:217–224

Mücher CA, Klijn JA, Wascher DM, Schaminée JHJ (2010) A new European landscape classification (LANMAP): a transparent, flexible and useroriented methodology to distinguish landscapes. Ecol Indic 10:87–103

Naveh Z (1995) Interactions of landscapes and cultures. Landscape Urban Plan 32:43–54

Naveh Z, Liebermann A (1994) Landscape ecology. Springer, New York

Nerlich K, Graeff-Hönninger S, Claupein W (2013) Agroforestry in Europe: a review of the disappearance of traditional systems and development of modern agroforestry practices, with emphasis on experiences in Germany. Agrofor Syst 87:475–492

Olmo HP (1995) The origin and domestication of the Vinifera grape. In: McGovern PE, Fleming SJ, Katz SH (eds) The origin and ancient history of wine, Luxemburg, pp 31–43

Otero I, Boada M, Tàbara JD (2013) Socio-ecological heritage and conservation of Mediterranean landscapes under global change. A case study in Olzinelles (Catalonia). Land Use Policy 30:25–37

Petanidou T, Kizos T, Soulakellis N (2008) Socioeconomic dimensions of changes in the agricultural landscape of the Mediterranean Basin: a case study of the abandonment of cultivation terraces on Nisyros Island, Greece. Environ Manage 41:250–266

Pinto-Correia T, Cancela d'Abreu A, Oliveira R (2002) Landscape units in Portugal and the development and application of landscape indicators. In: NIJOS/OECD Expert Meeting—Agricultural Landscape Indicators, Oslo

Pinto-Correia T, Ribeiro N, Sá-Sousa P (2011) Introducing the montado, the cork and holm oak agroforestry system of Southern Portugal. Agrofor Syst 82:99–104

Pinto-Correia T, Vos W (2004) Multifunctionality in Mediterranean landscapes—past and future. In: Jongman RHG (ed) The new dimensions of the European landscape. Springer, Wageningen

Sánchez IA, Lassaletta L, McCollin D, Bunce RGH (2010) The effect of hedgerow loss on microclimate in the Mediterranean region: an investigation in Central Spain. Agrofor Syst 78:13–25

Scazzosi L (2004) Reading and assessing the landscape as cultural and historical heritage. Landscape Res 29:335–355

Sereni E (1961) Storia del paesaggio agrario italiano. Laterza, Bari

Sestini A (1963) Il Paesaggio. Touring Club Italiano, Milano

Sicca U (1924) Grammatica delle iscrizioni doriche della Sicilia. Arpino

Sirami C, Nespoulousa A, Cheyland JP, Martya P, Hvenegaarda GT, Geniezh P, Martina JL (2010) Long-term anthropogenic and ecological dynamics of a Mediterranean landscape: impacts on multiple taxa. Landscape Urban Plan 96:214–223

Solymosi K (2011) Indicators for the identification of cultural landscape hotspots in Europe. Landscape Res 36:3–18

Stoate C, Bàldi A, Beja P, Boatman ND, Herzon I, van Doorn A, Snoo GR, Rakosy L, Ramwell C (2009) Ecological impacts of early 21st centuary agricultural change in Europe—a review. J Environ Manage 91:22–46

Terral JF (2000) Exploitation and management of the olive tree during prehistoric times in Mediterranean France and Spain. J Archeol Sci 27:127–133

Trischitta D (1983) Toponimi e paesaggio nella Sicilia orientale. Edizioni Scientifiche italiane, Napoli

Venturi Ferriolo M (1989) Nel grembo della vita. Le origini dell'idea di giardino. Guerini e Associati, Milano

Venuto A (1516) De Agricoltura opusculum, per Sigismondo Mayr Alemano, Napoli

Vicente AM, Alés RF (2006) Long Term persistence of dehesas. Evidences from history. Agrofor Syst 67:19–28

Vogiatzakis IN (2011) Mediterranean experience and practice in landscape character assessment. Ecol Mediterranea 37:17–31

Vos W, Klijn J (2000) Trends in European landscape development: prospects for a sustainable future. In: Klijn J, Vos W (eds) From landscape ecology to landscape science. Kluwer Academic Publishers, Dordrecht, pp 13–29

Vos W, Meekes H (1999) Trends in European cultural landscape development: perspectives for a sustainable future. Landscape Urban Plan 46:3–14

Watson AW (1983) Agricultural innovation in the early islamic world. Cambridge University Press, Cambridge

Zimmermann RC (2006) Recording rural landscapes and their cultural associations: some initial results and impressions. Eviron Sci Policy 9:360–369

Zohary D (1995) Olive. Olea europaea (oleaceae). In: Smartt J, Simmonds NW (eds) Evolution of crop-plants. Longmans, London, pp 279–382

Zohary D, Hopf M (1993) Domestication of plants in the old world. Clarendon Press, Oxford

Chapter 3
Connections Between Natural and Cultural Diversities in the Landscape of the Małopolski Vistula Gorge and the Nałęczów Plateau (Eastern Poland)

Barbara Bożętka

Abstract The study focuses on interconnections between natural and cultural spheres of the landscape constituted by the Małopolski Vistula Gorge and the Nałęczów Plateau, eastern Poland. The relations between a man-made impact, physical structure of the area and biological diversity are analysed. Additionally, the issues of traditionality and cultural richness of the area are explored. A special character of interrelations between natural and cultural spheres of the landscape is revealed in the present work. A distinctive landscape pattern showing high diversity has evolved as the reaction to natural conditions and historic, traditional land use. In consequence, the area is featured by a special combination of different habitats, including thermophilous grasslands and river islands, and it is characterized by high biodiversity. The analysis of spatial structure is supported by an insight into the theme of co-evolution of natural and cultural in the landscape. Furthermore, cultural implications involving the role of natural values for national heritage are regarded. In conclusion, the importance of knowledge about natural–cultural interspersion is emphasized.

Keywords Nature · Biocultural diversity · Values · Eastern Poland

3.1 Introduction

Relationships between man and nature imply the concerns of processes and their effects; of direction of changes; of values and their threats. The issues raise important questions, such as how culture transforms nature and how culture values

B. Bożętka (✉)
Institute of Geography, Gdańsk University, 1a Bażyńskiego Street, 80-952 Gdańsk, Poland
e-mail: geobb@univ.gda.pl

M. Agnoletti and F. Emanueli (eds.), *Biocultural Diversity in Europe*,
Environmental History 5, DOI 10.1007/978-3-319-26315-1_3

nature. Co-evolution of the human and the natural follows different paths, but it is always accompanied by an anthropogenic impact. Biotic elements response to changes in a relatively short time and thus indicate the state of the environment. In Europe vegetation has been under anthropogenic pressure for centuries, a steady impact on natural environment in Central Europe by agriculture and afterwards by forestry is dated at least on the Neolithic period (Ellenberg 1963). With no doubt, transformation of natural elements involves negative influences on biodiversity (including irreversible changes, loss and disturbance), which do have to be counteracted. However, there are also examples showing the human role as leading to an increase in diversity and strengthening ecological and landscape values. Though not being able to display the values of primaeval environments, several semi-natural systems can demonstrate important ecological assets. Presence of some semi-natural habitats considerably adds to biodiversity at different levels—at least from regional to continental.

Regarding the European context, it is evident that numerous ecological patterns we currently assess highly have evolved under a steady human influence. They not only require precise maintenance regime, but also demonstrate pronounced socio-cultural importance. The landscape of dehesas [Portuguese Montado, see for instance Costa et al. (2014), Pinto-Correia (1993)], Fennoscandian wooded meadows and pastures or Machairs in Ireland that exhibit inherent cultural values along with rich biodiversity can serve as examples. Characteristically, a large part of habitats protected under the Natura 2000 network (Interpretation manual of European Union habitats 2007) are of cultural origin (or strictly defining, having two inseparable sections: natural and cultural). This is, for example the case of the Fennoscandian lowland species-rich dry to mesic grasslands (a code 6270), the Lowland hay meadows *Alopecurus pratensis- Sanguisorba officinalis* (6510) and the Semi-natural dry grasslands and scrubland facies on calcareous substrates *Festuco-Brometalia* (6210). Calcareous grasslands can be easily found in the western part of the Lublin Upland, eastern Poland, in the area exposing distinctive natural conditions and demonstrating long-standing human influence.

Stretching close to the Vistula river in eastern Poland, the Małopolski Vistula Gorge and the Nałęczów Plateau show unique landscape features, including high biodiversity and remarkable cultural values. To investigate the links between natural and cultural diversity, and to trace the character of coexistence of the two primal elements became the main underlying issues of the work. These demanded an introduction to the processes of landscape evolution. A landscape approach, which closely adheres to the issue and which is capable of bringing the synthesis was employed.

3.2 The Area. Co-evolution of Man and Nature

3.2.1 The Area and Its Characteristic Features

A spatial extent of the analysis covers the area of the western part of the Lublin Upland (eastern Poland). It encompasses a central part of the Vistula river valley and an adjacent large fragment of a loess plateau, within the boundaries of two physico-geographical units (Kondracki 1994), mesoregions of the south-eastern Polish Uplands: Małopolski Przełom Wisły (The Małopolski Vistula Gorge or the Lesser Poland Vistula Gorge) and Płaskowyż Nałęczowski (The Nałęczów Plateau; Fig. 3.1). The town of Nałęczów to the east and Janowiec to the west demarcate the boundaries of a detailed study. Regarding natural landscapes classification (Kondracki and Ostrowski 1994), the area belongs to Lowland (valley floors) and Upland landscapes (a Loess Plateau). Both units are parts of the Lublin Upland geobotanical region (Matuszkiewicz 2001a, b) (Fig. 3.1).

In essence, two chief characteristics of the natural environment are water and relief. Proximity of the gorge with a 100-m steep slope to seemingly gentle hills, but in fact cut with innumerable sharp gullies, makes a strong impression and together with the changing force of the river's water expresses the dynamics of nature. A mixed vegetation structure noticeably increases diversity of the landscape. Potential vegetation is still present in the landscape, but its area was reduced by an anthropogenic factor. This is especially the case of natural forest formations [mainly subcontinental dry ground forest *Tilio-Carpinetum*, oak formations *Potentillo albae-Quercetum*, riparian assemblages of *Circaeo-Alnetum*, *Ficario-Ulmetum* and *Ribo nigri-Alnetum*, Matuszkiewicz W (1995), Matuszkiewicz JM (2001a, b), Kucharczyk et al. (1998)], which at present occur only locally. On the other hand, several associations, such as calcareous grasslands (e.g. *Festucetalia valesiacae*) occupying steep valley slopes (Kucharczyk et al. 1998) have shown more resistance to changes, finding acceptable conditions under human influence.

Noteworthy, though governed and in its cardinal, still easily observed features organized by nature, the area bears a considerable human impact.

The epoch of human expansion began in the Late Palaeolithic period. The first groups of people reaching the western part of Lubelskie region consisted of reindeer hunters moving after the herds of animals during recession of the Scandinavian ice sheet (Reder 2006). The central Vistula valley and, particularly, the Małopolski Vistula Gorge has fulfilled a role of a transit route, both for ancient tribes and further on, for national groups. Great prehistoric migrations from Trans-Carpathians regions to European Lowlands took place using the corridor. The Neolithic Period is dated here not earlier, than at the end of 4000 BC, when extensive, slush-and-burn land cultivation connected with grain production and animal breeding led to deforestation of relatively large areas (Nogaj-Chachaj 2000; Reder 2006). It is assumed that the turn of the Neolith and Bronze Age was the time when first considerable anthropogenic transformation of the Loess landscapes in southern Poland took place (Nogaj-Chachaj 2000; Reder 2006).

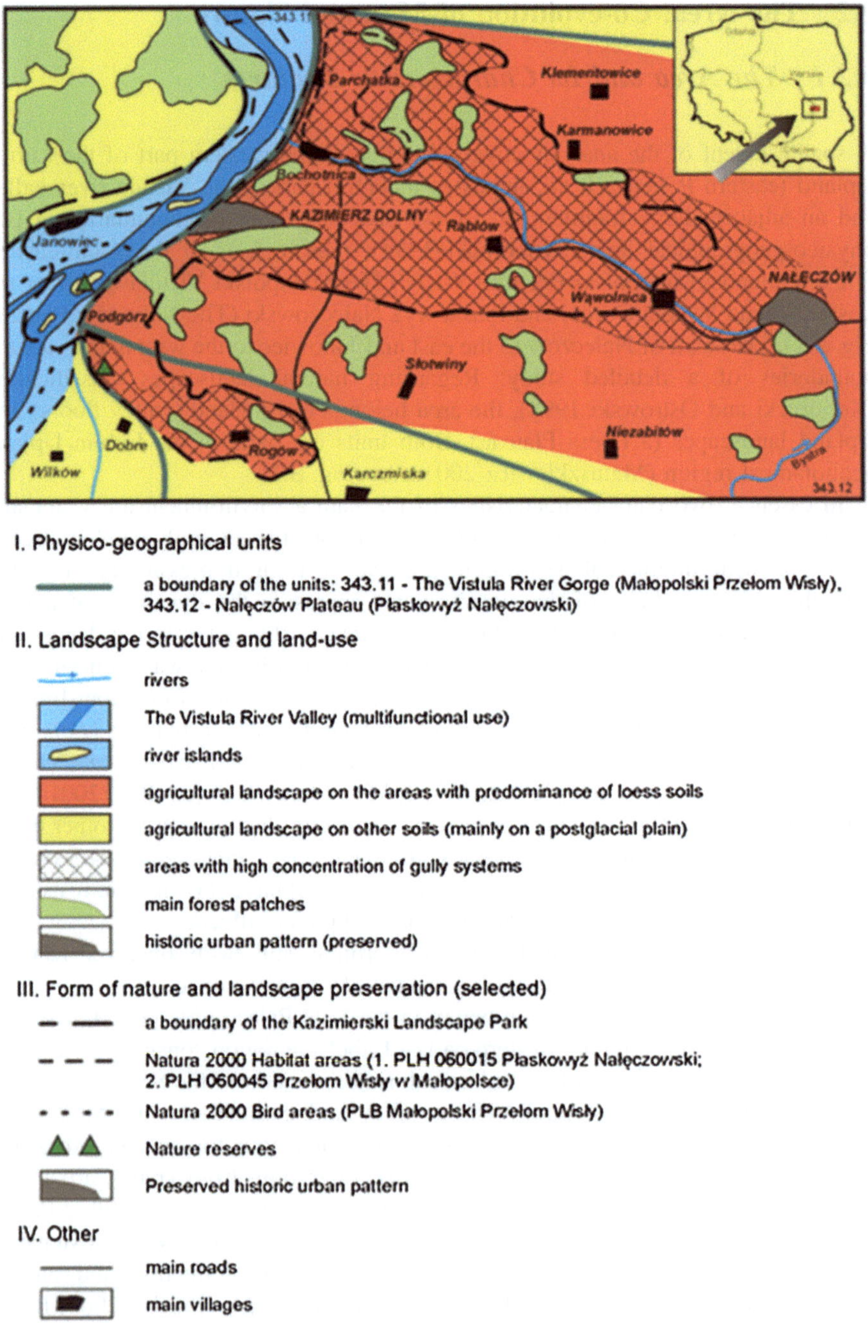

Fig. 3.1 Structure of the Małopolski Vistula Gorge and the Nałęczów Plateau [*Source* Author's own work, delimitation of physico-geographical units after Kondracki (1994)]

A network of permanent settlements spread mostly along rivers' valleys and its layout was formed in the Middle Ages. A particular attention in the context of settlement development and other sociocultural and economic influences should be paid at the Vistula valley. The river has been the 'focus point' in the history of culture and economy for a vast area (see, e.g. Pawłowski 2004, 2006a). Development of two important towns of the region, today historic and emblematic: Janowiec on the left bank and Kazimierz Dolny on the right bank was ensured by the river and economic opportunities it provided.

A long history of human migration and land management explains the presence of interesting artefacts, monuments and above all, multi-layered temporal structure of the landscape. It is also one of the reasons of species richness of the central part of the Vistula valley—many plants accompanied human groups coming from the south, and some species migrated along the river corridor from the Baltic region. In consequence, the analysed region shows one of the richest flora in Poland, exceeding in several places 600 species per 1 km^2 (Kucharczyk and Wójciak 1995).

3.2.2 *The Małopolski Vistula Gorge*

The valley of the Vistula river descends in this area to 115 m asl and embraces the Lublin Upland from the west. The Vistula river has cut in a neotectonic uplift in Pliocene (Kondracki 1994), and after the Oder glaciation period, about 260,000 years ago further considerable changes took place (Pożaryski et al. 1994).

The Gorge is 80 km long (from Annopol to Puławy) and can be divided into three parts, the most northerly and simultaneously the narrowest part, called the 'Kazimierzowski Gorge' ('Przełom kazimierzowski') is analysed in this study (Fig. 3.2). The valley bottom narrows here to almost 1 km, the river intersects Cretaceous rocks ('opoka') or Maestrichtian limestone. Slopes are steep and ascend to 100 m on the right bank (Maruszczak 1972). Evidently, this section of the gorge is strongly emphasized in the landscape.

As mentioned before, the Vistula river played a significant cultural, geopolitical and economical role. Important international trade routes linking the Black Sea with Western Europe and with the Baltic Sea ran along the Gorge, with the river crossing in today's Kazimierz Dolny. Benefits from location by its bank led the town of Kazimierz Dolny to a prosperous development observed till the beginning of the eighteenth century (Teodorowicz-Czerepińska 1981; Pawłowski 2006a). At the turn of the sixteenth and the seventeenth centuries it served as a main river port and a trade centre for Lublin, Radom, Chełm regions, and further, for Russia, Podole and Ukraine. Salt, wood, potash, and above all, grain were transported down the river to Gdańsk by the Baltic Sea, where spices, citrus fruit, alcohol and herrings were purchased to bring them up the river. Altogether with the progress in trade and commerce, shipbuilding and agriculture expanded. As a result, immense transformation of natural environment occurred—forest cover shrank rapidly, and a side effect: intensification of weathering and erosion took place. In response, the Vistula,

Fig. 3.2 The Małopolski Vistula Gorge near Kazimierz Dolny. In the background a large island ('The Cows' Island', a nature reserve) can be noticed (*Photo* Author)

not being able to transport sediments in its previous channel, was transforming into a more braided and convoluted system. Meanwhile, the main phase of gully erosion on the Nałęczów Plateau was induced (Zgłobicki et al. 2003; Warowna 2006; Pawłowski 2006a).

Contrary to the past, nowadays the Małopolski Vistula Gorge does not fulfil the role of a high-ranking transport route. An international position of the route ceased in the eighteenth century, local transport has receded in the twenty-first century. Several authors (e.g. Kondracki 1994) show the waterway's role in a context of regional development not as a linking corridor, but as a dividing line.

The Gorge exposes wilderness of the river, nevertheless, a question about the degree of naturality may be posed. Structure of vegetation cover informs about long-lasting human influence, patches of natural vegetation are found only in several places. Moreover, the Vistula river did not avoid regulation of its channel. Since the nineteenth century flood embankments have been formed, whereas first larger scale regulation works were introduced at the beginning of twentieth century (Warowna 2004, 2006). The 'Vistula Plan' aiming to regulate the whole river system was created in 1970s (Górecka 2004). However, it was not implemented and additionally, many of the undertaken hydrotechnical works did not succeed. Together with changes taking place in an entire catchment basin caused by alterations in land use, technical works enhanced the tendency of the river to form a sub-natural pattern with numerous braids and meanders (Warowna 2004). Paradoxically—against and owing to human attempts to organize the river, the

Vistula maintains its wild character, which is mirrored in morphology of the valley. Braids, meanders, shallows, sandbars and islands are widely distributed within the valley and form a very attractive set of wildlife habitats.

3.2.3 The Nałęczów Plateau

The Nałęczów Plateau is an undulated plateau rising above 200 m asl featured by a varied and changeable relief. It is covered by a thick layer—sometimes more than 30 m—of loess deposited during the Baltic (Würm) glaciation period, which buried Cretaceous rocks (Harasimiuk and Henkiel 1978; Gardziel et al. 2006). The loess cover is very vulnerable to changes and is intensively modelled by weathering and water erosion. In consequence, this region exposes extraordinary relief and land cover built by active systems of gullies. The northern part of the area exhibits extremely high density of loess gullies, fixing one of the highest ratios in Europe: 11 km/km^2 (Maruszczak 1973).

Loess gullies can be defined as deep, long valley forms within the loess deposits, generally dry and with steep slopes (Rodzik 2006). They usually consist of rows of hummocks and V- or U-shaped valleys, sometimes with streams or spring zones at the bottom level. Semi-natural and strictly anthropogenic (road gullies, called in Polish 'głębocznice') types of gullies are distinguished (Rodzik and Gardziel 2004). Characteristically, in the western part of the plateau these erosion incisions form complicated arrangements: semi-natural gullies compose a dendritic pattern and road gullies form rectilinear and fan-shaped structures (Rodzik and Gardziel 2004).

Noteworthy, development of the gullies of the Nałęczów Plateau is closely linked with evolution of the landscape (see, e.g. Zgłobicki et al. 2003; Zgłobicki 2006)—a human factor plays a significant role in both cases. Water erosion increases in a deforested area, and it has been confirmed that occurrence of high surface runoff coincided with intensification of agriculture and a rise in precipitation (Śnieszko 1995; Zgłobicki 2006). Additionally, it is assumed that Neolithic land cultivation could not result in high gully erosion (Maruszczak 1988). These are the Middle Ages which brought far-reaching landscape changes with them: large areas were deforested and an introduced field pattern actively contributed to erosion, in consequence, the development of gullies was enhanced. However, the turning point in landscape evolution was the 16th century—according to Maruszczak (1988) at that time cultural elements of the landscape had definitely dominated natural components, simultaneously, progress of gully systems had accelerated (Zgłobicki et al. 2003; Zgłobicki 2006). A complicated relief causes serious difficulties in land cultivation. Nowadays, the great part of gullies is excluded from agriculture, which concentrates on the hilltops. Gullies are affected by the process of reforestation.

Apart from large Loess Hills predominant in the landscape, the Nałęczów Plateau encapsulates two other types of environment: the valley of the Bystra river and historic urban areas.

3.2.4 *Urban Areas*

Four historic towns: Janowiec, Kazimierz Dolny, Nałęczów and Wąwolnica are situated in the discussed area. Three of them: Janowiec, Kazimierz Dolny and Nałęczów present high cultural values exceeding the local level. Interestingly, cultural importance (renowned for instance in architectural aspects)is accompanied by noticeable links with natural environment.

Janowiec, formerly a private town of the Firlej's family, was founded in 1537. Its location took advantage of the Vistula Gorge situated on the very height of the western slope, the town benefited from nearby agricultural land and from a water artery during the epoch of trade prosperity (sixteenth–seventeenth century). Its monumental renaissance castle on the top of the escarpment marks the boundary of the Gorge and provides a magnificent view of the valley.

Kazimierz Dolny, a town since 1370, stretching from the lower levels of the Vistula valley up the Loess Hills exhibits inherent connections with the nature. They consist not only in location, but also in the content and values of an urban tissue (see, e.g. Teodorowicz-Czerepińska 1981). An arrangement of an urban pattern adjusted to natural relief; technology of buildings constructed out of local stone (cream—white Cretaceous rock called 'opoka') and wood; abundance of traditional agri- and horticulture forms (gardens, orchards, field and even grasslands) in addition to cultural importance have always been distinctive features of the town. Kazimierz Dolny convincingly demonstrates deep codependency of people and nature and the culture-forming role of rivers, not excluding such issues as wealth and failure (e.g. Pawłowski 2006b). The town can also be treated as a record of life of different social groups and multicultural community, with the Jewish minority in particular. With regard to rich history and great values Kazimierz Dolny was granted in 1994 a status of the Historic Monument of Poland.

Comparing to Janowiec and Kazimierz Dolny, Nałęczów is a juvenile town, however, it is also of considerable art-historical interest. Situated by the Bystra river, it grew fast in the eighteenth century thanks to recognition of mineral springs (bicarbonate soda, ferruginous) and soon formed the most attended nineteenth century spa in Poland. Interestingly, lying in central part of the Nałęczów Plateau, it shows stunning density of wood cover—trees and bushes accompany historic residential quarters, linking with gullies in the outskirts.

3.3 Culture

Culture forms an incredibly wide area of research; regarding solely the perspective of cultural anthropology, it ranges from learned patterns of behaviour to culturally determined patterns of thought and perception (Scott and Marshall 2009). This is impossible to characterize all cultural phenomena of the region in the following study, therefore only some prevalent issues are highlighted.

At first, multiculturality of the society bringing new dimensions to local and regional cultures should be mentioned. Different communities and nationalities have been offered hospitality in the Lubelskie region for centuries. Regrettably, the Second World War destroyed the long epoch of coexistence of Polish, Jewish, Ukrainian and other nations and religions, but traces of these differing cultures are still widely represented in the landscape. Out of three towns investigated in this study, especially Kazimierz Dolny exhibits strong multicultural influences, mostly owing to the presence of a sizeable Jewish community. After the dramatic incidents during the war, when almost all the Jewish citizens were murdered by the German Nazi troops, this town became a symbolic 'shtetl', particularly for artists (Odorowski 1991). Multiculturality regards also cohabitation of different social groups, and again, Kazimierz is an example of many cultural facets within this perspective, with different attitudes, tensions and changeability of physical appearance being portrayed by painters and characterized by writers.

Second, the Lubelskie region is still the area of a lively traditional rural culture, though many serious changes affected the countryside during last two decades. Although boundaries of the Lubelskie ethnographic region have not been defined clearly (Tymochowicz 2014), traditional features of the area are not questioned.

In spite of many unfavourable conditions the rural culture has retained its individuality and dynamics, and it still provides commodities unprecedented in Poland (e.g. Regional Encyclopedia 2014). Traditional pottery, braiding (willow and cat's-tail), boat building, rafting, fishing, tanning and weaving belong to accustomed occupations here (Malawska 1980; Litwin 2007), whereas embroidery, painting and sculpture in wood nearly approached the art (Folk Artists of the Land of Puławy 2011). Traditional music distinguishes the region, since the highest number of rural ensembles in Poland comes just from the Lubelskie region (Mazur-Hanaj and Baliszewska 2014).

Nevertheless, the cultural significance of the region lies not only in ethnographic values. The area is rich in impressive architecture, literature and painting. The architectural works of Tielman van Gameren and Santi Gucci (Janowiec), Jakub Balin (Kazimierz Dolny), later of Stanisław Witkiewicz, Jan Koszczyc-Witkiewicz (several works in Kazimierz and Nałęczów), Karol Siciński (Kazimierz) visibly mark the landscape (Żurawski 1986). Importance of natural substrate: stone and wood; a line of ornamentation for architecture and design altogether with outstanding accommodation of buildings to natural conditions were the reasons opening the way to the protection of the area and rose Kazimierz Dolny to the rank of the National Monument.

It seems that the region is especially inclined to cultural and artistic influences. Kazimierz Dolny has fulfilled a role of an 'artistic colony' of painters and writers (Odorowski 1991). The town's prosperity has gone away, but a special landscape mood attracted artists and visitors. First open air was organized in 1909. After that painters, poets, writers and their works became a kind of inseparable element of the cultural landscape. A special motif incorporating the nature and the landscape of Kazimierz (Odorowski 1991) was established in art and entered the national culture. Additionally, Nałęczów, being strongly connected with literature and with such writers as Bolesław Prus and Stefan Żeromski should be mentioned in the context of art achievements.

Together with the assets of traditional rural culture, aforementioned artistic values make the area of exceptional cultural importance for national heritage. Remarkably the landscape, which wonderfully ties culture with nature is not only viewed as beautiful, but it also encompasses a symbolic meaning connected with historicity, naturality and familiarity (e.g. Kosiński 2010).

3.4 Landscape and People. Structure, Use and Values

Landscape structure plays a crucial role for the maintenance of diversity (see, e.g. Forman and Godron 2006; Forman 1995; Antrop 2005). Heterogeneity, which represents the very feature of the analysed area, involves itself an assumption of potential rise in diversity.

Taking under account the classification of cultural landscapes (World Heritage Cultural Landscapes 1992–2002 1992), the area of the Małopolski Vistula Gorge and the Nałęczów Plateau constitutes an organically evolved landscape, which has developed with close relations to natural environment. And indeed, natural conditions have always been framing the landscape. Cultural identity accompanies natural distinctiveness and a sense of cultural–historical continuity expressed by particular land uses, landscape arrangement and traditional architecture is pronounced strongly. A primary profile of the contemporary landscape structure was outlined in Middle Ages, when strong anthropogenic influences were set in motion and a typical field division was organized (Teodorowicz-Czerepińska 1981). Considerably, although many changes have taken place for the last decades, the notion of traditionality is characteristic for the whole landscape, regardless the 'rural–urban' contradiction.

The Vistula valley, exposing features characteristic of the wild river has been touched by traditional use for centuries. Scarps of the valley have been used as pastures; meadows in the valley bottom, and even some islands have been maintained for hay and grazing herds of cattle. Unfortunately, a popular scene of cattle and horses transported on wooden ships across the river to the meadows located on the other bank (Litwin 2007) is currently a rarity, similarly to floats of simple wooden boats. Limestone quarries exploiting 'opoka' stone and exhibiting rare fossils operated on both river banks, in Kazimierz Dolny and Nasiłów. Production of fruit, especially apples and plums has a long history here, the Vistula valley with the area surrounding Kazimierz Dolny has been an important orchard region since eighteenth century (Pawłowski 2004; Wojciechowski 2006).

As highlighted before, the Vistula valley has experienced anthropogenic influences, but it is still a river featured by a high degree of naturality, especially concerning its geomorphological and hydrological attributes (Warowna 2004, 2006). In the effect of a particular usage of natural habitats, valuable plant associations developed, for instance thermo-xerophilous communities on the river slopes. Changeability of water levels, diversity of land cover types, occurrence of diverse biotopes, multifunctional use incorporating traditional farming activities, with addition to the presence of beautiful old towns on the river's banks, all these create an extraordinary landscape pattern.

The evolution of the Loess Plateau is strictly connected with a human impact (part 1.3). Today, apart from a few small towns, this area of very fertile soils constitutes an agricultural landscape. It is characterized by traditional land use and considerable structural and functional differences between the higher elevation (plateau tops, 'wierzchowina' in Polish) and gullies. The undulated plateau comprises an open farming land, where varied crops are produced: grain (maize, barley), sugar beet, vegetables and fruit. The orchards of cherry and apple trees are frequently mixed with hops fields (Łoboda and Sobczyk 2006).

No doubt, gullies are a remarkable feature of the plateau. However, though being very attractive from the ecological and landscape points of view, they are not easy for agriculture. Previously often used as pasture areas, nowadays tend to be excluded from agricultural use and transform into woodlands (Baran-Zgłobicka and Zgłobicki 2006). Dynamic, arranging themselves in complicated systems, covered mostly by woods and forests, with diversity of relief and a tendency to renaturalisation, they constitute a characteristic landscape element and fulfil important ecological functions, forming wildlife refuges and ecological corridors.

The agricultural landscape of the plateau is featured by a distinct pattern. Gentle, undulated tops are occupied by long and narrow fields divided by grassy baulks, lines and groups of hedgerows, dispersed patches of woodlands and single trees. Homesteads are either scattered (can be present even in gullies) or grouped in villages characterized by a rather dense housing network. Villages and hamlets frequently accompany river valleys, for instance the valley of the Bystra river.

The area is still utilized as traditional family-based agriculture, where small holdings prevail. Since the region is quite densely populated, it abounds with human activities. Field works are featured by a great input of human labour, different cultural elements follow land organization. Wooden huts hidden in gardens, crosses and chapels appearing along country roads, old tools abandoned in fields in addition to traces of many customs are evident in the area. Traditional character of the landscape involves occupation of people and their lifestyles. A main task: agricultural production has been widened with many kinds of craft, such as smithery, weaving and pottery (Sect. 3.3). Although the rural areas have been experiencing tendency to modernize and an increase of emigration of young people recently, the '*genres de vie*', to evoke Vidal de la Blache (1911–1912) is traditional, based on time-honoured values of work, family and religion, predominantly Roman Catholicism. Worthy to note, the area is inhabited by people, who are not detached from the environment. Skills connected with land cultivation, in specific conditions of loess relief, knowledge of how to use natural resources (e.g. wood and stone for house building, clay for pottery, willow for braiding) decides when obtaining an income from local area is necessary. Currently, this society is not entirely dependent on local supplies, but use of relevant resources is indispensable. Characteristically, many aspects of traditionality are demonstrated not only by rural landscapes, but also by urban areas, for instance in Janowiec and Kazimierz Dolny. Even such typical rural works as pastoral activities and fruit production have frequently taken parts in the towns. One of the spatial consequences of this way of land management involves occurrence of a sequence of different semi-natural habitats.

Both natural conditions and long-lasting sociocultural and economic processes led to formation of a unique landscape characteristic of the Western Lublin Plateau. It belongs to the 'Poland's strip fields', which are recognized as a regional landscape in the typology of Pan-European landscapes (Meeus 1995). A settlement pattern originated in the Plateau in the Middle Ages introduced a 'belt-form' land division (Teodorowicz-Czerepińska 1981), and subsequent multiple dissections of fields (always parallel and within the belt) resulted in the appearance of a system of very narrow and long fields, present up today. This mosaic of arable fields, orchards, spontanic groups of trees and shrubs, meadows and pastures has been used extensively for centuries (e.g. Baran-Zgłobicka and Zgłobicki 2006).

Owing to presence of exceptional combination of natural and cultural values the landscape of this part of the Małopolski Vistula Gorge and the Nałęczów Plateau was given protection by a state law in 1979 as The Kazimierski Landscape Park, constituting the first landscape park in the Lubelskie region. The park comprises several nature reserves and Natura 2000 areas, and hundreds of historical monuments. Maintenance of these values, especially in the epoch of increasing tourism and leisure pressure (Wołoszyn and Skowronek 2007; Chmielewski et al. 2004) requires not only continuation of specific land use, but also steady involvement in local and regional policies. The engagement belongs for instance to main aims of one of the NGOs—the Association of the Friends of Kazimierz (Towarzystwo Przyjaciół Kazimierza). The organization gathers people engaged in protection of the landscape assets, unfortunately, participation of local community in these activities is low.

Importantly, the analysed landscape has been featured not only by high structural diversity and variety of values, but also by unity of its parts. In consequence, balance and harmony, the outcomes of many processes became its major structural and functional characteristics.

3.5 Biodiversity

3.5.1 Habitat Pattern and Vegetation Cover

Key anthropogenic processes affecting ecological values of the Małopolski Vistula Gorge and the Nałęczów Plateau consist in influence on habitats properties and on their spatial pattern. Advanced modification of natural ecotopes and introduction of anthropogenic elements led to a considerable rise in landscape diversity. However, this change can be identified with transformation of the entire ecological system and a simultaneous loss of its naturality. Contemporary, the ecological system is of semi-natural character, in which biotopes presenting a high degree of naturalness, such as fragments of riparian forests are scarcely distributed within the prevailing semi-natural matrix. Nevertheless, it exhibits important ecological values and rich biodiversity on the landscape, habitat and species level.

Human pressure exerted a widespread effect on vegetation cover and main alterations included the change of spatial structure, habitat conditions and species composition (Kucharczyk et al. 1998). Forests have been particularly influenced—a very strong fragmentation and sometimes attrition affected riparian communities stretching along the Vistula valley and dry ground forests of the Loess Plateau. According to the assessment of anthropogenic changes of vegetation in Poland conducted by Faliński (1991), the majority of the Vistula valley has transformed to a semi-natural pattern, whereas the plateau consists mostly of synanthropic communities.

On the other hand, traditional land use connected with agriculture and horticulture provided beneficial conditions for a set of new habitats (Table 3.1), such as pastures and meadows. Some of them, for instance xerophilous steppe grasslands with *Sisymbrio-Stipetum capillatae* constitute an ecological rarity in Poland and in this part of Europe. However, great ecological values of the area stem above all from the character of the entire ecological mosaic, which is characterized by diversity, species richness and strong interconnections of biotopes (Plan Ochrony Kazimierskiego Parku Krajobrazowego-operat generalny 1998; Kucharczyk et al. 1998). Extensive use of land became one of the main factors leading to the biotic richness of these agricultural areas (Gacka-Grzesikiewicz 2006).

Variability of relief and land use resulted in formation of a special ecological pattern, where patches of different character lie in close proximity to each other, for example on both slopes of the Vistula Gorge, which assembles steppe grasslands, willow plantations and *Tilio-Carpinetum* forests. A sharp ecological contrast is noted on the Loess Plateau—deep gullies overgrow with broad-leaved forests, whereas large farming areas form a traditional mosaic of an open agricultural landscape (Fig. 3.3).

A large part of existing plant communities establishes important floristic and ecological values. Associations of dry grasslands and scrublands: *Sisymbrio-Stipetum capillatae* (Dziub. 1925) Medw.-Korn. 1959, *Koelerio-Festucetum rupicolae* Kornaś 1952, *Inuletum ensifoliae* Kozł. 1925, *Thalictro-Salvietum pratensis* Medw.-Korn. 1959, which comprise those of steppe characterisitics are extremely rare in Poland. In Central Europe, outside their core range (i.e. Pontio-Pannonian zone), they are of semi-natural origin, formed thanks to extensive pasture and when abandoned, they turn quickly into scrubs and woodlands (Matuszkiewicz 2014; Barańska and Jermaczek 2009).

Numerous habitats occurring within the area are ranked as of high ecological importance with relevance to Council Directive 92/43/EEC (Annex I). The Nałęczów Plateau comprises, for instance *Galio-Carpinetum* oak-hornbeam forests (a code 9170) and alluvial forests *Alnion incanae*, *Salicion albae*: *Salici-Populetum* (91EO). The Vistula Gorge gathers mentioned before semi-natural dry grasslands and scrubland facies on calcacerous substrates (6210), rivers with muddy banks with *Chenopodion rubri* and *Bidention* vegetation (3270), steppic woods with *Quercus* spp.: *Potentillo albae-Quercetum* (91I0), oak-hornbeam forests (9170) and alluvial forests (91EO).

Table 3.1 Vegetation types and important flora species

Vegetation types	Plant communities (selected)	Particular flora species
1. Aquatic habitats and rushlands	Associations with *Alisma lanceolatum* and *Hippuris vulgaris*; *Bidention* communities	*Salvinia natans* L., *Wolffia arrhiza* L.
2. Meadows	– *Arrhenatheretum medioeuropaeum, Poo-Festucetum rubrae, Cirsietum rivularis, Lolio-Cynosuretum, Scirpetum silvatici* – *Molinion, Filipendulo-petasition, Magnocaricion*	*Cirsium pannonicum* (L.F.) Link, *Dianthus superbus* L., *Orchis latifolia* L., *Ononis spinosa* L., *Pulsatilla pratensis* (L.) Mill
3. Dry grasslands and scrublands on chalk and limestone	– *Sisymbrio-Stipetum capillatae* (Dziub. 1925) Medw.-Korn. 1959 – *Koelerio-Festucetum rupicolae* Kornaś 1952 – *Inuletum ensifoliae Kozł. 1925, Thalictro-Salvietum pratensis Medw.-Korn. 1959, Festuco-rupicolae-Brachypodietum pinnati, Origano-Brachypodietum pinnati* – Scrublands with *Ligustrum vulgare, Crataegus* sp., *Berberis vulgaris, Prunetum fruticosae Dziub. 1921*	*Adonis vernalis* L., *Clematis recta* L., *Inula ensifolia* L., *Iris aphylla* L., *Jovibarba sobolifera* (Sims) Opiz, *Linum flavum* L., *Scorzonera purpurea* L., *Stipa capillata* L., *Cerasus fruticosa* PALL
4. Forests	1. Riparian forests: (A) alluvial plains of the Vistula: *Salici-Populetum*; *Salicetum pentandro-cinereae, Salicetum triandro-viminalis*; *Ribo nigri-Alnetum*; (B) the Bystra valley: *Circaeo-Alnetum* 2. Loess Plateau forests: *Tilio-Carpinetum stachyetosum*; *Galio-Carpinetum*; *Potentillo albae-Quercetum* Libb. 1933 with adjacent *Peucedano cervariae-Coryletum, Quercoroboris-Pinetum*	*Adenophora liliifolia* (L.) Besser, *Asarum europaeum* L., *Cephalanthera damasonium* (Mill.) Druce, *Daphne mezereum* L., *Galanthus nivalis* L., *Lilium martagon* L., *Listera ovata* (L.) R.Br
5. Synanthropic and segetal	*Vicietum tetraspermae, Galinsogo-Setarietum, Lamio-Veronicetum politae* and others	*Adonis aestivalis* L., *Caucalis daucoides* L., *Ornithogalum umbellatum* L.

Source Author's own work, on the basis of: Kucharczyk and Wójciak (1995), Kucharczyk et al. (1998), Plan Ochrony Kazimierskiego Parku Krajobrazowego (1998), Zagadnienia ochrony środowiska w gminie Janowiec (1991a, Zagadnienia ochrony środowiska w gminie Kazimierz Dolny1991b, Zagadnienia ochrony środowiska w gminie Wąwolnica 1991c, Zagadnienia ochrony środowiska w gminie Wilków 1991d)

Fig. 3.3 The Loess gullies of the Nałęczów Plateau display a characteristic set of habitats (from the left: agricultural areas with many small biotopes, gullies covered with deciduous broad-leaved forest, *Lilium martagon*—a protected forest species; *Photo* Author)

Significance of ecotopes and species is appreciated by presence of areas attached to the Natura 2000 network. Two areas of protected habitats: The Vistula Gorge in the Lesser Poland Province (PLH 060045), The Nałęczów Plateau (PLH 060015), and the Area of Bird Protection: The Lesser Poland Vistula Gorge (PLB 140006) have been established in the examined landscape.

Although semi-natural habitats prevail in the landscape structure, it encapsulates also several units characterized by a high degree of naturality, where natural associations (in terms of Faliński's (1969) definition) are found. Ecotopes of high natural values include meadows and fragments of riparian forests in the Bystra river valley, some patches of dry ground forests growing in ravines and ecotopes occurring in flood embankments within the Vistula valley (Plan Ochrony Kazimierskiego Parku Krajobrazowego-operat generalny 1998; Kucharczyk et al. 1998).

Ecological functions of the analysed area result not only from the presence of individual ecotopes, but also from connections at a landscape level. However, concerning both the uniqueness and the species richness, two kinds of habitats have been indicated to have the greatest faunistic importance: steppe grasslands and sandy islands on the Vistula river (Plan Ochrony Kazimierskiego Parku Krajobrazowego-operat generalny 1998; Kucharczyk et al. 1998).

3.5.2 Diversity of Fauna

A varied landscape structure forms favourable conditions for numerous fauna species. Three dominant types of habitat patterns can be distinguished: (1) dependent on the Vistula river, (2) habitats of agricultural areas of the Loess Plateau, (3) habitats of loess gullies and other woodland or forest patches outside the Vistula valley.

The position of the Vistula river valley is especially important. The Vistula river constitutes one of the most important ecological corridors in Europe (see Gacka-Grzesikiewicz 1995) and a system of biodiversity hot spots with refuges of international significance. Several International Bird Areas (IBAs) are spread over the valley, one of them—The Lesser Poland Vistula Gorge (PL 096; Wilk et al. 2010) stretches out in the studied area. Key bird species of the refuge embrace: the black stork *Ciconia nigra*, Montagu's harrier *Circus pygargus*, oystercatcher *Haematopus ostralegus*, ringed plover *Charadrius hiaticula*, bluethroat *Luscinia svecica*, gulls *Larus* spp. and terns *Sterna* spp. (Wójciak et al. 2010). Rare fish species include the thunderfish *Misgurnus fossilis* and the rapfen *Aspius aspius*, amphibians the tree frog *Hyla arborea*, mammals the river otter *Lutra lutra* and numerous bat species (Kucharczyk et al. 1998; Wójciak et al. 2010). Species richness derives from a sequence of natural and semi-natural biotopes (ranging from remnants of alluvial forests to semi-natural pastures), where river meanders, sandy bars and islands play a great role. A big river island south of Kazimierz Dolny: 'The Cows' Island', formerly used as pastureland, nowadays a nature reserve is assessed to be the most valuable water-dependent fauna habitat in the analysed fragment of the valley (Plan Ochrony Kazimierskiego Parku Krajobrazowego-operat generalny 1998; Kucharczyk et al. 1998).

This is astonishing that the Vistula valley is featured not only by biotopes and taxa connected with wet environments, but also by those demanding dry conditions and large exposure to sun. Patches of dry grasslands located on the slopes are habitats of many rare species. A unique butterfly: the chequered blue *Scolitandides orion*, and some reptiles, for instance the smooth snake *Coronella austriaca* in addition to other thermophilous species occur here (Zagadnienia ochrony środowiska w gminie Janowiec 1991a Zagadnienia ochrony środowiska w gminie Kazimierz Dolny1991b, Zagadnienia ochrony środowiska w gminie Wilków 1991d).

Agricultural parts of the Nałęczów Plateau group fauna typical of farming areas, however, thanks to traditional land use and occurrence of interesting microenvironments such as rock walls and caves, they demonstrate an interesting species composition, quite well recognized with regard to bird species (Plan Ochrony Kazimierskiego Parku Krajobrazowego-operat generalny 1998; Kucharczyk et al. 1998). A mosaic of arable fields, pastures and hedgerows is densely inhabited for instance by the skylark *Alauda arvensis*, stonechat *Saxicola torquata*, quail *Coturnix coturnix*, ortolan bunting *Emberiza hortulana* or great grey *Lanius excubitor*. Several species nest in villages and homesteads: the barn swallow *Hirundo rustica*, house martin *Delichon urbica*, black redstart *Phoenicurus ochruros*, barn owl *Tyto alba*. A species rare in Poland: the bee-eater *Merops apiaster* settled within the rock walls. Vast stretches of ecotones are necessary for many rare insect species: of bumblebees *Bombus* spp. and butterflies, e.g. *Apatura ilia*. Noteworthy, slopes of the Vistula river and the Nałęczów Plateau expose many species of the *Lepidoptera* group, for instance the swallowtail *Papilio machaon*, red admiral *Pyrameis atalanta*, clouded yellow *Colias croceus*, while orchards are settled by the giant peacock moth *Saturnia pyri* (Plan Ochrony Kazimierskiego Parku Krajobrazowego-operat generalny 1998; Kucharczyk et al. 1998).

Ecotones and groups of trees tend to link patches of woods and forests located in the gullies with an open landscape. The fauna of mammals in these woods embraces the pine marten *Martes martes*, common dormouse *Muscardinus avellanarius* and roe deer *Capreolus capreolus*. Birds are represented by the black woodpecker *Dryocopus martius*, blackcup *Sylvia atricapilla* or chiffchaff *Phylloscopus collybita*. Humid and shady ravines form habitats of numerous amphibians, for example the moor frog *Rana arvalis*. Amphibians as well as fish species are frequent in the valley of the Bystra river, which runs across the plateau and demonstrates high ecological importance. Caves in Bochotnica village with at least 14 bat species and meadows with abundance of insect and amphibian species considerably rise the biodiversity of the plateau, although the river can be regarded as a separate, non-loess unit (Plan Ochrony Kazimierskiego Parku Krajobrazowego-operat generalny 1998). Noteworthy, a network of woods in gullies provides important ecological functions fulfilling a role of local and regional ecological corridors for many groups of species (Plan Ochrony Kazimierskiego Parku Krajobrazowego-operat generalny 1998).

Values related to biodiversity are highlighted by high density of nestling birds of prey. The honey bazzard *Pernis apivorus*, Montagu's harrier *C. pygargus*, marsh harrier *Circus aeruginosus*, goshawk *Accipiter gentilis*, sparrow hawk *Accipiter nisus* nestle here, whereas other species, such as the golden eagle *Aquila chrysaetos* occur occasionally (e.g. Plan Ochrony Kazimierskiego Parku Krajobrazowego-operat generalny 1998).

High floristic and faunistic diversity is accompanied by its values in the context of nature conservation. The area of the Małopolski Vistula Gorge and the Nałęczów Plateau assembles numerous species enlisted in the Polish Red Data Book of Animals (2001, 2004) and given a high rank by the Bird and Habitat Directives (Council Directive 92/43/EEC; Directive 2009/147/EC; Table 3.2).

3.5.3 Biodiversity, Culture and Society

As highlighted previously, the pattern of biodiversity of the region has been determined by several factors and human pressure belongs to the most prominent. Anthropogenic influence consists in transformation of the landscape leading to the growth of spatial diversity. Introduction of new elements, such as cultivated parcels was only one of the processes taking part. In consequence, an interesting landscape pattern with a characteristic mosaic of habitats has evolved. Additionally, many semi-natural habitats show distinct ecological values, such as grasslands on Loess and limestone. Rise in habitat diversity in turn affected biotic diversity, currently well marked for instance in the number of species.

A long-lasting co-evolution of man and environment brought to life many intriguing phenomena. Influence of transregional trade routes on flora structure is one of such issues. Kucharczyk (2003) attempts to resolve the question of appearance of steppe grasslands and xerophil species in this area. He assumes that

Table 3.2 Fauna species of the highest importance within the two Natura 2000 areas: the Małopolski Vistula Gorge and the Nałęczów Plateau

Type of species	The Małopolski Vistula Gorge	The Nałęczów Plateau
1. Vertebrates	The Eurasian beaver *Castor fiber*; greater mouse-eared bat *Myotis myotis*, Bechstein's bat *Myotis bechsteinii*, pond bat *Myotis dasycneme*, white-backed woodpecker *Dendrocopos leucotos*, oystercatcher *Haematopus ostralegus*, ringed plover *Charadrius hiaticula*, bluethroat *Luscinia svecica*, little tern *Sternula albifrons*, Montagu's harrier *Circus pygargus*, avocet *Recurvirostra avosetta*, black stork *Ciconia nigra*, smooth snake *Coronella austriaca*, tree frog *Hyla arborea*, the Amur bitterling *Rhodeus sericeus*, Aspius *Leuciscus aspius*, spined loach *Cobitis taenia*, *Sabanejevia aurata*, European weatherfish *Misgurnus fossilis*	The barbastelle *Barbastella barbastella*; Bechstein's bat *Myotis bechsteinii*, pond bat *Myotis dasycneme*, greater mouse-eared bat *Myotis myotis*; Eurasian beaver *Castor fiber*; white-backed woodpecker *Dendrocopos leucotos*, European bee-eater *Merops apiaster*, European pond turtle *Emys orbicularis*, smooth snake *Coronella austriaca*, European fire-bellied toad *Bombina bombina*, European weatherfish *Misgurnus fossilis*
2. Invertebrates	The large copper *Lycaena dispar*, dusky large blue *Maculinea nausithous*, large blue *Maculinea arion*, scarce large blue *Maculinea teleius*	The large copper *Lycaena dispar*, violet copper *Lycaena helle*, large blue *Maculinea arion*, dusky large blue *Maculinea nausithous*, scarce large blue *Maculinea teleius*, Danube yellow *Colias myrmidone*

Source Author's own work, data: Kucharczyk and Wójciak (1995), Plan Ochrony Kazimierskiego Parku Krajobrazowego (1998), Kucharczyk et al. (1998), Zagadnienia ochrony środowiska w gminie Janowiec (1991a, Zagadnienia ochrony środowiska w gminie Kazimierz Dolny1991b, Zagadnienia ochrony środowiska w gminie Wąwolnica 1991c, Standard Register… 060015 (2013), Standard Register… 060045 (2013), Evaluation of species importance with regard to: Council Directive (92/43/EEC), Directive (2009/147/EC), Polish Red Data Book of Animals (2001, 2004)

expansion of these neophytes was enhanced by historic trade roads and by landscape transformation that resulted in a decrease in forest cover and in a growth of agricultural areas. The direction of grain import in the sixteenth–seventeenth centuries from the south-east (Wołyń, Podole and Ukraine, where numerous xerophilous species occurring in Kazimierz originate from) to the Vistula river played a significant role in the process (Kucharczyk 2003; Pawłowski 2006b).

Natural resources have been constantly used by local communities. With respect to biotic values, traditional use and its extensive character are of much importance. Customary way of use is linked not only with terrestrial environment. Freshwater fishery, shipbuilding and rafting needed both the Vistula river and the wood from the Nałęczów Plateau to operate. Shipbuilding does not operate any more, but wood is commonly used as a construction and fuel timber. The Vistula banks provide a purple willow (*Salix purpurea*) for braiding. Meadows and pastures have been established on the areas previously covered by alluvial forests.

Notably, the agricultural landscape encapsulates many issues connected with traditional use and methods of maintenance of particular habitats. A special use must have been connected with thermophilous grasslands, today often under protection (Plan Ochrony Kazimierskiego Parku Krajobrazowego-operat generalny 1998). They constituted pasturelands, usually extensively used for cattle, sheep and goat grazing, however, sometimes a method of burning was also employed (Kucharczyk 2003; Pawłowski 2006b). Interestingly, even today historic tracks of breeding animals can be read from the landscape structure, since they are related to the network of ravines. As revealed before, land cultivation of loess areas requires the detailed knowledge of the area, its soil and vegetation. The knowledge and the way of use of farming assets have been elaborated by local communities for a long time. Thus, it may be argued that the habitats form not only a particular ecological structure, but also a kind of cultural heritage, which can be called after Wojciechowski (1996) the pattern of rational management.

The rural areas exhibit lots of close connections between culture and biodiversity. The most obvious is food production. Importantly, cereals, fruit, vegetables and products of animal breeding are not only trade commodities, but serve for different local purposes. The Lubelskie Region and the area of Puławy county have developed an own traditional cuisine depending both on ingredients acquired by farmers and fishermen and on wild plants collected in forests and meadows. Wild fruit, such as several kinds of berries, the Cornelian cherry *Cornus mas*, hawthorn *Crataegus* spp., barberry *Berberis vulgaris*, European elder *Sambucus nigra* and many other species (incidentally, frequently accompanying thermophilous scrublands) are widely used in gastronomy and traditional medicine. For example, the European elder *Sambucus nigra* is applied in cases of respiratory diseases, hawthorn *Crataegus* spp. in case of heart difficulties. Honey has been a local speciality for ages, used in many dishes, forming a beverage, as well as a recommended medicine. Not surprisingly, floristic motifs are frequently found in local art and decorations. Floral ornaments, such as tree branches and flowers on women's skirts and blouses, and hats made of different grasses or straw worn by men are typical for customary clothes (Gauda 2001). Remarkably, classic weaving patterns with long, thin, multicolour stripes seem to follow strictly a typical field pattern occurring in the area ('strip fields').

It is worth noting that geographical names occurring in this area emphasize correspondence between natural environment and society. Not trying to carry on an expertise, some uninvolved observations can be made. The geographical names of villages and their parts teem with direct relations to natural forms of the landscape and land use management. The role of relief is underlined by referring in a core of the word to 'góra' ('mountain' in English) in numerous spots: Podgórz, Góry Pierwsze, Góry Trzecie, Górki, or to 'dół' ('bottom', 'foot' in English), for example in Doły in Kazimierz Dolny. Numerous villages are named after forms of traditional land use, such as Łąki (near Nałęczów) after Polish equivalent to 'meadows', and comprising 'Las' ('Forest'): Las Stocki ('Stocki's Forest'), Nowy Las ('New Forest'). Further, several settlements are named after flora species, especially trees:

Grabówki after the hornbeam ('grab' in Polish), Jaworce after the sycamore tree ('jawor'), Za Dębem after the oak ('dąb').

However, the relationship between culture and biosphere should not be limited to the concern of traditional land use. Importance of natural values has a wider meaning here. It may be argued that an impact of natural elements on culture has surpassed the notion of local resources and a customary way of life. Natural values of the area (reflected in art and confirmed by art) considerably influenced the national culture and identity. Nonetheless, this contentious theme requires further exploration.

Moreover, the concern of social appreciation of the values in the context of present spatial planning and local decisions should be evoked as far as biodiversity and landscape are regarded. The analysis of post-war history of spatial planning in Kazimierz Dolny and its vicinities (Kopciowski 2013) shows that requirements of spatial order and the role of natural values (with biodiversity among them) have been frequently underestimated in local policy. Regrettably, this outstanding landscape is confronted with many threats. Shrinkage of ecologically important areas around spots of great touristic attractiveness, uncontrolled settlement sprawl and deterioration in many places, not excluding the centre of Kazimierz Dolny, are observed (Chmielewski et al. 2004; Kosiński 2010).

3.6 Concluding Commentary

Interconnections between nature and culture are extremely difficult field of enquiry. The undertaking to connect the natural and the cultural, the physical with the mental —which is a challenge—must meet obstacles, because the separation of natural and cultural elements of the world lays deeply in our system of knowledge. Nevertheless, nature and culture do converge on many levels and in many dimensions.

Studies on interlinkages between natural and cultural diversities tend to focus on material representations of environment. However, the notion of resource and its socio-economic importance for society is only part of the problem. Pretty et al. (2008) indicated four spheres (literally 'bridges'), where biodiversity and culture are in a very close contact. These are: (1) Beliefs, meanings and worldviews, (2) Environmental practices and activities, (3) Knowledge about nature, (4) Norms and regulations. The case of the Małopolski Vistula Gorge and the Nałęczów Plateau can contribute to this wide recognition. The work presented here reveals that relations between biodiversity and culture may go far beyond the aspect of economic subsistence, use and service. Values attached to a landscape and biosphere may play an even greater role. They are reflected not only in material and immaterial culture of local societies, but they rise to a national level. Furthermore, the way in which nature expresses itself, historicity and high aesthetic values considerably added to symbolic meaning of the landscape and influenced national identity.

Interspersion of natural and cultural values is unavoidable in many disciplines and in many (should we suggest nowadays: most?) places in the world. Naveh

(2000) formulating the Multifunctional Landscape Theory (2000) highlighted complexity of the natural–cultural interaction system. According to his formula, cultural, multifunctional landscapes are formed not only by the bioecological and socio-economic domains, but also by 'the third domain': socioecological and cultural. Many values attached to nature and biodiversity strongly depend on influences encapsulated in this third dimension. Changes in cultures, even their slight alterations may cause far-reaching consequences for individual natural elements as well as for the whole ecological system. Again, the area demonstrated in the following chapter confirms this thesis, just to bring to memory the process of landscape transformation taking part in the Renaissance. Great architectonic achievements, a splendid row of over sixty granaries on the Vistula banks in Kazimierz Dolny coincided with overwhelming changes of the surrounding (deforested) landscape. Without careful reflection on interweaving natural and cultural processes we would not be able to maintain values, areas, species and habitats nor understand our own history.

Acknowledgments I thank warmly the staff of The Head Office of the Lubelski Landscape Parks, in particular to Maria Straka and Piotr Deptuś, for generous help in conducting the study on biodiversity values of the Kazimierski Landscape Park. Special thanks should go to Tomasz Furtak for enriching the knowledge about the Lublin Upland, its landscape functions and amenities.

References

Antrop M (2005) Why landscapes of the past are important for the future. Landscape Urban Plan 70:21–34

Barańska K, Jermaczek A (2009) Poradnik utrzymania i ochrony siedliska przyrodniczego 6210-murawy kserotermiczne. Wydawnictwo Klubu Przyrodników, Świebodzin

Baran-Zgłobicka B, Zgłobicki W (2006) Geomorfologiczne uwarunkowania struktury użytkowania ziemi w krajobrazach lessowych Polski SE. Probl Landscape Ecol 18:483–490

Chmielewski TJ, Chmielewski S, Jakubaszek S, Krawczyk A (2004) Zmiany struktury użytkowania ziemi i cech fizjonomicznych obszaru Kazimierskiego Parku Krajobrazowego w II połowie XX wieku. In: Kucharczyk M (ed) Współczesne problemy ochrony krajobrazu. Zarząd Zespołu Lubelskich Parków Krajobrazowych, Lublin, pp 67–76

Costa A, Madeira M, Santos JL, Plieninger T (2014) Recent dynamics of evergreen oak-wood pastures in south-western Iberia. In: Hartel T, Plieninger T (eds) Ecological wood-pastures in transition, a socio-ecological approach. Routledge, Oxon, New York, pp 70–89

Council Directive 92/43/EEC (1992) of 21 May 1992 on the conservation of natural habitats and wild fauna and flora. Official J L206:7–50

Directive 2009/147/EC (2010) of the European parliament and the council of 30 November 2009 on the conservation of wild birds. Official J L20:7–25

Ellenberg H (1963) Vegetation Mitteleuropas mit den Alpen. In: Walter H (ed) Einfuhrung in die Phytologie. Ulmer, Stuttgart

Faliński JB (1969) Zbiorowiska autogeniczne i antropogeniczne. Próba określenia i klasyfikacji. Dyskusje fitosocjologiczne 4. Ekologia Polska 15(2):173–182

Faliński JB (1991) Kartografia geobotaniczna, cz. 3: Kartografia geobotaniczna i stosowana, PPWK. Warszawa, Wrocław

Folk Artists of the Land of Puławy (2011) Wydawnictwo informacyjne Starostwa Powiatowego w Puławach, Puławy
Forman RTT (1995) Some general principles of landscape and regional ecology. Landscape Ecol 10(3):133–142
Forman RTT, Godron M (2006) Land mosaics: the ecology of landscapes and regions. Cambridge University Press, Cambridge
Gacka-Grzesikiewicz E (ed) (1995) Vistula as an ecological corridor. State-functioning-threats. IUCN Foundation, Warsaw
Gacka-Grzesikiewicz E (2006) Ochrona walorów turystycznych krajobrazu doliny Środkowej Wisły szansą na rozwój regionu. In: Kucharczyk M (ed) Współczesne problemy ochrony krajobrazu. Zarząd Zespołu Lubelskich Parków Krajobrazowych, Lublin, pp 99–105
Gardziel Z, Harasimiuk M, Jezierski W, Pawłowski A, Zgłobicki W (2006) Erozja wąwozowa w zachodniej części Płaskowyżu Nałęczowskiego. Przegląd Geologiczny 54(9):768–776
Gauda A (ed) (2001) Dziedzictwo kulturowe Lubelszczyzny. Kultura ludowa. Polskie Towarzystwo Ludoznawcze, Lublin
Górecka M (2004) Możliwości rozwoju agroturystyki w Małopolskim Przełomie Wisły (jako forma rewitalizacji terenów nadwiślańskich). In: Kucharczyk M (ed) Współczesne problemy ochrony krajobrazu. Zarząd Zespołu Lubelskich Parków Krajobrazowych, Lublin, pp 117–125
Harasimiuk M, Henkiel A (1978) Wpływ budowy geologicznej i rzeźby podłoża na ukształtowanie pokrywy lessowej w zachodniej części Płaskowyżu Nałęczowskiego. Annales Univ. M. Curie-Sklodowska, sectio B., 30/31:55–80
Interpretation manual of European Union habitats, July 2007, EUR 27
Kondracki J (1994) Geografia Polski. Mezoregiony fizycznogeograficzne, PWN, Warszawa
Kondracki J, Ostrowski J (1994) Typy krajobrazu naturalnego i jednostki fizycznogeograficzne 1:2000000. In: Kondracki J (ed) Geografia Polski. Mezoregiony fizyczno-geograficzne. PWN, Warszawa
Kopciowski D (2013) Ochrona zabytkowych wartości Kazimierza Dolnego. Przestrzeń i forma 13:99–115
Kosiński W (2010) Zespół krajobrazowy Kazimierz Dolny-Janowiec. Ochrona *genius loci* wobec aktualnych zagrożeń. Architektura, Czasopismo Techniczne 3 5:149–177
Kucharczyk M (2003) Phytogeograpical roles of lowland rivers on the example of the middle Vistula. Univ. M. Sklodowska-Curie Press, Lublin
Kucharczyk M, Wójciak J (1995) Ginące i zagrożone gatunki roślin naczyniowych Wyżyny Lubelskiej, Roztocza, Wołynia Zachodniego i Polesia Lubelskiego. Ochrona Przyrody 52:33–46
Kucharczyk M, Kucharczyk H, Wójciak J, Rafałowski K (1998) Operat ochrony przyrody ożywionej, adaptacja "Planu zagospodarowania obszaru funkcjonalnego Kazimierskiego Parku Krajobrazowego" do „Planu Ochrony". Lubelska Fundacja Ochrony Środowiska Naturalnego, Lublin
Litwin J (2007) Szkutnictwo w okolicy Kazimierza Dolnego. In: Życie między dwoma brzegami, katalog wystawy fotografii (5.08.-31.10.2007). Muzeum Nadwiślańskie w Kazimierzu Dolnym, Muzeum-Zamek w Janowcu, Janowiec
Łoboda K, Sobczyk W (2006) Rozwój sadownictwa oraz jego wpływ na współczesne przekształcenia w krajobrazach rolniczych w zachodniej części Wyżyny Lubelskiej. In: Kałamucka W (ed) Krajobraz kulturowy zachodniej części Wyżyny Lubelskiej. Zakład Ochrony Środowiska Univ. M. Curie-Skłodowska w Lublinie, Lublin, pp 52–61
Malawska J (1980) Kultura materialna Kazimierza-tradycyjne zajęcia ludowe. Katalog wystawy, Galeria Letnia, Kazimierz Dolny
Maruszczak H (1972) Wyżyny Lubelsko-Wołyńskie. In: Klimaszewski M (ed) Geomorfologia Polski. PWN, Warszawa, pp 340–384
Maruszczak H (1973) Erozja wąwozowa we wschodniej części pasa wyżyn południowopolskich. Zeszyty Naukowe Postępów Nauk Rolniczych 151:15–30

Maruszczak H (1988) Zmiany środowiska przyrodniczego kraju w czasach historycznych. In: Starkel L (ed) Przemiany środowiska przyrodniczego Polski. Ossolineum, Wrocław, pp 109–135
Matuszkiewicz W (ed) (1995) Potential natural vegetation of Poland, general map 1:300000, sheet n° 7. PAN, IGiPZ, Warsaw
Matuszkiewicz JM (2001a) Geobotanical regions, 1:2500000. In: Matuszkiewicz JM (ed) Zespoły leśne Polski. PWN, Warszawa
Matuszkiewicz JM (2001b) Zespoły leśne Polski. PWN, Warszawa
Matuszkiewicz W (2014) Przewodnik do oznaczania zbiorowisk roślinnych Polski. PWN, Warszawa
Mazur-Hanaj R, Baliszewska M (2014) Rewizja raportu 2011 "Muzyka ludowa od tradycyjnej do folkowej". In: Grozdew-Kołacińska W (ed) Raport o stanie tradycyjnej kultury muzycznej. Instytut Muzyki i Tańca, Warszawa, pp 54–65
Meeus JHA (1995) Pan-European landscapes. Landscape Urban Plan 31:57–79
Naveh Z (2000) Introduction to the theoretical foundations of multifunctional landscapes and their application in transdisciplinary landscape research. In: Brandt J, Tress B, Tress G (eds) Multifunctional landscapes: interdisciplinary approaches to landscape research and management. Centre for Landscape Research, Roskilde, pp 27–43
Nogaj-Chachaj J (2000) W epoce kamienia. In: Banasiewicz-Szykuła E (ed) Archeologiczne odkrycia na obszarze Kazimierskiego Parku Krajobrazowego. Wojewódzki Konserwator Zabytków w Lublinie, Lublin, pp 15–16
Odorowski W (1991) Malarze Kazimierza nad Wisłą. Krajowa Agencja Wydawnicza, Warszawa
Pawłowski A (2004) Kształtowanie się krajobrazu kulturowego w dolinie Wisły Środkowej. In: Kucharczyk M (ed) Współczesne problemy ochrony krajobrazu. Zarząd Zespołu Lubelskich Parków Krajobrazowych, Lublin, pp 51–66
Pawłowski A (2006a) Kazimierz Dolny- symbol w krajobrazie (krajobraz historyczny, rozwój przestrzenny, funkcje miasta). In: Kałamucka W (ed) Krajobraz kulturowy zachodniej części Wyżyny Lubelskiej. Zakład Ochrony Środowiska Univ. M. Curie-Skłodowska w Lublinie, Lublin, pp 87–96
Pawłowski A (2006b) Okres prosperity szlaku wodnego jako apogeum znaczenia nadwiślańskich miast- konsekwencje dla krajobrazu Małopolskiego Przełomu Wisły. Probl Landscape Ecol 18:329–343
Pinto-Correia T (1993) Threatened landscape in Alentejo, Portugal: the 'montado' and other 'agro-silvo-pastoral' systems. Landscape Urban Plan 24(1–4):43–48
Plan Ochrony Kazimierskiego Parku Krajobrazowego-operat generalny (1998) Zakład Planowania Przestrzennego. Pracownia w Lublinie, Lublin
Polish Red Data Book of Animals, Invertebrates (2004) In: Głowaciński Z, Nowacki J (ed) Instytut Ochrony Przyrody PAN, Kraków
Polish Red Data Book of Animals, Vertebrates (2001) In: Głowaciński Z (ed) Państwowe Wydawnictwo Rolnicze i Leśne, Warszawa
Pożaryski W, Maruszczak H, Lindner L (1994) Chronostratygrafia osadów plejstoceńskich i rozwój doliny Wisły środkowej ze szczególnym uwzględnieniem przełomu przez wyżyny południowopolskie. Prace Państwowego Instytutu Geologicznego 147, Warszawa
Pretty J, Adams B, Berkes F, Ferreira de Athayde S, Dudley N, Hunn E, Maffi L, Milton K, Rapport D, Robbins P, Samson C, Sterling E, Stolton S, Takeuchi K, Tsing A, Vintinner E, Pilgrim S (2008) How do biodiversity and culture intersect? Plenary paper for conference 'sustaining cultural and biological diversity in a rapidly changing world: lessons for global policy'. IUCN, 2-5.04.2008
Reder J (2006) Środowiskowe uwarunkowania rozwoju osadnictwa prehistorycznego w zachodniej części Wyżyny Lubelskiej. In: Kałamucka W (ed) Krajobraz kulturowy zachodniej części Wyżyny Lubelskiej. Zakład Ochrony Środowiska Univ. M. Curie-Skłodowska w Lublinie, Lublin, pp 62–75
Regionalna Encyklopedia Kultury Ludowej. www.regiony.kulturaludowa.pl. Accessed on 10 May 2014

Rodzik J (2006) Wąwozy- naturalne czy kulturowe elementy krajobrazu? Probl Landscape Ecol 18:457–464
Rodzik J, Gardziel Z (2004) Układy krajobrazowe wąwozów kazimierskich (geneza i warunki rozwoju). In: Kucharczyk M (ed) Współczesne problemy ochrony krajobrazu. Zarząd Zespołu Lubelskich Parków Krajobrazowych, Lublin, pp 85–92
Scott J, Marshall G (eds) (2009) Dictionary of sociology. Oxford University Press, Oxford, New York
Śnieszko Z (1995) Ewolucja obszarów lessowych Wyżyn Polskich w czasie ostatnich 15000 lat. Prace Naukowe Uniw, Śląskiego w Katowicach 1496
Standard register of data for Natura 2000 area PLH 060015 Płaskowyż Nałęczowski, 2013 Generalna Dyrekcja Ochrony Środowiska w Warszawie
Standard register of data for Natura 2000 area PLH 060045 Przełom Wisły w Małopolsce, 2013 Generalna Dyrekcja Ochrony Środowiska w Warszawie
Teodorowicz-Czerepińska J (1981) Kazimierz Dolny. Monografia historyczno-urbanistyczna. Towarzystwo Przyjaciół Kazimierza, Kazimierz Dolny
Tymochowicz M (2014) Lubelszczyzna-kulturowe zróżnicowanie regionu. In: Regionalna Encyklopedia Kultury Ludowej. www.regiony.kulturaludowa.pl/publikacja/73. Accessed on 10 May 2014
Vidal de la Blache P (1911–1912) Les genres de vie dans la géographie humaine. Annales de géographie 20:193–212, 289–304
Warowna J (2004) Naturalna i antropogeniczna ewolucja Wisły Lubelskiej w aspekcie możliwości jej renaturyzacji. In: Kucharczyk M (ed) Współczesne problemy ochrony krajobrazu. Zarząd Zespołu Lubelskich Parków Krajobrazowych, Lublin, pp 93–97
Warowna J (2006) Naturalna i antropogeniczna ewolucja krajobrazu doliny Wisły w odcinku przełomowym. In: Kałamucka W (ed) Krajobraz kulturowy zachodniej części Wyżyny Lubelskiej. Zakład Ochrony Środowiska Univ. M. Curie-Skłodowska w Lublinie, Lublin, pp 29–34
Wilk T, Jujka M, Krogulec J, Chylarecki P (2010) Important bird areas of international importance in Poland. Ogólnopolskie Towarzystwo Ochrony Ptaków, Marki
Wójciak J, Szewczyk P, Urban M (2010) Małopolski Przełom Wisły (96). In: Wilk T, Jujka M, Krogulec J, Chylarecki P (eds) Important bird areas of international importance in Poland. Ogólnopolskie Towarzystwo Ochrony Ptaków, Marki, pp 331–332
Wojciechowski KH (1996) Krajobrazy-elementy dziedzictwa kulturowego i ich ewolucja. Annales Univ. M.curie-Sklodowska, Lublin, sectio B, vol LI, p 19
Wojciechowski KH (2006) Historia zagospodarowania Powiśla Chodelskiego. In: Kałamucka W (ed) Krajobraz kulturowy zachodniej części Wyżyny Lubelskiej. Zakład Ochrony Środowiska Univ. M. Curie-Skłodowska w Lublinie, Lublin, pp 48–51
Wołoszyn W, Skowronek E (2007) Tourism planning and management in Kazimierz landscape park, Poland. In: Roca Z, Spek T, Terkenli T, Plieninger T, Höchtl F (eds) European landscapes and lifestyles: the mediterranean and beyond. Edic. Universitârias Lusofônas, Lisboa, pp 375–383
World Heritage Cultural Landscapes 1992–2002 (2003) In: Fowler PJ (ed) World heritage papers 6. UNESCO World Heritage Centre, Paris
Zagadnienia ochrony środowiska w gminie Janowiec (1991a) In: Wilgat T (ed) Zarząd Zespołu Lubelskich Parków Krajobrazowych w Lublinie, Lublin, manuscript
Zagadnienia ochrony środowiska w gminie Kazimierz Dolny (1991b) In: Wilgat T (ed) Zarząd Zespołu Lubelskich Parków Krajobrazowych w Lublinie, Lublin, manuscript
Zagadnienia ochrony środowiska w gminie Wąwolnica (1991c) In: Wilgat T (ed) Zarząd Zespołu Lubelskich Parków Krajobrazowych w Lublinie, Lublin, manuscript
Zagadnienia ochrony środowiska w gminie Wilków (1991d) In: Wilgat T (ed) Zarząd Zespołu Lubelskich Parków Krajobrazowych w Lublinie, Lublin, manuscript
Zgłobicki W (2006) Wąwozy jako zapis działalności człowieka w zachodniej części Wyżyny Lubelskiej. In: Kałamucka W (ed) Krajobraz kulturowy zachodniej części Wyżyny Lubelskiej. Zakład Ochrony Środowiska Univ. M. Curie-Skłodowska w Lublinie, Lublin, pp 35–40

Zgłobicki W, Rodzik J, Schmitt A, Schmidtchen G, Dotterweich M, Zamhöfer S, Bork HR (2003) Fazy erozji wąwozowej w okolicach Kazimierza Dolnego. In: Waga JM, Kocel K (ed) Człowiek w środowisku przyrodniczym-zapis działalności, Polskie Towarzystwo Geograficzne, Oddział Katowicki, Sosnowiec, pp 234–238
Żurawski J (1986) Kazimierz Dolny. In: Zin W (ed) Zabytki urbanistyki i architektury w Polsce. Odbudowa i konserwacja 1: Miasta historyczne, Warszawa, pp 187–211

Chapter 4
Wooded Grasslands as Part of the European Agricultural Heritage

Csaba Centeri, Hans Renes, Michael Roth, Alexandra Kruse, Sebastian Eiter, Jutta Kapfer, Antonio Santoro, Mauro Agnoletti, Francesca Emanueli, Maurizia Sigura, Martina Slámová, Marta Dobrovodská, Dagmar Štefunková, Zdeněk Kučera, Dénes Saláta, Anna Varga, Salvador Villacreces and Johannes Dreer

Abstract Wooded grasslands have always played an important role in rural life with changing issues: They are of high importance for questions of biodiversity, soil, and water resources and in preserving agricultural heritage, but their maintenance is labor intensive. Abandoned wooded grasslands undergo succession, and food production alone does not support their survival. They require special attention

---

C. Centeri (✉) · D. Saláta
Faculty of Environmental and Agricultural Sciences, Institute of Nature Conservation and Landscape Management, Department of Nature Conservation and Landscape Ecology, Szent István University, Páter K. u. 1, 2100 Gödöllő, Hungary
e-mail: Centeri.Csaba@mkk.szie.hu

D. Saláta
e-mail: Salata.Denes@mkk.szie.hu

H. Renes
Faculty of Geosciences, Utrecht University, P.O. Box 80115, 3508 TC Utrecht, The Netherlands
e-mail: j.renes@uu.nl

H. Renes
VU University Amsterdam, Amsterdam, The Netherlands

M. Roth
School of Landscape Architecture, Urban and Environmental Planning, Nürtingen-Geislingen University, Schelmenwasen 4-8, 72622 Nürtingen, Germany
e-mail: Michael.Roth@hfwu.de

A. Kruse
Institute for Research on European Agricultural Landscapes e.V, Hauptstr. 48, 51491 Overath, Germany
e-mail: kruse@eucalandnetwork.eu

S. Eiter
NIBIO - Norwegian Institute of Bioeconomy Research, P.O. Box 115, 1431 Ås, Norway
e-mail: sebastian.eiter@nibio.no

M. Agnoletti and F. Emanueli (eds.), *Biocultural Diversity in Europe*, Environmental History 5, DOI 10.1007/978-3-319-26315-1_4

and at the beginning a well-established subsidy system can help to contribute to their survival. Their sustainable use in the present-day landscapes can only be conceivable in complexity where food production, reintroduction of their cultural values, biodiversity and landscape protection, and ecotourism are playing an

J. Kapfer
NIBIO - Norwegian Institute of Bioeconomy Research, P.O. Box 2284, 9269 Tromsø, Norway
e-mail: jutta.kapfer@nibio.no

A. Santoro · M. Agnoletti · F. Emanueli
Laboratory for Landscape and Cultural Heritage, University of Florence, Via San Bonaventura 13, 50145 Florence, Italy
e-mail: antonio.santoro@unifi.it

M. Agnoletti
e-mail: mauro.agnoletti@unifi.it

F. Emanueli
e-mail: francesca.emanueli@unifi.it

M. Sigura
Department of Agriculture and Environmental Sciences, University of Udine, 33100 Udine, Italy
e-mail: maurizia.sigura@uniud.it

M. Dobrovodská · D. Štefunková
Institute of Landscape Ecology, Slovak Academy of Sciences, Stefanikova 3, P.O. Box 254, 814 99 Bratislava, Slovak Republic
e-mail: marta.dobrovodska@savba.sk

D. Štefunková
e-mail: dagmar.stefunkova@savba.sk

Z. Kučera
Faculty of Science, Department of Social Geography and Regional Development, The Research Centre for Cultural and Historical Geography, Charles University in Prague, Albertov 6, 128 43 Praha 2, Czech Republic
e-mail: zdenek.kucera@natur.cuni.cz

A. Varga
Centre for Ecological Research, Hungarian Academy of Sciences, Vácrátót, Hungary
e-mail: varga.anna@okologia.mta.hu

S. Villacreces
Technical University of Madrid, Madrid, Spain
e-mail: salvavillacreces@gmail.com

J. Dreer
Hof und Leben GmbH, Unternehmensberatung für Landwirte, Johannes Dreer, Welfenstr. 16, 86825 Kirchdorf, Germany
e-mail: Dreer@hofundleben.de

M. Slámová
Technical University in Zvolen, Department of Landscape Planning and Design, T.G.Masaryka 24, 96053 Zvolen, Slovakia
e-mail: mslamova@hotmail.com

important role. This chapter gives an overview on the recent situation of wooded grasslands and their historical development, based on the work done by the Institute for Research on European Agricultural Landscapes (www.eucalandnetwork.eu). National pictures, definitions, history (including local names), threats, potentials, cultural values, spatial distributions, subtypes, and available databases have been collected, described, and analyzed. The main results of this survey are as follows: (1) Wooded grasslands are known to the public but mainly to local communities where they occur; (2) Many subtypes of wooded grasslands exist in various European countries; (3) Wooded grasslands underwent tremendous changes during the past centuries and lost their importance for various reasons; (4) There are many local and regional projects focusing on wooded grasslands, often as "lighthouse" projects to valorise cultural achievements.

Keywords Biodiversity · Rural countryside · Best-practice examples · Sustainable land use · Biocultural diversity

4.1 Introduction

Wooded grasslands belong to the richest and most beautiful landscapes in Europe (Alexander 1998; Fleming 2012; Rackham 1998, 2003, 2006, 2013; Rotherham 2013a, 2013b). They are, however, increasingly rare and under threat by modern farming and forestry (Green 2013). In this paper, an attempt is made to define and describe this landscape type, as well as to map its dispersal.

The paper is prepared by members of the EUCALAND Network[1] (European Culture Expressed in Agricultural Landscapes). EUCALAND is an expert network on the agricultural landscapes of Europe and their heritage (Printsmann et al. 2012; Pungetti and Kruse 2010). In 2010, the network published a glossary of 42 European agricultural landscape types and terms with translations into six languages (Kruse et al. 2010), aiming at future extension. In 2012, the Network organized a workshop on "wooded pasture," the traditional landscapes that combine trees with grazing lands.

The main research questions were:

1. What types of wooded grasslands exist in Europe and what is their dispersal?
2. What are the cultural values of the agricultural landscapes of wooded grasslands?
3. How to maintain the cultural values of this cultural landscape type?

[1]www.eucalandnetwork.eu.

4.2 Definitions

The landscape we describe in this paper, and for which we use the umbrella term of 'wooded grasslands,' combines trees with grassland. Such landscapes can be found in different European regions.

Natural wooded grasslands are defined by various authors. White (1983) gives a good example: "Wooded grasslands are lands covered with grasses and other herbs with woody plants [trees (≥7 m tall), bushes (3–7 m), dwarf trees, palm trees or shrubs (≤2 m)] covering between 10 and 40 % of the ground. Woody plants nearly always occur scattered."

Most definitions do not distinguish between man-made and natural wooded grasslands. Outside Europe, landscapes that combine grazing with trees are often described as 'natural' landscapes. Examples are the African savannas. In parts of Europe, there is an ongoing discussion on the 'natural' state of the landscape. Most authors take it for granted that, without human influence, most of temperate Europe would consist of mixed forest. Old ideas of natural open lands, such as Gradmann's 'Steppenheide' theory, (Gradmann 1933) have been revived in recent decades, particularly by the Dutch ecologist Frans Vera who claims that the natural landscape in temperate Europe would be a half-open, park-like landscape. However, most authors see these half-open landscapes as man-made.

Man-made 'wooded grasslands' probably first appeared along with other types of man-made 'grasslands' (although there is little evidence for that). Secondary grassland that appeared clearly due to man-made deforestation (Dirkx 1998) started with the spread of Neolithic husbandry (Bredenkamp et al. 2002) and other land management activities like mowing, slash-and-burn agriculture, arable farming, etc. (Coupland 1979; Knapp 1979; Pott 1995; Jääts et al. 2011). According to Pott (1995), man-made grasslands were obviously in use and wide spread since the Pleistocene. However, Vera declares that the European Atlantic period (8000–5000 BP) is the period before the introduction of the livestock (Vera et al. 2006).

The main sources for the existence of former (man-made) wooded grasslands are soil characteristics and pollen. As the prevention of the wooded grassland from reverting to woodland required burnings, the signs of these burnings can be found in the soil, as coal beds with their attendant fossil pieces or thin layers of charcoal and fusain can prove that fire was part of ancient ecosystems (Cole 1986; Schaetzl 1986; Komarek et al. 1973; Kozlowski and Ahlgren 1974). This fossilized carbon does not tell us if it has a natural or man-made origin, but if it is historically proven that grasslands were in use as part of the agriculture in the given area, these signs can prove the possibility of existence of a wooded pasture, too.

In Europe, the other evidence of former open or partly open (not fully covered forest areas) grasslands is the presence of light-requiring plant species, e.g., shrub hazel (*Corylus avellana*), pedunculate oak (*Quercus robur*), and sessile oak (*Q. petraea*) (Vera 2002; Vera et al. 2006). The possibility of using light-requiring species as indicators of openness is also used in other continents, e.g., in the USA, where red maple (*Acer rubrum*), black birch (*Betula lenta*), and yellow birch

(*Betula alleghaniensis*) were such indicator species (Whitney 1984, 1990; Rooney 1995; Abrams et al. 2000, 2001).

In Sweden, wooded grasslands, called "leaves meadows," are highly distinguished in a landscape change study where a 200-year period is examined and five principal land cover groups were made, one of which is "wooded grassland." This must have played an important role in the Swedish countryside (Skånes and Bunce 1997) and was combined with animal husbandry. These enclosed leaves meadows, often close to the villages, served for keeping ill or pregnant animals. Today, it remains only as relics, e.g., on the island of Öland.

In Italy, wooded grasslands are mentioned for various areas, normally as wooded pastures. In the mountainous areas of the Alps, the most frequent species are larch (*Larix decidua*) and red fir (*Picea abies*); on the higher Apennines are beech (*Fagus sylvatica*) and turkey oak (*Quercus cerris*), and in the hills can be found pubescent oaks (*Quercus pubescens*). In Mediterranean areas, the mainly used trees are holm oak (*Quercus ilex*), cork oak (*Quercus suber*), and carob trees (*Ceratonia siliqua*) (Del Favero 2008). The wooded pastures of Salten (Alps of Trentino Alto Adige, Italy) have groups of larch trees (*L. decidua*) creating a beautiful landscape (Agnoletti 2013). Here we have to mention that Agnoletti (2013) does make a difference between wooded grasslands and wooded pastures. Another interesting point of Agnoletti (2013) is that he suggests with regard to the situation of Sardinia the attribute of "treed" instead of "wooded" for grasslands with trees that might be a considerable suggestion for future use.

As our research aimed at agricultural landscapes, we excluded sylvopastoral systems (this is why forest pastures are not listed here) and included only those landscapes where grasslands were created for using the grass for forage or other purposes and trees were left there (e.g., from forest clearing) or planted on purpose. On the other hand, it was important to include all landscape types that look physiognomically similar. This is the main driving rule for defining wooded grasslands: their appearance. This is the reason why all grasslands where any type of tree appears as an important landscape characteristic have been included. The reason of its creation was not the sole organizing rule for defining a particular landscape as wooded grassland; however, it was important that it was identified as a result of a human activity rather than a natural environment.

There are some similar landscape types (e.g., transhumance) that might look physiognomically similar to wooded grasslands, but differ by the purpose of the use. They were, in fact, created to be used as grasslands, but developed naturally by common succession processes.

4.3 Agricultural Landscapes as Heritage

Wooded grasslands are a part of the European heritage (Laszlovszky and Szabó 2003; Rackham 2003; Howard 2003). Heritage can best be defined as 'those remains from the past that are seen as valuable for the present.' Another definition of heritage is "properties passed, according to law, from fathers and mothers to their children" (Choay 2005). It has been gradually expanded to become "a set of representations and attributes attached to a non-contemporary object (artwork, building (Ellenberg 1990), landscape, site, …) which recognized significance requires a protection" (Lazzarotti 2003).

This notion of heritage covers a much broader field (Poulot 1998; Neyret 2004): "the range of meanings attached to this formerly precise legal term has recently undergone a quantum expansion to include almost any sort of intergenerational exchange or relationship […] between societies as well as individuals" (Graham et al. 2000). It means that the broader notion includes immaterial elements (e.g., oral traditions, folklore, music, etc.), too.

We wish to follow this notion during the analyses and listing of the cultural heritage of the recently described agricultural landscape type: wooded grassland.

4.4 Materials and Methods

Country synopses were written using a standardized questionnaire (Fig. 4.1) with EUCALAND network members as key experts, collecting national perspectives on history, threats, potentials/chances, names, cultural values, spatial distributions, subtypes, other values (e.g., biodiversity, economy, etc.), key actors (history,

Description of Agricultural Landscape Types and Agricultural Landscape Elements – Questionnaire

Description of Cultural Landscape Types (one per year)
By the EUCALAND Network

Definition:

Theme 2013: Water meadows.
That include
That does not include

	Question	Sources / References	Explanation How to do it
	If you need more space for your answer, use additional paper apart by naming the number of the question you are referring to		
1.	Does it (or did it ever) exist in your country? □ Yes □ No		cross the right answer
2.	What is the used name in your country? __________ Describe it from your national perspective.		Give the national or regional name(s). Describe differences if any with the other, similar types in Europe. You can also describe the importance, meaning etc.
3.	Occurrence in a given cell raster (grid)[1]: ○It does not occur •It occurs ●It occurs and is important ●It is dominant /characteristic Describe the change in occurrence over time. (If you can fill in the same grid for a historical period, its fine)		Describe, as you know it from plants or animals, the occurrence. Try to be as precise as possible. In this way, we will receive the occurrence in Europe. You can produce several grids for different times in order to demonstrate the changes during time.
4.	How/why/by, whom was it created? Describe the genesis (history) of that landscape in your country. E.g.		Here you have the possibility to describe more in detail, what you know about the landscape type/element regarding your country. You can

Fig. 4.1 Extract of the annual EUCALAND questionnaire for gathering country information

present) for cultivation and management of wooded grassland, recognition in science, and public media (references on national situation, bibliography).

The questionnaire was elaborated by the EUCALAND network members on various workshops and finalized at the 5th EUCALAND Workshop, which took place at Szent István University (Hungary) from 18 to 20 of April 2012. Results were introduced and discussed at the 6th EUCALAND Workshop at Utrecht University (The Netherlands), from 24 to 26 of April 2013.

4.5 Results

In the following text, selected results are presented. Data from ten European countries have been gathered: Estonia, France, Germany, Hungary, Italy, Norway, Slovakia, Spain, the Netherlands, and the United Kingdom. Considering the unequal importance and variety of wooded grasslands in these countries, the following country synopses differ in their length and detail. Nevertheless, this summary is only a first step toward a European description of the national situations which will follow later in the project of describing the European agricultural landscapes.

4.5.1 Country Synopses

4.5.1.1 Estonia

Although Estonia did not fill in a questionnaire, Kukk and Kull (1997) published a description of wooded meadows in Estonia. As in the present work we consider wooded meadows as a type of wooded grasslands, a description of wooded meadows is useful for getting closer to a European perspective.

Kukk and Kull (1997) described the wooded meadows where—besides mowing—branches and twigs of trees and shrubs were collected for winter leaf fodder. The authors state that there are large regional differences in the method of this collection. In Sweden, trees were pollarded when collecting twigs but they were coppiced in Estonia. According to Kukk and Kull (1997), the wooded meadows of Estonia were classified based on their soil conditions as follows:

(a) meadows on calcium-rich soils (species rich);
(b) on acidic soils (species poor);
(c) on flooded river valleys (no specification on species richness added).

All types are divided to subtypes based on their moisture conditions (e.g., wet wooded meadows). It is interesting and gives a point to the European perspective that as there are breaks in regular mowing, there are a series of successive stages between wooded meadows and forests. This should be the case in the majority of European wooded grasslands, too.

4.5.1.2 France

Local name: *systèmes agro-sylvopastoraux méditerranéens, pâturage boisés.*[2]

The wooded grassland type exists in France in several different forms.

Type 1: The appearance is very much like the German "Streuobstwiesen" (see German national description), where the production of especially apples and pears plays an important role, e.g., in Normandy (north), Alsace, and Lorraine (east). In the Centre (center), the dominating trees are walnuts and peach.

Type 2: The agro-sylvopastoral system, e.g., in the Var region and in the Jura (both south-east), is a combination of crop production, meadows, and extensive forestry with the aim of acting against forest fires (Joffre et al. 1991). There exist four subtypes, depending on the coverage of trees. Many (or most) of them are located in mountain regions. We find it in many Mediterranean countries, especially where there are heavy problems with forest fire. The system exists since the end of land reclamation of the monarchy in the Middle Ages, although with the agricultural disconnection after World War II, the terrain was abandoned and/or reclaimed and the system declined. Only 15 years ago, it started to be promoted by the regional authorities (Conseil Général CG) of Var within the guideline of a sustainable forestry (DFCI: Défense de la Forêt contre les Incendies)—especially regarding forest fires. In the Var region, it is connected with the production of quality products (olive, chestnut, truffle, wine) and it also is a part of a know–how transmission of the ancestors. In the Jura the today's awareness and protection of wooded grasslands results especially from their function for recreation and tourism. Another subtype does exist, e.g., in the Vosges (north-east): half-open areas. But here, the use of tree fruits does not play any role.

Three systems of breeding are connected to the timber (fire wood) cutting (type 2):

1. winter transhumance: the animals spend the summer in the mountains, and during winter they stay at the coastal zone;
2. reorganisational pastoralism: one tries to establish a local transhumance, while inviting local breeder to send their animals to the common fields. A well-known and only lately introduced example is in the Puy-de-Dôme region

[2]The author thanks the students of the Master 1 "Dynamique des Territoires et nouvelles ruralités" (2012), Clermont Universités et VetAgro Sup, with their professor Yves Michelin (VetAgroSup et Laurent Rieutort UBP Clermont) for providing information on the recent situation in France.

(Clermont-Ferrand, center), where sheep from the Parisian region spent their summer;
3. installation of sylvo-pastoralism: breeder sends their animals to the clearing zones in order to keep the animals there and the zone open.

Regarding our definition for wooded grasslands, we see that type 1 and 2 are at the edge of the definition. But as they nearly always include (intensive) fruit production and animal husbandry, they are considered as within our topic of wooded grasslands, although in France itself, they are considered as a kind of forestry.

Type 3: *Paysage autoritaire* (authoritarian landscapes). Here, the landscape is the fruit, at once, the result of its own and of human activity. There, where the landscape is managed and is receiving "help," for its proper development, e.g., in the Jura Massif, it is determined for and by mankind: cutting and zoning (and protection) of the natural areas, clearings, etc.

The described systems, especially type 2, are of growing importance, developed by and with means of the public authorities. But the system depends completely on the financial support of the "Conseil Général" (CG).[3] The management is constantly developing, and among others there is research and consultancy done by CERPAM,[4] a research center in the department of the Mediterranean Alps. There are contracts with the animal producers. The national forest office (ONF) is one of the main important partners. Agro-sylvo-pastoralism is seen as *the* regional tool against the forest fires. It increases the worth and value of soil and regional products as it is the tool for forest fires prevention. Therefore, agro-sylvo-pastoralism can be considered having a double use. The system sets in relation to several agricultural activities: forestry, soil production and orchards, renewal of agricultural activities on former agricultural land (olive- and oil-production, truffle-, wine- and chestnut production, and others), alpine and pasture farming (sheep, goats), and crop production.

Only a few people know about the agro-sylvo-pastoralism in the Var region. But the CG communicates toward the inhabitants about the risks of fire and the measures that have been put in place. Other countries have taken notice of the program and are interested. Based on the Var experiences, a common Mediterranean project within the cadre of the European INCENDI program has been developed.

4.5.1.3 Germany

Local name: Streuobstwiese (national), regional: Bongert, Bitz

[3]This means the administration of a department, France has 101 departments.

[4]Centre d'études et de réalisations pastorales Alpes Méditerranée (CERPAM).

Wooded grasslands are very well known in Germany since the prehistory times, as already Stone Age people have used the fruits in order to enrich their nutrition. Planned or maintained, wooded grasslands are also very old. There are several records and examples for promotion by the Cistercian monks in the thirteenth century. New influences and cultivars have been (re-)introduced by the Huguenots in the sixteenth to seventeenth centuries. From the eighteenth to nineteenth centuries, Prussian kings promoted the planting of fruit trees by law (Kruse 1999).

There are several German names used to denominate wooded grasslands; however, "Streuobstwiese" or "Obstwiese" are the most frequent in comparison with regional names like "Obstgarten," "Bitz," or "Bongert." The meaning of these names is meadow with fruits or simply fruit garden. The Streuobstwiese has a high importance, ecologically and for regional identity and is very well known. Nevertheless, there is an important decrease in its extent since the 1950s because of the industrialization of farming. A second reason for the loss of Streuobstwiesen is their location at the fringe of rural settlements, a favored place for the development of new residential estates. As many rural settlements, especially in Western Germany, have been growing massively since the 1950s, many Streuobstwiesen have been cut down for new building land.

In Germany there are many initiatives—local, regional, national—in order to protect and to reinvest in Streuobstwiesen. One aim is to keep or to re-plant high standard trees (instead of low or half standard trees), because they have a many times higher ecological value and meaning. There are many different cultivars and types. In order to keep this knowledge and also the seeds, a garden was created with public financial support by DBU (German Federal Environmental Federation), where traditional, regional sorts are cultivated, including a seedbank.

The common appearance is:

1. A mixed use of fruit productions and grass, haymaking, and/or pasturing (Fig. 4.2), often combined with beekeeping or extensive pony or cattle rising.
2. Grazed land, wet, close to floating water: Ash tree, wych elm, aspen, goat willow, small-leaved lime, oak, and birch: Trees often in a row/line.
3. Very little "forests" (*Gehölz*) within (in the middle) of grazed land: different species, often combined with wild fruit trees, with or without water in the middle.

In the later two types the fruits are normally not harvested. They are anyway at the edge of our definition.

Streuobstwiesen are endangered in Germany. This fact is well recognized. They are listed in the Red List of Biotope types of the German Ministry for Nature Conservation. Also, in several federal states, they are listed as a protected biotope, e.g., Saxony. At the same time, the recognition of their cultural, aesthetical (Fig. 4.3), and last but not least ecological values is very high, though is the awareness.

There are many protections and maintenance programs, but also successful and well-known social (regional) events (Fig. 4.4), brands, and marketing (see paragraph on "use of the wooded grasslands today").

Fig. 4.2 Pasturing of cherry orchard meadows in the Swabian Alb, Germany, June 2010 (*Photo* Huber S.)

Fig. 4.3 Spring aspect of "Streuobstwiese" with flowering apple trees, April 2011 (*Photo* Habeck J.)

4.5.1.4 Hungary

Hungarian name: fás legelő.

Wooded grassland habitat complexes were dominant management types across centuries known from the medieval times in Hungary (Bartha 2003; Szabó 2005; Varga and Bölöni 2009). In Hungary, we can state that wooded grasslands are recognized as wooded pastures and wooded meadows (Bölöni et al. 2008). Their maintenance is basic part of the traditional ecological knowledge (Varga and Molnár 2014). Because of abandonment derived from the change of agricultural

Fig. 4.4 Teaching how to prune fruit trees on wooded grasslands as a mean of conservation, February 2008 (*Photo* Habeck J.)

structure—mostly 50'–80' in Hungary—and regulation (Saláta et al. 2009) the wooded pastures are in the 8th most endangered woody, semi-natural habitats (Molnár et al. 2008); moreover, no more than 5500 ha left from them (Bölöni et al. 2008). The abandonment is followed by large-scale reshrubbing and self-reforestation (Szabó et al. 2007; Varga and Bölöni 2009, 2011; Varga et al. 2014). Still existing wooded grasslands are recognized as important landscape and nature conservation values (Márkus 1993; Haraszthy et al. 1997; Saláta et al. 2011; Vityi and Varga 2014). There are several conservation areas dedicated to wooded pastures (Varga et al. 2014). One of them is the Hollókő Landscape Protection District, a "UNESCO World Heritage Site" (Fig. 4.5), where the wooded pasture is part of the world heritage site as the surrounding of the village, bearing an important landscape value (Harmos 2013).

4.5.1.5 Italy

Italian name: pascolo arborato, local names: salti, difese.

Wooded grasslands are widely used as wooded pastures in Italy. These can usually be found in areas with difficult climatic conditions and poor soils, where

Fig. 4.5 Wooded grassland near by the castle of Hollókő after renovation of the abandoned wooded pasture (some of the trees were cut in order to create the state of the former look of the pasture, regardless the fact that there is no grazing on the area at the time of writing, 2014) (*Photo* Centeri Cs., 2013)

other cultivation is not possible. The first information about Italian wooded pastures date back to the sixteenth century, when these land uses were listed in some maps of central Apennines where there was a pastoral system based on grazing and on the leaves coming from the pollarding of the trees that grew on the pastures (Afan De Rivera 1842; Moreno and Poggi 1995; Palumbo 1912; Tondi 1821; Zanzi Sulli and Di Pasquale 1993). These pastures have an anthropic origin, related to the historical and economic situation of the local people that modified the surrounding environment to obtain services and food. They had few economic resources and so they created systems that were able to provide all the goods they needed, such as food, pastures for the animals, firewood, and fruits. These agro-sylvo-pastoral systems were ecologically and energetically sustainable. Due to the various climatic and geographic situations that can be found in Italy, different kinds of wooded pastures can be found throughout the country.

Trentino Alto Adige: In the Alps of Trentino, Alto Adige can be found one of the biggest wooded pastures with larches of Europe, in a place called Salten. This word comes from the Latin word *saltus*, used for describing a landscape characterized by open spaces with trees used for grazing in the middle of large woodlands. The fact that the name of this place reminds to such a Latin word means that this landscape dates back to 2000 years ago (Biasi 2010).

Tuscany: In the Tuscan Cadastral Registry of 1832, a lot of woods were classified as pasture woods, representing the 45 % of all the land used for grazing in the mountainous areas. Since the beginning of the twentieth century, the 90 % of the woods of Tuscany were regularly grazed, and they supported up to 50 % of the livestock of the farms (Pontecorvo 1933). In the 14 areas surveyed for the Tuscan landscape monitoring system, wooded pastures, due to the abandonment of traditional activities, decreased by 88 % in the period 1832–2002. Most of the wooded pastures investigated (67 %) have been covered by shrubs and trees and became woodlands after they were no longer grazed, while 25 % were turned into cultivations. Traditionally, the trees founded on wooded pastures were oaks, pollarded at 2–3 m from the ground. This way of managing the trees allowed to keep the animals on the pastures, as they could not reach the new branches and leaves with their bite. The trees provided firewood, leaves, and acorns for the pigs that grazed on the pasture (De Ricci 1830; Merendi 1957; Pavari 1930), and they were about 8–10 m apart from each other so the ground could receive enough light from the sun (Gabbrielli 1980; Iacini 1801). This management system was already known among the ancient romans, as *silva fructifera* or *silva glandaria*, and it was also cited by Cato the Elder in 53 B.C. in his *De Agricoltura* as the most important way to increase the benefits for a farm (Biblioteca degli scrittori latini 1846). Usually, these pastures were very common until the beginning of the twentieth century in different areas of Tuscany, but the average surface was small (Brilli 1992; Pavari 1934). The density of the trees was between 50 and 170 trees per hectare (Oliva 1924). Every year one or two plants per hectare were cut, chosen among those trees that showed signs of decline and lower acorns production (Del Noce 1849). The lack of natural renovation was remedied with the seeding; the new plants were usually protected from the livestock bite by shrubs with thorns (Gautieri 1813).

Liguria: In the Ligurian Apennines between the eighteenth and nineteenth centuries local communities used a 12-year cycle, where wooded pastures with grey Alder (*Alnus incana*) and other species were alternated to cultivations, in mountainous areas with very poor soils (Bertolotto and Cevasco 2000; Moreno 1990). First, the grey alders on the pastures were coppiced and some of them used as firewood by the local people. Then the remaining wood was burned on the pasture and the ashes were dispersed on the ground to fertilize it; the fire, covered by soil to protect the trees, was also used for clearing the ground from shrubs. The land was then cultivated for 2–5 years, while the alders grew, and finally, when the alders were grown enough and resistant against game animals, sheep and goats were allowed to graze.

Central-southern Apennines: In Abruzzo the *difese* were the wooded pastures on public lands used by the local population for breeding or for collecting firewood and acorns. Traditionally, the trees on the pastures were pollarded at 2.5–3 m for keeping new branches and leaves apart from the bite of the animals.

Sardinia: The *salti* of Sardinia were wooded pastures for public use. The Latin word *saltus* comes from the word used for "gap." In fact it was used for open spaces used for grazing, in the middle of woods (Di Berenger 1859–1863). Since ancient times, the lands that legally belonged to the villages and that where close to it were used as agricultural areas (called *vidazzone*) or for the grazing of domestic livestock (called *paberile*). The public lands far from the villages were called *saltus*, and were used for the grazing of cattle or pigs and for collecting firewood (Beccu 2000), with the grazing that was allowed only when the acorns were on the ground (Alias et al. 2008; Beccu 2000).

Sicily: The wooded pastures in the Ragusa Province date back to the fourteenth century, when the lands of the *latifundio* were divided among the farmers. One of the main features of these wooded pastures is that there are circular dry stone walls surrounding the carob trees for protecting them from the animals (Fig. 4.6). Other dry stone walls are used for dividing the pastures. The trees are mainly used for the carob fruit as food for the cattle and for providing shade during the hot summer months.

Overall, the surface covered by wooded pastures in Italy has decreased during the last 150 years, mainly because of land abandonment. These land uses, in fact, were usually found in mountainous areas or in places with harsh climatic conditions and poor soils, and thus were the first places to be abandoned by local population after the World War II. The abandonment of pastures caused the increase of the forest areas and a loss of biodiversity at the landscape scale. Even if there are no specific statistics on this land use, pastures in Italy decreased from 6,113,000 to 3,346,951 ha in the period 1861–2010.

4.5.1.6 Norway

Norwegian name: *hagemark* (with reservations (only certain pastures): *beiteskog, skogsbeite, innmarksbeite).*

The most typical wooded grassland in Norway is the so-called “hagemark (skog)” (Fig. 4.7). This term is used for grassland with scattered trees and, sometimes, bushes. “Hagemark” tends to be more open than simply grazed forest, although there is no well-defined threshold to separate the two. “Hagemark” can be divided into different types according to its use (e.g., grazed gardens, hay meadows with trees). “Hagemark” is located close to farms, where forests have been used for grazing and for production of firewood and timber. An important component of “hagemark” is pollarded trees whose tops and branches have been cut to provide fodder for domestic animals or for tanning. Ash trees, wych elm, aspen, goat

Fig. 4.6 Wooded pasture with carob trees in Sicily, Italy (*Photo* Agnoletti M.)

Fig. 4.7 "Hagemarkskog" in Norway (*Photo* Hofsten J. / NIBIO, August 2008)

willow, small-leaved lime, oak, and birch are the tree species that are most often dominant in this landscape type (Direktoratet for naturforvaltning 2007; Hauge and Austad 1999; Rekdal and Larsson 2005; Skog og landskap 2012).

It is assumed that forests have been used for grazing since the beginning of agricultural land use in the Stone Age, but at least for 2000 years. Forests have also been utilized for hunting, firewood, timber, and production of tar and charcoal. The forests' function as grazing areas for domestic animals has probably been the most important one, at least until the start of coal mining in the 1500–1600s. Grazed forests have decreased or disappeared locally because of the increase of timber production, but have been common at least until World War II. Forests were used as grazing areas mainly during spring and autumn, whereas alpine pastures have been used in the summer months. Besides the development from forests through high grazing pressure, wooded pastures in Norway may also have developed from grasslands and meadows, whose use has changed from, for instance, hay production to grazing purposes (Hauge and Austad 1999; Kielland-Lund 1999; Kvamme 1999).

Wooded pastures occur in areas with animal husbandry where environmental conditions allow the growth of trees, and with vegetation that is of grazing value. They are not restricted to certain altitudes, neither to a specific substrate, slope, or aspect. The link between wooded pastures and certain environmental conditions is rather indirect. Their occurrence depends more directly on habitats comprising vegetation with high grazing potential, whose distribution is controlled by environmental conditions. Nutrient-rich and productive habitats with scattered trees are areas of high potential for grazing (Frislid 1999; Losvik and Hjelle n.d. b; Skog og landskap 2012).

Forests have historically been used for grazing by all kinds of domestic animals (sheep, goats, cattle, pigs, and horses). Today, sheep grazing dominates, changing the combined grazing by several grazer species to more selective grazing. As for cattle grazing, the trend today is that only young animals are out for grazing, whereas other cattle are kept indoors for milk production. Wooded pastures have decreased over time as farms have been abandoned, with regrowth of shrubs and trees as a consequence. This is due to changes in land-use policies and economy.

For instance, small farms with sheep have increasingly become abandoned, whereas bigger farms have increased livestock numbers and, hence, grazing intensity (Frislid 1999; Kielland-Lund 1999).

Wooded grasslands are historically often connected with small storage barns. Storage close to the area of fodder collection saved time during the short and busy summer, and transport back to the farm which might have even been easier in the snowy winter months, when sledges could be used. More recently, abandoned outfields in the vicinity of infields and abandoned hay meadows have increasingly been used as pastures, especially for cattle. This is due to government regulations stipulating that farm animals must be outdoors during the summer months. Locally, implementing new and modern technology may maintain or increase grazing. This applies to areas where the use of newer, larger machines is not practicable, such as steep slopes or dense forests. Grazing animals may make use of areas that are not accessible for mechanical operations (Losvik and Hjelle n.d. a).

Wooded pastures are important for different reasons, including both economic (e.g., milk and meat production) and ecological (e.g., specific biodiversity of endangered species) (Direktoratet for naturforvaltning 2007). They also have a high value as a semi-natural system that contributes to better health and welfare of domestic animals (e.g., exercise, fresh air, variation in forage plants), which may also be reflected in better quality of meat and milk (Sickel 2009). Moreover, open forests with scattered trees have an aesthetical and nostalgic value that is closely connected to pre-industrial farming (Skog og landskap 2012). They are depicted in paintings and photographs, although this may be due to their occurrence close to alpine summer farms, which play an important role in Norwegian history, rather than due to the interest of the painter for the particular landscape type. Nevertheless, wooded grasslands are rather little known by the general public compared with other landscapes.

4.5.1.7 The Netherlands

Dutch name: hoogstamboomgaarden and bosweide (relict)

The only type of wood pasture that still exists in the Netherlands is traditional orchards (*hoogstamboomgaarden*, with high trees), particularly in two regions (the riverine region and south Limburg). During the last half century, most of these orchards have been replaced by large-scale orchards, dominated by low fruit trees and no longer grazed. Remains of the old orchards are preserved by local initiatives, often with government grants.

Other systems of wood pasture (*bosweide*) have all but disappeared. However, since the 1980s, a number of ecologists claim that landscapes that are half-open by the effects of grazing animals are in fact the original natural landscapes. Therefore, such landscapes have been reintroduced. In practice, they need intensive management, proving their character as man-made landscapes.

4.5.1.8 Slovakia

Slovakian name: pasienkový les (grazing forest)

Grass-covered mosaics originally tended toward mixed system. We can distinguish two types of the grass-covered mosaics:

- Grass-covered former mosaics of arable lands, grasslands, orchards (sometimes vineyards) with or without balks, with or without fruit trees on balks. They occur in areas of dispersed settlements (mostly without balks in hilly country of moderate relief) or in highlands of steeper relief, mostly with various balks (terraces, mounds, heaps, etc.). From the natural condition point of view, they were uninteresting for intensive cultivation by socialistic cooperative farms. From the biodiversity viewpoint, its focuses are on the balks (many species of invertebrates and vertebrates, vegetation—besides herbs and greens, there are rosa, blackthorn, juniper, spruce, fir, beech, hawthorn, oak, hornbeam, etc., from the fruit trees there are cherry woods, apple trees, pear, nut trees, plum trees, raspberry, etc.) as well as on the grasses arable land.
- Grass-covered former mosaics of arable lands and grasslands with balks (terraces, mounds, heaps, etc.). They occur in mountain areas of higher altitude and steeper relief. The orchards and the fruit trees on the balks have not been found because of unsuitable climatic conditions.

Further wooded grasslands can be found on today abandoned former more intense utilized grasslands (meadows, pastures). They were originally intensively utilized large block grasslands used as meadows or pasture. From the wooded vegetation, there are individuals or groups of rosa, blackthorn, juniper, beech, spruce, fir, hawthorn, oak, hornbeam, etc. Here too, it is possible to distinguish two types:

- Wooded grasslands as a result of recultivation of former mosaics. They originated in socialism time as in mountain areas in moderate natural condition suitable for heavy agricultural machinery and cultivated by cooperative farms. After 1990 due to changes of political and economic systems, many of them were abandoned or less utilized. From the biodiversity point of view, they are poorer compared with grass-covered mosaics or residue of original grasslands (next type).
- Wooded grasslands as residua of original traditional meadows and pastures. They occur in highlands and they are situated above mosaics or arable lands as part of closed agricultural system in the past. Their biodiversity is high because of long-time cultivation as grasslands.

They exist mostly in highlands up to higher altitude (about 1000 m and more) generally on poor soil conditions (more stony and shallower cambisols, rankers, and rendzinas), steeper slopes, climate of longer winter with snow, less sunny days, and more rainfalls. Because Slovakia is a country of very heterogeneous substratum and relief conditions, it is difficult to state natural conditions more specifically at this stage.

Fig. 4.8 Wooded grassland (*dehesa*) in Spain (*Photo* Cañas I.)

Within original meadows and pastures we can find log hay barns. Parts of some grassed mosaics are isolated groups of traditional residence and farm buildings within dispersed settlements as well as isolated stone cellars and log cellars for agricultural products storage, coming from the period of existence of arable land, and old country roads with traditional crosses and chapels can be found within grassed mosaics.

Some areas of grassed mosaics with sheep grazing are typical for the production of special sheep cheese *bryndza*, of the drink *žinčica* and production of wood products from pastoral tradition: *valaška* (small wood axe), *črpák* (wood pot for drinking), etc.

4.5.1.9 Spain

Spanish name: dehesa (Ferrer et al. 1997).

Dehesa (Fig. 4.8), the most extensive wood pastures in Mediterranean Europe (Marañón and Ojeda 1998), is a traditional agro-forestry system that simultaneously has extensive (or semi-extensive) livestock grazing, forestry, and even agricultural production, maintaining high values of biodiversity (García Tejero et al. 2013). The tree component is made of different oak species: cork oak (*Q. suber*), holm oak (*Q. ilex*), quejigo oak (*Quercus canariensis*), melojo oak (*Quercus pyrenaica*), etc. Shrub pasture and orchards pasture are also very common, as a consequence of the human transformation of Mediterranean forests in Spain (López-Pintor et al. 2006). The *dehesa* is derived from progressive thinning-out of the Mediterranean forest ecosystem, and it is a traditional management practiced over centuries (Marañón and Ojeda 1998). The *dehesa* has important social and economic values in Spain, and plays an important role in maintaining rural population levels.[5]

[5]http://sigpac.mapa.es/fega/visor/.

4.5.1.10 United Kingdom

In the United Kingdom, wooded grasslands exist since many centuries, and they played an important role in the life of farmers and thus in the formation of the landscape in the countryside (Hooke 2012, 2013). We can consider wood pasture as an ancient system of management that developed as a multifunctional landscape in a period when woodland was plentiful and where there was very little need for formal coppice. The latter is a very intensive and well-managed system, intended to provide vital supplies of timber in a resource-limited landscape. Due to the decrease of the traditional use of these landscapes, between the nineteenth and twentieth centuries, they have gone through major changes regarding the structure of the trees and the overall appearance. Usually, in the wood pastures and in the Royal forests, the cover of the tree has increased, as a consequence of the decrease of the ungulates and of the cattle. Nowadays, their importance is mainly linked to the presence of monumental trees.

4.5.2 Types and Subtypes of Wooded Grasslands

Using information compiled from the literature review, the questionnaire analyses, and the EUCALAND workshops, we can conclude that there is a huge variety of subtypes of man-made wooded grasslands. Some examples where land use and plant species determine the cultural value of these subtypes are (related cultural background, traditional knowledge, etc. are different and determinative for the type and subtype):

1. Wooded pasture: grassland with trees, established predominantly for grazing. Different subtypes exist where the physical appearance and the cultural value differ by the animal grazed on the pasture:

   1a: cattle;
   1b: sheep;
   1c: horse;
   1d: goat;
   1e: pig, etc. (e.g.,, mix of sheep and goat).

2. Wooded meadow: grassland with trees, established predominantly for hay cutting/forage production. Here any kind of tree appears, according to region, location, biogeographical preconditions. It can be fruit trees for human use or for animal use only (Bargioni 1998).
3. Wooded grassland with "(mixed) fruit trees": wooded grassland, established predominantly for fruit production (can be used for grazing and/or hay cutting, too). Certain subtypes exist determined by the tree species used:

   3a: mixed fruit trees (e.g.,, apple, pear, plum, peach, etc.)—this is an extensive production type;

3b: mono fruit tree production (more intensive, les extensive but no plantation), grasslands with:

3ba: olive trees;
3bb: oak trees (cork oak, holm oak, turkey oak, pubescent oak, etc.);
3bc: chestnut;
3bd: walnut.

3c: other trees (e.g.,, Robinia groves in Slovakia were described by Šikrová-Bodnárová 1954).

All of these are cultural landscapes with a long history of human influence.

4.5.3 Common Cultural Values of Wooded Grasslands of Europe

Based on the research conducted so far, we can conclude that the main common cultural value of the wooded grasslands of Europe is the traditional knowledge related to these wooded grasslands, formulated over centuries, taught by a chain of local people in time (verbal communication between former and later land users, farmers, shepherds, etc.). This knowledge is the common umbrella over the physical appearance of cultural values, such as buildings (e.g., buildings for animal husbandry (Fig. 4.9), buildings for shepherds, etc.) and man-made artifacts (including stone walls to protect a single tree and the soil around its trunk in a harsh environment, clothing, tools, cooking facilities, traditional food and drink, etc.).

Fig. 4.9 Wooded grassland with protection stone walls and wooden gate for controlling animal movement, Hydra Island, Greece, 2013 (*Photo* Centeri Cs.)

4.5.4 Identification of the Threats that Wooded Grasslands Are Facing

The biggest threat to systems of combined land use is, on one side, the growing specialization in agriculture (Bergmeier et al. 2010) and, on the other side, the abandonment of traditional practices linked to animal husbandry.

The survival of the wooded grasslands greatly depends on the land users and on policy makers. Since wooded grasslands include grassy areas, their mowing or grazing is vital. Being extensive in most of the cases, management mostly includes handmade grass cutting or requires a shepherd for animal husbandry. These management activities are not really inviting in recent days.

4.6 Discussion

A well-organized subsidy system might help in motivating local people. It is extremely important to include locals in the maintenance. They have to be involved, otherwise they will not feel its importance, and the sustainable management of the wooded grasslands will be neglected after subsidy is ceased. If it is important to save these landscape types, land users have to be convinced that it is good for them and this kind of understanding has to be economical as well.

The evolution of human-induced extensive landscape types—like wooded grasslands—is a good example of people living in cooperation with nature, using the best practice to gain products from the land. A man-made extensive—low input—landscape type is normally following the given natural conditions, taking away as many resources from nature as is sustainable. This is one of the greatest messages of wooded grasslands, be they used only for pasture where single tree mainly provides shade during summertime, or also for other productions, such as fruit trees, olive, cork oak, etc. The evolution of a given landscape type is a learning process of local people. They learn how to use their land the best practical way to produce meat, fruits, forage, and other goods.

The evolution of wooded grasslands is variable, depending on the local geographical conditions and on the cultural background of the people creating them. People's understanding of the wooded grasslands is often contradictory. Some people, particularly those who are mapping them (e.g., botanists), know exactly what they are, while others do not even call them wooded, regardless of the visible trees, because there have always been trees in these places where a given land user (e.g., shepherd) has worked (cit. personal communication with Hungarian shepherds).

All landscape types are continuously evolving (Bender and Winer 2001) and wooded grasslands do as well. This is the reason of their vulnerability. Their evolution is open-ended so their presence is volatile (Bender and Winer 2001).

Future development of wooded grassland is complicated because they are all labor-intensive land uses, invented to achieve risk minimization by diversification of cultivation. Thus the discussion of its functions and meaning may become an important part of resilience discourse, especially regarding:

- Social aspects: voluntary landscape management, social work programs, young farmers programs;
- Sustainable food planning: local food, slow-food, farm-to-table movement, community engagement in food production;
- CAP: subsidy distribution.

Cultural values of agricultural landscapes are rooted in agricultural production, evolved as part of land management by people living in a specific area, and in an optimal process a climax can be reached where people find a way to live in symbiosis with their land.

Reaching this stage means sustainable use of natural resources that is economically viable, while at the same time people take good care of the land and do not create overuse. This sustainable and economically viable land use is already an important message for the present-day human society, suffering from unemployment, lost identity, uncertain future, etc.

Extensive landscapes are multifunctional landscapes where coexistence and integration of different uses and ecosystem services allow the balance between human activities and nature, and thus the sustainability of the socioecological landscape system.

Wooded grasslands are cultural landscapes whose existence is guaranteed by the perpetuation of managing human activity according to traditional techniques which allow land users living in harmony with the environment and preserving the identity of the people. The persistence of these landscapes will measure how alive are relationship between community and its own territory, and how alive are their sense of belonging.

The existence of categories of wooded grasslands around Europe described by a series of specific subtypes at local level shows how the diversity imposed by environmental and geographical constrains found in the way of resource utilization by humans an expression of the common cultural heritage for European communities.

4.7 Added Values of Wooded Grasslands

Wooded grasslands have faced a lot of attention from various aspects, e.g., biodiversity protection (Rackham 2000), provision of ecosystem services, socioeconomic aspects, tourism, and agricultural added values (Hæggström 1998).

One of the major concerns is related to nature conservation (e.g., some of the wooded grasslands are part of the Hungarian national nature conservation area system which means that their value has been recognized by the Hungarian authorities). In Germany they are listed among the landscapes which have to be protected (by law).[6] Nature conservation value of wooded grasslands is also mentioned in connection with high natural value (HNV) areas, where Iberian wooded grasslands are cited to have 135 plant species on a square meter compared with as few as one or two plant species on intensively managed grasslands.[7]

Specific landscape types with specific purpose, history, biodiversity, nature conservation value, and economical importance have evolved over time, thanks to the human activities. Landscape diversity, also appreciated by ELC, faces similar global threats (climate change, globalization of agriculture, and others).

4.8 Conclusions

We can state that wooded grasslands played in Europe in former times a more important role in many local landscapes than they play nowadays. Many sources prove that they have already a long history, with summit of their importance in the nineteenth and first 60–70 years of the twentieth centuries. Based on the literature review and the questionnaires, we can state that wooded grasslands are in use in the largest extent in Southern Spain.

Wooded grasslands throughout Europe face similar problems (decline, abandonment) regardless of their location, origin, situation, and subtypes. As they have common, typical problem, a comparable approach by policy, planning, and decision-makers has to be applied. Their exact coverage in Europe is not known and thus the risk of losing them, along with their cultural, biological, and agricultural values, is increasing. Mapping—at least the most valuable ones (mapping of hot-spots)—is needed in order to provide information about their presence for those who want to save them. There are different types and subtypes of wooded grasslands and their comparability and cultural heritage values have not been sufficiently discussed and investigated from a Pan-European point of view. Local studies and studies about single subtypes do exist. However, outlook and research are desirable.

Wooded grasslands can only be protected if they remain in use. Based on the discussions with experts, local people, scientists, and the examined questionnaires, we can state that single, short-time efforts for their renovation are not enough for their long-term preservation.

[6]http://www.bfn.de/0311_schutzw_landsch.html.

[7]http://www.birdlife.org/eu/pdfs/HNV_Policy_document_proof6_010910.pdf.

References

Abrams MD, van de Gevel S, Dodson RC, Copenheaver CA (2000) The dendroecology and climatic impacts for old-growth white pine and hemlock on the extreme slopes of the Berkshire Hills, Massachusetts, U.S.A. Can J Bot 78:851–861

Abrams MD, Copenheaver CA, Black BA, van de Gevel S (2001) Dendroecology and climatic impacts for a relict, old-growth, bog forest in the Ridge and Valley Province of central Pennsylvania. Can J Bot 79:58–69

Afan De Rivera C (1842) Considerazioni sui mezzi da restituire il valor proprio a'doni che la natura ha largamente conceduto al Regno delle Due Sicile. Naples

Agnoletti M (ed) (2013) Italian historical rural landscapes: cultural values for the environment and rural development. Springer, Heidelberg

Alexander KNA (1998) The links between forest history and biodiversity: the invertebrate fauna of ancient pasture-woodlands in Britain and its conservation. In: Kirby KJ, Watkins C (eds) The ecological history of European forests. CAB, Wallingford, pp 73–80

Alias S, Bianchi L, Calamini G, Campus S, Gregori E (2008) I pascoli arborati in Sardegna. Un caso in provincia di Nuoro. Sherwood 141:15–19

Bargioni E (1998) The production of fodder trees in Valdegno, Vicenza, Italy. In: Kirby KJ, Watkins C (eds) The ecological history of European forests. CAB, Wallingford, pp 43–52

Bartha D (2003) Történelmi erdőhasználatok Magyarországon (Historical forest uses of Hungary). Magyar Tudomány 12:90–102

Beccu E (2000) Tra cronaca e storia le vicende del patrimonio boschivo in Sardegna. C. Delfino Editore, Sassari

Bender B, Winer M (2001) Contested landscapes: movement, exile and place. Berg Publishers, Oxford

Bergmeier E, Petermann J, Schröder E (2010) Geobotanical survey of wood-pasture habitats in Europe: diversity, threats and conservation. Biodivers Conserv 19:2995–3014

Bertolotto S, Cevasco R (2000) The 'Alnocolture' system in the Ligurian Eastern Appennines: archive evidence. In: Agnoletti M, Anderson S (eds) Methods and approaches in forest history. CABI Publishing, UK, pp 190–202

Biasi R (2010) Prati e pascoli arborati di Salten. In: Agnoletti M (ed) Paesaggi rurali storici per un catalogo nazionale. Laterza, Bari, pp 230–232

Biblioteca degli scrittori latini (1846) Opere di M. Porcio Catone con traduzione e note (De Re Rustica I), Tip. Giuseppe Antonelli, Venice

Bölöni J, Zs Molnár, Bíró M, Horváth F (2008) Distribution of the (semi-) natural habitats in Hungary II. Woodlands and shrublands. Acta Bot Hung 50:107–148

Bredenkamp GJ, Spada F, Kazmierczak E (2002) On the origin of northern and southern hemisphere grasslands. Plant Ecol 163:209–229

Brilli A (1992) Lo spirito della campagna toscana. Istituto di credito agrario, Florence

Choay F (2005) Patrimoine (bâti urbain et rural, paysager ou naturel). In: Merlin P, Choay F (eds) Dictionnaire de l'urbanisme et de l'aménagement. PUF, Paris, pp 617–620

Cole MM (1986) The savannas: biogeography and geobotany. Academic Press, London

Coupland RT (ed) (1979) Grassland ecosystems of the world: analysis of grasslands and their uses, vol 18. Cambridge University Press, Cambridge

De Ricci L (1830) Delle scarse rendite dei querceti ghiandiferi e dei metodi per aumentarli. Giornale Agrario Toscano 4:320–326

Del Favero R (2008) I boschi delle regioni meridionali e insulari d'Italia: tipologia, funzionamento, selvicoltura. CLUEP, Padua

Del Noce G (1849) Trattato istorico, scientifico ed economico delle macchie e delle foreste del Granducato di toscana con un nuovo regolamento delle acque fluenti per i fianchi scoscesi delle montagne appennine e delle centrali. G. Ducci, Florence

Di Berenger A (1859–1863) Dell'antica storia e giurisprudenza forestale in Italia. Longo, Treviso–Venezia

Direktoratet for naturforvaltning (2007) Kartlegging av naturtyper. Verdisetting av biologisk mangfold. DN håndbok 13, Direktoratet for naturforvaltning, Trondheim

Dirkx GHP (1998) Wood-pasture in Dutch common woodlands and the deforestation of the Dutch landscape. In: Kirby KJ, Watkins C (eds) The ecological history of European forests. CAB, Wallingford, pp 53–62

Ellenberg H (1990) Bauernhaus und Landschaft in ökologischer und historischer Sicht. Ulmer, Stuttgart

Ferrer C, San Miguel A, Ocaña M (1997) Propuesta de un nomenclátor definitivo de pastos de España. Revistas Pastos 27(2):125–161

Fleming A (2012) Working with wood pasture. In: Turner S, Silvester B (eds) Life in medieval landscapes: people and places in the Middle Ages. Windgather, Oxford, pp 13–31

Frislid R (1999) Man in nature—cultural landscapes of Norway. Landbruksforlaget, Oslo

Gabbrielli A (1980) Selvicoltura toscana nel '700. Parte prima. Annali Accademia Italiana Scienze Forestali 29:211–242

García Tejero S, Taboada A, Tárrega R, Salgado JM (2013) Land use changes and ground dwelling beetle conservation in extensive. Biol Conserv 161:58–66

Gautieri G (1813) Quando e come abbiasi a permettere il pascolo nei boschi si resinosi che da fronda, si d'alto fusto che cedui. G. Silvestri, Milan

Gradmann R (1933) Die Steppenheidetheorie. Geographische Zeitschrift, 39(5):265–278

Graham B, Ashworth GJ, Tunbridge JE (2000) A geography of heritage. Power, culture & economy. Arnold, London

Green T (2013) Ancient trees and wood-pastures; observations on recent progress. In: Rotherham ID (ed) Trees, forested landscapes and grazing animals: a European perspective on woodlands and grazed treescapes. Routledge, Abingdon, pp 127–142

Hæggström C-A (1998) Pollard meadows: multiple use of human-made nature. In: Kirby KJ, Watkins C (eds) The ecological history of European forests. CAB, Wallingford, pp 33–41

Haraszthy L, Márkus F, Bank L (1997) A fás legelők természetvédelme (Nature conservation of wood pastures). WWF füzetek 12

Harmos K (2013) Eltűnt fáslegelők nyomában (In the wake of disappeared wood pastures). Zöld Horizont 26–27(3–4):4–5

Hauge L, Austad I (1999) Hagemark. In: Norderhaug A, Austad I, Hauge L, Kvamme M (eds) Skjøtselsboka for kulturlandskap og gamle norske kulturmarker. Landbruksforlaget, Oslo, pp 85–94

Hooke D (2012) 'Wealdbæra & Swina Mæst': wood pasture in early mediaeval England. In: Turner S, Silvester B (eds) Life in medieval landscapes; people and places in the Middle Ages. Windgather, Oxford, pp 32–49

Hooke D (2013) Re-wilding the landscape; some observations on landscape history. In: Rotherham ID (ed) Trees, forested landscapes and grazing animals: a European perspective on woodlands and grazed treescapes. Routledge, Abingdon, pp 35–50

Howard P (2003) Heritage: management, interpretation and identity. Continuum, London

Iacini S (1801) Atti della Giunta per la inchiesta agraria e sulle condizioni della classe agricola. Volume III, Fascicolo I, Forzani e C. Tip. Del Senato, Roma

Jääts L, Kihno K, Tomson P, Konsa M (2011) Tracing fire cultivation in Estonia. For Stud 53:53–65

Joffre R, Hubert B, Meuret M (1991) Les systèmes Agro-Sylvo-Pastoraux Méditerranéens: Enjeux et réflexions pour une gestion raisonnée. Dossier MAB 10, UNESCO, Paris

Kielland-Lund J (1999) Beiteskog. In: Norderhaug A, Austad I, Hauge L, Kvamme M (eds) Skjøtselsboka for kulturlandskap og gamle norske kulturmarker. Landbruksforlaget, Oslo, pp 95–102

Knapp R (1979) Distribution of grasses and grasslands in Europe. In: Numata M (ed) Ecology of grassland and bamboolands in the world. W. Junk, The Hague

Komarek EV, Komarek BB, Carlysle TC (1973) The ecology of smoke particulates and charcoal residues from forest and grassland fires: a preliminary atlas. Misc. Publ. No. 3. Tall Timbers Research Station, p 75

Kozlowski TT, Ahlgren CE (eds) (1974) Fire and ecosystem. Academic Press, New York

Kruse A (maiden name: Schmidt) (1999) Beitrag der historischen Landschaftsanalyse für aktuelle Fragen des Naturschutzes - Eine Untersuchung durchgeführt am Beispiel des Biosphärenreservates Schorfheide-Chorin (Contribution of the historical landscape analysis for recent questions of nature conservation - a research done at the example of the biosphere preserve Schorfheide-Chorin). GCA-Verlag, Herdecke

Kruse A, Centeri Cs, Renes H, Roth M, Printsmann A, Palang H, Benitojordá L, Velarde MD, Kruckenberg H (2010) Glossary on agricultural landscapes. Tájökológiai Lapok 8(Special Issue): 99–127. [open access: www.tajokologiailapok.szie.hu/pdf/2010SpecialIssue/15_Kruse.pdf]

Kukk T, Kull K (1997) Puisniidud/Wooded meadows. Estonia Maritima 2

Kvamme M (1999) Stølslandskapet. In: Norderhaug A, Austad I, Hauge L, Kvamme M (eds) Skjøtselsboka for kulturlandskap og gamle norske kulturmarker. Landbruksforlaget, Oslo, pp 183–192

Laszlovszky J, Szabó P (ed) (2003) People and nature in historical perspective. CEU Medieval Studies and Archaeolingua (Budapest CEU Medievalia 5)

Lazzarotti O (2003) Patrimoine. In: Lévy J, Lussault M (eds) Dictionnaire de la géographie et de l'espace des sociétés. Belin, Paris, pp 692–693

López-Pintor A, Gómez Sal A, Rey Benayas JM (2006) Shrubs as a source of spatial heterogeneity—the case of *Retama sphaerocarpa* in Mediterranean pastures of central Spain. Acta Oecol 29:247–255

Losvik MH, Hjelle K (n.d. a) Wooded hay meadows in Norway. In: CULTBASE—database on cultural landscapes and cultural landscape ecosystems in Europe. http://pan.cultland.org/cultbase/?document_id=132

Losvik MH, Hjelle K (n.d. b) Wooded pastures in Norway. In: CULTBASE—database on cultural landscapes and cultural landscape ecosystems in Europe. http://pan.cultland.org/cultbase/?document_id=133

Marañón T, Ojeda JF (1998) Ecology and history of a wooded landscape in Southern Spain. In: Kirby KJ, Watkins C (eds) The ecological history of European forests. CAB, Wallingford, pp 107–116

Márkus F (1993) Extenzív mezőgazdaság és természetvédelmi jelentősége Magyarországon (Importance of extensive agriculture and nature conservation in Hungary). WWF füzetek 6

Merendi A (1957) Il problema forestale nel territorio del Chianti. In: Convegno del Chianti. Atti Firenze 25–27 maggio. Accademia Economico-Agraria dei Georgofili, Tip. Vallecchi, Firenze, pp 228–261

Molnár Zs, Bölöni J, Horváth F (2008) Threatening factors encountered: Actual and endangerment of the Hungarian (semi-) natural habitats. Acta Bot Hung 50(Suppl):199–217

Moreno D (1990) Dal documento al terreno. Storia e archeologia dei sistemi agro-silvo-pastorali. Il Mulino, Bologna

Moreno D, Poggi G (1995) Storia delle risorse boschive nelle montagne mediterranee: modelli di interpretazione per le produzioni foraggere in regime consuetudinario. In: Cavaciocchi S (ed) L'uomo e la foresta secc. XIII-XVIII. Atti della ventisettesima settimana di studi, 8–15 maggio 1995. Le Monnier, Florence

Neyret R (2004) Du monument isolé au tout-patrimoine. Géocarrefour 3(79):231–238

Oliva A (1924) L'economia del bosco pascolivo dell'Italia centrale. L'Alpe 11:103–108

Palumbo M (1912) Boschi e selve. Provvedimenti di governo. Salerno

Pavari A (1930) Il leccio o elce. L'Alpe 17(special issue):390–398

Pavari A (1934) Il problema dei querceti in Toscana. Atti Regia Accademia dei Georgofili Firenze 5(31):3–31

Pontecorvo G (1933) Considerazioni sui rapporti tecnici ed economici fra agricoltura e selvicoltura in Toscana. Atti della Reale Accademia dei Georgofili 5(30):260–273

Pott R (1995) The origin of grassland plant species and grassland communities in Central Europe. Fitosociologia 29:7–32

Poulot D (1998) Le patrimoine et les aventures de la modernité. In: Poulot D (ed) Patrimoine et modernité. L'Harmattan, Paris, pp 7–67

Printsmann A, Kruse A, Roth M (2012) Introduction for living in agricultural landscapes: practice, heritage and identity. Eur Countryside 2:89–100

Pungetti G, Kruse A (eds) (2010) European culture expressed in agricultural landscapes—the Eucaland project book. Palombi Editori, Roma

Rackham O (1998) Savanna in Europe. In: Kirby KJ, Watkins C (eds) The ecological history of European forests. CAB, Wallingford, pp 1–24

Rackham O (2000) Prospects for landscape history and historical ecology. Landscapes 1(2):3–15

Rackham O (2003) The illustrated history of the countryside. Weidenfeld & Nicholson, London

Rackham O (2006) Woodlands. Collins, London

Rackham O (2013) Woodland and wood pasture. In: Rotherham ID (ed) Trees, forested landscapes and grazing animals: a European perspective on woodlands and grazed treescapes. Routledge, Abingdon, pp 11–22

Rekdal Y, Larsson Y (2005) Veiledning i vegetasjonskartlegging. NIJOS rapport 05/05. Norsk institutt for jord- og skogkartlegging, Ås

Rooney TP (1995) Restoring landscape diversity and old growth to Pennsylvania's northern hardwood forests. Nat Areas J 15:274–278

Rotherham ID (2013a) Grazed tree landscapes; overview and introduction. In: Rotherham ID (ed) Trees, forested landscapes and grazing animals: a European perspective on woodlands and grazed treescapes. Routledge, Abingdon, pp 2–10

Rotherham ID (2013b) Reinterpreting wooded landscapes, shadow woods and the impacts of grazing. In: Rotherham ID (ed) Trees, forested landscapes and grazing animals: a European perspective on woodlands and grazed treescapes. Routledge, Abingdon, pp 72–86

Saláta D, Horváth S, Varga A (2009) Az erdei legeltetésre, a fás legelők és legelőerdők használatára vonatkozó 1791 és 1961 közötti törvények (Laws for the use of forest grazing, wood-pastures and grazing forest between 1791 and 1961). Tájökológiai Lapok 7(2):387–401

Saláta D, Wichmann B, Házi J, Falusi E, Penksza K (2011) Botanikai összehasonlító vizsgálat a cserépfalui és az erdőbényei fás legelőn (Botanical comparison study on the wood pasture of Cserépfalu and Erdőbénye). Anim Welf Ethol Hous Syst 7(3):234–262

Schaetzl RJ (1986) Complete soil profile inversion by tree uprooting. Phys Geogr 7(2):181–189

Sickel H (2009) Fettsyresammensetning og antioksidantinnhold i melk fra utmarksbeite i fjellet. Bioforsk FOKUS 4(2):208–209

Šikrová-Bodnárová M (1954) Zahájený les alebo pasienkový les na svahových pasienkoch južného Slovenska. Poľnohospodárstvo 1(3):306–317

Skånes HM, Bunce RGH (1997) Directions of landscape change (1741–1993) in Virestad, Sweden—characterised by multivariate analysis. Landscape Urban Plan 38(1–2):61–75

Skog og landskap (2012) Innmarksbeite. Versjon 2.0, Norsk institutt for skog og landskap, Ås

Szabó P (2005) Woodland and forests in medieval Hungary. Archaeopress, Oxford

Szabó M, Kenéz Á, Saláta D, Szemán L, Malatinszky Á (2007) Studies on botany and environmental management relations on a wooded pasture between Pénzesgyőr and Hárskút villages. Cereal Res Commun 35(2):1133–1136

Tondi M (1821) La scienza silvana ad uso dei forestali. Naples

Varga A, Bölöni J (2009) Erdei legeltetés, fáslegelők, legelőerdők tájtörténeti kutatása (Landscape history research of forest grazing, wooded pastures, forest pastures). Természetvédelmi Közlemények 15:68–79

Varga A, Bölöni J (2011) Egy felhagyott fás legelő területének tájhasználat változása Olaszfalu (Veszprém megye) határában (Land-use change of an abandoned wood pasture at Olaszfalu – Veszprém County). Folia Musei Historico-Naturalis Bakonyiensis 28:55–66

Varga A, Molnár Zs (2014) The role of traditional ecological knowledge in managing wood-pastures. In: Hartel T, Plininger T (eds) European wood-pastures in transition. Routledge, Abingdon, pp 187–202

Varga A, Bölöni J, Saláta D, Biró M, Horváth F, Samu ZT, Bodor Á, Molnár Zs (2014) Magyarországi fás legelők és legelőerdők jelenlegi természetvédelmi helyzete és problémái

(Present conservational state and problems of wood pastures and grazing forests in Hungary). Abstract book of X Aktuális Flóra- és Vegetációkutatás a Kárpát-medencében nemzetközi Konferencia, p 225

Vera FWM (2002) The dynamic European forest. Arboricultural J 26(3):179–211

Vera FWM, Bakker E, Olff H (2006) The influence of large herbivores on tree recruitment and forest dynamics. In: Danell K, Duncan P, Bergström R, Pastor J (eds) Large herbivore ecology, ecosystem dynamics and conservation. Cambridge University Press, Cambridge, pp 203–231

Vityi A, Varga A (2014) Wood pasture in Hungary. WP2-initial stakeholder meeting report. AGFORWARD, p 13

White F (1983) The vegetation of Africa: a descriptive memoir to accompany the UNESCO/AETFAT/UNSO vegetation map of Africa. Natural resources research report XX, UNESCO, Paris

Whitney GG (1984) Fifty years of change in the arboreal vegetation of Heart's Content, an old-growth hemlock-white pine-northern hardwood stand. Ecology 65:403–408

Whitney GG (1990) The history and status of the hemlock-hardwood forests of the Allegheny Plateau. J Ecol 78:443–458

Zanzi Sulli A, Di Pasquale G (1993) Funzioni delle matricine nei cedui nella teoria selvicolturale del XVIII e XIX secolo. Rivista di Storia dell'Agricoltura 33(1):109–121

[illegible]

Chapter 5
A Comparative Study of Two Mediterranean Transhumant Systems and the Biocultural Diversity Associated with Them

Pablo Domínguez

Abstract The alliance between the natural and social sciences has proven to be a successful analytical approach to understand and conserve ecosystems worldwide, while seeing humans as key agents within these (1971 Man and the Biosphere Programme, 1972 Stockholm Declaration, 1992 Rio Conference). In this context, authors from various areas of expertise have stressed the importance of recognizing the inextricable link between biological and cultural diversity and the need to raise awareness of these interactions for global sustainability. Despite scientific research repeatedly insisting on the importance of such a link, there remains a gap calling to highlight the concrete ways in which this diversity of long-held biocultural relations manifests and is generated. In fact, many of the works demonstrating the aforementioned bond are focused on the bioecological consequences of human diversity. At the same time, when they introduce a more sociocultural focus, they most often make linguistic indexes, their main measure for culture and/or use quantitative and macro-geographical approaches. In this sense, the general trend of this type of works, although always valuable, seems somewhat reductionist or incomplete. A less hard science and more detailed ethnographic-humanist analysis of this diversity and its groundings are still lacking. In order to address the exposed problem, I will present my preliminary works comparing agro-pastoral transhumant systems of the High Atlas of Marrakech and the Central Spanish Pyrenees. The ultimate goal is to push for an increasingly holistic approach to biocultural analysis including the humanities to a greater extent, and a broader spectrum of the social sciences.

P. Domínguez
Laboratory of Socio-Ecological Systems in the Globalization (LASEG), Institut de Ciencia i Tecnologia Ambientals (ICTA), Universitat Autònoma de Barcelona, 08193 Bellaterra (Cerdanyola del Vallès), Spain

P. Domínguez (✉)
Departament de Antropologia Cultural i Història d'Amèrica i Àfrica, Universitat de Barcelona, Montalegre 6-8, 08001 Barcelona, Spain
e-mail: eco.anthropologies@gmail.com

M. Agnoletti and F. Emanueli (eds.), *Biocultural Diversity in Europe*,
Environmental History 5, DOI 10.1007/978-3-319-26315-1_5

Keywords Biocultural diversity · Agropastoralism · Transhumance · Commons · Pyrenees · High atlas

5.1 Introduction

The outcomes of biocultural diversity (Maffi 2005), particularly in traditional agro-sylvo-pastoral systems, have been critical throughout history for creating resilient landscape patterns (Gómez-Baggethun et al. 2010; JP-BiCuD 2010) and shaping ecosystems (Alaoui 2009; Fillat et al. 1995) as well as very particular cultural identities of different regions (Mahdi 1999; Pallaruelo 1988). This is especially true for the Mediterranean region (McNeil 2003; Vidal and Castán 2010). The particular geography of this area, defined by a sinuously shaped inland sea surrounded by an enormous variety of mountain ranges (Braudel 1949), favours in fact great ecological and cultural diversity, which implies an immense heritage of plants, animals and other living beings, as well as an enormously varied range of agro-pastoral techniques, customs, beliefs and social relationships. But this diversity is not only a result of the millenary derivatives and adaptations to local and particular conditions favoured by such compartmented systems. It is also the consequence of direct and indirect millenary exchanges between some of the most important farming civilisations in human history (Barbera and Cullota 2012). Due to this, the different agro-sylvo-pastoral management modes of the Mediterranean region share important similarities while they also have important specificities related to each particular biophysical conditions and each particular history. This makes them elements with (1) common grounds and (2) unique differentiated systems with their own biological and cultural diversity, intertwined through long-held biocultural transformations. Indeed, it is precisely through a specific process of trial and error instigated by the local communities, in direct or indirect connection with other populations over the centuries and millennia of Mediterranean relations, that the different and original communal management systems have been established in each case. Based on the scarcity and need for detailed ethnographic comparative studies on biocultural diversity, and the long-term field-experience accumulated by the author in Mediterranean transhumant systems, the present work will focus on the comparative analysis of two general models of such systems within two mountain areas of the Mediterranean bioregion where many different transhumant systems are still operating: the southern side of the Spanish Central Pyrenees and the northern rim of the Moroccan High Atlas of Marrakech, both covering an extension of approximately 100 km^2 (Fig. 5.1).

Transhumant modes of herding seem to have been widespread throughout the entire Mediterranean region (Chassany 2008), which has been subjected to long-term agro-pastoral uses since at least 8000 BC in the Fertile Crescent and 5000 BC in the Iberian Peninsula and the Maghreb (Rasse 2008). Nevertheless,

Fig. 5.1 Map with the two areas of study marked in red, the Spanish Central Pyrenees and the Moroccan High Atlas of Marrakech

transhumance has slowly faded away from most plains up to the point that today they are mainly present in its mountain systems, where industrialization and private/state appropriation of collective land has had a lower historical incidence than in the plains. Located in sites which are often considered hot spots of endemism at a global level (Mittermeier 2004), the highland pastures object of transhumance, many times managed commonly, host in general a significant biodiversity due to the isolation effect that mountain areas produce as well as to the consequence of pastoralist activity and transhumance (Auclair and Al-Ifriqui 2012; Dominguez and Hammi 2010; Fillat et al. 2007). In purely agronomic terms, the key element of these transhumant systems in the two regions consists of a regulation of access to altitude grassland that prevents shepherds from allowing their animals to graze for three or four months. This period mainly coincides with the respective springs, which is the moment of maximum growth and reproduction of pastoral vegetation. This regulation is established and implemented by an assembly of users, both in the case of the Atlas and the Pyrenees. The purpose of such regulation is to ensure the regeneration and sustainable use of the common meadows, as well as to provide equal access to the different shepherds as they all access the pasture at the same time, and they jointly decide their ways and rules of management. However, as we will see below, these mountain commons are also a *total social fact* in the sense defined by Mauss (2012), around which revolves a whole system of social, economic, political, ritual and symbolic relations driven through the figure of the Mediterranean saints (Muslim in the Atlas and Christian in the Pyrenees). This fact makes transhumance a faithful reflection of two "connected" (Chassany 2008) and at the same time "particular" mountain cultures (Dominguez 2010; Beltrán 1993).

In short, the two regions of study seem both to converge within a certain general framework, at the same time as to diverge due to socio-historical and bioecological differences, turning them into a Mediterranean common heritage and an example of difference and particularity. It is in fact this tension that will be the guiding thread throughout the entire text, the final objective of which is to give some preliminary steps towards gaining a more complete and holistic understanding of the complex essence of the biocultural diversity. In fact, since many of the works studying and demonstrating the biocultural bond are either focused on the bioecological consequences of human diversity (Hammi et al. 2007; Sirami et al. 2010), or they make linguistic indexes their main measure for culture (Sutherland 2003; Stepp et al. 2005), using mainly quantitative and macro-geographical approaches (Zent 2001; Loh and Harmon 2005), it seems necessary to include a broader spectrum of social sciences and humanities.

5.2 Convergences and Divergences

The northern watershed of the Marrakech High Atlas and the southern watershed of the Central Pyrenees have certain analogies in biophysical terms, concerning certain botanical formations that are partially similar, such as supra-Mediterranean forests and scrublands dominated by holm oak (*Quercus ilex*), prickly juniper (*Juniperus oxycedrus*) and gum rockrose (*Cistus Ladanifer*). But what probably brings them closer in terms of the biophysical substrate of their corresponding transhumant systems, are the highland pastures that exist above the line of tree growth and that are one of the common key elements for the sustainment of these systems. Nevertheless, this pastoral frontier where the presence of meadows becomes dominant, diverges somewhat. It can be located at around 1800 m above sea level in the case of the Pyrenees, when in the Atlas it is at about 1900–2100, depending on the intensity of human pressure. At the same time, although certain general biophysical similarities exist, the Marrakech High Atlas highland pastures receive a much lower annual pluviometric average (500–700 mm) that can be up to two or three times higher in the Pyrenean high pastoral plateaux (1000–2000 mm). There is also a considerable difference of almost 10 °C in the annual mean temperatures between the oro-Mediterranean pastures of the two areas. And nevertheless, they have both attracted the attention of shepherds for the same reason (cyclically available fresh summer pastures) since pre-historic times (Rodrigue 1999; Mazzucco 2012). In these high places indigenous populations saw a complementary, abundant and nutritive resource for their flocks during the dry periods of the Mediterranean temperate climate zone that usually left the lower lands dry and unproductive in comparison to these alpine meadows. The economic importance of such sites became obvious for these populations at very early stages of the Neolithic and even the Mesolithic. It became so important in fact, that these spaces became centres of great symbolic meaning for pastoralist and proto-pastoralist populations. The outstanding presence of rock art that exists in both regions bears witness to this importance. Nevertheless, the convergence in this arena seems to stop mostly in the generic symbolic importance given to these highland pasture areas, since this importance is expressed very differently according to where we place ourselves. In the High Atlas they take the form of rock carvings where they are the norm, whereas they are rather rare in the case of the Pyrenean high pastures (Clot 1974) being megalithic art much more dominant.

5.2.1 Rock Art

The abundance of rock carvings found in the high transhumant plateaux and on the paths of shepherds in the Marrakech High Atlas is the proof of to the antiquity of the transhumant use of highland pastures. These carvings date back to between 2500 and 3500 BC, although it is important to mention that the drying up of the Sahara between 2500 and 1200 BC would have accentuated the sense of refuge of these mountain pastures, leading to an increase in the production of these both ritual and artistic manifestations (Rodrigue 1999). This dating and these carvings reveal the existence of an ancient transhumant society, essentially based on pastoralism due to the great dominance of pictures linked to animal-rearing activities above agriculture, particularly cattle, which in any case already coexisted closely with agriculture (Sellier 2004; Pascon 1983; Bellaoui 1989). In the Oukaïmeden and the Yagur where Mahdi (1999) and myself (2010) described fertility rites associated with the opening of the highland pastures, authors such as Simoneau (1967) and Auclair and Al-Ifriqui (2005) defend a certain continuity of these practices in these same sites since the paleo-Berber shepherds from the Bronze Age (2000–500 BC), from whom the abundance of symbols carved on the red sandstone of these pastoral centres (fish, moons, fibula, anthromorphic scenes of procreation and childbirth) remain.

On the contrary, in the Pyrenees, rock carving is very scarce, and consequently, quite unknown (Clot 1974). However, megalithism, which is practically absent in the Atlas, can be stated to have an overwhelming presence in the case of the Pyrenees. It has even been pointed out as evidence for the existence of a single and powerful nucleus of cultural identity throughout the whole mountain range until the second millennium BC (Almagro 1942, p. 169). This has often been explained as the influences of a clear continental-European inspiration, which has always pushed towards the South and had a powerful influence on the pre and proto-Neolithic people of the Pyrenees, counteracting the influences from the rest of the Iberian peninsula, and decanting the link of the Pyrenees Mountains towards the influences of the North represented by megalithism (Jiménez 2006). Later on, from the Iron Age onwards, the growing influence and establishment of Celt populations during the last centuries of the second millennium BC is understood as the process that caused the end of the megalithic culture of the Pyrenees (Maluquer 1987, p. 44). Nevertheless, this influence, again from the North, injected new differences between the two models of cultural expression. Thus, although the African and European models continued to be based on various similar ecological typologies and uses of natural resources (mountain pastoralism in Mediterranean climates), the different cultural influences made Maghreb rock art last for a great time longer (until approximately 500 BC) and in a very different way than in the Pyrenees.

5.2.2 Transhumance, Between Agricultural Systems and Highland Pastures

In both cases, the Central Pyrenees and the Marrakech High Atlas, the main habitat and the farmland are located in what we could call "the villages", generally at mid-altitude between the high summer pastures and the lower winter valleys and plains. However, this means slightly higher in the High Atlas (on account of their lower latitude), and slightly lower in the Pyrenees, due to the limiting factor of the temperatures at this same altitude. Depending on the community observed, in the High Atlas of Marrakech "the villages" are located at between 1000 and 1800 m above sea level, while in the Pyrenees they are rarely higher than 1200–1300 and start even at 600 m, or even lower. Nevertheless, in all cases, through similarity in the type of natural resource and the exploitation techniques available, the traditional model of organization essentially coincides between the two areas and always revolved around the main village which became the centre of gravity of a double transhumance: one "normal" or "vertical" in summer, on the high pasturelands, and the other "inverse" or "horizontal" in winter, on the large steppe plains.[1] However, in both cases, today, transhumance towards the plain is maintained at a residual level. Only some owners of the biggest flocks are interested in doing it today, since the travel costs are only offset if a large number of animals is mobilized, and this generally concerns only sheep. The fundamental reason for this alteration of winter movement seems to coincide in this case. In both Morocco and Spain, plains are increasingly owned by landlords, notably by owners of large estates or by large cooperatives, and therefore are occupied by crops in winter and spring, often even blocking the paths to them, meaning that transhumance no longer finds its place in these spaces and cannot be exercised as it was before.

In any case, although transhumance is increasingly bereft of its winter movement, the two transhumance models continue to make part of a vast mountain agricultural system. This is based on two principles: on the one hand, the integrated complementarity of the agricultural and pastoral productions, and on the other, the vertical

[1]In the case of the southern watershed of the Central Pyrenees, the flat lands of the Ebro; in the case of the northern watershed of the Marrakech High Atlas, the plains of the Haouz.

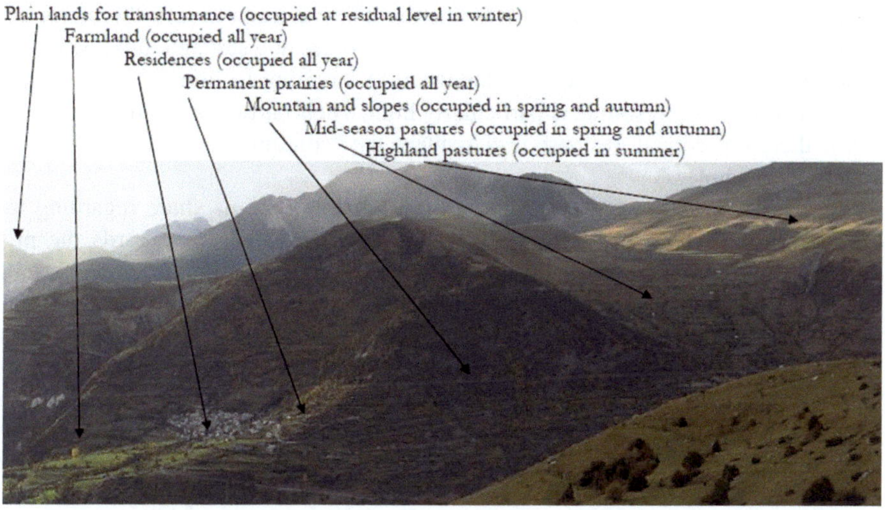

Fig. 5.2 General image of the use of space by agropastoralists in Gistain (Spanish central Pyrenees), in the Central Pyrenees and which corresponds directly to the model used in the High Atlas

complementarity of the spaces belonging to different bioclimatic stages. The integrated complementarity of animal and plant productions means that the animal-breeding provides manure for the crops and farm animals to carry out ploughing and threshing, and in return, the animals rely on crop products (e.g. barley, corn or fodder) for complementing their feeding, particularly in winter when pastoral resources are scarcer. At the same time, the vertical complementarity refers to farmers' use of the pasturelands belonging to diverse bioclimatic stages. As shown in the diagram below, shepherds combine the use of several spaces: (1) in winter, the plains and in the last decades increasingly the edges of the village; (2) in the intermediary seasons (spring and autumn), the mountains and their slopes; (3) in summer, the highland pastures (Fig. 5.2).

In both the Pyrenees and the High Atlas, transhumance highland pastures are located as a continuation of the main village lands, generally at a maximum distance of a half-day walking from the village houses. In fact, highland pastures which are used for transhumance find their place in a greater agro-pastoralist system. In other words, they are only part of a wider system and their use falls within a global and collectively rationalized use of the communal lands. And yet, highland pastures do not only contribute to the functioning of the local terraced systems by providing to the general production of the farm activity simply some "extra" fodder; but, in addition to providing very nutritional grasses for the animals (probably the richest of the whole annual cycle), these pastures provide food on site (i.e. without any need for transport or external inputs, and therefore at a very low cost), at an absolutely critical time in the pastoral cycle, and this is what is most crucial. That is, when the spring routes of the animals and the summer heat have finally exhausted the lowlands, leaving little to eat for the herds. It is important to note therefore, that the fodder contribution of the highland pastures comes at a strategic time, when the majority of other pastoral areas no longer have anything to offer, and the opportunity cost of such a resource is particularly high. Their participation in the animals' diet is therefore crucial, because it guarantees the "continuity" of the agro-pastoral system as a whole.

The above reveals what the different transhumant systems share regarding the economic role played by these highland pastures. However, as regards the productivity and composition of the highland pastures, they differ from one site to another as a result of the biophysical conditions and the different ways of management established in each site (Dominguez et al. 2012). In a first physico-technical definition, these highland pastures could be defined as homologous systems, altitude pastures with no trees, humid areas exposed to extreme conditions where grass grows in abundance in the spring thanks to the collection of water after the snow melts and the temperatures rise. But if we look in greater detail, more important elements related to the subject of biocultural diversity are at play.

In both cases, there is a practice of excluding animals from the pastures apart from some summer months, to maximize production and to guarantee the durability of the exploitation of the resource. This allows only the product of the year to be taken, and thereby saves the seeds for the following years. Several ecologists have underlined the particularly beneficial effects of this practice on the plant cover and the conservation of the local biodiversity (Auclair and Al-Ifriqui 2012; Fillat et al. 2007). But in fact, this exclusion of animals from the pastures, and the particular means of managing them in each case (essentially linked to the histories and needs of each socio-ecological context), not only enables species for pastoral use to be regenerated, but also "those" species which are more sought-after in each community, depending on the types of animals raised, and other additional resources used by each community, etc. For example, Alaoui (2009) confirmed this capacity of local ecological knowledge by demonstrating, through several tests that the majority of plants defined by the shepherds as most useful and beneficial for their animals on the pastures of Oukaïmeden in the High Atlas, reach maturity and drop their seeds after reproduction precisely a few days or weeks before the opening of this space to rearing, enabling the reproduction and conservation of a specific botanical community year after year. Human rationality and biological logics therefore meet as there is an intentional choice of species through a certain set of managerial habits that leads to very precise systems of plants. Table 5.1 shows the particular case of the pasture of the Yagur in the High Atlas, managed by the Mesioua tribe.

5.2.3 Architecture

As we have seen, the transhumance model affecting the different populations in the Pyrenees and High Atlas involves the splitting in two main areas: that where people usually live and that where they practice transhumance temporarily. The main

Table 5.1 Species sampled in three different sites of the Yagur High Atlas pastures managed by the Mesioua community (Dominguez and Hammi 2010)

ASSAGOUL	TAMADOUT	ZGUIGUI
Most external site of the Yagur	Intermediate site of the Yagur	Most internal site of the Yagur
Species: 54	Species: 59	Species: 66
Alchemilla atlantica	*Allium roseum*	*Allium roseum*
Allium roseum	*Alyssum serpylifolium*	*Alyssum serpyllifolium*
Alyssum serpylifolium	*Avena sterilis*	*Alyssum spinosum*
Asphodelus microcarpus	*Bromus madritensis*	*Androcymbium graminum*
Astragalus maroccanus	*Bromus tectorum*	*Asteriscus pygmaeus*
Avena sterilis	*Carlina involucrata*	*Bromus madritensis*
Bromus madritensis	*Carthamus lanatus*	*Bupleurum semicompositum*
Campanula dichotoma	*Catananche caerulea*	*Calendula aegyptiaca*
Campanula trichocalycina	*Catananche caespitosa*	*Campanula rapunculus*
Carduus pycnocephalus	*Centaurea sulphurea*	*Carex divisa*
Catananche caespitosa	*Cirsium casabonae*	*Catananche caerulea*
Cephalanthera longifolia	*Cirsium syriacum*	*Cephalanthera longifolia*
Cirsium syriacum	*Crepis vesicaria*	*Ceratocnemum rapistroïdes*
Coronopus procumbens	*Dactylis glomerata*	*Cirsium monspessulanum*
Dactylis glomerata	*Dianthus lusitanus*	*Cirsium syriacum*
Dianthus lusitanus	*Echium pustulatum*	*Crepis vesicaria*
Diplotaxis assurgens	*Erodium gruinum*	*Crucianella angustifolia*
Erodium gruinum	*Erodium praecox*	*Dianthus lusitanus*
Erodium praecox	*Erodium tordyloïdes*	*Echinops spinosus*
Eryngium ilicifolium	*Eryngium ilicifolium*	*Emex spinosus*
Festuca elatior	*Galium aparine*	*Erodium gruinum*
Festuca maroccana	*Genista hirsuta*	*Eryngium ilicifolium*
Filago pygmaea	*Gladiolus segetum*	*Festuca maroccana*
Galium aparine	*Hedypnois arenaria*	*Filago pygmaea*
Geranium dissectum	*Hordeum murinum*	*Galium aparine*
Geranium rotundifolium	*Koeleria vallesiana*	*Gladiolus segetum*
Helianthemum croceum	*Leersia hexandra*	*Hedypnois cretica*
Hordeum murinum	*Leontodon hispidilus*	*Helianthemum croceum*
Koeleria vallesiana	*Lotus creticus*	*Hierarcium pseudopilosella*
Koelpinia linearis	*Mantisalca salmantica*	*Hordeum murinum*
Lotus creticus	*Mentha pulegium*	*Hypecoum procumbens*
Mantisalca salmantica	*Nonea vesicaria*	*Koeleria pumila*
Ononis spinosa	*Oropetium africanum*	*Koeleria vallesiana*
Pallenis spinosa	*Paronychia argentea*	*Koelpinia linearis*
Papaver rheas	*Plantago coronopus*	*Lasiopogan muscoïdes*
Paronychia argentea	*Plantago lagopus*	*Lepidium sativum*

(continued)

Table 5.1 (continued)

ASSAGOUL	TAMADOUT	ZGUIGUI
Plantago coronopus	*Plantago major*	*Lotus arenarius*
Plantago ovata	*Plantago ovata*	*Mentha pulegium*
Poa bulbosa	*Poa bulbosa*	*Miniuartia tenuissima*
Reseda luteola	*Rumex crispus*	*Minuartia geniculata*
Rhaponticum acaule	*Salvia verbenaca*	*Paronychia argentea*
Rumex crispus	*Scorzonera laciniata*	*Phalaris minor*
Senecio vernalis	*Scorzonera pygmaea*	*Plantago coronopus*
Silene cucubalus	*Senecio gallicus*	*Plantago lagopus*
Silene gallica	*Senecio vulgaris*	*Poa bulbosa*
Sonchus asper	*Silene apetala*	*Ranunculus aquatilis*
Sonchus tenerimus	*Siybum marianum*	*Ranunculus spicatus*
Stipa retorta	*Sonchus asper*	*Reichardia tingitana*
Teesdalia nudicaulis	*Sonchus tenerimus*	*Resada battandieri*
Urginea maritima	*Stipa nitens*	*Rumex crispus*
Vulpia ciliata	*Stipa retorta*	*Rumex vesicarius*
Vulpia geniculata	*Teesdalia nudicaulis*	*Salvia verbenaca*
	Thymus pallidus	*Scorzonera undulata*
	Trifolium campestre	*Senecio vulgaris*
	Trisetaria pumila	*Sonchus tenerrimus*
	Urginea maritima	*Stipa retorta*
	Urginea undulata	*Thymus pallidus*
	Vicia onobrycoïdes	*Trifolium campestre*
	Vulpia ciliata	*Trifolium repens*
		Trifolium tomentosum
		Urginea maritima
		Urginea undulata
		Vallerianella microcarpa
		Vicia dasycarpa
		Vulpia ciliata

settlement is always in "the villages" as we pointed out before, where the main croplands for the families is located, as well as where the more solid buildings which are, this is, their houses. The secondary settlement is generally in higher lands, marked by much smaller and more precarious settlements known as *azibs* in Morocco and *majadas* or *bordas* in Spain. The existence of these two main conceptual spaces, involves the splitting of the group and the family between those who practice transhumance and those who stay close to the croplands. These constructions supporting transhumance range from restored cave-like structures, to sheepfolds or little houses, and small isolated dry stone constructions, the nature and quality of which depends on the materials available in the areas surrounding the fields in which they are built. Today, it increasingly depends on the ability to

Fig. 5.3 On the left a sheepfold with a horizontal roof in the High Atlas, and on the right a sheepfold with a stepped roof in the Pyrenees

purchase material from the outside (cement, iron, aluminium, etc.), which is more obvious in the Pyrenees where the shepherds are more capitalized, notably as a result of European subventions which support their way of life (even if always insufficiently), whereas in the Moroccan case they are absolutely inexistent (Fig. 5.3).

The construction differences are always linked to both, the biophysical and the sociocultural differences of each context. As can be observed in the two photos above, the Pyrenean example has a roof at an inclination of more than 45° which contrasts to the completely flat roofs of the High Atlas. This could be directly related to the pluviometric difference between the two areas, with much higher precipitation in the Pyrenees, therefore requiring a much more powerful water and snow runoff system. Nevertheless, unlike the explanation of the over 45° inclination of the roof in the Pyrenean case, mainly ecologically determined by the need to face high pluviometry and snow, the form of the roof of the Pyrenees sheepfold characteristically formed by "little steps" cropping up the roof, is at the same time related to the technical and aesthetic notions of the French valleys of the northern side of the Pyrenees. In fact, this sheepfold is located only a few kilometres from the French–Spanish border and evidently the shepherds from this region (Sant Juan de Plan) were influenced by the culture of their neighbours. In fact, the shape of these roofs is both identical to constructions from the north face of the Pyrenees and different from the high mountain Spanish sheepfolds in other more southern but much closer neighbouring regions. At the same time, the horizontal roofs of the sheepfolds and houses of the High Atlas, coming from the conditions and techniques of the much drier North African plains, and which have expanded upwards through the slopes of the High Atlas for centuries, up to altitudes at times above 2500 m and 700 mm of rainfall and snowfall, are probably partially poorly adapted to the conditions of the north side of the High Atlas of Marrakech, which is quite close to the humid influences of the Atlantic Ocean. Nevertheless, these still survive, in this case not precisely due to bioecological conditions but mainly as a heritage of a particular know-how and aesthetic tradition belonging to a much wider region, that of the greater Maghreb.

5.2.4 Spirituality of Euro-African Transhumant Systems

At the same time, and in their specific forms, approaching both transhumant models as a *total social fact* enables us to gain a better understanding of the essence of these systems. In other words, to understand all the aspects of the social life of the transhumants and to understand how they reflect the social system of the livestock breeders' communities of origin. In fact, the "transhumance period" sets all the communities in motion and projects all the aspects of the social system onto the space used by the shepherds. Thus, a reduced social and symbolic model of the livestock breeders' community of origin is projected onto the different spaces where transhumance takes place (Mahdi 1999). The highland pastures and the villages are a physical and sociocultural space alike. They are a productive space but also an annual meeting point where livestock breeders and village people socialize and reencounter each other year after year. The perenniality of the transhumant building depends on the answer to a fundamental question: How can a collective discipline be kept to manage these spaces and ensure that they are commonly used by different livestock breeders without becoming a tragedy of common resources (Hardin 1968)? How can the collapse of the common pool be avoided, while preventing or minimizing conflicts between shepherds? In other words, how can the competition between these different users be effectively regulated in this double biophysical and sociocultural perspective?

The results of my different theoretic and field works on the phenomenon of the transhumance conducted over the past decade, reveals a regulation model in which the management institutions quite greatly coincide in the two Mediterranean mountain areas, in the sense that they are at the same time fundamental and structured similarly in managerial assemblies of users, but were traditionally also never isolated from religion and the sacred. The spiritual and the immaterial world have always had a fundamental place in Mediterranean mountain transhumant systems. To portray this, the place of local saints referring to transhumance and highland pastures is a particularly sharp example of this. In fact, we generally find these spaces under the patronage of mythical figures and sanctities that centre the performance of pastoralist and transhumant rituals, in order to ensure a complete and balanced management of the villagers' common lands. Throughout the transition cycle between the arrival of the shepherds to "the villages" and the move up to the high pastures, individual and collective rites, celebrations and ceremonies are performed always in linkage or homage to a local saint, both in the Pyrenees and the High Atlas. There are several examples of hermitages and other places associated with saints where shepherds bring their animals to be blessed or where saints are honoured. Through these rituals the livestock breeders seek to safeguard the collective and individual assets, to protect the livestock and to bring prosperity and fertility to their animals and community.

The place of saints in tribal Moroccan and North African societies is well-known. The livestock breeders placed, and often still place their herds under the protection of a saint. The saint guaranteed fertility, protection (particularly against spirits and devils), and prosperity for the livestock, and in exchange, the livestock breeders gave thanks to the saint in the form of gifts and sacrifices. Gellner (1969) showed particularly well how, according to their physical location and that of their sanctuary, on the border between tribes, the saints and their families acted as arbitrators in conflicts, bringing about agreements, assisting traders in times of war and guaranteeing the safety of passages (Gellner 1981). For example, the pastures of Oukaïmeden and the Yagur are, respectively, placed under the patronage of the saints Sidi Fars and Sidi Boujemaa. By analysing the histories of different Maghreb saints, theoretically forbidden by the most orthodox Islam, and less present in the Machriq, we observe a sacralisation of pacts and discipline regarding the use of pastures being established, distributing rights to groups and establishing the tithes and sacrifices that the livestock breeders must make annually before, during and after transhumance. The regulation of the pastures, with their opening and closing dates, is thus linked to the saint and their descendants who are themselves a saint caste, announcing the day on which the pastures would open, thereby sacralising, legitimizing and reinforcing the respect for the regulations that pend upon these common pastures.

At the same time, although such a lineage of saints acting as arbitrators never seems to have existed in the case of the Pyrenees, at least not within historic times, after the period in which shepherds were traditionally excluded from the village, when they had left for the plains, and just before moving up to the highland pastures, the shepherds lived a time of celebration, joy and reunion between shepherds and their families, neighbours and friends that very often matched with key religious dates (Nadal et al. 2010). This is still celebrated today, on the anniversary of the specific saint of each community, giving a ritual and sacred nature to the entry of the animals to the collective highland pastures immediately after. Among other things this was set up to express that a lack of respect for the exclusion of animals before the key date was serious and dangerous. As in the High Atlas, the saints, in other words, God's emissaries, were always there to remind us of the rules and good pastoral conducts. Likewise, livestock breeders in the Pyrenees turn to and notably turned to a huge number of saints (San Urbez, San Visorio, San Victorián, Santa Orosia, Santa Elena, San Cosme, San Damián, San Martín, Santa Marina) and to different apparitions of the Virgin at some of the different key moments of their transhumant cycle. If honoured according to the correct rituals, these figures were able to ward off the dangerous action of sorcerers (Pallaruelo 1988, pp. 196, 194). In this sense, the opening of the pastures in the Pyrenees is often also lived as a festivity of homage to one of the patron saints of the village, and thereby structuring the whole Pyrenees pastoral calendar, in which the entertainment and socializing is nevertheless, often even more important than the religious sense for local populations (Villa 1988), as it seems to happen very often too in the Moroccan cases (Dominguez 2010). The passage from one cycle to another, that is, from "life in the lowlands" to "life in the highlands", is frequently initiated by a village celebration starting at the moment that the shepherds arrive, or simply just before moving up to the highland pastures if they do not migrate to the plains anymore. This is undertaken at a place of pilgrimage, like a hermitage or a place where we find the reference to a local saint, through the celebration of a collective meal in which several small ruminants, often sheep, are sacrificed and offered to the community with fresh cheese produced during these celebratory days. Thus, by bringing together all the representatives of the community around very precise dates linked both to the saints and the transhumance, by bringing children and adults together through a convivial collective activity of sharing and harmony, these meals and celebrations refer both to the social structures and functions of the community particular to each human and ecological environment, as well as to the rules and values of pastoral cohabitation and inter-solidarity, since this time also serves to recall them and to update them through the different rituals each new year (Beltrán 1993, pp. 364–366; López 2010, p. 153).

One of the points that seems most significant to me regarding the divergence between the two models of this saint and religious management of the transhumance and the plant communities that it affects in the highland pastures, is the different denominational traditions of each type of transhumants, one Catholic Christian and the other Sunni Muslim. In this case, the ethnographies available show us diverse forms of beliefs on highland pastoral areas and their management. In Morocco, if the collective rule is not respected, they manifest in particular in the omnipresent threat of spirits or devils, representations notably rooted in the Koranic tradition (Mahdi 1999). At the same time, in the case of the Pyrenees, it is in the form of witches serving Lucifer, which pastoral behaviour is threatened and conditioned, beliefs linked in turn to a different European-biblical tradition (Pallaruelo 1988).

At the same time, the most fundamental biocultural or eco-anthropological convergence that can be found between the High Atlas and the Pyrenees in this arena of magical-religious management of shared pastures, is that to a great extent, magic and/or the sacred pervaded the majority of precautions, management and remedies applied by shepherds in the management of their land and flocks. Since the transhumants were obliged to live in inhospitable, solitary and at times dangerous places, always threatened by wild animals, verticality and typical the storms of high Mediterranean mountains, as well as to ensure an increase in their flock and boost the fertility of these, on which their life and that of their community depended, the shepherds used their traditional ecological knowledge, passed, tested and constantly updated from generation to generation, but also their spiritual knowledge and religious beliefs. Through these shared characteristics of their biophysical and sociocultural means of production, which surpassed many times their capacity to control the different unexpected events at each time, particularly high in the oro-Mediterranean pastoral context, both Berber and European livestock breeders faced uncertainty and risk with a rich cosmology of the spiritual world.

All the components of these hagiographic transhumant models are not always found in each case studied, not when comparing the High Atlas and the Pyrenees, nor when comparing different communities within each of these two macro-units. Clearly they are never expressed in the same way because they always depend on the particular history and geography of each case. Even from one side of a valley to the other, the versions of one same hagiographic story can vary …. And so, from one mountain range to another, even more. Thus, variants meet passing from one situation to another, diversifying with the biophysical settings and the particular social histories of each group of transhumants. And nevertheless, they do not exclude the existence of astonishingly close elements, almost shared in their most essential nature, as we have seen. In short, in both cases, transhumance to pastures is an opportunity for the organization of social, economic, political and religious activity, which participates, consolidates and even rules great part of the collective discipline, accompanying the success of this model of management of natural resources, while co-determining with the local ecological conditions a rather particular socio-ecosystem built for and by each community. Transhumance to pastures is multidimensional; seeing it just as folklore is to overlook a type of social organization where spirituality, the sacred, has an important place and gives material activities a lot of their meaning. Likewise, only focusing on the ecological effects of pastoralism or on an antiquated mono-directional and functionalist determinism of the environment over human culture, would be forgetting other factors that determine or condition this same environment and that are born out of other type of interactions too, also out of human interaction and cultural exchanges.

5.3 Conclusions

With the view of contributing to deepen in the study of the biocultural bond and the different forms that it adopts, study which is still only emerging and is very rarely approached from the humanities or a broad perspective in social science, the present text has explored different links between natural and cultural attributes of two transhumant mountain systems, attempting to open towards such broader analytic lenses. At the same time similarly and differently, transhumance still exists in these mountain ranges of the Western Mediterranean region, the central Pyrenees in Spain, and the High Atlas in Morocco. In the Spanish case, like in other European countries, transhumance has undergone a very severe process of degradation and decline, particularly during the last one hundred and fifty years, mainly due to the communal land seizures and enclosure processes imposed by central states with the aim of assuring public control over natural resources or land privatization within global capitalist dynamics, feeding the industrialization of the nineteenth and early twentieth centuries, but which increasingly imposed obstacles on the fragile and costly mobility of the flocks through their transhumance, traditionally of several hundreds of kilometres. This favoured the abandonment of rural areas, the agricultural intensification of private corporations and the subsequent countryside re-urbanization in the more recent years (MacDonald et al. 2000; Stoate et al. 2001; Uhel 2006).

With the goal of implementing a particularly rich comparative approach of a world that has been for long time declining in the Pyrenean case, the inclusion of a study of Moroccan transhumant systems is related to the fact that it helps to understand the history of the former (the Pyrenean), since contrary to what happened in Europe, in Southern Mediterranean countries transhumance has not suffered such intense, systematized and widespread communal land seizures or enclosure processes as those in Europe. In fact, most High Atlas transhumant systems, although not free from increasingly external impositions and also intensely modified, are still highly traditional and a main agro-pastoral management tool for highland communities and economies (Auclair and Al-Ifriqui 2012). They follow demographic evolutions that are very different from the decay or abandonment of mountain areas in Europe and are in constant and rapid re-adaptation to the continuous social and environmental changes. Hence, even if the adaptive processes undertaken by the two different transhumant models are not in any case symmetrical nor fully predictable, the present text has shown how exploring comparatively the two cases from a detailed historical, ethnographic and eco-anthropological point of view among others, helps to further our understanding of certain aspects of their biocultural diversity and its genesis, while providing us with examples of organizational systems which we can still trace and which we should examine more holistically. This is, including the humanities and a wider spectrum of the social sciences in their interaction with the environmental sciences, when asking ourselves about the possible ways of assessing, conserving and/or adding value to biocultural diversity.

References

Alaoui S (2009) Les pelouses humides dans le haut Atlas: Biodiversité végétale, dynamique spatiale et pratiques de gestion coutumière. Dissertation, Université Cadi Ayyad

Almagro M (1942) La Cultura megalítica en el Alto Aragón. Ampurias 4:155–169

Auclair L, Al-Ifriqui M (2005) Les agdals du Haut Atlas marocain. Enjeux d'une recherche pluridisciplinaire. Cahiers de Recherche du Centre Jacques Berque 3:60–79

Auclair L, Al-Ifriqui M (ed) (2012) Agdals. Society and resource management in the Moroccan Atlas. Royal Institute of Amazigh Culture—Institute of Research for Development, Rabat

Barbera G, Cullotta S (2012) An inventory approach to the assessment of main traditional landscapes in Sicily (Central Mediterranean Basin). Landscape Res. doi:10.1080/01426397.2011.607925

Bellaoui A (1989) Les pays de l'Adrar-n-Dern. Etude géographique du Haut Atlas de Marrakech. Dissertation, Université de Tours

Beltrán O (1993) Es Aranesi: adaptació a l'entorn i organització social al Pirineu central. Dissertation, Universitat de Barcelona

Braudel F (1949) La Méditerranée et le Monde Méditerranéen a l'époque de Philippe II. A. Colin, Paris

Chassany JP (ed) (2008) Les paysages culturels de l'agropastoralisme, méditerranéen edn. Conseil général de l'Aveyron, Meyrueis

Clot A (1974) El arte gráfico prehistórico en los Altos Pirineos. MUNIBE (Sociedad de Ciencias Naturales ARANZADI) 1–2:57–62

Dominguez P (2010) Approche multidisciplinaire d'un système traditionnel de gestion des ressources naturelles communautaires: L'agdal pastoral du Yagur (Haut Atlas marocain). Dissertation, École des Hautes Études en Sciences Sociales/Universitat Autònoma de Barcelona

Dominguez P, Hammi S, (2010) L'agdal du Yagur, écologie et pastoralisme. In: Fernandez K (ed) Ecología y Pastoralismo, Koldo Michelena, San Sebastián, p 34–56

Dominguez P, Bourbouze A, Demay S, Genin D, Kosoy N (2012) Culturally mediated provision of ecosystem services: the agdal of Yagur. Environ Values 21:277–296
Fillat F, García-González R, Gómez D (1995) Importancia de la ganadería en la conservación del paisaje pirenaico. Quercus 107:24–26
Fillat F, García-González R, Gómez D, Reiné R (eds) (2007) Pastos del Pirineo. CSIC, Madrid
Gellner E (1969) Saints of the Atlas. Weidenfeld and Nicolson, London
Gellner E (1981) Muslim society. Cambridge University Press, Cambridge
Gómez-Baggethun E, Mingorría S, Reyes-García V, Calvet L, Montes C (2010) Traditional ecological knowledge trends in the transition to a market economy: empirical study in the Donana natural areas. Conserv Biol 24(3):721–729
Hammi S, Al-Ifriqui M, Simonneaux V, Auclair L, Montes N (2007) Évolution des recouvrements forestiers et de l'occupation des sols entre 1964 et 2002 dans la haute vallée des Ait Bouguemez (Haut Atlas Central, Maroc). Sécheresse 18(4):271–277
Hardin G (1968) The tragedy of the commons. The population problem has no technical solution; it requires a fundamental extension in morality. Science 162(859):1243–1248
Jiménez J (2006) La imagen de los espacios de alta montaña en la prehistoria: El caso de los Pirineos Catalanes Occidentales. Dissertation, Unversitat Autònoma de Barcelona
Joint programme on the links between biological and cultural diversity. 2010
Loh J, Harmon D (2005) A global index of biocultural diversity. Ecol Ind 5(3):231–241
López JD (2010) Las fiestas de la primavera: bienvenida a los pastores trashumantes. In: Vidal P, Castán JL (ed) Transhumancia en el Mediterráneo, Ceddar, Zaragoza
MacDonald D, Crabtree JR, Wiesinger G, Dax T, Stamou N, Fleury P, Gutierrez-Lazpita J, Gibon A (2000) Agricultural abandonment in mountain areas of Europe: environmental consequences and policy response. J Environ Manage 59:47–69
Maffi L (2005) Linguistic, cultural, and biological diversity. Annu Rev Anthropol 29:599–617
Mahdi M (1999) Pasteurs de l'Atlas. Production pastorale, droit et rituel, Ed. Fondation Konrad Adenauer, Casablanca
Maluquer J (1987) La població prehistórica a la Vall de Serradell. Collegats 1:31–44
Mauss M (2012 [1925]) Essai sur le don. Forme et raison de l'échange dans les sociétés archaïques. PUF, Paris
Mazzucco N, Gassiot E, Rodríguez D, García D, Obea L (2012) Les primeres comunitats ramaderes de la Vall de Sant Nicolau (5.000—2.300 CalANE). In: Aniz MM (ed.) IX Jornades sobre Recerca al Parc Nacional d'Aigüestortes i Estany de Sant Maurici, Ed. Generalitat de Catalunya & Parc Nacional d'Aigüestortes i Estany de Sant Maurici, Lleida, pp 221–233
Mcneil JR (2003) The mountains of the Mediterranean world. An environmental history. Cambridge University Press, Cambridge
Mittermeier R (2004) Hotspots revisited: Earth's biologically richest and most endangered terrestrial ecoregions. Cemex, Mexico
Nadal E, Igleias J, Estrada F (2010) Generalitat de Catalunya. Departament de Cultura, Barcelona
Pallaruelo S (1988) Pastores del Pirineo. Ministerio de Cultura, Madrid
Pascon P (1983) Le Haouz de Marrakch. CURS-CNRS-INAV, Rabat
Rasse M (2008) La diffusion du Néolithique en Europe (7000–5000 av. J.-C.) et sa représentation cartographique. Mappemonde. http://mappemonde.mgm.fr/num18/articles/art08205.pdf. Accessed 27 Jan 2015
Rodrigue A (1999) L'art rupestre du Haut Atlas marocain. L'Harmattan, Paris
Sellier E (2004) L'agdal du Yagour. Territorialités au pluriel pour la protection de la nature dans le Haut Atlas de Marrakech. Dissertation, Université de Provence
Simoneau A (1967) Les gravures du Haut Atlas de Marrakech. Revue de Géographie du Maroc 11:67–76
Sirami C, Nespoulous A, Cheylan JP, Martya P, Hvenegaarda GT, Geniezh P, Schatza B, Martin JL (2010) Long term social and ecological dynamics of a Mediterranean landscape: impacts on biodiversity. Landscape and Urban Plan 96:214–223
Stepp R (2005) Mountains and biocultural diversity. Mt Res Dev 25(3):223–227

Stoate C, Boatman ND, Borralho RJ, Rio Carvalho C, de Snoo GR, Eden P (2001) Ecological impacts of arable intensification in Europe. J Environ Manage 63:337–365

Sutherland WJ (2003) Parallel extinction risk and global distribution of languages and species. Nature 423:276–279

Uhel R (coord) (2006). Urban sprawl in Europe: the ignored challenge. EAA Report 10(2):56

Vidal P, Castán JL (eds) (2010) Transhumancia en el, Mediterráneo edn. Ceddar, Zaragoza

Villa J (1988) Una reflexión sobre la fiesta, el rito y la danza en las Comunidades Pirenaicas. Revista El Gurrión 32:8–12

Zent S (2001) Acculturation and ethnobotanical knowledge loss among the Piaroa of Venezuela: demonstration of a quantitative method for the empirical study of traditional ecological knowledge change. In: Maffi L (ed) On Biocultural diversity: linking language, knowledge and the environment. Smithsonian Institution Press, Washington

Chapter 6
Apicultural Spaces as Biocultural Places: A Comparative Temporal and Spatial Examination of Beekeeping Practices and Their Contextual Landscapes in the Northwest Apennines

Robert Hearn and Rebekka Dossche

Abstract The historical rural landscapes of the Val Borbera (Piedmont) and the Val di Vara (Liguria) were characterized by intensive agro-silvo-pastoral systems. This paper explores the comparative bio-cultural history of apiculture, beekeeping and the production of honey and wax in these areas of the northwest Italian Apennines during the past 200 years. In 1798–99, an enquiry (*inchiesta*) into the quality, production and territorial needs of the Republic of Liguria included a question concerning the nature, extent and practices associated with apiculture and beekeeping, but also on the landscapes where this rural practice were performed. Through contextualizing these written and oral sources on a landscape scale alongside historical cartography and later aerial photography, this paper discusses the temporal and spatial evolution of apiculture, beekeeping and the production of these practices from the late eighteenth century to the modern day. In doing so, apiculture and beekeeping are shown to be important components within and indicative of the bio-cultural diversity and heritage in these parts of the northwest Italian Apennines and indeed elsewhere across Europe.

Keywords Apicultural and beekeeping practices · Historical approaches · Landscape research · Interdisciplinarity · Mediterranean basin · Bio-cultural diversity

---

R. Hearn · R. Dossche (✉)
Laboratory of Archaeology and Environmental History (LASA),
Department of Antiquities, Philosophy and History (DAFIST),
University of Genoa, Genoa, Italy
e-mail: rebekka.dossche@ugent.be

R. Hearn
e-mail: robert.hearn@edu.unige.it

R. Dossche
Institute for Agricultural and Fisheries Research (ILVO), Department of Agriculture and Society, Merelbeke, Belgium and Ghent University, Department of Geography, Ghent, Belgium

M. Agnoletti and F. Emanueli (eds.), *Biocultural Diversity in Europe*,
Environmental History 5, DOI 10.1007/978-3-319-26315-1_6

6.1 Introduction

European landscapes are complex manifestations of representation of centuries of continuous and interactions between the land, how it is used, and whom it has and is used (Agnoletti 2011, 2013). Rural mountainous areas are amongst the landscape types in Europe that have changed most drastically through time, especially those in the Mediterranean basin. Many of these were historically characterized by an intensively used rural landscape where inhabitants lived from the exportation of local produce in combination with self-sufficient farming and herding. The physical and social conditions in these areas necessitated the creation and maintenance of an interesting balance between human management, restrictive environmental conditions and biological diversity the aim of which was to optimize annual fluctuations in production and productivity (Naveh and Lieberman 1994). This long history of these balances, interchanges and relationships are generally not thought to have caused ecological degradation. Moreover, these facets created a landscape with a specific cultural value (Pinto-Correia 1993; Agnoletti 2013).

However, many highly valuable rural mountain landscapes have undergone significant land abandonment processes since the mid-twentieth century, particularly during the 1970s. This initiated a process of extensification that led to the gradual disappearance of an environment and landscape closely related to by the rural communities. The unbalancing of previous temporally and spatially specific socio-cultural systems historically adapted to natural resource and local environmental constraints are considered in literature to pose a degradation threat, potentially leading to the disappearance of highly valuable landscapes (Pinto-Correia 1999; Naveh and Liebermann 1994).

In Europe, the switch from 'traditional' to postmodern landscapes has resulted in the loss of many valuable landscapes during the last 200 years (Cevasco and Moreno 2013). This turnover was characterized by significant differences in the dynamics, speed, and scale of landscape change and the subsequent changes in perceptions, values and the related behaviours of their users (Antrop 2005). The driving forces behind these changes can be classified into three main categories (Antrop 2005); namely (i) a more dynamic evolution of areas based on infrastructural accessibility, (ii) the urbanization and industrialization of areas and the resulting exodus of remote areas, and (iii) the globalization of decisions and actions with an impact on the local level.

However many areas of Europe are currently undergoing processes of planned or spontaneous 'rewilding'. The rapid spread of spontaneous woodland has and is occurring in many parts of Mediterranean Europe, and is seen by some as a welcome return to idealized notions of landscapes before they were shaped by human activities (Monbiot 2013). As with many areas of the Italian Apennines, in Liguria rural depopulation and abandonment have led to the unintentional regeneration of dense woodland thereby constituting an example of unplanned 'rewilding' (Balzaretti et al. 2004; Cevasco 2007, 2013; Watkins 2014). This secondary, semi-natural vegetation has created the ideal habitat for the return of species, particularly the wild boar, deer and wolves (Hearn et al. 2014; Hearn 2015). However, whilst regarded positively by many, 'the new woodland disguises the loss of traditional woodland knowledge, the disappearance of thousands of small agricultural terraces and the reduction of biodiversity associated with the pastures and meadows of the cultural landscapes it replaces' (Hearn et al. 2014: 54). In short, 'rewilding' can have a significant impact on bio-cultural heritage and can serve to disrupt notions of 'naturality'.

In Mediterranean agricultural areas where, due to the natural environment and its agricultural productions and structure, mixed land use systems (mainly based on permanent crops and integrating open wooded areas) were developed during the past, such driving forces have had significant impacts. The high variety of crops means that they cannot compete with the productivity levels in areas of specialized agriculture, particularly when contextualized within a global urbanization and the industrialization of European agriculture, with the strengthening of market forces and an increased competition with the highly productive agriculture in northwest Europe. The inability for competition by remote Mediterranean areas combined with free market strategies and the homogenisation of agricultural rules (Common Agriculture Policy) in northwest Europe, puts Mediterranean areas in a weakened position, leading to the further global marginalization of these regions (Pinto-Correia 1993).

In northwest Italy, the processes of land abandonment have been described by several authors (cfr Cevasco et al. 2013). Vos and Stortelder (1992) describe the land abandonment in the Tuscan Apennines as a consequence of the decrease in the number of people active in agriculture and forestry during recent decades. The combination of hard work with declining demand for local produce has resulting in a non-sustainable economic situation for traditional activities and, as such, constitutes a key factor in explaining why people have emigrated. As stated by Pinto-Correia (1993), the rural management in the marginal areas most affected by land abandonment processes and subsequent landscape changes resulting from an aging population and depopulation of young and active people has led to socio-economic depression in many areas.

Research on landscapes and the related actors and past practices are vital components in the creation of resilient and sustainable future landscape research, and as such the adoption of an historical approach is crucial in the valorisation of bio-cultural landscapes (Cevasco et al. 2015). This chapter explores the rural practice of apiculture and beekeeping, because of both its cultural (beekeeping techniques change during time) and biological diversity (bees are dependent on their biological environment for pollen and nectar) through time. As such, these practices are densely connected with the bio-cultural history of a landscapes. Through an interdisciplinary examination of two empirical case studies, drawing on an historical approach, this chapter discusses the temporal and spatial evolution of beekeeping and its resultant production from the late- eighteenth century, concluding with a regressive analysis of the bio-cultural history of the specific landscapes. In order to do so, this research posed two fundamental questions; namely (i) how beekeeping was executed in the past and (ii) what is the relation between the honey production of today and the landscapes of the past?

6.2 Apiculture and Beekeeping Practices in the Ligurian Apennines

The research is situated in two case studies in the northwest Italian Apennines, in the regions Piedmont and Liguria. The landscape histories of both valleys, the Val di Vara (Liguria) and the Val Borbera (Piedmont), are characterized by intensive agricultural use, abandonment and later transformations into more extensively used landscapes with rural traces and other functions, such as hiking, hunting, and some food tourism (Cevasco 2011; Moreno 2011). Both areas were known for ancient mercantile trails which connected the Ligurian Sea with the Po plain (Balzaretti 2013). Until the 1950s, the rural landscapes of the Apennines, between Piedmont and Liguria, were characterized by terraces, cultivated with wheat, vineyards and small-scale gardening around the mountain villages (Balzaretti et al. 2004). On the other hand, there were grass and pasture land with herds of sheep and cows more uphill. Farmers of the lower valleys brought their cattle up unto the highlands, using the ancient trails, and met on the summer pastures (Moreno and Raggio 1990).

In his bibliography of the Italian works concerning apiculture and beekeeping, Francesco Beltramini de' Casati reflected that 'there are many periodicals containing articles on bees; so far one can count 700 [periodicals] in which there are numerous articles on beekeeping' (Beltramini de' Casati 1890). Over a century later, Accorti (2000) had 48 apicultural and beekeeping bibliographies at his disposal, and based on these listing compiled an extensive bibliography with several thousand references, omitting reprints. In the fourteen years since Accorti's work, this number has without doubt swollen significantly. However, the works on apiculture are only the 'ears of the hippopotamus', and the growth of literature on beekeeping must be viewed in relation to the increase in works on apian and applied entomology, which in turn are imbued in the flowering of natural history, particularly from the mid-eighteenth century (Smith et al. 1973). In Italy, Genoese aristocrats and institutions played a significant role in the development of natural and entomological history (Raggio 1998; Poggi 2012), such as Marquis Massimilliano Spinola (1780–1857), who identified the 'Ligurian' or 'Italian' honeybee (*Apis mellifera ligustica*) in (1806). Examination of the personal archives of Spinola held at the dynastic castle in Tassarolo, Piedmont, reflect the circulation of entomological knowledge in Italy and throughout Europe during the late-eighteenth and nineteenth centuries (Passerin D'Entreves 1980). Many insights resulting from scientific entomological study concerning bee behavior and biology were in turn applied to apicultural and beekeeping practices and production, and linked to efforts to 'rationalize' or 'modernize' practices in Italy. Amoretti's work on beekeeping in Lombardy (1788; 1811) is widely considered to be the first examples of 'modern' apiculture (Crane 1999), 'modernizing' serving as shorthand for innovations geared towards increasing hive harvests and greater economic returns. During the course of the nineteenth century, a number of initiatives and institutions were created to stimulate this process, such as the Associazione Centrale per l'incorraggiamento dell'Apicoltura in Italia, founded in Milan in 1867 (Figs. 6.1 and 6.2).

Fig. 6.1 Geographic location of case study areas: Val di Vara, Liguria [1] and Val Borbera, Piedmont [2]: map produced by Dossche (2014)

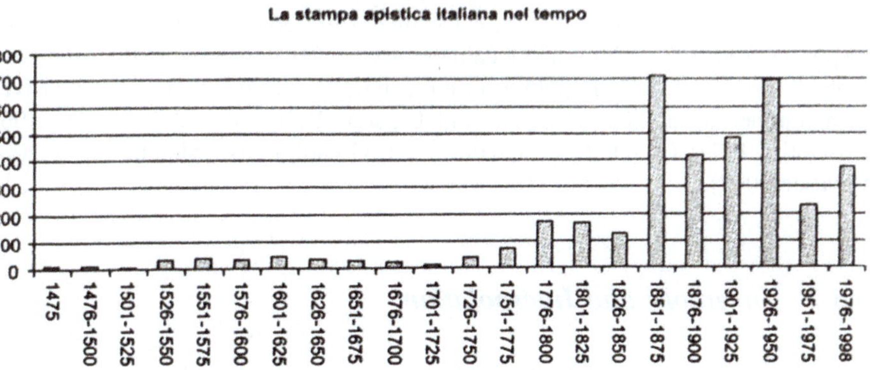

Fig. 6.2 Italian apistic and apicultural publications (1475–1998). In: Accorti M (2000) Le api di carta: bibliografia della letteratura italiana sull'ape e sul miele. Leo S. Olschki Editore, Florence, pp. xxix–xxx

6.3 Methodology

The research was divided into two main parts; (i) the historical analysis of the contextual documentation of beekeeping in northwest Italy, and (ii) the landscape analysis of the environment within which the beekeeping practices took place. The combination of these different data sources results in a multidisciplinary and integrated conclusion.

6.3.1 *Inchiesta Istituto Nazionale della Repubblica Ligure (1798–1799)*

From the end of the eighteenth century, 'modern' forestry techniques and schemes devised by enlightened agricultural 'improvers' who sought to replace many historical, traditional agro-silvo-pastoral land management and usage practices in Liguria. In doing so, 'the data collected by contemporary agricultural 'improvers' can provide evidence of old land-management practices' (Moreno 2004: 131). In Liguria, a questionnaire (*inchiesta generale*) circulated by the Istituto Nazionale (Scotti 1979)—'a group made up of a group of doctors, naturalists, agriculturalists, and lawyers from the emerging Genoese bourgeoisie who were concerned with propagating French enlightenment ideology' (Moreno 2004: 131)—is one such example. Comprised of 35 questions concerning various elements of parish social, cultural, political and economic life, in late 1798, the *inchiesta* was circulated to parish priests and mayors throughout the newly-created Repubblica Democratica Ligure (est. 1797). Whilst many were lost in the early nineteenth century, two hundred and forty responses from 193 parishes and localities survive and are housed in the Archivio di Stato di Genova (A.S.G., Repubblica Ligure, n.610) and have been used extensively in a number of studies on Genoa and Liguria (Bulferetti and Costantini 1966; Costantini 1973; Grendi 1973; Quaini 1973; Grendi 1996; Moreno 1990, 2004; Cevasco 2007).

6.3.2 *Inventory and Interpolation*

Based on the *inchiesta* of 1798–1799, the responses to three key questions were examined: The first two -[Q14] and [Q17]- were used for the landscape analysis, containing information concerning the territory, extension and land use, and the third -[Q28]- in order to gather precise information concerning apicultural and beekeeping practices within these landscapes. These questions sought to gather information concerning [14] the extension of the territory, the extent of cultivation and non-cultivation on the plain, hilly and mountainous areas, [17] whethter the mountains and non-cultivated areas were covered by pasture or by trees and shrubland, and the extent and quality of the woodlands, and finally [28] if bees were kept, how the honey and wax were collected, whether the bees were killed during the extraction and the annual production.

Those three questions were analyzed and the results were interpolated on both a historical map of the area (*Minute di Campagna*: Gran Carte degli Stati di Terraferma, Corpo di Stato Maggiore 1828) and on a recent military aerial photo (2010).

6.4 Results

6.4.1 Apicultural and Beekeeping Practices and Production in the Late-Eighteenth and Early-Nineteenth Centuries

Question [28] of the *inchiesta* sought to gather information concerning apicultural and beekeeping practice and production, asking respondants (i) whether beekeeping was conducted, (ii) the manner in which the honey and was were collected, (iii) whether the bees were killed during the collection, and (iv) the annual production. Alongside more general historiographical issues relating to historical documentation like the *inchiesta*, including those relating to diverse levels of literacy (Cevasco 2004; Moreno 2004), Question [28] reveals a number of shortcomings in the manner in which the information was requested and the phraseology employed by the Istituto Nazionale in doing so, both of which have implications for the responses generated and the subsequent contemporary and historical use of the documentation. Firstly, the Question [28] would only elicit information concerning 'domestic' apiculture and beekeeping, thereby potentially omitting details concerning the collection of honey and wax of 'wild' bees. Secondly, no information concerning the number if hives and/or those involved in their husbandry were sought. Whilst some offered this information, where found, only a handful offered figures concerning hives and/or beekeepers and, where found, it is not clear whether these figures detail 'professional' beekeepers or simply the keepers of bees. Thirdly, various adjectives were frequently employed *in lieu* of precise numerical information concerning hives and/or beekeepers and annual production, the use of 'few' or 'very few' being particularly common, many respondents stating that they (hives and/or beekeepers) were too few to count. However, it is unclear what they may have meant by 'few', 'very few', 'little', or 'very little', and whether these descriptions were 'absolute' or 'relative' observations, reflections that would have been particularly relevant in light of the intentions of the *inchiesta* to identify areas of rural production that could be 'improved' or 'modernized'. Finally, there are a number of issues relating to the request for information concerning 'annual production', in that it is unclear whether this was measured in weight and/or monetary value, an issue further complicated by the use of inconsistent measures and currencies. By extension, it is unclear whether the figures, where provided, relate to the quantities of honey and was that were sold or existed in non-monetary economies. Such historiographical considerations relating to the information sought and the way in which it was collected could therefore have had significant implications in examinations of the nature and scale of historical apiculture and beekeeping practices and production.

However, putting these historiographical issues aside, a number of trends can be gleaned from the information provided in response to Question [28]. The majority responded positively to the question concerning whether beekeeping was an activity practiced in their community (Fig. 6.3a), most responding that they had ‘a small amount’ of them, be it bees, hives or keepers. Looking at other responses, for example, ‘1 or 2’, ‘a pair’, ‘2 or 3 people have hives’, we classified this ‘small amount'' into a quantities of less than ten hives. Only a few communities responded with higher quantities, such as more than 20, or even 32 hives (San Giorgio—Val di Vara). The amount of hives in the Val Borbera (Albera: Table 6.1) were generally higher than in the Val di Vara, most of them responding with amounts higher than 20.

The *inchiesta* responses provided interesting regarding the techniques people used to extract the honey and the wax (Fig. 6.3b). Most of the communities killed the bees in order to collect the honey and the wax, particularly in the middle and lower Val di Vara and Val Borbera a technique that was widespread in the Apennines at that time (cfr Table 6.1). However, at this time it would appear that new techniques to extract the honey and wax without killing the bees were being developed and being assimilated, albeit very gradually, into rural apicultural and beekeeping practices. This technique resulted in lower production, however apiculturalists and beekeepers slowly began introducing it so as to maintain the bee-family for the next year, whereas the killing-technique obliged them to search for new families to repopulate the hives, be these ‘wild’ or those from other domestic sources in the parish. The *inchiesta* shows that this change in extraction method can first be seen in the Upper Val di Vara (Fig. 6.3b).

Considering the annual production quantities (Fig. 6.3c), the *inchiesta* shows that honey was the main product, but also that wax had an important output. Wax has a number of uses, from the basics for lighting through to preparatory uses in carpentry and other practical applications. For example, Chabrol de Volvic’s *Statistique* (1824), concerning the fleeting French occupation of Liguria (1805–1814), reveals that there were large wax factories in several locations, producing wax on what would today be described as an 'industrial' level. However, what is both clear and inferred from the historical literature is wax production remained on a localized level well into the nineteenth century. Echoing sentiments explicitly stated and gleaned from responses to the *inchiesta* and other archival documentation concerning the Repubblica Ligure (1799–1805), the statistics concerning the French département of Montenotte produced by Chabrol de Volvic (1773–1843) reflect that: 'apiculture could form a productive branch of the rural economy in the department, particularly in the two northerly *arrondisement*, where the meadows and heaths could offer abundant nutrition for the hives…however the keeping of bees is rudimentary, the inhabitants leaving this production almost to chance' (Chabrol de Volvic 1824: 233), reflections that find parallels in works on agriculture produced later in the nineteenth century (Zuccagni-Orlandini 1838).

(a)

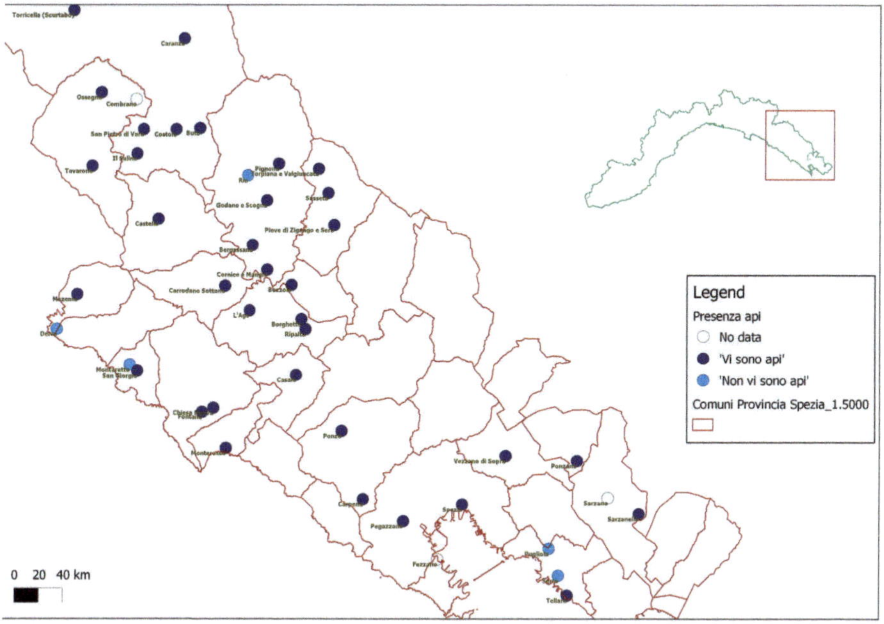

(b)

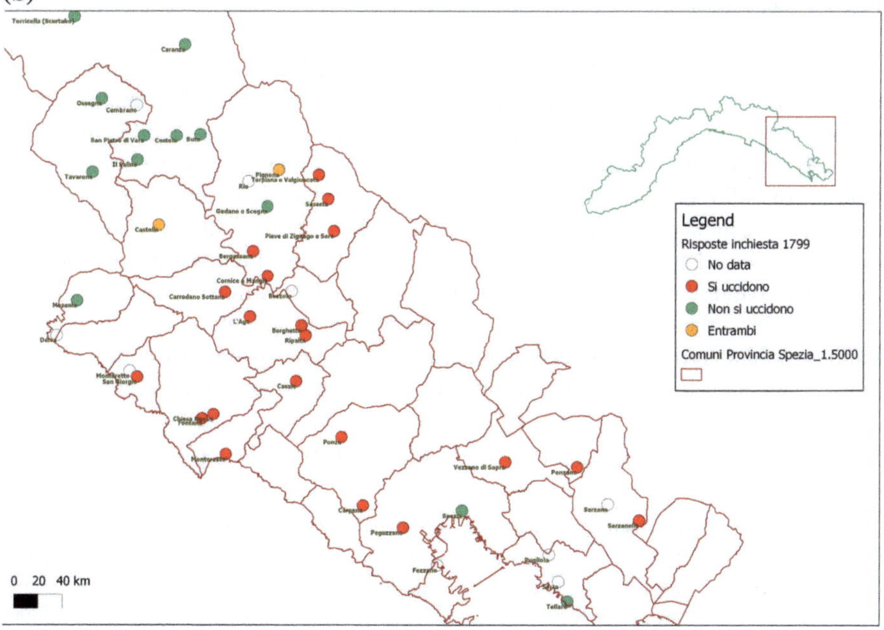

Fig. 6.3 Major trends of beekeeping practices in the case study area 'Val di Vara' in 1798-99. **a** Presence (dark blue) or absence (light blue) of beekeeping practice. **b** Way of collecting honey by killing bees (red), not killing bees (green), or both (orange). **c** Different products of beekeeping: honey (yellow), honey and wax (pink), wax (orange). Map produced by R. Dossche 2014. A.S.G., Repubblica Ligure, n. 610

(c)

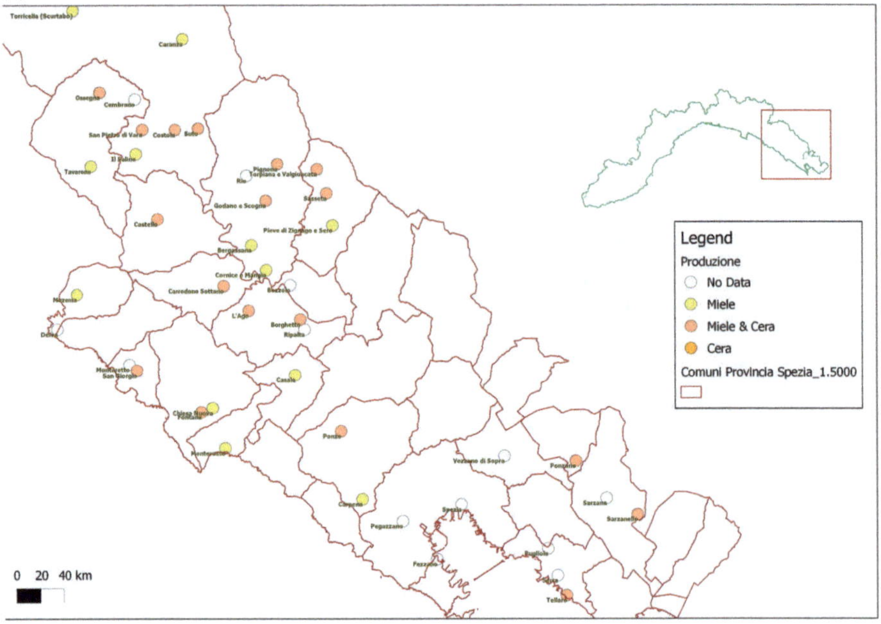

Fig. 6.3 (continued)

The ‘rudimentary’ methods employed by inhabitants in this *département* included digging hives into the trunks of chestnut trees rather than making them, placing the hives close to inhabited areas rather than in open or wooded areas, and annually killing the swarms with straw smoke in order to remove the wax and honey. In addition to the ‘methodological’ shortcomings employed by rural inhabitants, Chabrol de Volvic recorded that 'in the canton of Savona (Liguria) were there is a lack of meadows and flowered fields, the inhabitants fear bees because they believe that they [the bees] will reduce the number of flowers on the fruit trees and so reduce the amount of fruit.' However, whilst apicultural production was seemingly hindered by rudimentary practices and the persistence of popular beliefs, innovations and developments were taking place elsewhere in the *département* of

Table 6.1 Responses to Question [28] for Albera (Val Borbera): A.S.G., Repubblica Ligure, n.610

Quantity of hives	Beekeeping techniques	Products	Annual production
There are around 30 small hives	The honey and the wax are collected after the bees have died	Honey and Wax	The product is not of a considerable entity, as in this country the winter is much longer than the rest of the seasons
In the parish there are 30 hives in poor condition	The honey and wax are collected when the bees have been killed	Honey and Wax	The production is scarce because the spring arrives very late and the winter comes very early

Montenotte, notably in the territories of Acqui (Piedmont) where 'some beekeepers had begun using hives with draws that could be removed when removing the honey and wax without killing the bees'. Citing the example of a single beekeeper in the hamlet of Saint-Bernard et de Legino (Piedmont) who annually profited 800Fr from 40 hives, Chabrol de Volvic sought to estimate the potential yield and value of hives in his *département*, and in doing so still accounted for the impact of rudimentary methods on production. Based on the statistics collected from the individual communes, Chabrol de Volvic calculated that there were 5272 hives in the *département* of Montenotte. However, 'half of these swarms are kept for re-population, and the other half being harvested and the bees generally killed.' Based on these figures, Chabrol de Volvic therefore calculated that 'every hive can produce 16 kg of honey and 3 kg of wax [...] the regular price of honey is 80 cents per kg and 4Fr per kg of wax [...] each hive therefore has a value of 24Fr and 80 cents [...] when multiplied by 2636, half the number of productive beehives, the total value of apiculture is 55,372Fr and 80 cents.' In concluding his treatise, Chabrol recorded that whilst 'in the past, honey, and wax were imported from Sardinia, Corsica, Sicily, and Spain [...] the political circumstances have interrupted this commerce.' In order to reduce the impact of the cessation in trade 'the department could easily reduce costs by giving these precious products the attention that they deserve' (Chabrol de Volvic 1824: 233–234). The nature of the information imparted by the *inchieta* and the *Statistique* characterizes the documentation produced following the restoration of the Kingdom of Sardina following the Congress of Vienna in 1814 until the unification of Italy documentation that, in short, reveals that honey and wax production was mainly conducted on a small, local scale, with successive administrations seeking to expand this industry into a profitable endeavour (Fig. 6.3; Table 6.1).

6.4.2 Apicultural and Beekeeping Landscapes in the Early-to-Mid-Nineteenth Century

When combined with the analysis of the 1828 historical map (*Minute di Campagna*: Gran Carte degli Stati Sardi di Terraferma, Corpo di Stato Maggiore Sardo), questions [14] and [17] of the *inchiesta*, provide a detailed insight into the characteristics of the nineteenth century landscape (Fig. 6.4a).

The brown circles on Fig. 6.4a (C: *Campo*) refer to the amount of cultivated land indicated in the responses to the *inchiesta*. In Albera (Val Borbera), the *campo* cultivated lands occupies only one third of the whole territory of the community. On the other hand, there is also non-cultivated land that occupying two thirds of the territory, consisting out of pastureland (P: *Pascoli*), woodland (B: *Boschi*), vineyards or fruit gardens (CV: *Campo e vigno*) and/or shrubland (g: *Gerbido*). All of these types of land use estimate a certain land use, and are considered as those under cultivatation, even if the respondents of the *inchiesta* classifed them as 'non-cultivated'. The results of the *inchiesta* are therefore interesting when

(a)

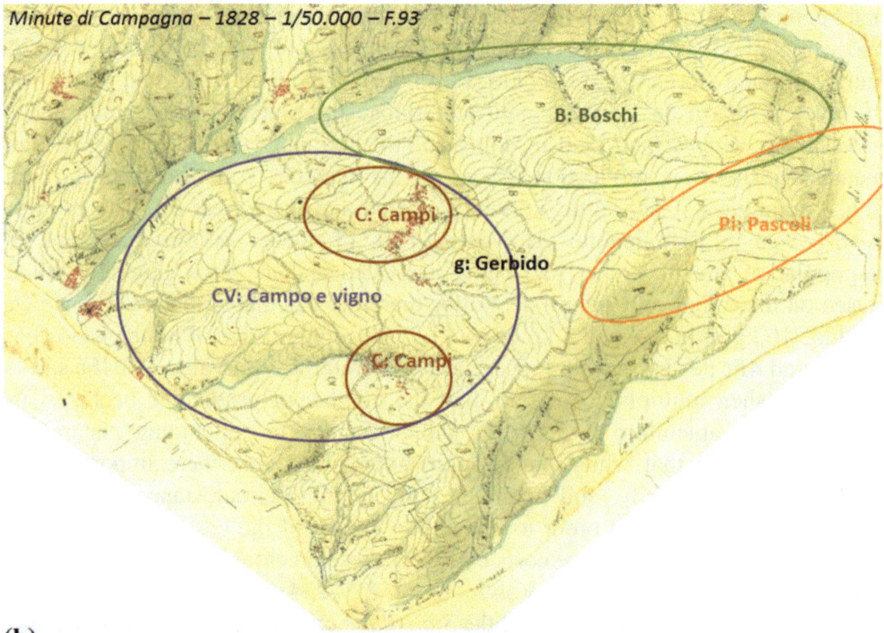

(b)

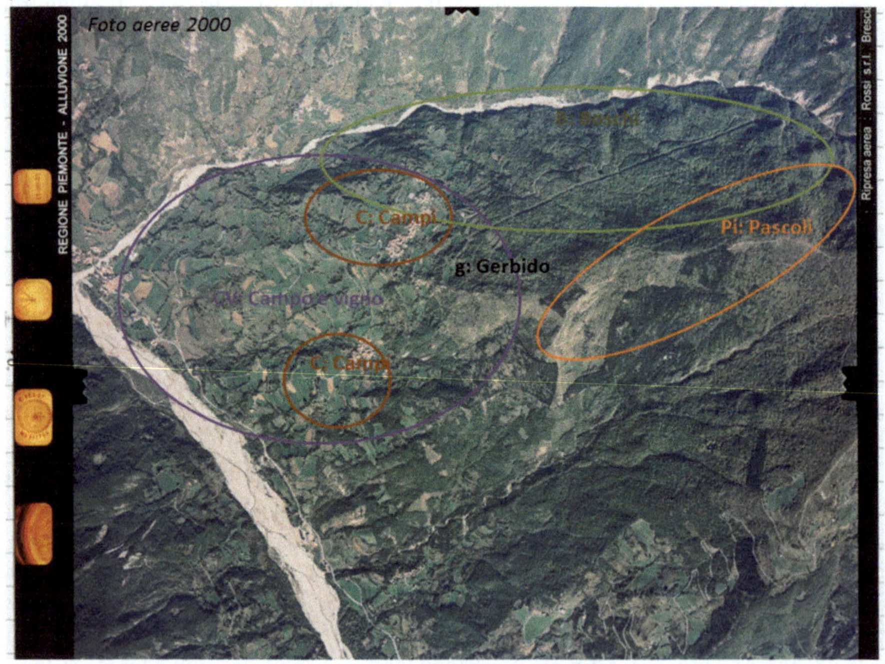

Fig. 6.4 Interpolation of **a** historical land use (Minute di Campagna: Gran Carte degli Stati Sardi di Terraferma, Corpo di Stato Maggiore Sardo 1828- 1:50,000 F93) and **b** the current land use actual ones (Military aerial photo 2010). Produced by R. Dossche 2014

Table 6.2 Interpretation of *Inchiesta Istituto Nazionaledella Repubblica Ligure* (1798–99): A.S.G., Repubblica Ligure, n.610

No.	Community	Surface	% cultivated	% uncultivated	Geomorphology	Land cover
394-1	Albera	Around 4 miles from east to west and around 2.5 miles from south to north	A third cultivated	Two thirds uncultivated	Few plain areas, one third part is hilly, the rest more mountainous	Partially uncovered, grass for good pastures, places with woodlands, chestnut woods, fruit trees, and various types of oak
394-2	Albera	For one part 3 miles, and around 2.5 for the other	A third cultivated	The rest uncultivated	Small plain areas, one third hilly, the rest mountainous	Partially uncovered, grass for pasture, woodlands for wood, chestnut woods, wild and domestic chestnuts, various types of oak, mulberries, moroni, apple trees, pear trees

combined with historical maps because they are complementary and provide a relatively clear indication of the nature of the landscape and its usage in the late-eighteenth and early-nineteenth centuries (Table 6.2).

In order to detect the nature and dynamics of the landscape transformations, and to understand the evolution of land use practices during the past 200 years, snapshots of the historical map of 1828 with the aerial photos of 2010 were compared thereby visualizing the different trajectories of each land use type within the landscape (Clark 2004). This comparison revealed the following changes and characteristics:

(i) From historical arable land to meadow or shrub and woodland
The ancient cultivated land, mostly oriented on terraces in the area and surrounding the villages, are still recognizable in the field and on the recent orthographic photography. Here, there has been an increase in the amount of hedgerows and laterally more shrubby areas on the terraces that have become more invasive in the last 30 years. Some terraces, mostly the ones situated far from the villages, have become completely dominated by secondary vegetation and converted into shrub and woodland. Other, less abandoned terraces are now used as meadows, or in a handful of cases, are cultivated for the production of grasses and fodder in relation to current livestock husbandry.
(ii) From historical plantations to shruband woodland:
Former chestnut plantations have mostly been invaded by shrubs and secondary vegetation. They are currently not maintained and suffer from several diseases. Several plantations suffer from the lack of maintenance causing lower production from the plantations. Some plantations also used for grazing are still recognizable in the field. However, many of these have become invaded over time by mixed shrub and woodland.
(iii) From historical pasture land to shrub and woodland:
The higher pastures and meadows have visibly lost territory over time and are now smaller and more invaded by shrubs. Since the arable land around the villages is no longer used for cultivations, there has been a shift in the meadows from *outfield* (higher situated historical pasture land) to *infield* (historical arable land, close to the villages) systems.
(iv) Continuous villages with concurrent decrease in road infrastructure:
Many mercantile trails disappeared. Ancient trails have been invaded by shrubs and secondary vegetation but some have survived because they were asphalted, or led to a small group of houses and therefore privately maintained.

The preview of the landscape evolution visualization shows that the area close to the settlements (*infield*) has undergone greater changes than the more uphill or downhill (*outfield*) areas. The *outfield* areas might have undergone changes with the same frequency, but the changes resulted more from the use of the land rather than the land cover. For example, the upper hill areas are mostly grasslands, but have changed from meadows to pasture since the agricultural practice of hay-production has been more concentrated in the *infield* than the *outfield*. The upper hill areas are nowadays only used for the grazing of cattle.

6.4.3 Historical Apiculture and Beekeeping in the Bio-Cultural Landscape

At the end of the eighteenth, beginning of the nineteenth century, beekeeping was common if not intensive rural practice in the Ligurian Apennines (Fig. 6.4), albeit more present in the mountain areas than those closer to the Gulf of Liguria. In addition, newer techniques for the extraction of honey and/or wax first appeared in the intensively used mountain areas. Figure (6.3) provides an overview of the larger amount of hives in the Val Borbera and the Upper Val di Vara in comparison with the Lower Val di Vara, lying closer to the Ligurian Sea. This culture of small-scale beekeeping was conducted in a landscape with a large biodiversity, as confirmed by the map of 1828 together with the responses to questions fourteen and seventeen of the *inchiesta*. The combination of these sources means that, since honey production was performed once a year, it is could be suggested that the type of honey produced was probably what is currently refered to as '1000-Flower honey (*millefiori*).

Apiculture and beekeeping is still performed in these areas, albeit with a much smaller diversity and on a much larger scale. Apiculturalists and beekeepers are less numerous but have a much larger number of hives due to the professionalization of the practice. This professionalization has also caused honey production to concentrate on certain flowerings, particularly chestnut, linden and acacia. Beekeepers working in the Apennine mountain landscape are mostly interested in the chestnut flowering, yet express concern at the perceived reduction in the biodiversity of the landscape owing to the presence and recurrent outbreaks of several diseases and the 'rewilding' of the landscape (Hearn 2015).

6.5 Conclusion

Beekeeping in the nineteenth century had a large dispersion and was a common rural practice, especially in the Apennines. At the same time, it was performed on a small scale and in a landscape with a large biodiversity. Whilst extremely limited in terms of quantity, the existing historical documentation detailing beekeeping practice reveal that during the study period, production was generally limited and with negligible market value. This of course does not negate the possibility that hive production was sufficient for daily, small-scale use, meeting the needs but nothing more. Nevertheless, there is a clearly perceptible desire for the increase in apicultural production with a large number of publications and initiatives geared toward 'modernizing' beekeeping. These desire were articulated by all successive administrations from 1797 to 1860, however, documentation concerning the success is largely absent and so the success hard to establish. This research therefore highlights the importance of interdisciplinary research and the combination of approaches from historical geography, environmental history, physical geography and the natural sciences. In fact, beekeeping as a bio-cultural activity occurring in bio-cultural landscapes, that bees as a species are fundamental in creating, is an ideal academic landscape for interdisciplinary investigations.

If we look at the way in which apiculture and beekeeping are conducted nowadays, there are large differences in the amount of people engaged in apiculture and beekeeping and the scale on which it is performed. The professionalization of apicultural and beekeeping techniques from the second half of the twentieth century onwards is one of the reasons explaining these changes.This professionalization has material form, but also in terms of the types of honey produced such as chestnut, acacia, and linden; in short unifloral rather than multifloral.

However, what is the relationship between contemporary honey production and the historical, bio-cultural landscape? Several types of honey that are produced today, are done in a landscape with a historical value. Chestnut honey, for example, is only possible in landscapes that have a culture of chestnut production of the last 200 years, and as such beekeeping of today is linked strongly with the bio-cultural heritage of our landscape, and its future scale and success is intricately tied to the future management and conservation of rural landscapes of historical importance (Agnoletti 2011, 2013).

References

Accorti M (2000) Le api di carta: bibliografia della letteratura italiana sull'ape e sul miele. Leo S. Olschki Editore, Florence

Agnoletti M (ed) (2011) Paesaggi rurali storici: per un catalogo nazionale. Editori Laterza, Bari

Agnoletti M (ed) (2013) Italian historical rural landscapes: cultural values for the environment and rural development. Springer, Dordrecht

Antrop M (2005) Why landscapes of the past are important for the future. Landscape Urban Plann 70(1–2):21–34

Balzaretti R, Pearce M, Watkins C (eds) (2004) Ligurian landscapes: studies in archaeology, geography and history. Accordia, London

Balzaretti R (2013) Dark Age Liguria: regional identity and local power, c. 400–1020. Bloomsbury, London

Beltramini de'Casati F (1890) Vocabolario apistico italiano e dizionario d'apicoltura. Tip. Guigoni, Milan

Bulferetti L and Costantini C (1966) Industria e commercio in Liguria nell'eta del Risorgimento (1700-1861). Banca commerciale italiana, Milan

Cevasco R (2007) Memoria verde: nuovi spazi per la geografia. Edizioni Diabasis, Reggio Emilia Italia

Cevasco R (ed) (2011) Piemonte. In: Agnoletti M (ed) Paesaggi rurali storici: per un catalogo nazionale. Editori Laterza, Bari, pp 157–179

Cevasco R (ed) (2013) La natura della montagna. Scritti in ricordo di Giuseppina Poggi. Oltre Edizione, Sestri Levante Spinger, Dordrecht, pp 141–152

Cevasco R, Moreno D (2013) Rural landscapes: the historical roots of biodiversity. In: Agnoletti M (ed) Italian historical landscapes: cultural values for the environment and rural development

Cevasco R, Moreno D, Hearn R (2015) Biodiversification as an historical process: an appeal for the application of historical ecology to bio-cultural diversity research. Biodivers Conserv 24:3167–3183

Chabrol de Volvic, Gilbert Joseph Gaspard (1824) Statistique des provinces de Savone, D'Oneille, D'Acqui et de partie de la province de Mondovi, formant l'ancien département de Montenotte. Par le comte de Chabrol De Volvic, conseiller d'État, préfet de la Seine. Imprimerie de Jules Didot ainé, imprimeur du Roi, Paris

Clark J, Darlington J, Fairclough G (2004) Using Historic Landscape Characterisation. English Heritage and Lancashire County Council, London

Costantini C (1973) Comunità e territorio in Liguria: l'inchiesta dell''Istituto nazionale' (1799). Miscellanea storica ligure 5(2):291–363

Crane E (1999) The world history of beekeeping and honey hunting. Taylor and Francis, London

Felloni G (1961) Popolazione e sviluppo economico della Liguria nel secolo XIX. I.L.T.E, Turin

Grendi E (1973) Introduzione alla storia moderna della Repubblica di Genova. Fratelli Bozzi, Genoa

Grendi E (1996) Storia di un storia locale : l'esperienza ligure 1792-1992. Marsilio, Venice

Hearn R, Watkins C, Balzaretti R (2014) The cultural and land use implications of the reappearance of the wild boar in northwest Italy: a case study of the Val di Vara. J Rural Stud 36:52–63

Hearn R (2015) Gains and Losses in the European Mammal Fauna. In: Kirby K, Watkins C (eds) Europe's Changing Woods and Forests: from wildwood to managed landscapes. CAB International, Wallingford, pp 193–206

Moreno D (1990) Dal documento al terreno: storia e archeologia dei sistemi agro-silvo-pastorali. Il Mulino, Bologna

Moreno D (2004) Escaping from 'landscape': the historical and environmental identification of local land-management practices in the post-medieval Ligurian mountains. In: Balzaretti R, Pearce M, Watkins C (eds) Ligurian landscapes: studies in archaeology, geography and history. Accordia, London, pp 129–140

Moreno D (ed) (2011) Liguria. In: Agnoletti M (ed) Paesaggi rurali storici: per un catalogo nazionale. Editori Laterza, Bari, pp 180–203

Moreno D, Raggio O (1990) The making and fall of an intensive pastoral land-use system. Eastern Liguria, 16–19th centuries. Riv Studi Liguri 56:193–217

Monbiot G (2013) Feral: searching for enchantment on the frontiers of rewilding. Penguin Books, London

Naveh Z, Lieberman AS (eds) (1994) Landscape ecology: theory and application. Springer, Dordrecht

Passerin D'Entreves P (1980) La collezione Spinola di Tassarolo. Museo Regionale di Scienze Naturali, Turin

Pinto-Correia T (1993) Threatened landscape in Alentejo, Portugal: the 'montado' and 'agro-silvo-pastoral' systems. Landscape Urban Plann 24:43–48

Pinto-Correia T (1999) Contribution to the extensification/intensification debate: new trends in the Portuguese montado. Landscape Urban Plann 46:125–131

Poggi R (2012) Entomologia in Liguria. Storie di insetti, uomini e collezioni. Erredi Grafiche Editoriali, Genoa

Quaini M (1973) Per la storia del paesaggio agrario in Liguria : note di geografia storica sulle strutture agrarie della Liguria medievale e moderna. Camera di commercio industria artigianato e agricoltura, Savona

Raggio O (1998) Collecting nature in Genoa 1780–1870. From aristocratic patronage to civic patrimony. J Hist Collect 10(1):41–59

Scotti P (1979) L'Istituto Nazionale (Genova 1798–1806). Atti della Accademia Ligure di Scienze e Lettere XXXV:330–339

Smith RF, Mittler TE, Smith CN (eds) (1973) History of entomology. Annual Reviews Inc, Palo Alto

Vos W, Stortelder A (1992) Vanishing Tuscon landscapes: landscape ecology of a submediterranean-montane area (Solano Basin, Tuscany, Italy). PUDOC, Wageningen

Watkins C (2014) Trees, woods and forests: a cultural history. Reaktion, London

Zuccagni-Orlandini A (1838) Corografia fisica, storica e statistica dell'Italia e delle sue isole, corredata di un atlante, di mappe geografiche e topografiche, e di altre tavole illustrative. Firenze

Chapter 7
History and Traditional Management of Italian Wood Pastures

Francesca Emanueli and Mauro Agnoletti

Abstract Wood pastures are semi-natural systems of elevated biological and cultural value that are by now recognised at international level. There are many reasons for the current interest in such woods: the diversification of the forest landscape, the rehabilitation of abandoned land through the rearing of indigenous breeds yielding quality products with an elevated level of food safety and the conservation of biodiversity. The historic origin of these forest formations is very ancient and is linked to the deforestation carried out by man to obtain land for cultivation. The traditional management of these woods over centuries has allowed them to survive up to the present. Unlike other countries such as Spain, for example, in Italy forestry science has not codified any form of silvicultural management for the wooded pastures. The purpose of this article is to offer a synthetic bibliographical overview of the historic origins of these forest systems in Italy and the management methods applied to them.

Keywords Italian wood pasture · History · Silvicultural management · Biocultural diversity

7.1 Introduction

Wooded pastures are considered as wooded areas that have traditionally and at length been managed through sustainable grazing of both domesticated and wild animals, combined with certain forms of use of the trees and shrubs, such as felling, pollarding and coppicing (Sanderson 1996). The structure of these forest formations

F. Emanueli (✉) · M. Agnoletti
CULTLAB—Laboratory for Landscape and Cultural Heritage,
University of Florence, Via San Bonaventura 13, 50145 Florence, Italy
e-mail: francesca.emanueli@unifi.it

M. Agnoletti
e-mail: mauro.agnoletti@unifi.it

M. Agnoletti and F. Emanueli (eds.), *Biocultural Diversity in Europe*,
Environmental History 5, DOI 10.1007/978-3-319-26315-1_7

consists both of low-density populations with sparse trees and of woods inserted within a mosaic of pastures and shrubland. In Europe the historic origins of the wooded pasture are dated to the start of the Neolithic with the advent of agriculture and the first thinnings of the woods to make way for arable land. However, Vera (2007) has recently developed a theory that explains the formation of the wooded pastures not as a degenerative phase of the primordial climax forest caused by human intervention, but rather as a primordial formation brought about by the effects on the development of the vegetation of the grazing of wild ungulates in prehistoric times. At present, traditional wood pasture practices have survived in the Alpine areas of Austria, Switzerland and the French Jura, in the Vosges and in the Massif Central, in the south-west of Spain (Humpherey et al. 1998; Smit et al. 2006) or in several nature reserves such as Junner Koeland in Holland, the New Forest in England or the Börkener Paradies in Germany (Tubbs 1986; Potte and Hüppe 1991). According to the database *Vegetation Heritage of Hungary* (Bölöni et al. 2008) in Hungary the surface occupied by this particular type of forest formation amounts to around 5500 ha. In Italy the wood pastures were already widespread from the sixth century BC along the Apennine chain (Cevasco 2007). Currently, their presence is concentrated in Sardinia, while they have largely disappeared from the rest of the country. The lower rate of survival of wood pastures in Italy compared with other European countries is partly the result of the modern silviculture approach (Rackham and Grove 2003). Starting from the eighteenth century, based on the German, Austrian and French forestry schools, the importance of timber production began to be stressed. The wood was no longer considered as a multifunctional resource for the local populations, but as a means for converting "wood into silver" (Ernst 1998). More recently, the importance of timber production has diminished in favour of a multiple management of the forest resources: this attempt to simultaneously meet economic, social, environmental and aesthetic landscape demands is defined as multifunctionality. The term "multiple use" was introduced in the United States in the early 1960s, whereas these practices had already been widespread in Europe for many centuries (Farrell et al. 2000). The wooded pastures are comprised among the multifunctional systems since they are associated with important values such as the diversification of the forest landscape, the development of typical productions (from meat to dairy products) which feature greater food safety and are therefore highly appreciated on the market, and the recovery of marginal areas by implementing zootechnical activities in territories where the obstacles of an environmental nature do not permit other forms of production in the primary sector.[1] Furthermore, the wooded pastures contribute to land use associated with a high level of biocultural biodiversity (Dennis et al. 1998; Haraszthy et al. 1997; Manninig et al. 2006; Olff and Ritchie 1998; Rozas 2004; Salata et al. 2011; Van Wieren 1998). In view of the multiple values of the wooded pastures as mentioned above, the identification, restoration and conservation of

[1]Through the insertion of the wood in the husbandry forage chain, the production costs can be cut by between 30 and 60 % compared with modern stable breeding.

these forms of land use have been developed in several European countries such as England, Germany, Switzerland (Bürgi and Stuber 2001; Bürgi and Gimmi 2007) Spain, Portugal and Belgium (Van Uytvanck et al. 2008). In Italy there are no historic reports of local practices that make it possible to retrieve information about the restoration and management of wooded pastures. Silvicultural and ecological information is instead necessary in order to establish management criteria for the woods where the classic models of high forest or coppice cannot be applied. It is therefore necessary to find a compromise between the traditional production objectives (firewood, acorns, chestnuts, cork, grazing), conservation of biodiversity and the cultural value of a traditional management system that has been practised for centuries and that now contributes to the characterisation of the historic rural landscape and the beauty of these territories (Farrel and Piussi 2000).

7.2 Wood Pastures in Italy: History and Management

In Italy the combination of trees and pasture was expressed in a number of different systems partially comparable to mixed crops, such as chestnut groves or fruit trees mixed with vines, where it was also possible to sow seeds or in any case to graze. The practice of pruning the trees to permit grazing is still to be found in various parts of Italy, but is hardly ever recognised in technical and scientific literature even though it gives rise to extensive landscapes—such as those in Sardinia—frequently considered as aspects that degrade the original forest vegetation.

There is evidence of the presence of the wooded pastures in Italy from as far back as the Roman period. The Romans referred to these forests using two terms: *saltus* and *silva glandaria*. The *saltus* could also indicate wooded areas, but unlike the *silva*, these were located in the vicinity of the villages; they were used as wood pastures and could also include temporary crops. The *silva*, on the other hand, generally indicated woods set at a distance from the settlements in which the silvicultural practices were implemented for procuring firewood and timber (Clèment 2008). The *silva glandaria* (or *silva pascua*) was maintained as a low-density high forest (6–18 m); the species normally subject to such treatment were oaks, chestnuts and beeches: in addition to acorns, the oaks also yielded timber for building. In order to foster the natural regeneration of these high forests, the Romans were in the habit of introducing pigs into the woods to break up the surface layers and loosen the earth, thus creating the optimal conditions for the germination of the seeds; they were introduced into the fields after sowing for the same reason. Nevertheless, the first documentary traces date back to the sixteenth century, in the form of several maps relating to the wooded pastures of Molise, Abruzzo and the Tuscan-Emilian Apennines. These show traces of a system designed for the production of hay and of leaves, obtained by the pollarding or lopping the trees (Cevasco 2007).

Recent pollen analyses carried out on centuries-old wooded pastures of beech in the northern Apennines backdate the origin of these stands to the Early Middle

Ages (around 800 AD) and illustrate the high level of biodiversity of the floral and herbaceous cover, correlated to the normal silvopastoral practices. The decline of the herbaceous resources during the twentieth century can be directly correlated to the suspension of the grazing (Cevasco and Moreno 2010). The presence of wooded pastures is still recorded in certain mountain areas, comprised within the system of monitoring of the Tuscan landscape, by the data in the Agricultural Land Register of 1832 Analysis of the data collected showed that wooded pasture covered 11.2 % of the territory and represented 44.5 % of the grazing land (Agnoletti 2010a). The contraction of this surface area became increasingly more marked starting from the years following the Second World War, particularly as a result of the reforestation of these areas connected with the abandoning of the mountain territories by the rural population. As sustained by Piccioli (1908), the term *wooded pasture* is considered to refer to the grazing areas in which the trees tend to be conserved, isolated or in groups, and more or less regularly cover the land in such a way that the number of trees is never detrimental to the herbaceous production. According to the Agricultural Land Register definition (1929) the distinction between the different types of crop is as follows:

Wooded pasture (*and meadow*): pasture (or meadow) where there exist aligned or scattered woody plants, the product of which considerably increases the soil rent. The degree of cover of the tree canopy projection on the soil must be between 5 and 50 %. The tree stand should not be too extensive and the trees, whether they are scattered or in groups, have to be distanced in such a way as to ensure that the grazing surfaces are never completely in the shade but have ample sunshine for several hours a day (Spampani 1910). The surface of the wooded pasture has to be divided into sections (known as parks): enclosed areas of a size proportionate to the productivity of the herbaceous pasture and the number of livestock to be introduced. The enclosures may be of a temporary or permanent character. The temporary enclosures consist of wooden fences or barbed wire between posts. The permanent enclosures can be made up of hedges or trees set along the perimeter of each plot operating as windbreaks and as shelter for the animals. The permanent enclosures are preferable because they keep the pasture divided up into a number of plots in which the animals can graze in turn, to the great benefit of the herbaceous development.

The replanting or formation of wooded pastures has been supported by a number of scholars (Bernardini 1962; Briganti 1914; Carrante 1909; D'Errico 1970–1971; Gori Montanelli 1937, 1939; Perona 1893; Piccioli 1908, 1913; Spampani 1910; Speranzini 1927; Voglino 1923), who underscore the numerous advantages of an environmental and economic nature as compared with bare pastures. The area currently occupied by the wooded pastures in Italy can be extrapolated from the data in the National Inventory of 2005 which makes a distinction between the wood and other "wooded land". The wood is considered as a population of trees and shrubs simultaneously featuring the following characteristics: a surface area of more than 5000 m^2, a level of land cover of the trees and shrubs greater than 10 % and a width of more than 20 m. The "other wooded land" corresponds to areas of a minimum surface area of 5000 m^2, a minimum width of 20 m and a level of tree cover between 5 and

10 %, made up of species capable of reaching a mature height in situ of 5 m, or alternatively formations with a cover of more than 10 % not capable of reaching a mature height in situ of 5 m. Under this new classification, the Italian forest surface amounts to 10,528,000 including 1,708,333 (approximately 17 %) classified as "other wooded land": low woods (2–5 m in height), low-density woods, scrub and shrubs. The "wood" classification comprises not only the new categories, but also the shrub and tree formations that are frequently wooded pastures or shrub pastures. Only the cork wooded pastures of Sardinia have been distinguished from the rest. According to these statistics, therefore, the wooded pastures have practically disappeared; this could lead territorial planners and managers to confirm that the wooded pastures no longer exist and that—in accordance with the current logic of environmental improvement—the existing low-density woods ought to be steered towards the ecological optimum, or that following the abandonment of cultivation practices they will evolve naturally into dense stands. A certain reversal of trend has begun to appear in Italian forestry policy at local government level in the north-east of Italy, where incentives for the maintenance and conservation of these traditional forms of land use were distributed through the Rural Development Programme 2007–2013.

7.3 Thin High Forests as Cultural Model of Pastured Woods

7.3.1 Even-Aged High Forests

The arrangement of the trees as single individuals set at regular or irregular intervals appears to be what, in a strict sense, characterises the wood pastures or wooded pastures at least for Briganti (1914), Carrante (1909), Perona (1893), Spampani (1910). They are high forests that are sufficiently thin to allow the grass to grow, giving rise to a pasture that is more or less abundant but continuous. Briganti (1914) enthusiastically described the undertaking of a private landowner of Altamura (province of Bari) who in 1896 enclosed a vast plot of land of about ten hectares by building a dry stone wall of little more than 1 m in height. Then he dug holes at distances of 6–8–10 m from one another and planted sycamore maples (*Acer pseudoplatanus*), or field elms (*Ulmus campestris*), black locusts (*Robinia pseudoacacia*), Aleppo pines, ash trees and a few oaks. By 1910 the trees were already large and healthy and in the less stony stretches the undergrowth was made up of lush grass which was cut and turned into hay. Again in the Bari Murgia region, Carrante (1909) suggested recreating the wooded pastures through reforestation by planting out 1000–1200 trees per hectare of species such as the *English oak* (*Quercus robur*) which produces acorns that are extremely well used by pigs, and when it is cut provides a highly prized timber, and is perfectly suited for creating excellent wooded pastures in not too shallow soil. The pedunculate oak (*Quercus peduncolata*) also yields acorns that are good for pigs and can adapt to conditions of poorer quality than

Fig. 7.1 Wooded pastures in Salten (province of Bolzano) represent a Roman period landscape (Agnoletti 2010b)

the English oak; the Valonia oak (*Quercus Aegilops*) too, and even more the holm oak (*Quercus Ilex*) in lower regions and the Turkey oak (*Quercus cerris*) in higher parts. The subsequent cultivation operations were to consist of a series of successive thinnings, starting from the fifteenth year of age, which would reduce the number of trees to 600–700 per hectare; in the 75th year a first thinning would have been carried out, halving the number of trees. The remaining stand would have undergone the first regeneration felling at 150 years of age. Other cultivation treatments were instead to be applied below the tree stand. To control the excess of organic substance and the invasion of the nitrophilous shrubs, periodical loosening of the soil was required, and the spreading of lime to neutralise the acidity of the soil and prepare it for the planting of legumes, excellent foraging crops. In thin planting the trees should consist of heliophilous species; in the Alps the larch is used, giving rise to the formation of the so-called thin larch wood pasture, as in the Alpine areas of Piedmont and the Aosta Valley. Among the largest larch wood pastures of Europe we can mention the wooded pastures of Salten which extend over 1000 ha on the Mazzoccolo upland (province of Bolzano) (Fig. 7.1).

The name *Salten* derives from the word *saltus*: regular mowing of the meadows and animal grazing is still carried out here today, ensuring the conservation of the wooded pasture. In view of its sparse crown, its rapid growth and the fact that it bears up well to isolation, the larch is perfectly suited for the creation of this thin high forest (Perona 1893). The number of trees to keep in the pasture ranges from a minimum of 40–50 per hectare, set at intervals of between 8 and 15 m (considering the trees distributed on a square plantation layout), whereas if the distance between the trees is 3 m the tree cover has to be reduced to permit the development of the herbaceous cover, with an initial thinning at ten years of age, leaving a distance between the trees of not less than 5 m (Spampani 1910). For similar reasons the Scots pine and the black pine can be maintained in isolated conditions; in the same way the Swiss stone pine, which in view of its slow growth ought to be used only in pastures at an altitude at the extremes of forest vegetation, while the European spruce reacts badly to the state of isolation and provides too much shade (Perona 1893). Among the deciduous trees, the beech was used for the planting of the pastures, especially in the central and southern Apennine areas. However, considering the ecological and edaphic requirements for the regeneration of the species,[2] it is obvious that the stand has to be sufficiently dense, with a semi-open cover. The release of a few isolated individuals (from between 80 and 120 up to 300 per hectare, depending on age and the development of the crown) has prevented the trees from being regenerated, to the point of creating over time thin or patchy high forests with just a few individuals of colossal dimensions with no possibility of reproducing themselves and destined to disappear. Consequently, the high forest system using beech trees does not appear to be the most suitable one for the wooded pasture; in the past these woods in fact applied the coppicing system, like the wooded pastures of Abruzzo, for example. The pollarding means that it is possible not to interrupt the grazing within the wood, since the coppice shoots grow at a height that makes them immune to the bites of browsing livestock. The pollarding/lopping also provided wood, and leaves for the animals. In the central and southern Apennine areas, the term *difesa* was used to indicate the woods and wooded pastures that belonged to the municipality and over which the local people enjoyed rights of use, including collecting wood, collecting acorns and grazing. The Bosco di Sant'Antonio (Fig. 7.2), extending over 550 ha in the municipality of Pescocostanzo (province of Aquila), represents a well-conserved example of this and is possibly one of the most significant of the Abruzzo *difese*. As the term indicates, these pastures enjoyed particular protection from the local people, and the grazing of 'foreign' or non-local livestock was prohibited. Grazing was indeed restricted to the horses and cattle of the local community, excluding sheep. It was forbidden to cut down the trees at the foot, while pollarding was permitted to obtain firewood and leaf. The constant practice of lopping allowed the reserved plants to reach colossal dimensions: the lopped plant is better able to defend itself against the attacks of pathogens and also has fewer mechanical problems than the old trees, since the coppice shoots are smaller than the trunk (Lonsdale 1996). The circumferences measured at a height of 1.30 m can reach 5.7 m for the beech, and 4.7 m for the sycamore maple and the linden. Cereal crop growing was carried out in the wooded pasture, albeit not continuously. The Bosco di Sant'Antonio was the subject of one of the first ecology battles in Italy, between the municipal authority

[2]The need for moist and porous soil with a thick layer of litter to reduce evaporation at ground level, climatic conditions without sudden changes in temperature, tolerance to shade at the seedling stage which, however, cannot compete with the tall nitrophilous grasses typical of the clearings and the large empty spaces.

Fig. 7.2 Impressive beech pollard in the Bosco di Sant'Antonio. The absence of management practices leads to the progressive disappearance of historically significant elements in the Italian forest landscape (Agnoletti 2010b)

which owned the land and the citizens of Pescocostanzo. The citizens objected to the proposal to perform forest utilizations that would have transformed the wooded pasture into a wood for timber production, thus completely distorting its structure. They submitted the matter to the Member of Parliament Gaetano Salvemini and to the President of the Republic Luigi Einaudi who, through a decree issued in January 1953, recognised the notable public interest of the wood and classified it as an area of natural beauty to be subjected to landscape protection pursuant to the 1939 law. From 1991 it was annexed to the Majella National Park. The discontinuance of lopping and the reduced grazing is jeopardising the original structure of the pasture: the progressive closing of the open spaces and the greater tree canopy protection are favouring the tree component to the detriment of the grass component. From a historic and landscape aspect these trees (Fig. 7.2) have assumed an important role both because these are particularly beautiful and picturesque landscapes in terms of the variety of aspects and the majesty of the individual trees—which are generally gnarled with many forks, with numerous excrescences in correspondence with wounds—and as ancient relics of the pastoral and rural civilisation (Bortolotti 1985).

7.3.2 Uneven-Aged High Forests

The wooded pastures, usually managed on the selection system, were made up of oak-type species. In his work *De agri cultura*, 160 BC Cato the Elder recommended that for the farm to obtain the maximum profit a portion should be set aside for the *glandaria silva*, a high forest pasture traditionally connected with the wild breeding of pigs. Such was the importance of this breeding that in the oak woods of Tuscany, the estimates of value were made not in terms of the surface area of the wood but of the number of pigs that it could feed. In the eleventh century, the oak woods belonging to monasteries in the northern Apennines were valued in the accounts by the number of pigs whose nutriment they could support, with the measurements varying from one to one and a half hectares per animal (Gabbrielli 2006). De Ricci (1830) recorded that the income of the Tuscan oak woods annexed to the farms was linked exclusively to the sale of the pigs fattened on acorns, and that 4000 trees were sufficient to feed 300 pigs. Up to the middle of the twentieth century the value of the timber production of the Tuscan oak woods was equalled or surpassed by that of the grazing (acorns, grass, leaf, branches). This is why the oak high forest has to be considered as a wood pasture rather than as a wood in which the production of timber is predominant (Pavari 1934). In the pre-war period in a Chianti farm the number of pigs introduced into the oak woods from October to May was 300 every hundred hectares. 52 % of the gross sellable production was represented by the products of grazing, and only 48 % was attributed to wood products (Marinelli 1980). The selection system management was very probably applied to the oak species by the owners of these woods for the purpose of distancing the trees and thus increasing the production of acorns (Merendi 1927). The cultivation techniques normally applied also included the cutting back of the undergrowth to facilitate the access of the livestock, and to permit the growth of the pasture and the manual collection of the acorns. Indeed part of the acorn crop was collected by the owners to be stored and fed to the animals in years when the yield of fruit was low. The density of the high forest could vary from 40 to 50 trees per hectare (Oliva 1924; Parente 1931) up to a maximum of 165–170 per hectare when they were of dual production. The selective felling at the foot of the tree usually took the form of the removal of 1–2 trees per hectare every year (Del Noce 1849). The trees selected for felling were usually the older ones that showed greater signs of deterioration and drop in fruit production. When natural regeneration did not take place the problem was solved by sowing or planting out (Iacini 1880). In the unanimous opinion of French foresters, the technique of selective felling at the foot of the tree leads to modest density, a notable irregularity of growth habit, an abundance of epicormic branches, the invasion of auxiliary species and, more often than not, difficulties of regeneration. There are also those who advocate the selected high forest created though group felling, which gives rise to a multilayer stand connected by small surface clearings. In the Adour region of France, patch cutting was carried out in an unusual manner in the pedunculate oak woods, previously managed with a sort of selected felling of the larger oaks by individual tree; the wood, which was constantly traversed by livestock, was thin, with sparse trees. Starting from the eighteenth century, this population was managed using the regular stand system. The frequency of the years of acorns, the good resistance of the seedling production under canopy cover, meaning that there is almost always vigorous early regeneration and the rare recurrence of spring frosts, make it possible to proceed to regeneration in a brutal manner, both through successive fellings within the space of 5 or 6 years, and through the clear felling of small areas. The latter is the system that tends to prevail, yielding good results. To favour the seedling production and development, the plots under regeneration are banned from grazing for several years before felling, and at the time of felling the total coppicing of the pre-existing seedlings, thorny shrubs and briars is performed, which moreover spring up again very rapidly; at times the grazing of the pigs is permitted, under careful surveillance, to ensure the burial of the acorns. At the time of the first thinning (15 years) the local inhabitants come to clear the wood of shrubs, which is then restored to pasture (Perrin 1954). In theory the normal state of a selected high forest pasture, of average fertility and for diametric categories from 20 to 70 cm, is expressed by the following figures: 183 trees per hectare, 13 m^2 of basal area and 121 m^3 of growing stock (De Philippis 1985). It is more convenient to establish the density of the trees based on the degree of cover of tree canopy projection:[3] this parameter is important for regulating the relations that are established between tree stand and pasture. To permit good forage production, the stand ought to have a density such as to permit the tree cover to regulate the microclimate in a significant manner through the so-called parasol effect: the influence of the tree on the lengthening of the period of standing availability of the forage supply (Bellon 1995). For the herbaceous cover the optimal level is between 40 and 70 % of crown canopy depending on the forest species (Heiter and Lilin 1989). In general, a degree of crown density of 60 % is considered the maximum limit; beyond this level major competitive phenomena develop between the trees and between these and the grass layer. The lower threshold is usually considered 10 %; beneath this minimum value of tree cover, the number of trees is so drastically reduced that both their contribution to the nourishment of the livestock and their beneficial microclimatic influence on the pasture become negligible (Champion and Vieillendt 2001; Rodriguez Estevez et al. 2007). It is no simple matter to translate the degree of crown density into a number of trees per hectare; the different dimensions that the crowns can reach has to be assessed on the basis of the diametrical dimensions of the trees, the site conditions, the state of plant health, etc. Table 7.1 shows the modular values of the crown area of a deciduous oak (Di Berenger 1887). To synthetically assess the area of crown of a seedling of turkey oak and downy oak, we can consider that it has a diameter 25 times greater than the diameter of the trunk at 1.30 m. The basal area of a single seedling (A) will prove to be $\pi/4 \times D^2_{\text{crown}} = 500\ D^2_{\text{trunk}}$.

The natural regeneration of the selected high forests is connected with the management of the grazing and the undergrowth vegetation. The number of animals has to be commensurate with the production capacity of the wood (grass pasture, undergrowth and acorns). The seedlings then have to be protected from the bites of browsing animals using the so-called nursery plant until they become resistant to the livestock bites. The importance of the protective plants, above all the thorny shrubs, in protecting the seedlings from drying out in the sun and from the bites of

[3]The degree of crown density (θ) of a wood is defined as the ratio between the sum of the areas ($\sum s$) of the crowns of the trees of a wood complex ($\sum s = \pi/4 \times \sum D_c^2$), for simplification of calculation treated as circumferences (D_c crown diameter), and the surface area of the wood S. In symbols, $\theta = \Sigma s/S$.

Table 7.1 The lower values of basal area correspond to trees grown in denser stands; the higher values refer to trees grown in isolated conditions (Di Berenger 1887)

Normal diameter of the trunk (cm)	Diameter of the crown (m)	Crown area (m^2)
5	2–2.5	3–5
10	2.5–3	5–7
15	3.5–4	10–13
20	4.5–5	16–20
25	5–5.5	20–24
30	5.5–6.5	24–33
35	6–7	28–38
40	7–8	38–50
45	8–9	50–63
50	9–10	63–79
55	10–11	79–95
60	10.5–11.5	37–104
65	11–12	95–113
70	12–13	113–133
75	12.5–14	123–154
80	13.5–15	148–177
85	14–16	154–201
90	15–17	177–227
95	16–18	201–254
100	16.5–19	214–284

animals (both domestic and wild) was already recognised in scientific literature as far back as the early**nineteenth century** for example in Gautieri (1813) and Gori Montanelli (1937). Recent researches have confirmed that the thorny shrubs can act as "safety zones" for the seedlings (Callaway and D'Antonio 1991; Callaway 1992; Callaway and Davis 1998; Rousset and Lepart 1999; Gomez et al. 2001; Kuiters and Slim 2003; Bakker et al. 2004; Smit et al. 2005, 2007). Research carried out in Holland (Bakker et al. 2004) and in Italy (Alias et al. 2010) has demonstrated how decisive the degree of shrub cover is in creating the microenvironmental conditions suitable for the germination of the seeds. The seedlings can escape the bites of sheep after 5–10 years, after 10–15 those of horses and goats, and after 20–30 years the damage caused by cattle. In the case of pigs, once the confirmation of the regeneration is obtained, there is nothing to prevent their grazing (Montoya 1983).

Barròn del Pozo (2006), on the other hand, maintained that the regeneration (both natural and artificial) can be considered confirmed, when coexistence with livestock or wild animals is possible without the need for individual protection. This condition is established depending on the circumference of the trunk measured 1.30 m from the ground:

- >10 cm for sheep or pigs if the pasture is abundant (diameter at 1.30 m > 4 cm).
- >20 cm for sheep, pigs and wild animals (diameter at 1.30 m > 6.4 cm).
- >25 cm for cattle (diameter at 1.30 m > 8 cm).

7.4 Conclusions

The origin of the wooded pastures in Europe dates back to very remote times. Generally, their creation is connected with the work of deforestation and clearing of the ground to permit the survival of the populations. Livestock was the inevitable choice for making available for human nourishment the scarce natural resources available in the territory. The deforestation was carried out in the forest stands which had a function linked only to the production of wood, such as the conifer woods of the Jura and French Savoy. In view of the dual production of wood and acorns—fundamental to the nourishment of pigs—the stands consisting of oak species were maintained as pastures. The tree component present in the pasture surface plays a very important role in forage production, and it is for this reason that the density and distribution of the trees have to adhere to very precise criteria established based on the environmental conditions and the types of tree present. The arrangement by single individuals is appropriate for heliophilous species with light crowns, such as larches and oaks, since their canopies do not overshadow the growth of the herbaceous vegetation. The larch is well-suited for use in the thin high forest pastures of the mountains, since it has a sparse crown, rapid growth and stands up well to isolation. The thin high forest of oaks, however, is an outstanding wooded pasture, since the production connected with the fattening of the pigs with acorns is greater than the income yielded by the production of timber. The treatment of the oak wood pasture is essentially based on selected felling. The parameters on which the selected felling is based are the absence of a real exploitable diameter (the tree is cut down when it shows symptoms of deterioration or drop in fruit production). The normal diametric distribution is established according to the degree of cover of tree canopy projection and not on a pre-established standard of wood provision. The management practices in the stand and the control of the undergrowth vegetation are indispensable to guarantee the production of forage and, at the same time, the regeneration of the tree component. The regeneration method consists essentially of the restriction on grazing and the recruitment of seedlings and coppice shoots to maintain the desired density. The density, expressed as the degree of cover of tree canopy projection, is the most important parameter to establish and maintain in order to guarantee the persistence of the wooded pasture and to define the structure of the selected high forest. The lower threshold is taken as a degree of tree canopy projection of 10 %, beneath which the ecological, environmental and productive role of the tree stand becomes negligible. Above 60 % of canopy projection there emerge phenomena of competition for nutritional elements, light, water and root space between the trees and pasture beneath.

References

Agnoletti M (2010a) Paesaggio rurale. Strumenti per la pianificazione strategica. Edagricole, Milano

Agnoletti M (ed) (2010b) Paesaggi rurali storici. Per un catalogo nazionale. Editori Laterza, Bari

Alias S, Bianchi L, Calamini G, Gregori E, Sioni S (2010) Shrub facilitation of Quercus ilex and Quercus pubescens regeneration in a wooded pastures in central Sardinia (Italy). Forest 3:16–22

Bakker ES, Olff H, Vandenberghe C, De Mayer K, Smit R, Gleichman JM, Vera FWM (2004) Ecological anachronism in the recruitment of temperate light-demanding tree species in wooded pastures. J Appl Ecol 41:571–582

Barròn del Pozo JL (2006) Ordenacion del arbolado de las dehesas. El tramo en densificacion con repeticiones. Aplicaciòn a una dehesa fosil de encinas. In: Gestion ambiental y economica del ecosistema Montado/Dehesa en la Penisula iberica. Junta de Extremadura Consejeria de Agricoltura y Medio Ambiente, pp. 145–152

Bernardini F (1962) Nozioni di selvicoltura e apicoltura. Edagricole, Bologna

Bellon S (1995) Stratègies sylvopastorales en règion mediterrannèenne. Cahiers Option Mediterrannèenne 12:195–198

Bölöni J, Molnar Z, Birò M, Horvath F (2008) Distribution of the (semi-) natural habitats in Hungary II. Woodlands and shrublands. Acta Bot Hung 50:107–148

Bortolotti L (1985) Problemi di conservazione dei pascoli arborati («difese») d'Abruzzo. L'Italia Forestale e Montana 45:65–72

Briganti G (1914) Rapporti tra boschi e pascoli nel Mezzogiorno (3° Tema). In: I demani comunali e gli usi civici in rapporto con la legislazione forestale. 3° Congresso Forestale Italiano e 1° Congresso per l'irrigazione. Napoli maggio-giugno 1914. Edizioni Della Torre, Portici, pp. 73–89

Bürgi M, Gimmi U (2007) Using oral history and forest management plants to reconstruct traditional non timber forest uses in the swiss Rhone Valley (Valais) since the late nineteenth century. Environ Hist 13:211–246

Bürgi M, Stuber M (2001) Agrarische Walnutzungen in der Schweiz 1800–1950. Waldweidw, Waldheu, Nadel und Laubfutter. Schweiz Z Forstwesen 152:490–508

Callaway RM (1992) Effect of shrubs on recruitment of Quercus douglasii and Quercus lobata in California. Ecology 73:2118–2128

Callaway RM, D'Antonio CM (1991) Shrub facilitation of coast live oak establishment in central California. Madrono 38:158–169

Callaway RM, Davis FW (1998) Recruitment of Quercus agrifolia in central California: the importance of shrub-dominated patches. J Veg Sci 9:647–656

Carrante A (1909) Il miglioramento dei pascoli murgiosi. La propaganda agricola: bollettino quindicinale della Cattedra ambulante di agricoltura della provincia di Bari 12:182–185

Cevasco R (2007) Memoria verde. Nuovi spazi per la geografia. DIABASIS, Perugia

Cevasco R, Moreno D (2010) Rural landscapes: the historical roots of biodiversity. In: Agnoletti M (ed) Historical rural landscapes. Editori Laterza, Bari, pp 121–131

Champion J, Vieilledent G (2001) Selvicultura mediterranea. La Encina Q.Ilex. ETSIAM, Universidad de Cordoba

Clement V (2008) Spanish woodpasture: origin and durability of an historical wooded landscape in Mediterranean Europe. Environ Hist 14:67–87

Del Noce G (1849) Trattato istorico, scientifico ed economico delle macchie e delle foreste del Granducato di toscana con un nuovo regolamento delle acque fluenti per i fianchi scoscesi delle montagne appennine e delle centrali. G. Ducci, Firenze

D'Errico P (1971) Alpicoltura 1 e agricoltura montana: anno accademico 1970–1971. Facoltà di Agraria, Bari

Dennis P, Gordon IJ, Young MR (1998) Distribution and abundance of small insects and arachnids in relation to structural heterogeneity of grazed, indigenous grassland. Ecol Entomol 23:253–264

De Philippis A (1985) Lezioni di selvicoltura speciale. CUSL, Firenze
De Ricci L (1830) Delle scarse rendite dei querceti ghiandiferi e dei metodi per aumentarli. G Agrario Toscano 4:320–326
Di Berenger A (1887) Selvicoltura: trattato scritto per uso degli agenti forestali, ingegneri e possidenti di boschi. R. Marghieri, Napoli
Ernst C (1998) An ecological revolution? The Schlagwaldwirtschaft in western Germany in the 18th and 19th centuries. In: Watkins C (ed) European woods and forests. CAB International, pp. 83–92
Farrel EP, Führer E, Ryan D, Andersson F, Hüttl R, Piussi P (2000) European forest ecosystems: building the future on the legacy of the past. For Ecol Manage 132:5–20
Farrel EP, Piussi P (2000) Interactions between society and forest ecosystems: challenges for the near future. For Ecol Manage 132:21–28
Gabbrielli A (2006) Le vicende storiche e demografiche italiane come causa dei cambiamenti del paesaggio forestale. Anal Accad Ital Sci Forestali 55:133–166
Gautieri G (1813) Quando e come abbiasi a permettere il pascolo nei boschi si resinosi che da fronda, si d'alto fusto che cedui. G. Silvestri, Milano
Gomez JM, Hodar JA, Zamora R, Castro J, Garcia D (2001) Ungulate damage on Scots pines in mediterranean environments: effects of association with shrub. Can J Bot 79:739–746
Gori Montanelli L (1937) Contributo alla conoscenza delle condizioni silvo pastorali dell'alto Appennino. Tip. P. Galeati, Imola
Gori Montanelli L (1939) Selvicoltura e alpicultura. Lega, Faenza 1939
Haraszthy L, Markus F, Bank L (1997) A fäs legelok termeszetvedelme (Nature conservation of woodpastures). WWF füzetek 12:1–23
Hètier JP, Lilin C (1989) Elevage et espaces boisès mediterranèens. Forêt mediterranèenns XI:187–202
Humphrey JA, Gill R, Claridge J (1998) Grazing as management tool in European forest ecosystem. Forestry commission, Edinburgh
Iacini S (1880) Atti della Giunta per la inchiesta agraria e sulle condizioni della classe agricola. In: Fascicolo I. Forzani e C. Tip, vol III. Del Senato, Roma
Kuiters AT, Slim PA (2003) Tree colonisation of abandoned arable land after 27 years of horsegrazing the role of bramble as a facilitator of oak wood regeneration. For Ecol Manage 181:239–251
Londsale D (1996) Pollarding success or failure; some principles to consider. In: Read H (ed) Pollard and veteran tree management II. Corporation of London, pp. 100–104
Manning AD, Fisher J, Lindenmayer DB (2006) Scattered trees are keystone structures. Implications for conservation. Biol Conserv 132:311–321
Marinelli A (1980) Aspetti agro-forestali e zootecnici del pascolamento in bosco. Atti del Convegno Monticiano 18 novembre 1979. Associazione provinciale Dottori in Scienze Agrarie e Forestali, Firenze, pp. 57–60
Merendi A (1927) I querceti del Chianti. L'Alpe 5:108–113
Montoya JM (1983) Mètodo para la ordenaciòn silvopastoral. Forêt mediterranèenne 5:73–82
Montoya JM (1989) Encina y encinares. Ediciones Mundi-Prensa, Madrid
Olff H, Ritchie ME (1998) Effects of herbivores on grassland plant diversity. Trends Ecol Evol 13:261–265
Oliva A (1924) L'economia del bosco pascolivo dell'Italia centrale. L'Alpe 11:103–108
Parente E (1931) La ricostituzione delle fustaie degradate di rovere. L'Alpe 18:357–379
Pavari A (1934) Il problema dei querceti in Toscana, vol XXXI. Atti Regia Accademia dei Georgofili Firenze quinta serie, pp. 3–31
Perona V (1893) Economia forestale: trattamento dei boschi. Stab. Tip. Della Casa Edit. Dott. F. Vallardi, Milano
Perrin H (1954) Sylviculture (Tome II). Les traitement des forêts. Theorie et pratique des techniques sylvicoles. Ecole Nationale des Eaux et Forêts, Nancy
Piccioli L (1908) Boschi e pascoli. Unione Tipografica Editrice, Torino
Piccioli L (1913) Note di apicoltura. Tip. M. Ricci, Firenze

Pott R, Hüppe J (1991) Die Hudelandschaften Nordwestdeutschlands. Westfälisches Museum für Naturkunde, Münster

Rackham O, Grove ATG (2003) The nature of Mediterranean Europe: an ecological history. Yale University Press

Rodriguez Estevèz V, Garcia A, Perea J, Mata C, Gomez AG (2007) Revision bibliografica. Production de bellotta en la dehesa: factores influents. Archivios de zootecnia 56:25–43

Rousset O, Lepart J (1999) Shrub facilitation of Quercus humilis regeneration on calcareous grassland. J Veg Sci 10:493–502

Rozas V (2004) A dendroecological reconstruction of age structure and past management in a old growth pollarded parkland in northern Spain. For Ecol Manage 195:205–219

Salata D, Penksza K, Malatinszky A, Kenez A, Szentes A (2011) Woodpastures of Hungary, a disappearing silvopastoral system. http://grassland-organicfarming.uni-kiel.de. Accessed 4 Feb 2015

Sanderson NA (1996) The role of grazing in the ecology of lowland pasture woodlands with special reference to the New Forest. In: Helen R (ed) Pollard and veteran tree management II. Corporation of London, Londra

Smit C, Beguin D, Buttler A, Müller-Schärer H (2005) Safe sites for tree regeneration in wooded pastures: a case of associational resistance? J Veg Sci 16:209–214

Smit C, Den Ouden JAN, Muller Scharer H (2006) Unpalatabale plants facilitate tree sapling survival in wooded pastures. J Appl Ecol 43:305–312

Smit C, Vandenberghe C, Den Ouden J, Müller-Schärer H (2007) Nurse plants, tree saplings and grazing pressure: changes in facilitation along a biotic environmental gradient. Oecologia 152:265–273

Spampani G (1910) Coltura montana con speciale riguardo all'apicoltura. U. Hoepli, Milano

Speranzini F (1927) Lezioni di elementi di botanica, silvicoltura generale, silvicoltura speciale, geologia e mineralogia, alpicoltura, zoologia, entomologia, patologia, topografia, estimo, assestamento forestale: raccolta del dott. F. Speranzini (milizia nazionale forestale, scuola allievi militi, città ducale). Tip. F. e M. F.lli Faraoni, Rieti

Tubbs CR (1986) The new forest. Collins, London

Van Uytvanck J, Maes D, Vandenhaute D, Hoffmann M (2008) Restoration of woodpasture on former agricultural land: the importance of safe sites and time gaps before grazing for tree seedling. Biol Conserv 141:78–88

Van Wieren SE (1998) Effect of large herbivores upon the animal community. In: Wallis De Vries MF (ed) Grazing and conservation management. Kluwer academic publishers, Dordrecht, pp. 168–214

Vera F (2007) The woodpasture theory and the deer park: the grove-the origin of the deer park. In: Rotherham ID (ed) The history, ecology and archeology of medieval parks and parkland. Landscape archeology and ecology, vol 6. Sheffield Hallam University, UK, pp. 107–110

Voglino E (1923) Boschi e pascoli alpestri. Ottavi, Casale Monferrato

[illegible]

Chapter 8
The Making of Olive Landscapes in the South of Spain. A History of Continuous Expansion and Intensification

Juan Infante-Amate, Inmaculada Villa, Eduardo Aguilera, Eva Torremocha, Gloria Guzmán, Antonio Cid and Manuel González de Molina

Abstract The objective of this work is to make an additional contribution that reveals new evidence about the nature of Mediterranean landscapes, their historical construction and the consequences of their transformation process. For this, we want to study the case of what is perhaps the most representative crop of the region: olive groves. The study area is the south of Spain which, for more than a century, has had the main concentration of olive trees in the Mediterranean and currently has the highest concentration of cultivated trees in the continent, with a continuous wood of more than 200 million trees. In the study period, between 1750 and 2010, we will outline the change in the social function of the crop, the management applied, the resulting landscapes and the socio-ecological consequences of the change. As we will see, historically, olive groves have not presented a constant image, but have changed from being a widespread crop, similar to exploitation

J. Infante-Amate · I. Villa · E. Aguilera · E. Torremocha · G. Guzmán · A. Cid · M. González de Molina (✉)
Agro-ecosystems History Laboratory, University Pablo de Olavide, Seville, Spain
e-mail: mgonnav@upo.es

J. Infante-Amate
e-mail: jinfama@upo.es

I. Villa
e-mail: ivilgil@upo.es

E. Aguilera
e-mail: emagufer@upo.es

E. Torremocha
e-mail: etorbou@upo.es

G. Guzmán
e-mail: giguzcas@upo.es

A. Cid
e-mail: ajcid@upo.es

M. Agnoletti and F. Emanueli (eds.), *Biocultural Diversity in Europe*,
Environmental History 5, DOI 10.1007/978-3-319-26315-1_8

systems such as dehesas and montados, to the industrial monoculture that they are today. The reconstruction of the geography of their expansion and the change in the morphology of olive grove landscapes and its causes, enable us to rebuild one of the most representative fragments of traditional Mediterranean landscapes, the olive groves of the south of Spain, and, with it, to participate in some of the debates that still persist about this matter.

Keywords Olive · Olive history · Olive landscapes · Mediterranean landscapes · Environmental history · Biocultural diversity

8.1 Introduction

For several millennia the Mediterranean basin has been home to different civilizations which have populated it continuously. Consequently, its landscapes are the paradigm of what is understood today as cultural landscapes: lands resulting from human intervention in natural landscapes in order to satisfy their needs (Sauer 1925; Antrop 2006). Since the first civilizations and until industrialised farming, their morphology has been changing for centuries in a process that has been studied at length by many authors (Lepart and Debusche 1992; Grove and Rackham 2003; Blondel 2006; Agnoletti 2007) and that has generated interesting debates. For some researchers it uncovers a history of failure, understanding that the human impact has led to increasingly impoverished landscapes (Attenborough 1987; McNeill 2002), continuing a secular perspective that is today known as the "Ruined Landscapes" theory. Other authors point out that the Mediterranean landscape is an example of historical resilience, having maintained its productive capacity through many centuries of human action (Grove and Rackham 2003; Blondel 2006). In any case, there is room for little doubt when critically highlighting the recent processes of abandoning the rural environment (Agnoletti 2014; Marull et al. 2014) or the industrial intensification of its farming (González de Molina and Guzmán 2006; Parcerisas et al. 2012) as evidence of its increasing unsustainability, as well as the loss of its secular cultural inheritance (Petanidou et al. 2008).

In any case, the account of the history of traditional Mediterranean landscapes, which conditions our interpretations of same and the proposals for their future organisation, was written on the basis of research that has reconstructed fragments of their evolution in different places in the region and in specific periods. That is, on the basis of local and regional studies which, together, help us to improve our understanding of their history. In this regard, the objective of this work is to make an additional contribution that reveals new evidence about the nature of Mediterranean landscapes, their historical construction and the consequences of their transformation process. For this, we want to study the case of what is perhaps the most representative crop of the region: olive groves (Grigg 2001). The study area is the south of Spain which, for more than a century, has had the main concentration of olive trees in the Mediterranean and currently has the highest concentration of cultivated trees in the continent, with a continuous wood of more than 200 million trees (Infante-Amate 2014).

Although a lot has been written about the history of olive groves in this region (Zambrana 1987; Guzmán 2004; Infante-Amate 2014), this work presents some unpublished contributions concerning the reconstruction of the landscapes. To begin, in the following section, the surface area of the olive grove at a municipal level is reconstructed in four historical moments, between 1750 and 2010; information unknown to date which stems from a wider research project in which we have documented the surface area at a local level, taken from various historical sources[1]. However, in addition to documenting the expansive process of the olive grove landscapes and their geography, this text seeks to examine their morphological change and their ecological dynamics. We understand that landscape is the result of humans' alteration of ecosystems in their constant struggle to satisfy their needs. Throughout history, agroecosystems have changed their productive function in the same way that man's relationship with olive groves has. This process has given rise to different landscape formulas which, on the contrary, reveal different levels of sustainability and resilience. In the study period, between 1750 and 2010, we will outline these aspects for the case of Andalusian olive groves: the change in the social function of the crop, the management applied, the resulting landscapes and the socio-ecological consequences of the change.

As we will see, historically, olive groves have not presented a constant image, but have changed from being a widespread crop, similar to exploitation systems such as *dehesas* and *montados*, to the industrial monoculture that they are today. The reconstruction of the geography of their expansion and the change in the morphology of olive grove landscapes and its causes, enable us to rebuild one of the most representative fragments of traditional Mediterranean landscapes, the olive groves of the south of Spain, and, with it, to participate in some of the debates that still persist about this matter.

8.2 Andalusia and the Olive Expansion

The cultivation of olives has been present in all Mediterranean civilizations. To our knowledge, it began some five thousand years ago (Zohary and Hopf 1994) as a result of the domestication of the wild variety of olives. However, there is no consensus about the geography of its first expansion (Besnard et al. 2013). There is abundant evidence that it played a fundamental role in the economy and culture of the region in its most bygone societies (Loumou and Giourga 2003; Kaniewski et al. 2012). In fact,

[1]The main challenge was the reconstruction of 1750, which required consulting the files of each Andalusian town in the *Catastro de Ensenada*, information that is generally dispersed and difficult to access. For the other years we availed of better information but it was not always complete. We consulted provincial files or performed estimations for the reconstruction (Infante-Amate 2014).

the importance of its trade in the old world is well known, when one of the main routes was that which took oil from the south of Spain to Rome (Remesal 1998). However, we only have aggregated data of its production and surface area, that is the real dimension of its presence in Mediterranean landscapes, for very recent periods. In the first decade of the twentieth century, for which we have the first estimations of the global production, only three countries, Spain, Italy and Greece dominated more than 80 % of the global production (Zambrana 1984). Today still, 98 % of olive oil production is concentrated in the Mediterranean basin. Although some countries in South America, as well as regions such as California, Australia and South Africa have developed the cultivation of olives since the colonial era, their dimension in global terms remains insignificant. Olives were, and continue to be, a Mediterranean affair (Grigg 2001).

There is evidence suggesting that throughout the nineteenth century Italy was the main oil producer but also, that from the beginning of the twentieth century, Spain exceeded it considerably (Fig. 8.1a). Italian oil had entered a period of crisis, to a large extent due to the competition of Spanish oil which was experiencing what was known as its "Golden Age", a period of expansion referring to the first three decades of the twentieth century in which the surface area of olive groves and the

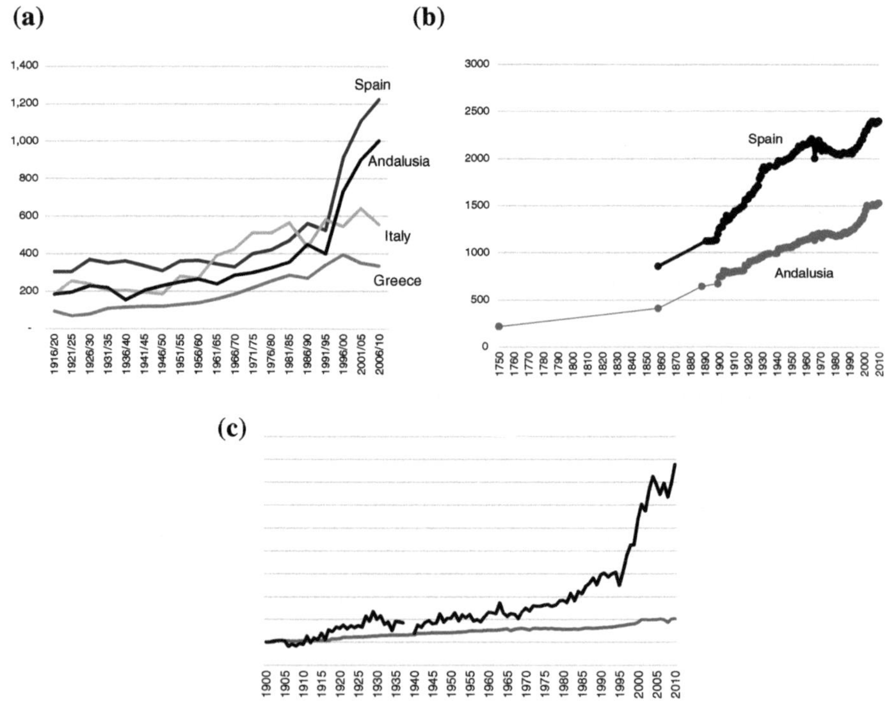

Fig. 8.1 **a** Olive oil production in thousands of metric tonnes. **b** Olive grove land in thousands of hectares. **c** Olive oil production and olive grove land in Spain (Index, 1900 = 1). *Source* Infante-Amate (2014)

production and exportation of oil multiplied by 1.61, 1.91 and 2.19, respectively (Zambrana 1987; Ramón-Muñoz 2007) (Fig. 8.1b). Spain dominated between 25 and 50 % of the global production of olive oil throughout the twentieth century. Without its progress, therefore, it would be difficult to explain the great olive oil expansion worldwide.

In Spain, the great growth of olive groves has mainly focused on one region: Andalusia, which comprises eight provinces in the south of the country (Fig. 8.2.). The history of Andalusian olive groves, which takes up the following pages of this work, has two particular features which make it an attractive context to study olive grove landscapes: as we have already stated, it has the highest concentration of olive trees in the Mediterranean and, secondly, it has a constant history of expansion that has not happened in other areas. In Italy, Greece, and the rest of Spain, olive groves have experienced times of considerable growth but also of decline (Infante-Amate 2014). In the case of Spain, other than the southern area of the country, the other main olive-producing focus was concentrated in the eastern provinces: Catalonia, Valencia and the Balearic Islands. They have all lost surface area in the last half century: the Balearic Islands, since the nineteenth century, Valencia, since 1930, and Catalonia, particularly, in the second half of the twentieth century. In 1858 the east had more than 25 % of the country's surface area of olive groves; today it does not reach 10 %.

Bearing in mind that the surface area of olive groves has decreased in these areas, how can the increase in Spanish olive groves, as shown in Fig. 8.1b, be explained? The key lies in the advance of the crop in the south of the country, especially in the region of Andalusia which, in the last century and a half has absorbed 70 % of new olive grove plantations in a continuous upward trend (Fig. 8.1b), a fact that informs us straight away of the first change in the agricultural landscapes of the region: an increasingly prominent presence of olive cultivation. In the mid-eighteenth century some 225,000 ha of olive groves were growing in Andalusia. Throughout the nineteenth century the crop continued to grow until it reached, in 1888, 643,000 ha, 7.37 % of the total surface area of the region and 56 % of the national olive-producing surface area. The effects of the fall of the Old Regime are often referred to, as it removed the obstacles to its expansion from the beginning of the nineteenth century, as well as the increasing demand for oil from other countries, as a result of industrial development and the emerging global economy(Zambrana 1987; Guzmán 2004; Garrido 2005). From then on, when olive groves were starting out in other parts of the country, they continued to grow in Andalusia until reaching, today, 1.5 million hectares: 17.37 % of the total surface area, 62 % of the Spanish surface area and 15 % of the global surface area (Table 8.1).

In this region, however, olive groves have historically been concentrated in three provinces: Jaén, Córdoba and Seville, which in the eighteenth century already dominated 80 % of the surface area of Andalusian olive groves and had more than 10 % of the total planted surface area of olives (Table 8.1). Its expansion, as well as that of other provinces such as Granada and Málaga, explains the tremendous dimension of olive groves. Jaén is today the quintessential olive-producing

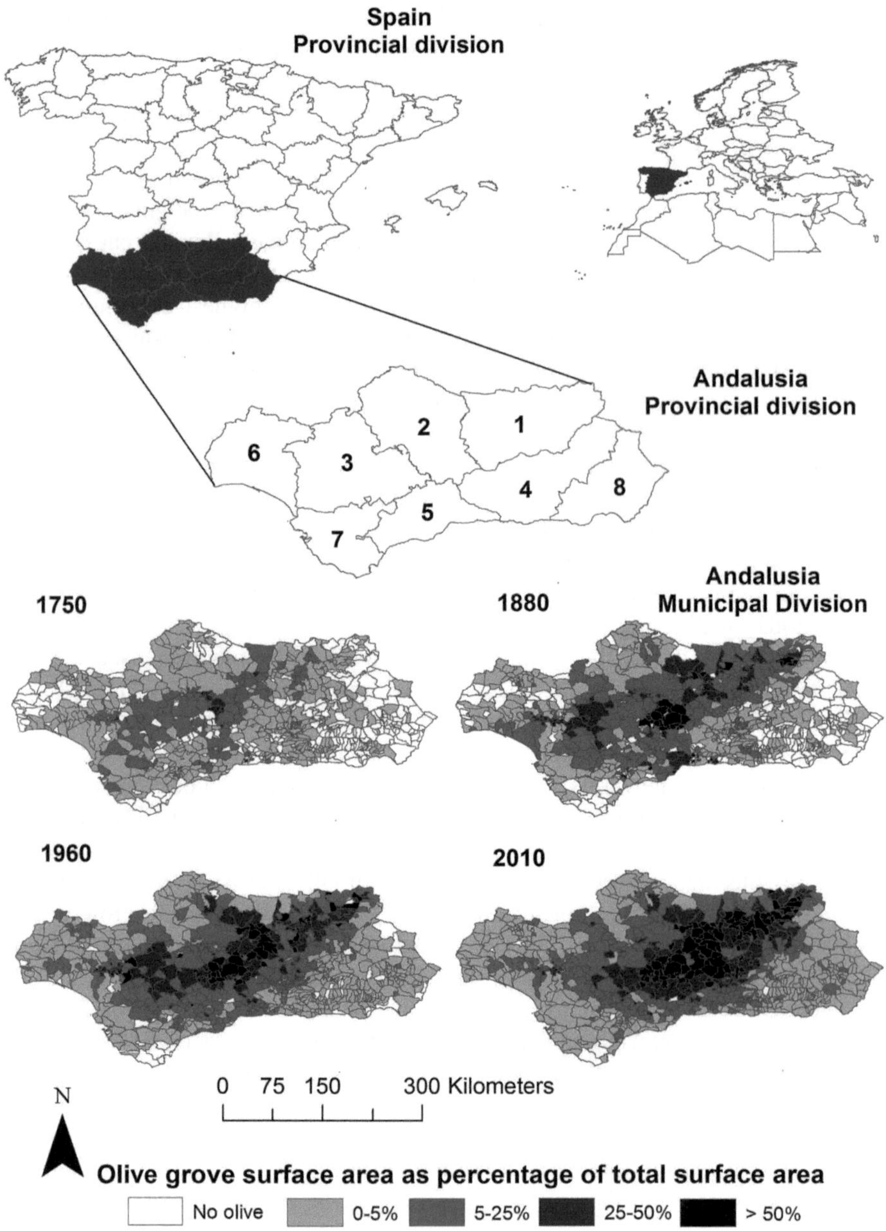

Fig. 8.2 Geography of olive expansion in Spain. *Source* See text

province. With some 570,000 ha taking up more than 42 % of the total surface area and 80 % of the planted surface area, it has become the leading administrative unit in olive groves. No other region in the world has such a prominent monoculture of olive groves.

Table 8.1 Olive surface area

Map code	Hectares					% of the total surface area	
	1750	1888	1930	1960	2010	1888	2010
Jaén	43,691	193,143	300,350	378,130	570,965	14.31	42.31
Córdoba	63,073	191,045	240,800	270,560	343,825	14.1	25.37
Seville	78,525	169,263	218,710	300,800	221,761	12.05	15.79
Granada	4844	12,838	53,305	78,020	178,534	1.02	14.25
Málaga	17,925	42,478	86,203	112,620	125,729	5.81	17.2
Huelva	4,427	19,675	26,848	31,140	33,692	1.94	3.32
Cádiz	12,603	12,577	20,293	32,470	22,850	1.69	3.07
Almería	724	1955	5960	8245	18,987	0.22	2.16
Andalusia	225,813	642,974	952,469	1,211,985	1,516,343	7.37	17.37
Spain		**1,138,851**	**1,882,289**	**2,316,425**	**2,449,828**	**2.11**	**4.53**

Source See text

Continuing with the geography of the expansion of olive groves in Andalusia, we can go into further detail if we analyse the process at a municipal level. In Fig. 8.2 their surface area is reconstructed in Andalusian municipalities between 1750 and 2010. In the mid-eighteenth century, 40 % of the almost 800 municipalities in Andalusia did not have olive groves. Today, they are present in 98 % of the municipalities. The second element explaining the expansion of olive groves is that the farms that already had olive groves now have a much greater quantity. For example, in 1750, only two municipalities had more than 10,000 ha of olive groves (Écija and Andújar, due to the wide extension of their municipal areas). Today, more than 30 municipalities have more than 10,000 ha. Those alone make up nearly a half million hectares.

As can be seen in Fig. 8.2, historically, olive groves have tended to be concentrated in the central part of the region, along the valley of the Guadalquivir River. We can observe that these same areas in which olives groves were already present in the eighteenth century have continued to abound in the olive oil specialisation. However, greater growth is observed in the southern part of this expansive area, corresponding to various regions located in the Subbaetic mountain range as well as in the northeast part, throughout the province of Jaén which, as we have pointed out, is the administrative unit with the highest concentration of olive groves and in which the lowland area is smaller than in regions on the west. In other words, it seems credible that olive groves have tended to increasingly occupy mountain areas. In this regard, the information generated for this work enables us to analyse the orography of the expansion of olive groves, bearing in mind that we have data on the surface area and on the average incline of each municipality. Performing a simple exercise, we differentiate between three groups: olive groves growing in lands with an incline of less than 7 %, which we could consider lowlands; those growing in lands with average inclines of between 7 and 15 %, mid-mountain; and, lastly, lands in which the incline is higher than 15 %.

Olive groves grew between 1750 and the present day in the three groups. That is, today there are much more olive groves both in lowland and mountain areas. However, we can observe that in relative terms historically olive groves used to occupy a greater place in lowland areas. In 1750, 76 % of olive groves grew in the lowland. However, today only 54 % of olive groves grow there, or, in other words, growth has been greater in mountain areas. This is due to two factors. First, in the mid-eighteenth century, olive groves already occupied a considerable surface area in the lowland and residual space in mountain areas, in such a way that possibilities of growth in mid-mountain areas were greater. Second, Seville is one of the few areas in which olive groves have declined. Throughout the nineteenth century, as a result of the loss of American trade and following the 1960 recession, olive groves in Seville declined. In other words, the province with the most lowland olive groves has seen the number of hectares dedicated to this crop fall.

In short, one of the main facts of the olive-producing history of the south of the country is its constant expansion, both in mountain and lowland areas. Today, more municipalities than ever have olive groves and those that have olive groves have a much greater surface area than before. This is the first element of the change of the agricultural landscapes of the south of the country: the monopoly exercised by olive groves in the landscapes of the majority of the region. However, beyond the mere analysis of the uses of the ground, this work seeks to document the morphological change in the landscapes. Figure 8.1b reveals the tremendous increase in the surface area of olive groves that shows the increasing presence in the landscapes of the south of the country. However, in Fig. 8.1c, we can observe that in the second half of the twentieth century, production increased at a much greater speed than the surface area, revealing an intensification process that decisively altered the region's landscapes. This transformation process began, however, several centuries previously.

8.3 Historical Changes in Management and Productive Functionality, and the Driving Forces at Play

If by agroecosystem we understand an ecosystem altered by human action to satisfy their needs (Gliessman 1990), cultural agricultural landscape can be understood as the visible part resulting from the transformation of said agroecosystems. They have fulfilled different functions throughout their history. In the preindustrial era they had to provide a considerable range of goods and services that disappeared in the industrial societies (González de Molina and Toledo 2014). On the contrary, humans' ability to transform agroecosystems has also changed throughout history. Industrialisation, for example permitted a greater degree of alteration of these (Pelletier et al. 2011). In any case, a cultural agricultural landscape changes as a result of human action—the management applied—which in turn varies according to different factors such as the technology available or the specific needs.

Agricultural and environmental history and historical and rural geography, among other disciplines, have documented the transformation processes of agricultural landscapes and the factors determining same, using examples from all over the world, as well as how humans have intervened to bring about such changes. The case of olive groves in the south of Spain is no exception and many pages have been written about its history, helping us to document the process of the factors determining their expansion or decline, their uses, their changing functionality and resulting landscapes. A brief account that helps us to understand their rise could be the following:

The climate of the region is completely adapted for the development of olive groves. Although we have evidence of their importance in antiquity—the massive exportation of olive oil from Hispania Baetica, the south of Spain, to Rome, as shown by the amphora found on Mount Testaccio (Remesal 1998)— what is certain is that it is difficult to reconstruct its productions, surface areas or management up to 1750, when a land registry was published in the Crown of Castile. It calculated, among other things, the agricultural surface areas at a municipal level, and their production, for the whole country. The evidence obtained in other works that have used this source is that olive groves appeared as a multifunctional exploitation, with low capitalisation and no involvement in large trade dynamics. In some way it is a comprehensible phenomenon. One of the characteristics of preindustrial societies is their almost complete dependence on agricultural production to satisfy the majority of human needs (Fischer-Kowalski and Haberl 2007; González de Molina and Toledo 2014). In addition to human food, the cultivation areas provided fodder, fuel, raw material… Andalusia, in the mid-eighteenth century, was empty; it was one of the regions with the lowest population density in the country with only 20 hab./km^2. This, as well as the institutional hurdles of the Old Regime that restricted the free production or trade of many products—among them, olive groves—explains its weak expansion and its extensive nature. Olive groves received very little work and their exploitation was integrated into other crops or livestock.

In subsequent periods, the statistical information is much more abundant and has helped to reconstruct the variables relating to the sector, as well as the main changes. A process of strong expansion can be observed throughout the nineteenth century; many authors have linked this to the fall of the Old Regime and its institutional hurdles (Zambrana 1987; Guzmán 2004; Garrido 2005). From this date onwards the land and labour market became free and some monopolies on the processing and trade of fruit were abolished. In other words, there were no environmental constraints on its development and, from the mid-nineteenth century, nor were there any institutional constraints. These facts, however, did not encourage the rise of the olive-growing sector; they simply inform us that there were no barriers to its expansion. The literature reveals two main nonexclusive theories which help us to understand the rise in the olive-growing sector in the nineteenth century.

The first and most widespread theory points to the growing demand for olive oil from other lands and also from the domestic market (Ramón-Muñoz 2000, 2007; Garrido 2005; Zambrana 2006; Mataix and Barbancho 2008). This occurred in the context of the industrial revolution and the consequent transport revolution which, together with the liberalisation of world trade, gave rise to what was known as the first wave of globalisation. Railways, steamboats and other types of transport connected world trade in an unprecedented manner. Olive groves were part of this inertia.

A second more recent theory relates the rise of olive groves from perspectives of Environmental History or Rural Studies. In the nineteenth century, Andalusia continued to be a preindustrial region and as a result, was dependent on agricultural exploitations to fully meet needs. However, there was one particular characteristic: during those years, it began to be increasingly populated. The availability of land was constantly decreasing and was conditioned, in turn, by a very low primary productivity. In this regard, the promotion of olive groves has been pointed out as a land-saving strategy via which peasants could stock up on oils for consumption, lighting, fuel from wood and pomace as well as food for the livestock, in a much more efficient way in terms of land cost (Infante-Amate 2012a, b, 2015). Thanks to this strategy, the pressure on pasture and mountain lands caused by the expansion of agricultural crops eased.

In relation to this last point, in Fig. 8.3a we portray the complex productive system of an olive grove agroecosystem in a preindustrial economy. Olive groves had extended so extremely in the region that in many areas they were the only purveyor of fuel and oils. As has been documented in other places, small landowners were responsible for the majority of this expansive process. It has been explained that olive groves produced a top range of products, and required little work (Infante-Amate 2013). In fact, the model of olive grove found in the region during the eighteenth and nineteenth centuries had a very low fruit production but nevertheless provided very high productions of other types of biomass (wood, leaves, branches…).

Thus, olive groves from the nineteenth century became less extensive and the production of fruit tended to increase. The theories of greater commercial demand and of greater ecological intensification to guarantee rural sustenance explain that olive groves would receive increasingly more work, would increase their plantation density and, with this, would multiply the production of fruit.

In Table 8.2 we reconstruct the work applied in 1750 and 1900 and the derivative products for two municipalities that represent the Andalusian olive-growing expansion. In Baena, a lowland municipality that specialised early, olive groves received 17.5 days' wages/day and 6.3 *obradas*[2]/day which increased to 37.6 and 15.2 respectively. In Montefrío, a mountain location, with less olive grove surface area and, in addition, a much more extensive nature, the trend was the same although less work was invested than that in the lowland.

A process of increasing domestication of the crop was taking place, mediated by the reasons outlined above and leading to an increasingly more intensive use which explains the increase in the production of the fruit: in Baena it went from 328.9 kg/ha of dry material (d.m.) to 727.7 kg/ha d.m. and, in Montefrío, from

[2]An *obrada* is the work of a couple of animals.

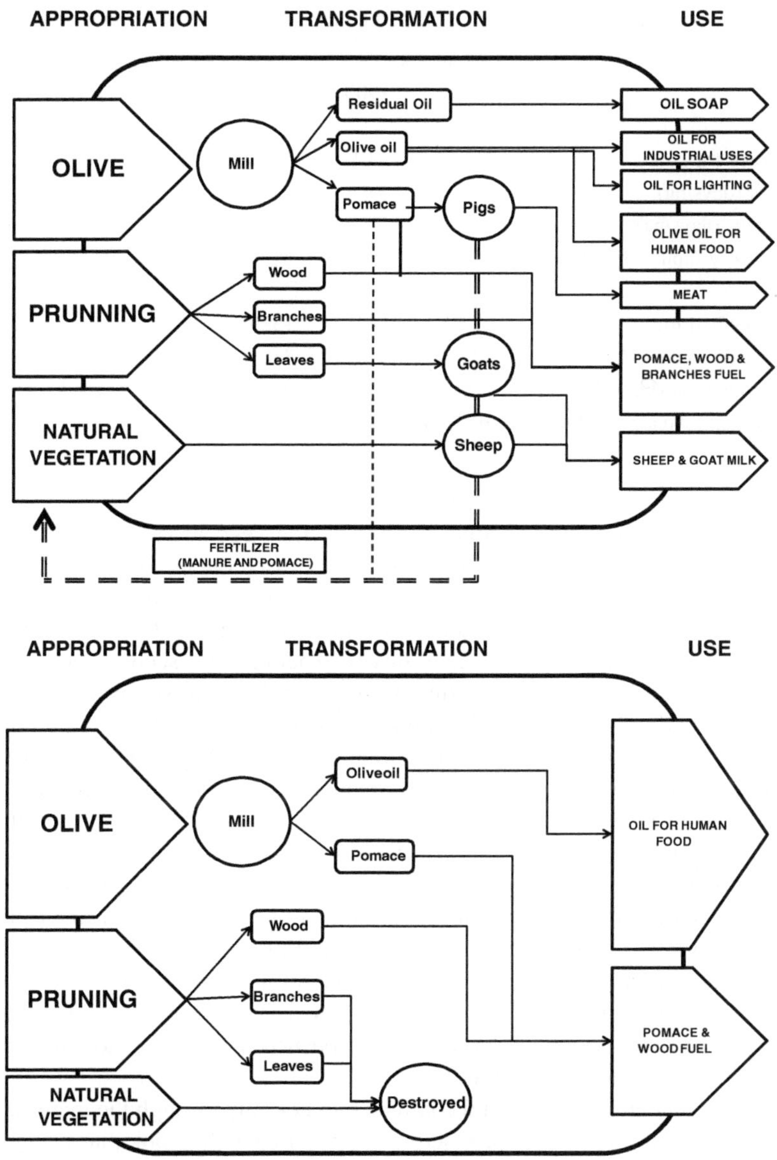

Fig. 8.3 System of appropriation and use of olive grove products in **a** preindustrial societies **b** industrial societies. *Source* Infante-Amate and González de Molina (2013)

108.7 to 534.2. Other research (Zambrana 1987; Infante-Amate 2014) and agronomical reports of the time (JCA 1891, 1921) as well as information from the land register highlight that olive groves required very little irrigation and no fertiliser, not even manure. In this way, it has been shown that the production of olives during

Table 8.2 Work applied to olive groves and resulting production in two local case studies

	Montefrío			Baena		
	1750	1900	2000	1750	1900	2000
Labour						
Workforce (days/ha)	11.0	28.1	18.4	21.2	37.6	10.8
Animals (days/ha)	4.4	8.6	–	6.3	15.2	–
Machinery (days/ha)	–	–	6.8	–	–	6.5
Total energy input (Gj/ha)	4.0	7.9	26.5	5.7	13.8	40.8
Production						
Olives (kg/had m)	109	534	1339	329	728	2641
Other used biomass (kg/had m)	2444	1961	710	2212	936	533
Total production (kg/had m)	2553	2496	2049	2540	1663	3174

Source Infante-Amate and González de Molina (2013)

said period was closely related to the quantity of preparatory work applied, essentially in the use of the ground, controlling the vegetation that competed with the crop in the summer, and pruning, now being carried out as a practice to improve production and not as a mere task to extract wood (Infante-Amate 2014).

The intensification under organic management continued to grow until the 1950s. However, from then on, olive groves experienced a second transformation process mediated by industrialisation and the global expansion of the markets. In the second half of the twentieth century, essentially from the 1970s onwards, various factors explain a new rise in the region's crop. They could be summarised as follows: after the Franco regime the Spanish economic policy was verging on openness again in the 1960s (Tió 1982). Olive groves had to compete with other cheaper oils, which began a crisis period. To tackle it, the government started various projects to help the sector, which were completed with the access to the European Union and its Common Agricultural Policy which, from then on, subsidised the production of olive oil (De Graaff and Eppink 1999). During these years, various studies related the consumption of olive oil with health benefits at the same time that lobbies related to the sector promoted its consumption. The aids for production and the growing external demand pushed olive groves to a new path of growth (Scheidel and Krausmann 2011).

Moreover, the growing demand could be met not only through the increase in the surface areas, but through the productive improvements that took place (Fig. 8.1c). Following the technological blackout of the Franco regime, Spain experienced an accelerated industrialisation process that also affected agriculture and, consequently, olive groves. Since the 1970s the consumption of synthesis fertilisers and chemical treatments multiplied, and the mechanisation of labour became widespread (Naredo 1983; Guzmán and Alonso 2008).

In Table 8.2 we can observe how the work added, on average, to one hectare of olive groves, fell between 1900 and 2000. This drop can be explained by the mechanisation process: today less labour is required. However, the production of fruit has grown: in the case studies cited, it has multiplied by 3–4. In the entire region, productivity has gone from 210 kg of oil per hectare in the 5-year period from 1961/65 to 670 kg/ha in 2006/2010.

Such a transition process has generated an increasing production of fruit, explained by the increase in consumables, essentially chemical fertilisers, and by the growth of the irrigated surface area, which went from taking up 5 % of the surface area of olive groves to 22 % (Infante-Amate et al. 2013). However, the nature of the labour applied to the crop also shows us another important change in its production: the intensive tilling of the soil, generalised to avoid vegetation from competing with the tree during the dry period, led to a loss in the total production of biomass, in addition to the fact that the new trees tend to be smaller and focus the production of biomass in the fruit.

The industrial transition process in the cultivation of olives transcends the mere industrialisation of its management because in addition it falls within the context of Socio-Ecological Transition via which olives, today, have lost their traditional multifunctionality (Infante-Amate and González de Molina 2013). The massive importation of fodder, the electrification and consumption of fossil fuels explain how the tree has lost some of its traditional functions such as producer of soap, lighting, fuel or animal food which characterised its traditional use and was a key factor for rural sustenance (Infante-Amate 2013) (Fig. 8.3b).

In short, olive groves have gone from being an extensive, multifunctional crop with little capitalisation, to experience a process of growing intensity under organic management which reached the mid-twentieth century. Following this phase, the crop experienced a period of productive industrialisation and simplification, focusing its production on providing good quality oil for the market. All the factors that defined this process of change as well as the uses applied to the tree, have led to different landscape morphologies that we will explain in the next section.

8.4 The Resulting Landscapes

The typology of olive grove landscapes could be rendered very complex taking into account orographic variables, types of exploitation, varieties, historical origins…. However, from a long-term perspective, a somewhat more simple taxonomy relating the great changes of nature's modes of appropriation can be established.

8.4.1 Landscapes in the Preindustrial Era

In the mid-eighteenth century Andalusia had one of the lowest population densities in Spain, with 20 hab./km^2. This, alongside the institutional restrictions that slowed down the agricultural colonisation process, meant that more than 50 % of the land was still dedicated to pasture and mountain. Furthermore, we know that the majority of the cultivated area referred to areas of cereal worked with light rotations. In the province of Seville, for example two-thirds of the cereal area was worked with a light 3-year rotation. What place did olive groves have in such a non-intensive type of agriculture?

Olive groves in the Old Regime, which were not very capitalised and had a notable multifunctional nature, as described above, appeared as an exploitation similar to agroforestal uses, such as *dehesas* or *montados*, than an intensive olive grove organised into rows. In other works, on the basis of the review of olive grove letting contracts in the region, between 1740 and 1850, the morphology of their landscapes has been documented (Infante-Amate 2012a, 2015). This source often provided details of the structure of olive grove landscapes in Andalusia. The results obtained indicate that the olive groves from mountain areas which, as we have observed in Fig. 8.2, were where the crop spread the least, appeared dispersed across the territory without forming organised groves and, in other cases, were associated with other crops. In fact, the land register sources from 1750 reveal more than one hundred municipalities in which there were olive groves in the municipal area but in which no olive grove farm was registered as a monoculture. Olive groves, therefore, were dispersed; peppering the landscapes of the mountain areas, integrated into other farming areas, or joined with livestock exploitations.

In countryside areas, where olive groves had a more notable presence, it was more common to find them organised into rows, where the main exploitation was the production of olives. However, in these places too, a considerable part of the olive groves appeared to be dispersed across the land, in cereal farms, irrigated gardens and forest and pasture areas. The review of the agricultural land registers from the nineteenth century shows us that in some counties with a considerable olive-producing specialisation, a very high percentage of the olive groves are registered as dispersed, of poor quality, or low production. They are characterised as olive grove farms with very few feet per hectare—between 40 and 50—in which the production of olives is combined with the exploitation of its plant cover as pasture.

Although it is difficult to quantify the dimension of this model of exploitation, it seems clear that until well into the nineteenth century the most recurring olive grove was a multifunctional crop, in which the extractions of wood could be greater than the appropriation of fruit. They grew in no particular order in other types of exploitations, and animals grazed on them when they were not part of fields or vines. In the municipality of Baena, the epicentre of the specialisation of Andalusian olive oil, in 1858, 40 % of the olive grove surface area was described as "dispersed", producing between 100 and 200 kg/ha compared to 1220 kg/ha of the better quality. However, the latter only occupied 6 % of the land. Likewise, in intensive olive grove areas close to the city of Seville, in 1900, it was insisted that the olive grove pastures were used (Infante-Amate 2014).

Somehow, as highlighted by the ecologist González Bernáldez (1981), olive groves appeared as a landscape resulting from Mediterranean woodland, the fruit of human action; similar, therefore, to systems such as *dehesa* or *montado*. This description of the landscape of the olive grove is not surprising insofar as other authors have indicated that its expansion occurred in many places, from Majorca (Morey and Molina de Dios 2013) to Italy (Bevilacqua 1989), on the basis of the grafting of olive varieties in wild olive. Likewise, other works have described dispersed and multifunctional olive grove landscapes in other areas of the Mediterranean, such as Cyprus (Makhzoumi 1997).

In the other agricultural colonisation processes in the Mediterranean, various strategies have been pointed out, which differentiate between the promotion of crops, often incorporated into complex tiles, or associative systems, and often of an agroforest nature (Blondel 2006). Andalusian olive groves described by the sources from 1750, and the majority of the nineteenth century, respond more to the second point than to a typically agricultural management.

8.4.2 Intensification Under Organic Management

Throughout the nineteenth century, as we have described above, the population density increased in the region at the same time that the institutional hurdles of the Old Regime collapsed. Agricultural colonisation advanced and with this, the surface area of olive groves grew (Barciela et al. 2005). But, furthermore, we pointed out that an intensification process took place in the works of the olive groves. Various case studies highlight that tilling the soil or pruning, in addition to requiring more work, became focused on the productive improvement of the fruit (Zambrana 1987; Garrido 2005; Infante-Amate 2014). In this way, land register information from the end of the nineteenth century reveals a process via which dispersed olive groves, associated with agro-silvo-pastoral management, tended to disappear, giving way to increasingly more organised olive groves. There were an increasing number of farms with olive groves in rows, or, in other words, with an olive-producing monoculture and increasingly larger feet per hectare densities.

The growing demand for oil on the one hand, and its multifunctional nature in contexts of growing population pressure on the other, meant that farmers in the region concentrated on olive groves, working them more intensively. A review of testimonies from the time shows that they insisted on describing the olive as a crop that was capable of producing a wide variety of products for rural sustenance.

In the mid-nineteenth century it was described in the region as follows:

> "This plant is the constant happiness of our fields; restorer of our strength; innocent relief from the heat, and the cold as fuel; (…) the plant that provides the cabinetmaker with excellent wood; that makes up for and condemns our destroyed and abandoned mountains" (Esponera 1851).

And in a similar fashion at the end of the nineteenth century:

> "Olive oil is an essential item for food for humans and as a condiment. It can also be used for lighting and in medicine, the arts and in industries in which it plays an important role (…). Its leaves can be used as food for goat livestock; a large quantity of nitrate and potassium carbonate can be extracted from its pomace, and mixed with bran this can be

used to feed pigs and as an excellent fertiliser. Lastly, its wood and roots are used for cabinetmaking, to make luxurious furniture and as one of the best fuels" (Serra 1878).

In 1924, in the international congress on olive growing, the same criterion abounded: "It is a great truth that the olive tree is the first of all trees; everything can be taken from it and obtained from it" (Rodríguez 1924). Such comments only abound in a context already highlighted by Columela and Virgillio in antiquity. What is interesting is seeing how, until well into the twentieth century, in preindustrial contexts, the olive tree maintained its nature of productive integration, although, as we have pointed out, it was tracing an evident landscape transition towards organisation of the plot.

Olive groves, therefore, throughout the nineteenth century, followed the transition described by Boserup: they went from a context of low land productivity and high labour productivity, to increasing the land productivity and that of labour employed (Infante-Amate et al. 2014). In short, they went from having an agroforestry nature to looking like a typical agricultural exploitation. Nevertheless, they continued to coexist on many occasions with the livestock, the tree size was still high, and the plantation density rarely exceeded 100 trees/ha. In fact, agronomic reports from the time indicated that in Andalusia the values were situated between 75 and 113 trees/ha, with 100 trees/ha being the most repeated, present in five of its eight provinces(JCA 1891). In short, they underwent a process of landscape transformation mediated by the productive and institutional revolutions of the nineteenth century (Antrop 2005).

The transition process, naturally, took place unevenly across the whole Andalusian territory. In Fig. 8.4 we make an approximation on the basis of local information available, which invites us to think that in the eighteenth century agroforestry olive groves must have extended in the same or a greater proportion than olive groves organised into rows. We have evidence that in counties with a strong olive-growing tradition olive grove pastures still occupied half of the surface area in 1858, as we have pointed out above. It is reasonable to think that both models extended equally until the mid-nineteenth century. What is certain is that in 1900, and to a much greater extent, in the mid-twentieth century, the dispersed olive grove, with low production and associated with livestock, stopped being a common exploitation. Furthermore, those that were preserved in such conditions were more the result of abandonment and marginalisation (Guzmán 2005). In 1975, Agricultural Statistics counted almost 1.7 million dispersed trees in Andalusia, which together could take up a virtual surface area of 17,000 ha. On this date, olive groves as a companion crop slightly exceeded 50,000 ha. However, the surface area of those olive groves that were not associated or dispersed exceeded 1.2 million hectares. In other words, the typology of the agroforestry olive tree, or the olive tree integrated with other exploitations tended to disappear at the end of the nineteenth century, becoming clearly residual in the mid-twentieth century.

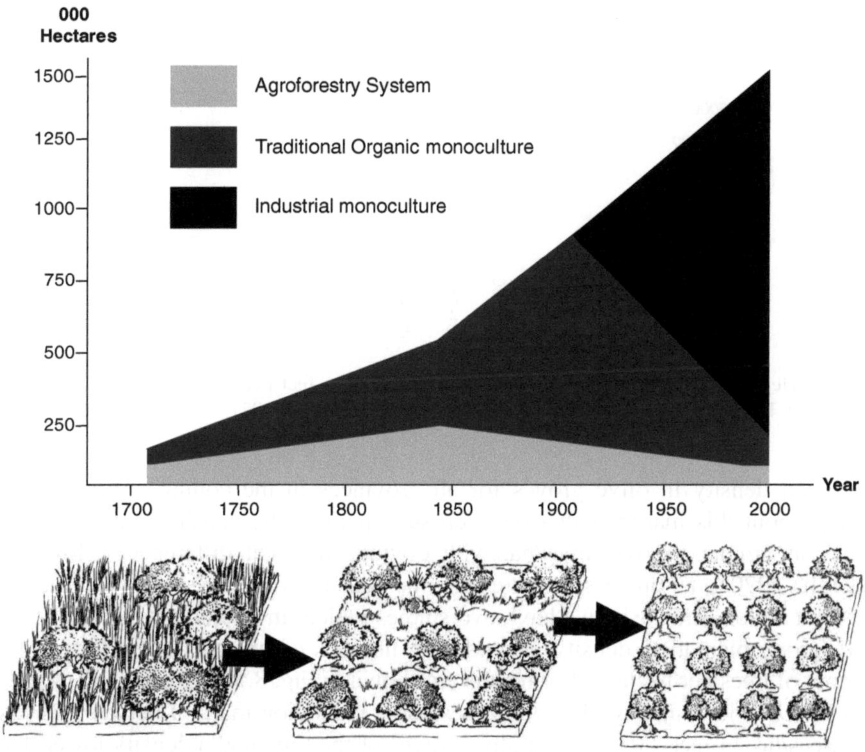

Fig. 8.4 Surface area occupied by each olive grove landscape in Andalusia. *Source* Compiled by author

8.4.3 Landscapes of the Industrial and Global Era

Agricultural industrialisation, the change of productive functionality and the integration into global markets outline the greatest and fastest olive-growing transformation process in history. As demonstrated above, since the 1960s a transformation process began which would extend to the majority of olive groves in the region in barely two decades and which would end up altering their landscapes. What were the landscapes resulting from the industrial and global era? How did it affect their morphology?

The mass addition of chemical fertilisers and the growing artificial irrigation helped to increase the plantation density, driving what was perhaps the most obvious process of change. After all, traditional olive groves, which rarely received fertilisation, could barely exceed production of 1000 kg/ha to a large extent due to such fertility replacement systems not being able to sustain more than 100 trees/ha (Naredo 1981). In fact, there is ample information about this. Both for 1888 (JCA 1891) and for the decade of the 1970s (MA 1970), we have data about the

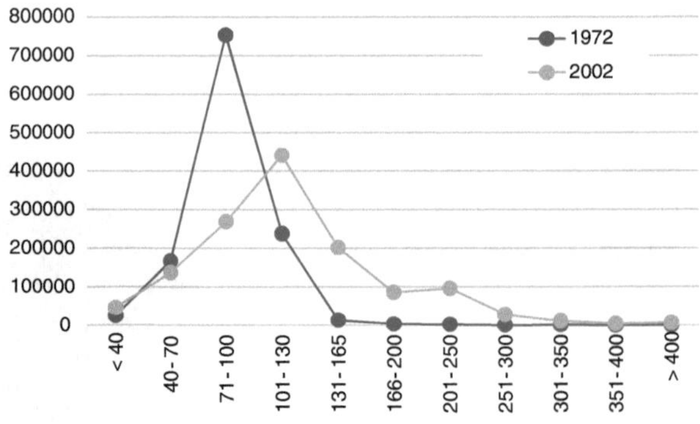

Fig. 8.5 Hectares of olive groves in Andalusia according to feet per hectare in 1972 and 2002, *Source* For 1972, MA (1975) and for 2002 (Junta de Andalucía 2002)

plantation density in olive groves for all provinces in the country. What is surprising about this matter is that between such periods the density remained quite stable, at around 90–100 trees/ha, with obvious regional differences. However, since the 1970s, the process of increasing the density of the trees has been radical. For the case of Andalusia, we have two studies which, in addition to providing the olive grove plantation density, provide the number of hectares according to the number of trees per hectare. The results are shown in Fig. 8.5 where we can observe how in 1972 the majority of olive groves had a plantation marc of between 71 and 100 trees/ha. In fact, 80 % of the olive groves of the time had a density lower than 100 trees/ha. Data from 2002 show an increasing distribution trend of feet per hectare. In fact, in that year, only 34 % of olive groves had less than 100 trees/ha. The rest exceeded said figure, in such a way that almost 150 thousand hectares already had more than 200 trees/ha.

While agricultural industrialisation brought an increase in the plantation density, the growing commercialisation of the crop and the subsidy policy of the Common Agricultural Policy (CAP), which associated aids with greater productive levels, ended up getting rid of the less productive (in terms of oil) olive grove models: dispersed trees and the associated crop disappeared completely. In Fig. 8.6 we show the evolution of dispersed olive trees according to estimations from the agricultural statistics: they went from 1.7 million in 1975, to more than half a million in 1995, to completely disappearing at present, when there are only 10,000 trees.

Other elements of the landscape change in recent years can also be described. The first is associated with the growing use of agricultural machinery and chemical treatments in the management of the grass cover. While in traditional systems it was difficult to remove spontaneous vegetation during the majority of the year, today, available technology means that the majority of olive groves in the region apply very intensive management that completely removes the vegetation, turning olive groves into a type of tree-lined desert in which loss by erosion has multiplied. It is estimated that since 1980 only, some olive groves in Andalusia have lost more soil, due to this reason, than in the previous two centuries (Gómez et al. 2014). Moreover, this fact is

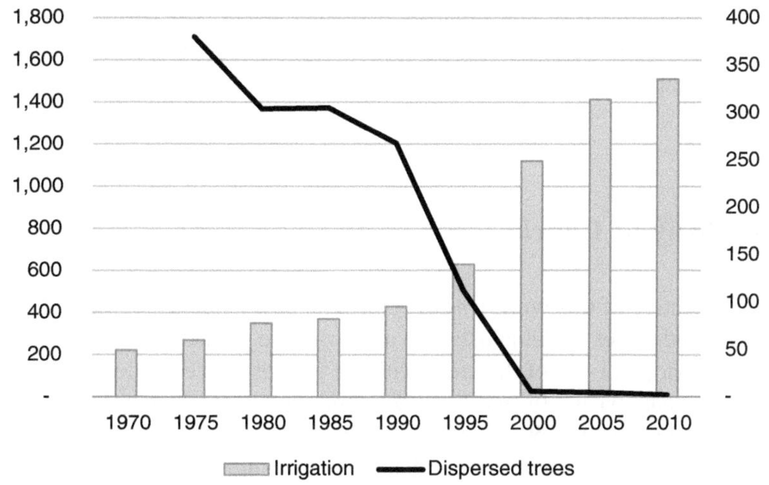

Fig. 8.6 The axis on the left represents thousands of dispersed trees; the axis on the right represents thousands of irrigated hectares. All data refer to Andalusian olives orchards. *Source* Infante-Amate et al. (2013)

also related to the loss of biodiversity of the olive groves; the flora and fauna present in traditional models have disappeared on account of the intensive work (Hermosín et al. 2009). This point is related to the change in the management of certain remains of pruning such as smaller leaves or branches. Traditionally they were eaten by animals—the leaves—or used as fuel—the branches. Today the majority are burned at the grove, generating a landscape which for several months is taken over by columns of smoke. Both the loss of the cover and the burning of plant remains reduce the ability of the olive grove soils to act as a carbon drain (Aguilera et al. 2013). It is also important to highlight that the expansion of the irrigated surface area (Fig. 8.6), as well as leading to an increase in the number of trees per hectare, has meant a plastic net has been spread over the base of the olive trees, since 94.5 % of the irrigation systems are drip irrigation. Lastly, it is essential to remember that during this period a process of variety simplification could be detected (García Brenes 2006), as well as a change in the morphology of the plant, which was increasingly small and only had one support base.

Today, however, other landscape models can be observed, which could mean a fourth—or fifth—typology in the future. In barely 10 years both superintensive olive groves and organic olive groves (modern) have gone from being inexistent to occupying 50 thousand hectares. The first case is a model of olive grove with hedges and plantation densities that can exceed 4000 feet/ha, similar to intensive vine cultivation, where trees are replaced with several hedges in trellises. Organic olive groves—modern, to differentiate them from the traditional organic groves described in these pages—are also increasing in surface area and a considerable number of these are using traditional management—proliferation of cover, integration with livestock, etc.—although many simply appear as an intensive organic model with replacement of consumables (Guzmán 2011). The surface area of both is still residual out of the total Andalusian olive groves, and more time is needed to analyse the real dimension of their significance in large Andalusian landscape models.

8.5 By Way of Conclusion

The main objective of this work was to contribute new evidence on the evolution of olive grove landscapes in Andalusia, the area with the greatest concentration of this crop in the world, and one of the areas with the greatest olive-producing tradition in the Mediterranean basin. In addition to documenting the history of its constant expansion, visible since 1750, when there were some 220,000 olive grove hectares, to the present day, with more than 1.5 million, the aim of the work was to characterise its different landscape typologies. This work has described three general models of olive grove landscape which to a large extent reflect a process of growing intensification and are characterised by the change in the social functionality of the tree.

Therefore, it is a story of growing expansion as well as constant intensification and, if the expression is allowed, new degrees of domestication. Agroforestry olive groves were present in Andalusian fields for millennia, without their sustainability being compromised. In fact, some analyses of the mid-eighteenth century reveal sustainable management that did not compromise their future. After all, it was a management with a very low anthropic impact. The organic olive grove model as a monoculture, in turn, started to reveal some environmental problems, essentially related to the replacement of nutrients (García-Ruiz et al. 2012), their erosion rates (Vanwalleghem et al. 2011) and, also, on a social plane, the growing dependence on foreign labour, generating seasonal migration processes (Infante-Amate 2011). However, it was a model that has also been present in the fields of the area for centuries and that has been the main feature in these fields since 1850. It has been suggested that the organic models of traditional olive groves were sustainable systems that were only compromised when certain levels of intensification in the working of the soil, fertility replacement and promotion in steep areas were exceeded.

The industrial transition shows the new mode of appropriation of the nature of industrial societies which, in the case of olive groves, has brought considerable environmental problems such as loss of biodiversity, contamination of aquifers, emission of greenhouse gases, erosion and dependence on external consumables which, for many authors, may compromise the future of the crop in the region (Beaufoy 2001; Gómez 2008; Alonso 2011). This landscape model has only been in existence for a half-century and, although it has expanded more than any other in the region in the past, everything suggests that over the course of the history of Mediterranean landscapes it will be the shortest-lived model on account of its lack of sustainability.

References

Agnoletti M (2007) The degradation of traditional landscape in a mountain area of Tuscany during the 19th and 20th centuries: implications for biodiversity and sustainable management. For Ecol Manage 249(1):5–17

Agnoletti M (2014) Rural landscape, nature conservation and culture: some notes on research trends and management approaches from a (southern) European perspective. Landscape Urban Plan 126:66–73

Aguilera E, Lassaletta L, Sanz-Cobeña A, Garnier J and Vallejo A (2013) The potential of organic fertilizers and water management to reduce N 2 O emissions in Mediterranean climate cropping systems. A review. Agriculture, Ecosystems and Environment, 164:32–52

Alonso A (2011) El olivar ecológico. Junta de Andalucía & Mundi-Prensa, Sevilla

Antrop M (2005) Why landscapes of the past are important for the future. Landscape Urban Plan 70(1):21–34

Antrop M (2006) Sustainable landscapes: contradiction, fiction or utopia? Landscape Urban Plan 75:187–197

Attenborough D (1987) The first Eden: the Mediterranean world and man. Collins, London

Beaufoy G (2001) EU policies for olive farming. Unsustainable on all counts, BirdLife Internacional-WWF

Besnard G et al (2013) The complex history of the olive tree: from Late Quaternary diversification of Mediterranean lineages to primary domestication in the northern Levant. Proc R Soc B: Biol Sci 280(1756):2012–2833

Bevilaqua P (1989) Clima, mercato e paesaggio agrario nel mezzogiorno, en Bevilaqua P (ed) Storia dell'agricultura italiana in etá contemporanea, Vol. I. Spazi e paesaggi, Marsilio Editori, Venecia, pp. 643–676

Blondel J (2006) The 'design'of Mediterranean landscapes: a millennial story of humans and ecological systems during the historic period. Hum Ecol 34(5):713–729

De Graaff J, Eppink LAAJ (1999) Olive oil production and soil conservation in southern Spain, in relation to EU subsidy policies. Land use policy 16(4):259–267

Esponera P (1851) Del olivo y su cultivo. Memoria presentada a la Junta de Agricultura de la Provincia de Jaén, Madrid

Fischer-Kowalski M, Haberl H (eds) (2007) Socioecological transitions and global change: Trajectories of social metabolism and land use. Edward Elgar Publishing, Cheltenham

García Brenes MD (2006) El olivar en Andalucía y el sistema de protección de la Unión Europea. Problemas del Desarrollo. Revista Latinoamericana de Economía 145:153–176

García-Ruiz R, González de Molina M, Guzmán G, Soto D, Infante-Amate J (2012) Guide lines for constructing nitrogen, phosphorus and potassium balance in historical agricultural systems. J Sustain Agric 36:1–33

Garrido L (2005) Olivar y cultura del aceite en la provincial de Jaén. Jaén, Instituto de Estudios Giennenses

Gliessman SR (1990) Agroecology: researching the ecological basis for sustainable agriculture. Springer, New York

Gómez JA (2008) Olivar sostenible. Prácticas para una producción sostenible de olivar en Andalucía. Córdoba

Gómez JA, Infante-Amate J, de Molina MG, Vanwalleghem T, Taguas EV, Lorite I (2014) Olive cultivation, its impact on soil erosion and its progression into yield impacts in Southern Spain in the past as a key to a future of increasing climate uncertainty. Agriculture 4(2):170–198

González Bernáldez F (1981) Ecología y paisaje. Barcelona

González de Molina M, Guzmán G (2006) Tras los pasos de la insustentabilidad: agricultura y medio ambiente en perspectiva histórica (siglos XVIII-XX) (Vol. 71). Icaria Editorial

González de Molina M, y Toledo VM (2014) The Social Metabolism. A socio-ecological theory of historical change. New York, Springer

Grigg D (2001) Olive oil, the Mediterranean and the world. GeoJournal 53(2):163–172

Grove AT, Rackham O (2003) The nature of Mediterranean Europe: an ecological history. Yale University Press, Yale

Guzmán JR (2004) El palimpsesto cultivado. Historia de los paisajes del olivar andaluz, Sevilla, Consejería de Agricultura y Pesca, Junta de Andalucía

Guzmán JR (2005) Territorio y medio ambiente en el olivar andaluz. Consejería de Agricultura y Pesca

Guzmán Casado GI, y Alonso AM (2008) A comparison of energy use in conventional and organic olive oil production in Spain. Agricultural Systems, 98(3):167–176

Guzmán G (2011) El olivar ecológico. Mundi Prensa, Madrid

Hermosín MC, Rodríguez-Linaza LC, Cornejo J, y Ordóñez-Fernández R (2009) Efecto de uso de agroquímicos en el olivar sobre la calidad de las aguas, en Gómez JA, Sostenibilidad de la producción de olivar en Andalucía, Sevilla, Consejería de Agricultura y Pesca, Sevilla, pp. 127–160

Infante-Amate J (2011) Los temporeros del olivar. Una aproximación al estudio de las migraciones estacionales en el sur de España (siglos XVIII–XX). Revista de Demografía Histórica 29:87–118

Infante-Amate J (2012a) Cuántos siglos de aceituna. Sobre el carácter de la expansión olivarera en el sur del España (1750–1900). Historia Agraria 58:39–72

Infante-Amate J (2012b) The ecology and history of the Mediterranean olive grove. The S-Spanish great expansion: 18th–20th Centuries. Rural Hist 23–2:161–184

Infante-Amate J (2013) ¿Quién levantó los olivos? La expansión olivarera decimonónica como estrategia campesina (1750–1930)", Historia Social 76(II):25–44

Infante-Amate J (2014) ¿Quién levantó los olivos? Historia de la especialización olivarera en el sur de España (ss. XVIII–XX), Ministerio de Agricultura, Alimentación y Medio Ambiente, Madrid

Infante-Amate J, González de Molina M (2013) The socio-ecological transition on a crop scale. The case of olive orchards in S. Spain (1750–2000). Hum Ecol 41(6):961–969

Infante-Amate J, Soto D, Cid A, Guzmán G, González de Molina M (2013) Nuevas interpretaciones sobre el papel del olivar en la evolución agraria española. La gran transformación del sector (1880–2010), XIV Congreso Internacional de Historia Agraria, Badajoz, 7–9 de noviembre de 2013

Infante-Amate J, Soto D, Iriarte I, Aguilera E, Cid A, Guzmán G, García-Ruiz R, González de Molina M (2014) La producción de leña en España y sus implicaciones en la transición energética. Una serie a escala provincial (1900–2000), Documento de Trabajo de la Asociación Española de Historia Económica

JCA (1891) Avance estadístico sobre el cultivo y producción del olivo, Junta Consultiva Agronómica, Madrid, Dirección General de Agricultura, Industria y Comercio

JCA (1921) El aceite de oliva. Junta Consultiva Agronómica, Dirección General de Agricultura y Montes, Madrid

Junta de Andalucía JA (2002) El olivar andaluz, Sevilla, Junta de Andalucía

Kaniewski D, Van Campo E, Boiy T, Terral JF, Khadari B, Besnard G (2012) Primary domestication and early uses of the emblematic olive tree: palaeobotanical, historical and molecular evidence from the Middle East. Biol Rev 87(4):885–899

Lepart J, Debussche M (1992) Human impact on landscape patterning: Mediterranean examples. In: Landscape boundaries. Springer, New York, pp 76–106

Loumou A, Giourga C (2003) Olive groves: the life and identity of the Mediterranean'. Agric Hum Values 20(1):87–95

MA (1970) Anuario de estadística agraria. 1970. Ministerio de Agricultura, Madrid

MA (1975) Inventario agronómico del olivar (varias provincias), Ministerio de Agricultura, Madrid

Makhzoumi JM (1997) The changing role of rural landscapes: olive and carob multi-use tree plantations in the semiarid Mediterranean. Landscape and Urban Planning, 37:115–122

Marull J, Tello E, Wilcox PT, Coll F, Pons M, Warde P, Ollés A (2014) Recovering the landscape history behind a Mediterranean edge environment (The Congost Valley, Catalonia, 1854–2005): The importance of agroforestry systems in biological conservation. Appl Geogr 54:1–17

Mataix J, Barbancho FJ (2008) El aceite de oliva. Jaén, Universidad de Jaén, Alma del Mediterráneo

McNeill JR (2002) The mountains of the Mediterranean world. An environmental history. Cambridge University Press, Cambridge

Morey A, y Molina de Dios R (2013) La pérdida de competitividad del aceite mallorquín y su pervivencia como cultivo marginal (1860-1960). XIV Congreso Internacional de Historia Agraria, Badajoz, 7–9 de noviembre de 2013

Naredo JM (1981) La incidencia del « estraperlo » en la economía de las grandes fincas del Sur. Agricultura y Sociedad 19:81–129

Naredo JM (1983) La crisis del olivar como cultivo biológico tradicional. Agricultura y Sociedad, 26:167–277

Parcerisas L, Marull J, Pino J, Tello E, Coll F, Basnou C (2012) Land use changes, landscape ecology and their socioeconomic driving forces in the Spanish Mediterranean coast (El Maresme County, 1850–2005). Environ Sci Policy 23:120–132

Pelletier N, Audsley E, Brodt S, Garnett T, Henriksson P, Kendall A, Kramer KJ, Murphy D, Nemecek T and Troell M (2011) Energy intensity of agriculture and food systems. Annu Rev Environ Resour 36:223–246

Petanidou T, Kizos T, Soulakellis N (2008) Socioeconomic dimensions of changes in the agricultural landscape of the Mediterranean basin: a case study of the abandonment of cultivation terraces on Nisyros Island, Greece. Environ Manage 41(2):250–266

Ramón-Muñoz R (2000) Specialization in the international market for olive oil before World War II, In: Pamuk S, Williamson JG (eds) The Mediterranean response to globalization before 1950. Routledge, London, pp 159–198

Ramón-Muñoz R (2007) La producción y el comercio de aceite de oliva en los países del Mediterráneo (1850–1938): competencia y especialización". Mediterráneo Económico 7:329–354

Remesal J (1998) Baetican olive oil and de Roman economy. In: Keay S (ed) The archaeology of early Roman Baetica, Rhode Island, Portsmouth, pp 183–199

Rodríguez J (1924) Actas del VII Congreso Internacional e Oleicultura, Sevilla, diciembre, Edición de Sucesores de Rivadeneyra, Madrid, 5–9

Sauer C (1925) The morphology of landscape. Publications in Geography 2, Universityof California

Scheidel A, Krausmann F (2011) Diet, trade and land use: a socio-ecological analysis of the transformation of the olive oil system. Land Use Policy 28:47–56

Serra M (1878) Elementos de agricultura, Jaén

Tió C (1982) La política de aceites comestibles en la España del siglo XX. Ministerio de Agricultura, Pesca y Alimentación, Madrid

Vanwalleghem T, Infante-Amate J, González de Molina M, Dy Soto, Gómez JA (2011) Quantifying the effect of historical soil management on soil erosion rates in olive orchards over the last 250 years. Agric Ecosyst Environ 142:341–351

Zambrana JF (1984) El aceite de oliva y su dependencia del mercado internacional de grasas vegetales. Un análisis histórico 1861–1935. Agricultura y Sociedad 33:159–196

Zambrana JF (1987) Crisis y modernización del olivar. Ministerio de Agricultura, Pesca y Alimentación, Madrid

Zambrana JF (2006) El sector primario andaluz en el siglo XX. Instituto de Estadística de Andalucía, Sevilla

Zohary D, Hopf M (1994) Domestication of plants in the Old World, 2nd edn. Clarendon Press, Oxford

[illegible]

Chapter 9
Historical Landscape of Lun Olive Trees at Pag Island, Croatia

Jadran Kale

Abstract Narrow and isolated landstrip of Lun village at the island of Pag is home to a substantial number of olive trees. Now a vast majority of wild olive trees (*Olea oleaster* sp.) are grafted, but there is also a biological reserve of ungrafted forest. Its inception is interpreted according to historical sources. Acquired preservational status of cultural landscape is compared with the activities of cultivators themselves, experiencing integrational period of Croatian EU-membership candidature as the second most important modern motion after post-feudal agrarian reforms. In such a way supralocal and supranational genesis of a local phenomenon is being confirmed also in the modern age. Agrogenetically diverse trees adapted to harsh karstic terrain grow from fruit stones and bear mutational stock of possible biotechnological importance. The case for a biocultural registry and adjusted conservation for development is argued. European and more generally implicated theoretical issues are also discussed.

Keywords Agrogenetic diversity · Agropastoral complex · Community-driven conservation · Biocultural diversity

9.1 Introduction

Cultural landscapes are widely recognised as dynamic and multifaceted phenomena. Their inclusion among UNESCO's heritage categories in 1992 had offered loosely delineated definitions, reaching from highly structured fields to purposely untouched natural spaces. All the administrative motion occured simultaneously with the nomination of periodically rebuilded Grand Shrine of Ise for the World Heritage List. Eligibility discussion that enfolded around these simple wooden

J. Kale (✉)
Department of Ethnology and Anthropology, University of Zadar, Novi kampus, F. Tudjmana 24 i, 23000 Zadar, Croatia
e-mail: jkale@unizd.hr

M. Agnoletti and F. Emanueli (eds.), *Biocultural Diversity in Europe*, Environmental History 5, DOI 10.1007/978-3-319-26315-1_9

Fig. 9.1 Olive trees and drystone walls denoting functional boundaries of agropastoral economy and commons. Position of grafting upon wild olive tree trunk can be clearly seen

buildings, cosmologically and ritually connected to its forest environment, led to the redefinition of authenticity from material integrity to processual function with an international legally binding act concerning collective expressions and production of culture (Smith and Akagawa 2009).

Along with particular practises of national legislations, a vast number of previously instigated preservations had to reconsider their terminologies and theoretical implications. Internationally minted category of intangible cultural heritage became also observational apparatus and occasional qualifier for the World Heritage List's nominations of cultural landscapes (Maurano 2005). From the opposite part of research spectrum nearly at the same time came scrutinised biotechnological expertise upon yet another invisible part of the landscape. After elaborating CBD-imposed mandate for legislative respect of collective creators' benefit through the Nagoya Protocol in 2010, two invisible perennial hands shaping cultural landscapes had legally met in the open. Cultural landscapes out of the strict associative sub-category now can be seen as biocultural terrains apt for interdisciplinary attention under biological, agronomical and cultural competences.[1] Continuously evolving cultural landscapes are also agrogenetically important terrains (Brush 2004; Brookfield 2001). Here we shall present a case in need of an integrated biocultural preservation.

[1]About cultural connotations of demonstrational feat for this new category in the World Heritage List, Ifugao rice terraces, see Brookfield (2001: 16) and Kwiatkowski (2013).

Fig. 9.2 Top of the vegetational zone. Olive shrubs are characteristic for the pastureland and trees growing out of fruit stones

Fig. 9.3 Wild olive trees biological preserve at Dudići pastureland, near Lun

9.2 The Environment and Orchard Inception

Landscape under consideration comprises the most distant and distinctively elongated strip of the Pag Island in Croatia, where the village of Lun and smaller neighbouring settlements are located. The opposite end of the island is the closest to the mainland, with its northeastern shores being gradually more and more distant from the coast, so Lun effectively figures as an isolate.[2] The island's endpoint is still under full-blown exposure of strong and dry mountainous wind. Land strip is steeply profiled towards mainland, with all vegetation growing at its milder southern exposures. Terrain is karstic with scarce and heavily conditioned arable fields. Water sources are rare and dependent of rainfall. Under given climate conditions, insolation is abundant and favourable for the growth of resistant vegetation. Grapes are absent. Sole prosperous fruit tree is olive. Number of olive trees is estimated at some 80,000, with a great number of multicentennials and individual plants cautiously dated with sixteen and seventeen centuries old age (Fig. 9.1).

Local Liburnian population of pre-historic times, archaeologically attested with several graves, four hilltop enclosures and one settlement, pursued a sheep herders' economy. Such pattern is confirmed at the opposed mainland coast, where Roman termination inscription from early first century CE delimited pastures of two Illyrian tribes with still visible drystone boundary. Insular environment, dry wind and certain shallow coasts had allowed for several saltworks, essential resource for sheep husbandry. At the joint base of the Lun landstrip the biggest pre-historic hilltop settlement of Košljun is situated, overseeing two saltwork locations in two out of three divergent bays. Between third century BCE and first century CE Roman forces subdued Illyric peoples, paving ways for its traders, ship owners, money lenders and plethora of citizen services aimed at new garrisons, colonies and settlements along roads and routes. According to the sources from either west or east provinces, profit extractors were rather ruthless. During the so-called Baton's Uprising at the beginning of first century CE, first to haunt and massacre were Roman traders. Once defeated and asked why they mutinied against Rome, Baton accused Romans for sending wolfs instead of shepherds or guardian dogs (Šašel 1963; Wilkes 1969).

In the first century centre of population activity settled at the coast downhill the Košljun drystone enclosures. Settlement change is attested with coinage found only at the lower sites. The name of enlarged settlement in Košljun and Caska area was Cissa, eponymic for the island of Pag in total. Epigraphic monument connects the new settlement with influential senatorial Roman family of Calpurnii. Long imposing walls can be clearly seen still today, but the main centre of activities was relocated to the third bay, exposed towards the open sea. Inner coastal navigational route flanked by hilltop guards was substituted by open-sea route between Salona and Aquileia. Brand new settlement served as a service port with waterfront installations and double aqueduct clearly oversized for a settlement alone, one being constructed over pillars and other being cut through limestone landmass in arduous and technologically advanced manner. In later centuries sign of service overproportion is recognised in oversized churches, obviously serving ship crews as well as

[2]The island of Pag is the fifth largest in the Eastern Adriatics, covering 284 km^2. Lun landstrip is protruded for 18 km, some 2 km in width.

inhabitants. Contemporary town name of Novalja stems out from Latin Navalia, denoting ship services (Oštarić and Kurilić 2013).

For Lun landscape genesis it is important to note possible component of Calpurnian possessions. Members of old Republican nobility and one among the mightiest Roman families which owned lucrative lands across the Empire established its property in the Balkans exclusively along the west Istrian coast and in south Liburnia. Two toponyms of *Krpinjan* (*Carpignano* in Italian) in Istria owe their forms to Calpurnii. Most known Istrian product was olive oil, used even as a cure and reputed highly enough to be faked with spices. It is very unlikely that olive trees were introduced at Pag as a planned economical effort, because several sites of Roman antiquity (some in remote provincial peripheries) reveal highly regular cultivation grid interspersed with oil production facilities (Matingly 1996). Locations of the oldest wild olive trees around Lun resemble no such intentions. Also, archaeological findings of olive mills, which are plentiful all over western Istria, are absent. We suppose that olive trees were introduced as side activities between Calpurnian land possessions, but preliminary genetic analysis among Istrian and Lun landraces gave no determinant connections (Matijašić 1998).

Summarising antique heritage of Lun olive trees, it could be stated that they owe their initial cultivation to prosperous trade from first century CE when an enormous amount of manufactured goods were imported from northern Italy to Salona and minor cities and garrisons. Previous trade was reserved for refined manufacturing like luxury weapons for chieftains, but new markets emerged for dozens of thousands of colonists, urbanised locals, stationed legions and veterans awarded with land allotments. Time window of aggrandized maritime transport and trade opportunities diminished in second century when provincial production had matured. Meanwhile, veteran settlements were established, like Pola (today Pula) in Istria where a Calpurnian was among founding leaders. A centuriated hinterland accompanied new colony, just as had happened in Iader (today Zadar) where Romans enlarged older Liburnian settlement. City of opportunities gave work also for locally attested vocation of *negotiator olearius*, trader of olive oil, so entrepreneurs were clearly aware of associated profit. Autochthonous communities could combine new culture with their own sheep herders' economy, but, as Cicero said for Galia, not a single coin changed hands without financial interest of a given Roman citizen. Coins represented signs of exchange out of subsistence proportion. Liburnians lacked central authority, capital city or distant surplus trade. One among symbols of their entering into wider commonwealth was olive trees (Suić 1981).

9.2.1 Feudal Heritage

The importance of antique past could be followed along pasture delimitations of the former Calpurnian possessions south of Caska and Novalja, which are observed through long drystone walls until nowadays. Boundary was confirmed as delimitation of Rab and Nin episcopies in eleventh century, regarded also as the borderline between communes of Rab and Pag, and serving today as the border between Lika and Zadar Counties which makes Pag as the sole Croatian island divided between two regional administrative units. Such a rigid observance of an antique trait can make understandable that Cissan husbandry traditions were practised over millennia, even after destruction of Cissa in 1203. Venetian order from 1441 treated it separately from the pasture regulations in the communal statute of Rab.[3] Commune of Rab, municipal settlement at the neighbouring island of the same name, used northern Pag pastures from the early Middle Ages and eventually encouraged growth of Novalja as their administrative outpost. Medieval inhabitation of Lun area was seasonal, bringing peasants from Rab as labourers of their feudal masters. Locally practised type of economy comprised mere utilisation of seasonal pastures.

Rab communal pastures went into state ownership in 1441 and became private possessions in 1640. New feudal masters (Borgo and later Dominis) replaced seasonal herdsmen with hinterland refugees during the War of Candia, which resulted in substantial migrations from middle and western Bosnia to Dalmatian coast, cities and islands. First immigration of sheep herders' families is documented in 1653, with their family name of Badurina denoting mainland nomadic Vlach population.[4] At 1711 there is still scant population of five families and 25 inhabitators. Drystone cottages were substituted with mortar builded from 1781 on. With permanent inhabitants, experienced mainland sheep breeders, Lun landscape had entered to a new phase of its shaping.

Decisive impact upon destiny of forests of wild olive trees, antique sideline import left to adapt under terrain, climate and pasture conditions, was implied by the arrangement of feudal obligations of the new inhabitators. Land owners persisted in seasonal pasturing which enabled forest growth and reliable wood supply. Olive trees were disseminated without active cultivation, just from their fruit stones. The absence of dissemination by cloning is obvious at the nearby islets, where olive trees are dispersed through seagulls' digestion. As the result, Lun olive trees became highly adapted and agrogenetically diverse.[5] Wild olive tree forests were owned by feudal masters as wood, annual taxes upon inhabitants included sheep husbandry products and any cultivational effort made by villagers enhanced value of the land which should be repayed to its cultivators if feudal contract had to be terminated. In time Lun villagers succeeded in prohibition of further seasonal employment of herders outside the island, and, as a sole cultivators, became fully aware of grafting. First, unsuccessful grafting initiative is dated in 1730, starting in earnest during nineteenth century. Villagers had to ask for grafting permissions, transforming their inclination for untaxed olive oil made from richer, grafted fruits, into a struggle with feudal

[3]"*Sed pastores Novaliae et Loni de satisfactione suae pastoriae accepiant bestias ad pascendum, secundum antiquam consuetudiem.*" Archival analysis is presented in Crnković (1988: 72).

[4]Vernacular Latinity formed characteristic *-ur* personal name with a common Slavic augmentative suffix *-ina*.

[5]I am indebted for unpublished agronomical data to Prof. Đani Benčić from the Faculty of Agronomy at the University of Zagreb.

masters. Owners employed foresters to cut off grafted branches, but villagers argued for their grafting rights well into 1930s. Along with provincial articulation of the first law enabling cultivators to buy cultivated land from the owners in 1870s, the owners started to cut the forests in massive proportions with large swaths of denuded karst left and still exposed today. Pastures of Lun and its daytime shelters, now separate settlements, were divided into 1847 and drystone-walled delimitations amassed over simple boundaries of feudal pasture units (Crnković 1988).

Dynamic nature of landscape management can be observed upon Dudići, hamlet of nicknamed Badurina branch. Their section of the Lun area landstrip borders the pasture of Lun in proper. After division of pastures they remained with commune of Rab as their feudal master, leaving every citizen of Rab as an ideal co-owner of feudal rights at one side and inhabitants of Dudići as their labourers at the another side. In terms of real life this meant paralysed chances for resolving villagers' economical emancipation and delineation of individual families' parts inside of commonly managed terrain. Because of this every improvement came to Dudići as the latest. About 1880 at Dudići only seven olive trees were grafted, all of them owned by one family from Rab. Other villagers ceased to pay their feudal tributes in 1920s but last feudal tax observed by Dudići happened in 1945, last year before final legal termination of feudal relations. Last oxen and cow were in Dudići, used still after 1945.[6] In landscape, being legally prevented to divide big pasture into individual enclosures, Dudići managed it as a commons with undivided flock (husbandry manner called *skupnô*, "doing together").[7] Because of milk processing and cheese production, they invented a multicellular stone enclosure in their hamlet (*Dudićev osik*), serving the purpose of daily milking and temporarily dividing individually owned sheep, letting them loose again in integral flock. Effectively they structured drystone-walled device for landscape management out of dire necessity, serving needs of the ageing inhabitants up to 1995. Even at a smaller scale of cultural signs, simple sling could be recognised as the tell-tale sign of their landscape management, because their neighbours with individualised narrower enclosures did not need them for directing the flock over open spaces any more (Kale 2011).

9.2.2 Actual Status

Because of this reason last forest of ungrafted wild olive trees survived at Dudići pasture, being protected in 1963 as biological reserve. Greater interaction with local inhabitants occured during fieldwork resulting in official proclamation of the cultural landscape ("ethnographic zone" of the time) in 1975, extending over all Lun

---

[6]Presence of oxen for cultivation of scarce arable fields made their mark in grafting, because older grafts had to be situated high enough and out of their reach (Badurina 2012).

[7]Mediterranean commons were typically connected with rearing animals, not with forests (Bourbouze and Rubino 1993).

landstrip, with zoning of enclosure at Dudići as the first inclusively protected particular drystone structure in the Eastern Adriatics. All this time olive trees were shadowed behind meat and cheese production of sheep herders economy (Fig. 9.2), practised in depopulated villages whose young and vital part of population had left for cities and abroad. When olive trees came into forelight in 1980s Lun was populationally depleted but connected with a paved road. Elaborately sculptured trunks emerged as aesthetically pleasing objects suitable as tourism attraction as well as sentimental reminder of the local origins.

After legal termination of the last remnants of feudal relations in 1946/1947, "the Lun issue" continued its divergent life as an obscurely recognised legal anomaly.[8] Individualization of ownership started from the base of the landstrip but never reached its tip, leaving Dudići out of the ownership even of the landplots under their houses and Lun inhabitants as the owners of grafted trees but without the land beneath them. Trees could be grafted separately, so owners happened to be divided even upon the same tree. First, engined oil mill came with state-driven collectivization in 1947, but privately bought cold-processing device has to wait the next war to finish, in 1997. Seasonal olive fruits' celebration started just at the time of Croatian EU-membership candidature in 2005, paving way for olive cultivators' civil association and cooperative after 2008. Croatia adopted subsidy scheme under regulation of the Common Agricultural Policy, leaving Lun cultivators disqualified because they do not own the land under cultivation. Because of this, villagers of Lun collectively sued state in 2009 over ownership entitlement upon land. Next spring started with the incident, when two multicentennial trees were sold by an impoverished villager, supposedly as a decoration for a shopping mall. Affair made the Lun olive heritage widely known, with experts at service for the local community, scientific symposium and public sessions held at Lun, state President invited to pick a first seasonal fruit under behemoth located at the new tourist route through olive trees orchard, and a big new tourist project established with municipal authorities. As the most important achievement came tacit understanding among villagers themselves, unrelated to their diverse affiliations, not to sell big trees any more.[9]

9.3 Discussion

It did not went for a long time until became clear that the initial research framework for an ethnologist, impressed with the quality and abundance of drystone-walling in the area, fell short of depicting the phenomenon. Luckily, thoroughfully executed

[8]Authoritative historical analysis of "The Lun Issue" ("*Lunjsko pitanje*") is published in Crnković (1988).

[9]For a value-driven understanding of cultural landscapes, see Buchecker et al. (2007). An example of religiously obligated agrogenetic conservation, an issue not dissociated from the cultural meanings of olive tree cultivation, can be seen at Iskandar and Ellen (1999).

projects concerned with toponomastic and archaeological heritage were finished at the same time, just as Lun descendants grasped strength to aggregate and publish personal memories, chronicles and ethnographies in series of publications (Badurina 2012). Young Lun descendant of mixed Italian–Croatian ancestry had specialised in olive tree expertise. It turned out that weakening of village population could offer an unexpected resource, combining competences and care of people sharing same local sentiments in achieving common goals. In a vibrant atmosphere of debated issues a redefinition of research objectives became a readily accomplished task.

Features of the Lun landscape had fermented the finest capacities of expert services concerned with cultural heritage. During 1970s a pioneering concept of "ethnographical zones" was introduced in two dozens of heritage denominations from the Rijeka Office of cultural heritage conservation. Lun was the only denominated "ethnographical zone" outside of the inner Rijeka region, following the wider administrative delimitation mentioned earlier in the text. Also, establishment of the Lun cultural landscape was not the first one with zoning included, but unusual multicellular structure of *Dudićev osik* made way for the first specific zoning of a drystone-walled building—practice to follow and advance with particular preservations no earlier than 1997. When faced with the crisis of selling fundamental phenomena from the landscape, awesome trees transported away in 2010, implementation of cultural landscape protection remained powerless because biological matter is not included in cultural heritage. Biological preserve *Dudićke krune* covers 24 ha with some 1500 ungrafted wild trees (Fig. 9.3). Institutional preservation of fundamental heritage phenomena proved impotent and effectually dependent on local activism.

Two distinctively European moments spring up. First one is concerned with longlived archival and other historic sources, so researched phenomenon can be approached for interpretation through centuries and even millennia. Second one is more important for its contemporary management. Local inhabitants as protagonists of their ownership struggle achieved literacy and fought with skills learned elsewhere, for example as priests and monks competent in guarding and interpreting archival sources.[10] First judicial appeal against feudal ownership in 1815 was articulated by a peasant. Today, it means outside expert advocacy being limited to a local assistance, and no foreigners to local community lead their causes. No matter how decimated, still multivocal and stratified, local community emerged vital and capable to define its priorities.

Next specific moment is connected with globalisation. Local amenity has an explicitly supralocal genesis, confirming vernacular traditions as a matter of exchange, dialogue and presentation. One could say that behind a picturesque tree a history of exploitation is being hidden, but such constatation does not displace its

[10]One among the finest Croatian art historians, late Franciscan Anđelko Badurina, grew up behind a windows looking directly at *Dudićev osik*. Later, he wrote en passant about the enclosure as the prime example of "pastoral urbanism" (Badurina Dudić 2006: 31).

contemporary cultivator as the rightful representational voice articulating all accumulated capacities. Wild olive trees lived here through history as a circumstance of a minor importance. Redefinition of heritage object promotes olive trees into apt issue for integrated conservation and development project, rather than community-based natural resource management (Maginnis et al. 2004). Biotechnological advance that dissects them as a potential agrogenetic database of mutation particularities, bringing them finally in the foreground, must find a common interest with the community of cultivators. Both sides are uttering voices of a global genesis, one belonging to the commonwealth of antiquity and the other aspiring to make sense out of biotechnological commons. Local effect is expected from both voices.

9.4 Conclusion

It has appeared that it is hard to say is it a cup of Lun olive oil half empty, or half full. People had left their village, but stayed alert and mobilised through the crisis. They also sold out a number of houses and engaged in tourist economy away from solely peasant work, but in effect they enlarged circles of symbolic influence and sentimental appeal. Two multicentennials were sold out of the landscape, but remaining number enjoyed waves of public attention and interests of investors in the industry of authentic experience. Even a cup half empty still became important because of oil, not any more for its wooden material. Critical was recognition of local initiatives and creativities to overcome inadequate legal frameworks regulating ownership and heritage management. Beside established biological and cultural preservation status, agricultural subsidies introduced during EU-integrational years have proved themselves as the only administrative tool capable to induce motion for change. That moment underlines capacity of cultural and agricultural creator, i.e. villager and cultivator.

Heritage registry dividing biocultural features into disciplinary domains lacks force to save grafted multicentennial tree or associated characteristically profiled dry stone walling, but it is even more important to combine it with productive, creative motives like socially responsible marketing, market niches or agrienvironmental schemes. Informed and educated biocultural creator, with his collective intellectual rights respected, is the most acceptable heritage guardian of its own domain because he can engage in sustainable adaptation, its responsible presentation, evaluation and further modification. Modern conservation had ceased to define itself as a restoration of stylistic integrity, and now it pursues research of "the social process by which the materials and values associated with objects, buildings and sites are transmitted through time" (Bracker and Richmond 2009: XIV). Lesson of a creative response like elaborated stone enclosure at Dudići, first Croatian drystone walls under explicit protection, is that heritage did not existed before certain age and social regulation in recent history. If we succeed in keeping and nurturing collective creative capabilities of culture creators, common good of productive human environment has a chance. At Lun, yesterday it was a matter of a Roman ship and today it is an another expensive device from a global technological commonwealth, a genetic sequencing machine—each one with plentiful strings attached.

References

Badurina D (2012) Lunjski libar. Ogranak Matice hrvatske i Grad, Novalja

Badurina Dudić M (2006) Povijest Luna. Gušć, Dudići

Bourbouze A, Rubino R (eds) (1993) Terre collettive nel Mediterraneo. Istituto Nazionale di Economia Agraria, Roma

Bracker A, Richmond A (2009) Introduction. In: Richmond A, Bracker A (eds) Conservation: principles, dilemmas and uncomfortable truths. Elsevier and Victoria & Albert Museum, London

Brookfield HC (2001) Exploring agrodiversity. Columbia University Press, New York

Brush SB (2004) Farmers' bounty: Locating crop diversity in the contemporary world. Yale University Press, New Haven

Buchecker M et al (2007) Value systems: drivers of human-landscape interactions. In: Kienast F et al (eds) A changing world: challenges for landscape research. Springer, New York

Crnković N (1988) Prilog istraživanju agrarnih odnosa na pašnjacima sjevernog dijela Paga i na otoku Rabu. In: Kaloper-Bakrač J (ed) Prinosi za gospodarsku povijest otoka Paga. Historijski arhiv, Rijeka and Pazin

Iskandar J, Ellen R (1999) In situ conservation of rice landraces among the Baduy of West Java. J Ethnobiol 19(1):97–125

Kale J (2011) Dudićev osik kao krajobrazni ključ. In: Skračić V (ed) Toponimija otoka Paga. Sveučilište u Zadru, Zadar

Kwiatkowski L (2013) Globalization, environmental change, and coping strategies among the Ifugao of the Philippine Cordillera Mountains. In: Lozny LR (ed) Continuity and change in cultural adaptation to mountain environments. Springer, New York

Maginnis S et al (2004) Conservation landscapes: whose landscapes? Whose trade-offs? In: McShane TO, Wells MP (eds) Getting biodiversity projects to work. Columbia UP, New York

Matijašić R (1998) Gospodarstvo antičke Istre. Žakan Juri, Pula

Matingly DJ (1996) First fruit? The olive in the Roman world. In: Shipley G, Salmon J (eds) Human landscapes in classical antiquity. Routledge, London

Maurano C (2005) La costiera amalfitana: il patrimonio intangibile di un paesaggio culturale. Centro Universitario Europeo per i Beni Culturali, Ravello

Oštarić I, Kurilić A (2013) Arheološka karta otoka Paga. Ogranak Matice hrvatske i Grad, Novalja

Smith L, Akagawa N (2009) Intangible heritage. Routledge, London

Suić M (1981) Zadar u starom vijeku. Filozofski fakultet, Zadar

Šašel J (1963) Calpurnia L. Pisonis auguris filia. Živa antika 12(2):387–390

Wilkes JJ (1969) Dalmatia. Routledge and Kegan Paul, London

Chapter 10
Rewilding the French Pyrenean Landscape: Can Cultural and Biological Diversity Successfully Coexist?

Tony Knight

Abstract Transhumant pastoralism has moulded the Pyrenean landscape for thousands of years. Ancient forests have ceded their dominance to verdant pastures that, today, symbolise the mountains; 'wild' life has ceded its historical presence to domesticated livestock. Now, the mountains are undergoing a process of rewilding: charismatic large predators have returned. A contested (re)introduction programme has reinforced the brown bear population, and natural agency has encouraged grey wolves to remake the Pyrenees their home. Conservationists and environmentalists argue that Pyrenean pastoralism has historically coexisted with large predators, and should simply (re)adapt its methods and practices to revalorise and reinvigorate a broader return to a once-present, but suppressed, level of biological and cultural diversity. On the other hand, far from imagining a more resilient natural and cultural landscape, pastoralists view these changes as a threat to their livelihood and identity. Drawing on ethnographic fieldwork, I explore these complex and conflicting interspecific relationships. I argue that conservation management programmes must seek interdisciplinary collaboration with social science in order to more profoundly understand the human and cultural implications of biocultural diversity.

Keywords Anthropocene · Biocultural landscapes · Human–animal conflicts · Large carnivore interactions · Pastoralism · Pyrenees · Rewilding · Biocultural diversity

10.1 Bio*cultural* Diversity

> The sustained concern with the modes of use and representation of the natural environment which has united structuralists, Marxists, ethno-scientists and cultural ecologists is obviously not reducible to mere fashion, and although it has generated countless misunderstandings, this unplanned convergence of interest reveals a common, albeit rarely explicit,

T. Knight (✉)
School of Anthropology and Conservation, University of Kent, Canterbury, UK
e-mail: ajk34@kent.ac.uk

M. Agnoletti and F. Emanueli (eds.), *Biocultural Diversity in Europe*,
Environmental History 5, DOI 10.1007/978-3-319-26315-1_10

> assumption: the principles of the construction of social reality are primarily to be sought in the relations between human beings and their natural environment.
>
> (Descola 1992: 109)

In 1948, then President of the New York Zoological Society, said

> It is man's earth now. One wonders what obligations may accompany this infinite possession (Osborn 1948).

Of course, since 1948, we have come to the realisation that the Earth's resources are far from 'infinite'. Furthermore, some recognise that humanity is facing its greatest challenge ever: socioeconomic growth driven by capitalism, globalisation, technology, and modern consumerism, is radically affecting the environment. Climate change and the degradation of biodiversity have become global concerns. At the same time, though, *social* inequalities are becoming a dominant feature of modern civilisation. The intersections between the *naturally* biological and the *culturally* social are fraught with challenges, and yet we clearly need to find equitable ways to be just to both components of what we are calling biocultural diversity. This is becoming increasingly difficult as we enter what some are calling the anthropocene, an age in which anthropogenic activities are becoming the principal driver of change—mostly adverse—to the Earth system

> The so-called Anthropocene – does not just imply conflation of the natural and the social, but also a 'radical' change in perspective and action in terms of human awareness of and responsibility for a vulnerable earth (Palsson 2012).

Most of the essays in this collection are written by natural scientists who are willingly incorporating human or social aspects into their work in an attempt to address these challenges.

I am an environmental anthropologist: my research focuses on the *relationships* that exist between humans and nature: the role of *culture*, for me, is central. In this chapter, then, I present a bio*cultural* focus on human–predator interactions in the French Pyrenees. Notwithstanding that neoliberal economic forces have challenged Pyrenean pastoralism for years, its actors perceive that the re*wilding* of their landscape through the reintroduction of large predators, in particular, brown bears (*Ursus arctos*) and wolves (*Canis lupus)* is an even greater threat to their livelihoods and identities. Drawing on *qualitative* ethnographic fieldwork in the Pyrenees conducted between 2013 and 2014, I explore how nature and culture are often perceived as being in mutual opposition, and within such a dichotomy, whether or not biological diversity and cultural diversity can successfully coexist.

'Culture' is a vast and complex concept, and despite well over a century of anthropological thought, has yet to be adequately defined. Nevertheless, as this book is largely based on scientific epistemologies, I feel it important to offer a simple framework for how I understand the term.

From Tylor (1871) to Geertz (1957), culture was variously said to include the knowledge, beliefs, art, customs, values, and anything else that gave humans the power of judgement. The focus of culture, then, was in the intangible and abstract. In the early 1960s, Goodenough synthesised culture within the complex realm of shared ideas that creates the foundation for the rules and meanings which allow us to make sense of the way societies live. Culture, he said, was the

> pattern of life within a community – the regularly recurring activities and material and social arrangements (1961: 521).

Even as the last century was closing, this was still the prevailing view

> culture consists not of things and events that we can observe, count and measure: it consists of shared ideas and meanings (Keesing 1997: 70).

Keesing believed we should be concerned with the systems of knowledge and beliefs through which the human group structures its experience and perceptions, formulates acts, and chooses between alternatives. For me, this is at the core of understanding human–nature relationships within a biocultural realm, but it is still missing an important point alluded to, but not elaborated on, by Goodenough. The *patterns* that make societal life what it is are clearly observable and, as such, are potentially tangible: the *material arrangements* further imply those *objects* and *things* essential to creating the shared *meanings* of culture. In Pyrenean pastoral society, the sheep raised by farmers and the predators who threaten them also have very deep symbolic meanings based on long collective memories. An understanding of culture which accepts that objects, things, and even nonhuman *beings*, symbolise its meaning(s), is essential if we are to make sense of the differing constructions of social realities, spoken of by Descola in my epigraph, that make up the biocultural space in which pastoralism and large predators are expected to coexist.

10.2 The Great Divide: The "Othering" of Wilderness

A common premise of 'Western' thought is that at some time in our history, humans lived in a unified socio-ecological environment in which 'wild' nature was not something apart from them. However, during the long transition from prehistory into 'civilised' times, this relationship changed. Little by little, the least hospitable components of nature came to be perceived as something that was not intrinsically 'human'. Humans *culturally* constructed a nature that was 'out there', separate from themselves, and potentially opposed to humanity. This new version of nature became thought of as wild(er)ness.

Wilderness is a Foucaultian *heterotopian* space: it is 'outlined in the very institution of society', yet intrinsically opposed to it (Foucault 1967: 332). It 'lies outside all places and yet is actually localizable' (ibid.): it is both isolated and, at the same time, penetrable. Over time, based on its desired use-value, society continuously reinvents the *function* of heterotopian wilderness: it becomes more, or less, 'wild'.

Wilderness, then, is an imaginary 'in which the abstracted essence of what society is not can take shape' (Whatmore and Thome 1998: 436). Society is the *civilised* 'Us', and is opposed to the 'Other' wild(er)ness. And yet, as Foucault implies, society *defines* wilderness and it is, thus, inherently a part of society while paradoxically remaining outside of it. Within this wild nature are unknowns that can become knowable through (superior) human reason,[1] and dangers that can be overcome through (superior) human technologies.[2] The wild can thus be domesticated or managed by humans. Nonhuman animals which do not belong in a civilised, human space such as a pastoral landscape, define this Other wild(er)ness.

Without this Other, how can we be an Us? This dichotomy has become known as the Great Divide (Latour 1993): culture versus nature; domestic versus wild; and human versus animal. To be human is to be not-animal (Haraway 2008: 11). In its reified 'Western' sense, then, human identity is dependent upon being 'on the opposite side of the Great Divide from all the others' (ibid.).

It is this Great Divide which has allowed 'Western' civilisation to grow and prosper through its domination and exploitation of a nature relegated to the Other (Ikerd 2005; Williams 2010). This nature includes all the earth's geological and biological resources: from rivers to rocks, oceans to oil, arboreal to animal. There is nowhere left on the planet still untouched by human activities: the deepest oceans, the tallest mountains, the remotest deserts, the coldest polar regions, and even the highest layers of our atmosphere are now tainted by the pollution from our agriculture, factories, transportation, energy sources, and weapons. We have localised and penetrated this heterotopia, violated its integrity.

The chasm separating nature from humanity continues to widen and deepen. Douglas (1975: 289) observed that the greater the separation between humans and nature, the more nature becomes alienated as an 'outsider'. The farm and the pasture are 'human' spaces: the encroaching predator is 'nature', 'wild'.

If we are to understand 'biocultural diversity', then, we need to recognise the complex and multiple cultural interactions between *biological* diversity and *cultural* human activities. In effect, we are trying to bridge the Great Divide between nature and culture in order to redefine a biocultural space in which human activities can coexist more harmoniously with biodiversity.

It is thus fitting that natural scientists and social scientists are coming together to seek solutions that can benefit both biological and cultural diversities. But to achieve this, we need to dissolve the ontologically false Great Divide.

---

[1]From an anthropological perspective, wilderness is also knowable through mystique and magic which interact with otherwise 'natural' elements in ways that challenge nature-culture dichotomies. A discussion on such alternative epistemologies is beyond the scope of this chapter.

[2]This particular aspect is very disturbing to me, for it is becoming the de facto essence of dominant climate change and conservation discourse.

10.3 Of Bears and Men … and Wolves and Vultures

10.3.1 The Decline of Brown Bears in Europe

It is easy to argue that the brown bear populations of Europe declined due to the simple expansion of agricultural development. But this minimises the deeper cultural meanings which allowed bears to be subsumed by civilisation. Indeed, the bear was revered by many European societies: while pastoralism has coexisted with brown bears in the Pyrenees for thousands of years, for much of that time, the bear held a powerful cultural symbolism. Pyrenean myth reveals that the bear 'fell' from grace: the massif was created by Hercules to protect the body of his lover, Pyrène, killed by bears who had become wild.

In fact, it was the Catholic Church that paved the way for the bear to become reviled. Ancient Pagan bear cosmogony persisted until well into the Middle Ages and threatened the essential role of Jesus within the Holy Trinity dogma. During more than a thousand years, the Church demonised the bear and encouraged its hunting in order to extirpate this fearsome symbol of the Devil (Conan 2009).

Memory, of course, rarely extends that far back into history. Today, then, when pro-conservation actors claim that pastoralism and bears have always coexisted, pastoralists respond that their ancestors 'always' refused peaceable coexistence with the Pyrenean bear, and were largely responsible for its extirpation.

Certainly, the relationship between pastoralists and predators has never been easy. Pastoralists were ready to abandon their 'territories' and 'cultures' because of the constant attacks by '*bestes salvatgines*' ('savage' beasts), as long ago as 1392: a royal decree was issued to the people of Espousouille en Capcir in the Eastern Pyrenees, allowing them to set fire to forests suspected of harbouring bears responsible for their losses (Berlic and Berlic 1987).

Nevertheless, Bouchet (1990), a specialist in Pyrenean history, argues that, *bergers* (farmer/shepherds) were not financially or practically capable of taking time away from their flocks to hunt bears, and personally killed few if any. Certainly, the Industrial Revolution had a far greater impact on the bear population. Gómez-Ibáñez (1975: 110ff.), describes how the emerging iron industry required massive deforestation to satisfy its needs for wood to fuel the foundries. Demand grew dramatically to provide the French navy with tall masts for its warships: within a decade 'most of the tall firs in the Aspe valley had been cut' (ibid.). The impact on the ecosystem, and in particular, the bear population, was devastating.

During the nineteenth century, the state offered hunters 'bounties' to kill large predators. These hunters sought out and killed many female bears to capture their cubs, who were sold to *montreurs*: the *montreurs* trained the cubs and toured the world with their 'show' bears (Bouchet 1990). The price for a bear cub was up to eighty times greater than the bounty for killing a single bear (ibid.). These activities took a terrible toll on the brown bear population of the Pyrenees.

By the beginning of the twentieth century, only 150 bears were estimated to be still alive in the Pyrenees (MEDD 2006). It is difficult to judge how significant predation was at this time, but Lamazou (1988), in his unique autobiography of life as a Pyrenean *berger* during the first six decades of the twentieth century, tells us that he lost twenty-five sheep to bears during his career. However, he never tried to kill the predator: 'we aren't, we've never been, and will never be hunters' (ibid.: 193–4). He added that he and his fellow *bergers* cursed the bear's attacks, but were "the first to regret" the near disappearance of the animal, for 'he [the bear] was an integral part [of our mountains], and without him, they will never be the same' (ibid.). Lamazou—a modern 'ancestor' of today's pastoralists, coexisted with bears, more-or-less peaceably.

Nevertheless, by the end of the twentieth century, just 16–20 brown bears still lived in the Pyrenees (Camarra and Dubarry 1992). Whichever 'reality' of the bear's decline is more to blame—expansion of general anthropogenic activities, the demonisation of bears, or pastoral refusal for coexistence via lethal protectionism—each reflects the Great Divide in which culture dominates nature, and the collective memories live on in today's cultural relationships with nature.

10.3.2 Anthropogenic Rewilding of the Pyrenees

The latter part of the twentieth century marked a shift in public consciousness towards a recognition that 'nature' needed to be protected. In 1984, the Bern Convention formalised the protection of brown bears in Europe and it was recognised that without urgent human intervention, the Pyrenean population would disappear (Camarra 1990). An EU-supported Franco-Spanish programme, LIFE Nature 'Conservation and restoration of threatened vertebrates in the Pyrenees', successfully reintroduced three Slovenian brown bears during 1996/97 (MEDD 2006).

However, the overall success of the programme was dependent on its local *human* acceptance, and that was not forthcoming due to the lack of appropriate consultation with the pastoral community, and the fact that depredations caused by bears were believed to be responsible for the 'excessive human-caused mortality and… extirpation of this species from the central Pyrenees' (Quenette et al. 2001: 120).

During the subsequent years, hunting, accidents, and the absence of continued coherent proactive conservation efforts meant that by 2005, the bear population remained essentially unchanged. By this time, however, a powerful environmental consciousness had transformed the bear into a symbol of the Pyrenees and, indeed, of 'nature' itself. Furthermore, the bear was now seen as a focal species, essentially a nonhuman *spokesperson* for biodiversity (Jeo et al. 1999). Indeed, the state reintroduction plan claimed that bears were:

> Fundamental to the functioning and equilibrium of the Pyrenean ecosystem (MEDD 2006: 79).

Through the way bears present themselves and their state of wellbeing, they 'tell' us how well—or not—the entire ecosystem is holding up. Successfully reintroduced indicator species such as bears 'tell' us that ecosystem services and resilience are improving (Jeo et al. 1999). Protecting the Pyrenean bears, then, gives a crucially important *voice* to the nonhuman biocultural actors.

However, conservation has become far more sophisticated than existing simply to protect certain animals, and it is beneficial to examine the sociopolitical context of Pyrenean rewilding as it is situated within the wider, global context. Anthropogenic climate change and massive biodiversity loss have become global concerns. The Nagoya Protocol goal of restoring 17 % of degraded land by 2020 (CBD 2010) affirms the European parliament's decision to improve protection and funding for wilderness areas (EC 2011). European 'wilderness' tends to be in marginalised, heavily depopulated European mountain regions which have traditionally been 'maintained' by pastoralists, themselves able to survive global market pressures thanks only to extensive financial subsidies provided under the EU Common Agricultural Policy (CAP). Ongoing CAP reform, however, strives to impose strict standards for increased productivity based on environmental compliance. This will likely accelerate the abandonment of pastoral practices and land, thus precipitating *natural* reforestation and further augmenting the scope for wilderness (Moriamé 2004). However, different actors have very different views on what wilderness means. For some, it might represent a temporary escape from everyday pressures of the 'civilised' world; for others, including pastoralists, it can represent a threat to their livelihoods.

In February 2009, the European Parliament Resolution on Wilderness in Europe (EPRWE 2009) was adopted as a framework to develop a wilderness and rewilding strategy. Areas such as the Pyrenees might not technically qualify as 'wilderness', but can be designated as 'wild areas', acknowledging their greater fragmentation and human impact (Wild Europe 2012). Potentially problematic is that wilderness and wild areas are not considered just on the basis of their intrinsic qualities, but are predicated on possessing underlying *material* benefits:

> In addition to their intrinsic and spiritual worth, wilderness and wild areas can generate important economic, social and environmental benefits (Wild Europe 2012: 3).

The emphasis on *economic*, as well as ecological and social benefits, is explicit: with its considerable tourism potential, wilderness areas 'can produce a range of often financially quantifiable benefits' (Wild Europe 2012: 4). As opposed to intrinsic ecological value, it is this socioeconomic concept of wilderness which appears to be the model actively being sought in Europe. However, in the context of late (or post) modernity and hegemonic globalisation, I find this very worrisome. As users of this 'wilderness', pastoralists are also concerned. They are wary of the 'real' objectives of conservation projects, and understand that the global capitalist political economy is predicated on significant financial returns from *any* investments, even those which are ostensibly ecologically altruistic.

Tourism is now the world's *largest* 'industry', and has an intrinsic need to maximise profits (Kearney 1995). In the guise of a collective commons to be protected, wilderness inevitably becomes profitised: it falls under the 'tourist gaze' (Urry 1990) and cannot escape being reduced to Marxian use-value in order to become a commodifiable experience (MacCannell 1989).

Tourism is not new to the French Pyrenees, but today, it has significantly replaced the economic importance of 'traditional' activities: tourism generates over 34 % of Pyrenean revenues compared to just 15 % for industry, agriculture, and pastoralism combined (CMP 2006). Boissevain (1994: 43–52) has called this modern phenomenon 'cultural tourism', where people strive to escape the stresses of modern society and find a more 'authentic' Other with which to compare themselves, and perhaps add meaning to their urban, and presumably, 'inauthentic' lives. The rewilding of the Pyrenees promises 'wild' 'natural' landscapes and the possibility of seeing 'wild' animals. Tourists can also witness the cultural activities of 'traditional peasants': shepherds working in partnership with their dogs tending their flocks on the mountain pastures.

Rewilding often involves animal reintroductions as part of a process in which humans consciously attempt to change a landscape, usually with the goal of recreating a specific version of what the landscape was *perceived* to be in the past. In the Pyrenees, this perception is of a landscape in which pastoralism happily coexisted with large predators. It is within this overall context that a new reintroduction programme was announced. The earlier lessons went unheeded, though, and inadequate attention was given to the fact that rewilding invariably entails changing people's relationships with the landscape in order to imagine desired biocultural and socioeconomic futures.

These imagined futures are based on differing perceptions of the same physical landscape. There are multiple ontologically distinct realities superimposed one on the other: pastoral, ecological, environmental, and developmental. Following a year of unilateral decision-making by the state, and a perceived avoidance of open consultation with the famers and shepherds most likely to be negatively impacted by the programme, it should come as no surprise that there was significant friction between each of the possible 'realities' (Tsing 2005). The result was that five Slovenian bears were 'successfully' reintroduced to the Pyrenees between April and August 2006; and pastoral activists refused this enforced cohabitation.

10.3.3 Friction

Pyrenean pastoralists are adamant that the Bear Reintroduction programme was imposed on them without ever seeking their consensus. Without this consensus, they say, a reintroduction programme was never feasible; but more importantly, is unallowable under Article 22 of the Habitats Directive (Europa 1992), which legally requires '*proper consultation of the public concerned*'.

What exactly is meant by 'the public concerned'? The entire population of France—who overwhelmingly support the bear reintroduction? Or does it mean those people who are the most *directly* concerned, in this case, the pastoralists?

There is a strong argument that France consciously bypassed this question, and even, perhaps, mislead the EU to win their original financial support for the programme (Chétrit 2012). Had they been properly consulted as required by European law, the pastoralists say, Europe would have denied the programme.

Nevertheless, the programme's 'legality' was far from the collective pastoral mind in 2006 in the lead-up to the first announced reintroduction. On 1st April, several hundred demonstrators sacked the small Pyrenean town of Arbas, planned to be a major reintroduction site. They attacked the buildings with firecrackers and sheep's blood, then tore down a large wooden sculpture of the bear before setting it alight. Death threats were made against the town's mayor.

During my ethnographic fieldwork I spent significant time with 'Jean-Marc' (actors' names have been changed to protect identities). His family has been raising sheep for several generations in Ariège: he was one of the Arbas demonstrators.

> We were angry. And at each step of the way, the anger grew. Suddenly, it exploded. We couldn't hold it back any longer. We were just like the *casseurs*[3]… we wanted to *break* the town. If anyone had tried to stop us, we'd have *broken* them too. I know it's not right. I'm quite a shy person and wouldn't hurt anyone. But that day… ?
>
> Now it's happening again. But we'll do anything to stop them rewilding our mountains. This is a *paysage totalement humanisé*. It's been humanised ever since pastoralism began here. That's why we wiped the predators out in the first place. We'll do it again – it's us, or them. (Fieldwork June 2014).

Although the physicality of Pyrenean pastoral anger did eventually quieten down, there has been a resurgence in the level of rhetoric during the last year or so. Many of the pastoralists with whom I spoke are beginning to lose hope for their future; there is a tangible mood of desperation. There is increasing talk of *abandonnement* of the *estives* (summer pastures), and even of pastoralism itself.

Planned rewilding is designed to create a desirable 'wilderness', but this is, itself, another form of human domination over nature, something the pastoralists are aware of: it is just a different way of 'domesticating' the wild than their own practices. But if pastoralists abandon their pastures, there can be undesirable affects.

'Pierre' has been a shepherd for some 40 years. A *Groupement Pastoral* of around 10 farmers employs Pierre to take care of their 1500 or so sheep every summer on a pasture in the heart of the rewilded 'bear country'. His job is to protect the family livelihoods of ordinary, small-scale *paysans* (peasant farmers):

> My father, grandfather, his father… were all shepherds in the Couserans. They had no predators.
>
> When I started as a shepherd, forty years ago, I used to lose between one and two per cent of my flock every year to accidents or disease… but no predation.
>
> For the last three years, each year, [in addition] the bear has taken over 50 sheep [3.34%]. I can no longer sleep at night. The slightest noise wakes me up… Every morning I wonder… "What am I going to find today"? A dead ewe? Forty dead ewes, their guts exploded all over the ground?

---

[3]"Casseurs" (from the verb "casser" which means "to break") is the name given to radical protesters who often join peaceful demonstrations with the single purpose of precipitating riot conditions, which often result in cars being burned and widespread looting of local businesses.

> You know? I've trained young shepherds for years now. These new shepherds… they come here with different ideas than us. We've made this nature, made it work for us, for the good of our sheep and for us, for generations.
>
> They dream of spending their lives working with animals in nature… and bears, well, they're a part of it. That's the way they see it. They come here thinking that the bear has a right to be here, too.
>
> But after just a few years, they can't take it any more… the stress, the fear, the fatigue, the sick feeling when they find a beast with her throat ripped out… So after a few years, they quit.
>
> When the wolf comes, we'll have to abandon the *estives*. We'll keep the sheep at our farms, which is not good for them. The *estives* will become overgrown, full of thorns and bad woods… and the mountain will no longer be beautiful. Tourists don't want that. They come here for pleasant and easy walks in the beautiful *estives* that we, and our sheep, maintain for them. (Fieldwork June 2014).

10.3.4 Natural Rewilding of the Pyrenees

The wolf is coming; and the subject invariably surfaces in most discussions about the future of Pyrenean pastoralism. Wolves were extinct in France by 1937, but in 1992, *Canis lupus italicus* made its way naturally to the southern French Alps from northern Italy. Today, there are about 300 wolves throughout the south and east of France. In the regions in which pastoralism is practiced, there have been devastating levels of depredations.

Interestingly, the first samples of wolf DNA were confirmed in the Pyrenees as far back as 1999 (Benhammou 2004), and yet today, there are officially still only a few 'lone' wolves in the territory. The state has not yet acknowledged any wolves who have settled in Ariège, but as one farmer told me:

> This suits us fine. If they're not here, then what dies in the woods must just be stray dogs, right? (Fieldwork June 2014).

The indifference that this farmer showed towards the protected status of wolves reflects the dominant pastoral discourse. An executive member of one of the official Ariège pastoral organisations indicated to me that during a meeting with the highest ranking departmental officials, there was a clear consensus: as long as the state does not recognise the presence of wolves in Ariège, then wolves cannot possibly be killed on Ariège soil. 'Everyone' knows this and 'everyone' shares the responsibility of assuring that wolves will not settle in Ariège.

10.3.5 Predations

The current territory inhabited by wolves is home to around three million sheep. So far this year, there have been just over 7500 depredations imputed to wolf attacks (Le Loup en France 2014). This is a rate of approximately 0.25 %.

In the Pyrenees, there are around 570,000 sheep and the number of bear depredations this year was 135. This rate of approximately 0.024 % is some ten times less than that for wolves (DREAL 2014). By contrast, up to 20,000 sheep are lost each year on the Pyrenean *estives* due to disease, accidental falls (in ravines for example), lightning, theft, and predation by bears and stray dogs (MEDD 2006: 26).

Such statistics are often used by environmentalists to argue that the problem of predation is extremely minor: pastoralists should therefore have no problem with coexistence, especially as they receive significant indemnifications. However, statistics reflect only a specifically desired reality, one that can be vehemently contested by actors experiencing a different reality.

The predations do not affect the entire livestock distribution: they are very localised. The bear-predation statistics provided above are for the entire Pyrenean massif, but the bears are almost all located in the department of Ariège, in the Central Pyrenees. Even there, they spend most of their time in the very small area of the Couserans, whose *estives* account for just 15,000 or so sheep, and where the great majority of predations occur. A level of predation that better reflects reality would be around 100 sheep killed out of 15,000, or 0.67 %. This is far higher than the Pyrenean-wide rate cited above, which tends to dominate public discourse. Furthermore, most of the depredations in the Couserans typically occur on just five *estives*. As such, the actual rate of predation can sometimes attain 10 % or even more of an individual farmer's flock.[4] At levels such as this, financial compensation does not even begin to ease the pain felt by the farmer.

One larger-scale farmer I have come to know quite well lost more than 40 % of his family's flock in a single bear attack. The sheep were grazing above a particularly rocky escarpment, and when the bear attacked, they panicked: many fell to their deaths or were never found. As with many pastoral families, this flock had been cultivated over multiple generations. Several years after this attack, the flock has still to regain its earlier 'quality'. Despite the years that had passed since the incident, the farmer's eyes filled with tears as he described the attack. Afterwards, he tried to explain the general feelings of pastoralists relative to their opposition to 'peaceable coexistence' with large predators:

> We don't want blood money [indemnifications]. We just want to be able to raise our animals in a dignified manner and make enough income to support our families. Our beasts don't deserve the suffering they go through because of the bear. They don't deserve to miscarry their lambs, year after year, because of their emotional stress. They don't deserve to be afraid when they graze in certain areas because the bear has visited them before. Saving the bears is killing pastoralism (see Fig. 10.1). So it's a choice, isn't it? Bears or pastoralism? Which has the greater right to survive? (Fieldwork 2014).

---

[4]Imagine a flock of 1500 sheep, owned by ten different farmers. The dominant farmer own one-third of the total, but a very small-scale farmer has only 60 sheep. A bear attack results in 30 sheep being killed, of whom, 15 belong to the small-scale farmer. In this case, recounted to me by a shepherd, this particular farmer lost 25 % of his flock.

10.3.6 Current Situation

At the European scale, brown bears are considered of Least Concern: however, the IUCN Red List considers the Pyrenean brown bear population as Critically Endangered (Djuro 2007). Despite decades of conservation efforts in France, the autochthonous Pyrenean brown bear is extinct, and today there are just 24 bears alive in the Pyrenees, all of Slovenian descent. The French state recently commissioned a scientific study which concluded that if imminent extinction due to consanguinity is to be avoided, then four males and thirteen females will have to be introduced over the next three to four years (Le Maho et al. 2013). The state hinted at a new 'bear plan' leading environmentalists to anticipate an imminent reintroduction. There was a predictable increase in pastoral anti-reintroduction rhetoric.

In parallel, news of significant wolf predations in Aude—the bordering department to the east—increased anxiety about the imminent arrival of wolves in Ariège. Bizarrely, this coincided with increasingly common reports of healthy young calves and lambs being predated by, of all creatures, vultures (*Gyps fulvus*). On 8th June, 2014, the Director of Veterinary Services for Ariège publicly confirmed that a ewe in good health had been killed by vultures. Apparently, the vulture had 'evolved' from a carrion-eater to a predator. Despite being ridiculed in the dominant environmental discourse, the official state and departmental positions were that vulture predation should be investigated.

An Ariège farmer who had 'accidentally' shot a vulture while attempting to deter a colony from attacking his livestock was taken to court for killing a protected animal. He refused to plead guilty, claiming it was 'legitimate defence'. Another farmer told me that he had been attacked three times in as many days:

> First the bear… and we know the wolf is at our door. Now the vulture… how can we carry on when protected predators are allowed to take our animals? (Fieldwork 2014).

On 10th June, 2014, the only male bear thought capable of taking over from the dominant male who is now the father or grandfather of almost the entire Pyrenean population, died in a natural accident. The problem of consanguinity took on an even greater urgency, and environmentalists appealed for his immediate replacement. Given the official status of the bear population, environmentalists could not imagine that such a reintroduction would not be rapidly forthcoming.

All of these events swelled to a new breaking point for the Central Pyrenean pastoral community who, in a rapid ad hoc action, joined forces with politically right-leaning hunting associations and farming unions to warn the newly appointed minister of ecology, Ségolène Royal, that they would not tolerate any further bear conservation, and neither would they allow wolves to settle in the region. In a matter of days, a large demonstration was organised in Foix, the departmental prefecture. Over 4000 protestors marched through the town to the prefecture building on 28th June, 2014 in a show of solidarity against policies which are perceived as protecting large predators against a vulnerable pastoralism. Their message was simple: we will not accept the continued rewilding of our mountains.

Fig. 10.1 Protesting: Livestock farmer in danger (Photo by Tony Knight, 2014)

Immediately following the Foix demonstration, Ségolène Royal authorised pastoral departments to hunt wolves, despite their strictly protected status. It was an attempt to appease the agricultural community in a year where the major thought on every government member's mind was the pitiful approval ratings and (re)election votes. Royal visited Ariège on 26th July, at which time she stated:

> The Pyrenean landscape is not adapted to the reintroduction of bears… When there are reintroductions of wildlife that are on the verge of disappearing, it should occur only in spaces where there is *no land-use conflict* (La Dépêche 2014; my emphasis).

She went on to declare that there would be no further bear reintroductions where this might threaten pastoralism. Ségolène Royal, the French Minister of Ecology, incredulously brushed aside decades of conservation research and work. She has shown that she is truly a minister for the new anthropocene era in that she has asserted politically that humans can and should take priority over nature, no matter what international legal obligations otherwise oblige.

For the first time in many years, the pastoral community is content in the knowledge that coexistence is off the table for a while, and happy just to wait for the collapse of the existing bear population. As for the wolves, they have made it abundantly clear that they will not allow them to settle in the Central Pyrenees.

10.4 Conclusions

It is extremely challenging to be able to draw any conclusions about the breakdown of biocultural relationships in the Pyrenees. It is clear to me that the pastoral community will continue to refuse any coexistence with large predators, no matter how much pressure is placed on them: many will simply abandon their *estives* and adopt a farm-based small-scale intensive animal husbandry if predation increases. This could be catastrophic for the desired aesthetics of the Pyrenean landscape and would have major financial implications for tourism as well as for pastoralism itself. The local economy of Ariège would be at significant risk.

There are competing ontological ideals of what the Pyrenees should represent. For many, they seek the natural aesthetics of today, but with the knowledge that the ecosystem is being 'improved' by the proliferation of wild animals once more, especially iconic large predators such as wolves and bears. For these people, the verdant expansive pastures combine with the rugged 'wild' mountain ravines and peaks to produce a nature as close to any wilderness that they would ever wish to experience. However, this biodiversity that they admire is a 'wild' biological nature—but one which is totally domesticated by cultural pastoralism (see Fig. 10.2). For these pastoralists, though, 'nature' is what they have made it to be. The 'wild' has no place in their lives. It is not the bear or the wolf who improve ecosystem services and resilient biodiversity, who are 'fundamental to the functioning and equilibrium of the Pyrenean ecosystem'—No, that role belongs to pastoralists and their sheep, and their culture and traditions. We are not dealing with just an ecosystem—it is an agro-ecosystem: humans are also part of this biodiversity.

The real issues are deeply rooted in socio-cultural values which can be very difficult to understand, and yet such an understanding is crucial if we are to successfully change attitudes and behaviours. Therefore, we should not presume that we can solve complex human–animal interactions and conflicts just though science-based management. Furthermore, even if budgets allow the responsible science teams to seek ad hoc help from social workers—survey questionnaires for example—this is wholly inadequate. What is really needed is a long-term open collaboration, in which a team of *interdisciplinary* researchers does not simply try to copy the methodological or epistemological frameworks of others, but actively engages in them; together.

It cannot be stressed strongly enough that even such a truly interdisciplinary approach can only hope to be successful within a unified and coherent political framework, within which, there is no room for personal biases and objectives: total honesty and transparency on the part of state policy makers is critical. It is essential that they (re)build a foundation for trust not by simply listening to, but by *hearing* all stakeholders. Such a 'round-table' should clearly involve pastoralists and environmentalists, but it must additionally include the 'voice' of the *nonhuman* actors who ultimately have the most to lose. Maintaining such a small consanguineous bear population for decades, for example—some of whom never even see another bear—is simply inhumane.

Fig. 10.2 Landscape domesticated by pastoralism (Photo by Tony Knight, 2014)

My research highlights the need for a truly *neutral* mediator or mediatory group to work from the outset to facilitate this safe round-table space in which opposing actors can air their views, their concerns, and maintain an ongoing dialogue. Most Ariégeois pastoralists are convinced that continuing to defend predator biodiversity will result in the extinction of cultural diversity. And yet, if they do not find ways to adapt to the inevitable presence of wolves, they indeed risk losing their livelihoods and unique identity. And the Pyrenees risks losing its socioeconomic foundation.

This seemingly unsolvable, but relatively minor problem of culture-nature conflict serves as a warning for what it will take to resolve the far greater issues facing us today in the anthropocene. Pastoralism is actually a model for such an anthropocenic mindset—human control over nature, where everything wild is seen as a contradiction to the domesticated ideal that is the dominant agricultural—and indeed, wider social—perception of nature. If we wish any kind of sustainable biocultural future, we must once and for all surmount the dominant mind set of culture versus nature which is intrinsic to capitalism's boom-or-bust structure.

References

Benhammou F (2004) Analyse stratégique et territoriale du retour du loup dans les Pyrénées: modalité, réalité et perspectives. Recherche Naturaliste 14

Berlic G, Berlic MF (1987) En 1985, l'ours dans les Pyrénées-Oriental. Naturalia Ruscinonensia: Revue de la Société d'Histoire Naturelle de Perpignan et des Pyrénées-Orientales 1:9–70

Boissevain J (1994) Towards an anthropology of European communities. In: Goddard VA, Llobera JR, Shore C (eds) The anthropology of Europe: identities and boundaries in conflict. Berg, Oxford, pp 41–56

Bouchet J-C (1990) Histoire de la chasse dans les Pyrénées françaises de XVIe au XXe siècle. Marrimpouey, Pau

Camarra J-J (1990) L'ours brun des Pyrénées. Suivi de population de 1979 à 1988. Bull Mensuel ONC 142:18–22

Camarra J-J, Dubarry E (1992) The brown bear in the French Pyrenees: distribution, size, and dynamics of the population from 1988 to 1992. In: Bears: their biology and management. A selection of papers from the ninth international conference on bear research and management, vol 9, issue 2. Grenoble, France, October 1992 (1997), pp 31–35

CBD (2010) Strategic plan for biodiversity 2011–2020 and the Aichi targets: "living in harmony with nature". Available via http://www.cbd.int/doc/strategic-plan/2011-2020/Aichi-Targets-EN.pdf. Accessed 1 May 2013

Chétrit D (2012) La réintroduction de l'ours: L'histoire d'une manipulation. Editions Privat

CMP Comité de Massif des Pyrénées (2006) Schéma interrégional d'aménagement et de développement des Pyrénées. Available via http://www.pyrenees.datar.gouv.fr/images/stories/autres_doc/Schema_de_massif.pdf. Accessed 20 Dec 2006

Conan C (2009) L'Ours. In: Des animaux, des dieux et des homes. Exhibition from 19 September 2009 to 28 March 2010 at the Musée d'Aquitaine, Muséum d'Histoire naturelle, and the Musée Goupil, Bordeaux

Descola P (1992) Societies of nature and the nature of societies. In: Kuper A (ed) Conceptualizing society. Routledge, London

Djuro H (Large Carnivore Initiative for Europe/Bear Specialist Group) (2007) Ursus arctos. The IUCN Red List of Threatened Species. Version 2014.2. www.iucnredlist.org. Downloaded on 13 Nov 2014

Douglas M (1975) Implicit meanings, essays in anthropology. Routledge and Kegan Paul, London

DREAL Midi-Pyrénées http://www.midi-pyrenees.developpement-durable.gouv.fr. Accessed 20 Oct 2014

Europa (1992) Council Directive 92/43/EEC of 21 May 1992 on the conservation of natural habitats and of wild fauna and flora Available via http://eur-lex.europa.eu/legal-content/EN/TXT/?uri=CELEX:01992L0043-20070101. Accessed 1 March 2012

EPRWE European Parliament Resolution on Wilderness in Europe (2009) Available via http://www.europarl.europa.eu/sides/getDoc.do?pubRef=-//EP//TEXT+TA+P6-TA-2009-0034+0+DOC+XML+V0//EN. Accessed 1 Mar 2012

Foucault M (1967) Of other spaces: Utopias and heterotopias. In: Leach N (ed) (1997) Rethinking architecture: a reader in cultural theory. Routledge, New York, pp 330–336

Geertz C (1957) Ritual and social change: a Javanese example. Am Anthropol 59:32–54

Gómez-Ibáñez DA (1975) The Western Pyrenees: differential evolution of the French and Spanish Borderland. Clarendon Press, Oxford

Goodenough WH (1961) Comment on cultural evolution. Daedalus 90(3):521–528

Haraway D (2008) When species meet. University of Minnesota Press, Minneapolis

Ikerd J (2005) Sustainable capitalism: a matter of common sense. Kumarian Press, Bloomfield

Jeo RM, Sanjayan MA, Sizemore D (1999) A conservation area design for the central coast region of British Columbia, Canada. Round River Conservation Studies, Salt Lake City

Keesing RM, Strathern A (1997) Cultural anthropology: a contemporary perspective, 3rd Revised edn. Wadsworth Publishing, Boston

Kearney M (1995) The local and the global: the anthropology of globalization and transnationalism. In: Annual review of anthropology, vol 24, pp 547–565

La Dépêche (2014) Ségolène Royal dit "non" à l'ours. Available via http://www.ladepeche.fr/article/2014/07/20/1921584-pyrenees-segolene-royal-dit-non-a-l-ours.html. Accessed 20 July 2014

Lamazou E (1988) L'ours et les brebis: Mémoires d'un berger transhumant des Pyrénées a la Gironde. Editions Seghers, Paris

Latour B (1993) We have never been modern (trans: Porter C). Harvard University Press, Cambridge

Le Loup en France (2014) http://www.loup.developpement-durable.gouv.fr. Accessed 20 Oct 2014

Le Maho Y (2013) Expertise collective scientifique "L'Ours brun dans les Pyrénées". Available via http://www1.nina.no/lcie_new/pdf/635379157217795378_2013-09-26-Expertise-Collective-Scientifique-Ours-Pyrenees-Museum-National-Histoire-Naturelle-MNHN.pdf. Accessed 1 Dec 2013

MacCannell D (1989) The tourist: a new theory of the leisure class. Schocken, New York

MEDD—Ministre de l'écologie et du développement durable (2006) Plan de Restauration et de Conservation de l'Ours Brun Dans les Pyrénées Françaises 2006–2009. Available via http://www.midi-pyrenees.developpement-durable.gouv.fr/IMG/pdf/OURS-plan_cle05c23a-1.pdf. Accessed 1 May 2013

Moriamé B (2004) Crise du pastoralisme: avec ou sans loup. Available via http://www.reportage.loup.org/html/elevage/crisepasto.html. Accessed 7 Mar 2011

Osborn F (1948) Our plundered planet. Faber and Faber, London

Quenette PY, Alonso M, Chayron L, Cluzel P, Dubarry E, Dubreuil D, Palazon S, Pomarol M (2001) Preliminary results of the first transplantation of brown bears in the French Pyrenees. Ursus 12:115–120

Tsing AL (2005) Friction: an ethnography of global connection. Princeton University Press, Princeton

Tylor E (1871) Primitive culture. J. Murray, London

Urry J (1990) The tourist gaze: leisure and travel in contemporary societies. Sage, London

Whatmore S, Thorne L (1998) Wild(er)ness: reconfiguring the geographies of wildlife. Trans Inst British Geogr 23(4):435–454

Wild Europe (2012) A working definition of European wilderness and wild areas. A discussion draft. Available via http://wilderness-society.org/wp-content/uploads/2014/02/116693719-Definition-of-Wilderness-in-Europe-Draft.pdf. Accessed 15 Mar 2013

Williams C (2010) Ecology and socialism: Solutions to capitalist ecological crisis. Haymarket Books, Chicago

Chapter 11
Traditional Agriculture as Cultural Heritage. Forgotten Agroforestry Practices Recorded in Textual Part of Nineteenth Century Tax Records

Jana Krčmářová and Martin Arnold

Abstract Agroforestry, integration of trees and agriculture, is an example of traditional European land use which became disadvantaged in the modernization. In the Czech Republic nowadays both agroforestry research and practices are virtually non-existent. Was agroforestry ever in use here? Analysis of Franciscan cadastre (1824–1845) revealed that it was present throughout the Bohemia province irrespective of landscape type, land fertility, altitude, population density and nationality (Krčmářová and Jeleček, in prep.). The following chapter will shed light on the main features of Bohemian agroforestry through content analysis of 166 cadastre evaluative protocols (*Schätzung elaborate*) of the aforementioned Franciscan cadastre. It will also discuss the interesting fact that even though the agroforestry tradition was clearly still alive it was not adequately documented in this record. Results showing how common rural practices were not recorded bring better understanding of the process of active forgetting of traditional agricultural knowledge during modernization.

Keywords Agroforesty · Franciscan cadastre · Agricultural modernization · Central Europe · Nineteenth century · Biocultural diversity

J. Krčmářová (✉)
Department of Common Anthropology, Faculty of Humanities,
Charles University, Prague, Czech Republic
e-mail: jana.krcmarova@yahoo.com

M. Arnold
Institut für Sächsische Geschichte und Volkskunde, Dresden, Germany
e-mail: martin.arnold1980@yahoo.de

M. Agnoletti and F. Emanueli (eds.), *Biocultural Diversity in Europe*,
Environmental History 5, DOI 10.1007/978-3-319-26315-1_11

11.1 Traditional Agriculture in Modern and Postmodern Times

Premodern or preindustrial European landscapes could be characterized by diverse mosaic of agroecosystems. These were formed by millennia of locally specific, multifunctional, extensive yet labour intensive management techniques combining various crops, procedures and time schedules. Spatially these landscapes reached their peak in second half of nineteenth century (Green and Vos 2001).

The management of these landscapes was based on knowledge about local environment and how to best adapt the commonly used set of techniques, crops and time schedules. It was key not only to long-term welfare of the households yet sometimes also to their survivor (Mitterauer 1995). This tradition based knowledge was characterized by slow changes (Malanima 2009) and careful sharing and transfer to following generations together with the crops varieties or technologies. Until the establishment of modern agricultural schooling system, the knowledge transfer was mostly local family, village or region based and took oral or written form. The time schedule of agricultural works was further strengthened by a system of occult or religious rituals.

Traditional land uses are commonly defined in opposition to modern practices as "*not...being...part of modern agriculture*" (e.g. Bignal and McCracken 2000). The changes of agriculture in the process of industrialization were so profound, that scholars often talk about "revolution" comparable with the beginning of farming during Neolithic times (Leaf 2003). The process of industrialization in Europe was slowly enfolding from eighteenth century and escalated in the Czech lands in the second half of nineteenth century (Jeleček 1985). As the innovations like machines, chemicals, new crops, technologies and time schedules naturalized in the European landscapes so did the modern "ways of doing things". The agricultural production was to be intensified, rationalized, made more efficient and controllable (Richards 2003).

With the enactment of the modern innovations many of the traditional small scale, extensive and seasonal practices and their intertwined complexes came into disuse, became relic or locally extinct. Some of the preindustrial practices were recorded, thanks to work of ethnographers, some even locally survived, but in general the majority was forgotten.

During industrialisation the mode of production and transfer of agricultural knowledge changed (Skolimowski 1994). Local land use and agricultural decisions were no longer to be grounded in the locally originated and transmitted traditions yet based on modern scientific conclusions. With the new generally applicable techniques and materials bringing unforeseen yields the intergenerational handover of local specific old style information was perceived as unnecessary or moreover heretic to the new belief in progress, rationality and control over nature. New educational institutions arose preparing cohorts of farmers thinking according to the modern standards and scientific disciplines of were created in agriculture as well as other economic sectors (Van Dijk 1989).

During the modernization and later on as the scientific and technical development escalated after the Second World War the importance of agricultural production in European landscape started to be limited to the most productive areas. Elsewhere the land use extensified or ceased et al. As a result of both intensification and abandonment the traditional agroecosystems shrank considerably (Plieninger et al. 2006).

A need to reconceptualize the rural landscape development came forward as the intensive agriculture (and forestry) had lost its prominence in European landscape. The hegemonic modern paradigm of duality of man and nature and the need of human domination and control (Eldredge 1995) was questioned by new paradigms. These are for example paradigm of ecosystem health in the sciences (Rapport 2007), biocultural diversity in the social sciences and the sustainable development on the policy level (Maffi 2007). The new conceptualization of agriculture can mostly be characterized by the plurality of possible approaches from the large scale monoculture intensive and industrial, production and yields-centered to the smallscale, diverse, extensive and conservation or even cultural heritage dedicated.

Preindustrial European landscapes due to high agroecological diversity can from a development point be perceived as a contemporary ideal or target landscape state. As a result the so called traditional ecological knowledge, the "*cumulative body of knowledge, know-how, practices and representations*" arising from long interaction of people with their local natural environment (abbreviated as TEK, ICSU 2002) is beginning to be studied, protected and restored (Maffi 2007). Besides its inherent value it is also appreciated because it can inform us about potentially long-term sustainable local resource management while protecting biocultural diversity (Plieninger et al. 2006; ICSU 2002: 5; Bignal and McCracken 2000; Bennett 1992). The term TEK is sometimes used as an interchangeable term with indigenous knowledge, folk biology or local ecological knowledge (Krech et al. 2004). In the case of temperate Europe TEK can be represented by pre-industrial local environmental knowledge (Bruchac 2014).

11.2 Agroforestry

Agroforestry, integration of trees with agricultural crops and/or livestock simultaneously or sequentially on the same unit of land (e.g. Nair 1993; Mosquera-Losada et al. 2008), is a classical example of traditional agricultural practice. Due to long interaction of European farmers with forested lands and also the rich climatic and microclimatic diversity, Europe is thought to be a region with the historically richest variety of agroforestry practices (Mosquera-Losada 2008). Trees are thought to have been a part of European agriculture from its very beginning (Magyari 2010). Involvement of animals in forests or the inclusion of trees in the fields or meadows can be even thought of as of first technique of landscape management (Hodder et al. 2009).

We can distinguish several basic types of agroforestry like agrisilviculture (crops with logged trees), silvoarable (crops with trees), silvopastoral (pasture/animals and trees), agrosilvopastoral (crops, pasture/animals and trees) and other (Sinclair 1999). Agroforestry systems can be found across climatic and edaphic gradients in all biogeographical regions of Europe (see Table 11.1). In all cases however, these systems rely on continuity of the ground management (grazing, hay collection, cultivation or other) preventing the tree canopy to close.

Agroforestry systems are disadvantaged in the current agricultural and forestry development paradigms. Eichhorn et al. (2006) identifies seven possible factors affecting the distribution of this system from mechanization, demand for more productive, less labour intensive land uses and bigger simpler and monocultural plantations and others. As a result the extensive agroforestry in Europe is on decline (Rois-Díaz et al. 2006; Kirby et al. 1995; Mitlacher et al. 2009; Dahlstrom et al. 2009) and so are the semi-natural open forests (SoEF 2011).

Agroforestry shrinkage is problematic in several aspects. Agroforestry systems possibly achieve the highest degree of agrodiversity worldwide (Geist 2006) hosting numerous species with sometimes great conservational value (lichens, beetles, terrestrial orchids). The cultural semi-closed tree systems are also known to be richer in species when compared to most European forests (Wulf 2003). With the abandonment of agroforestry allotments the landscape and species diversity drops (Rigueiro-Rodríguez et al. 2008). Moreover the cultural diversity of local practices and varieties is lost too (Geist 2006).

As the agroforestry systems become scarce or extinct their ecological, cultural, socio-economic and historical value is starting to be recognized (McAdam et al. 2008). They are studied on the theoretical level and policy incentives for their preservation and restoration are done. Agroforestry presence in the postindustrial landscape briefly stated helps to "*... mitigate the environmental problems caused or exacerbated by commercial agricultural and forestry production enterprises*" (Nair 2008). Trees planted throughout the agricultural ecosystems enhance (a) productive potential of the parcels providing fruits, oil, nuts, timber, firewood, cork or fodder in addition to the agricultural goods, (b) habitat and species biodiversity, (c) regulation functions of essential ecological processes and life support systems (soil and water conservation, reduced nutrient leaching, reduced fire risk and carbon sequestration) and (d) cultural functions (e.g. cultural heritage, recreation) (McAdam et al. 2008).

The techniques represented by agroforestry are also those on which the contemporary strategic planning is focusing in search of a long-term sustainable developmental plan for European semi-forested landscape (Gustavsson 2004).

11.2.1 Agroforestry in the Czech Republic

Agroforestry is nowadays a nearly forgotten phenomenon in the Czech Republic. It is neither recognized on the land use policy level nor studied and the state of

Table 11.1 Temperate agroforestry types and biogeographic distribution

Types of agroforestry (Mosquera-Losada et al. 2008 as enriched AFTA 1997 and Alavalapati and Nair 2001 classification)	Silvoarable agroforestry	Alley cropping (trees planted in single or grouped rows with agricultural or horticultural fields with crops growth in the wide alleys between the tree rows)
		Scattered trees (trees in low density with an annual cropping pattern)
		Line belts (hedgerows, shelterbelts, windbreaks and forest belts)- lines of trees around crop plantations, farms, roads, pastures protecting from wind or blown snow or hedgerows from thorny shrubs creating walls to separate crop areas from grasslands
	Forest farming	Forested areas used for production or harvest of natural standing specialty crops for medicinal, ornamental or culinary uses
	Riparian bufferstrips	Strips of perennial vegetation (trees/shrubs/grass) natural or planted between cropland/pastures and water sources such as streams, lakes, wetlands, and ponds to protect water quality. Act as corridors for flora and fauna; are often remnants of former river plain forests with willows, alder and variety of hard wood species
	Improved fallow	Fast growing, preferably leguminous woody species planted during the fallow phase of shifting cultivation; the woody species improve soil fertility and may yield economic products
	Multipurpose trees	Fruit and other trees randomly or systematically planted in cropland or pasture for the purpose of providing fruit, fuelwood, fodder and timber, among other services, on farms and rangelands. Streuobst in German, pres vergers in France, fruit tree meadows and orchards in English; dehesas in Spain (with e.g. *Castanea sativa*, *Fraxinus* spp., *Betula* spp., *Quercus* spp.)
	Silvopasture	Combining trees with forage and animal production. It comprises forest or woodland grazing and open forest trees
Agroforestry in biogeographical perspective (Rois-díaz et al. 2006)	Atlantic area	Wood pastures in northern Spain, living fences or hedgerows in Great Britain (and examples found also in France)
	Continental area	Windbreaks in Russia and tree belts to protect pastures in Ukraine
	Alpine area	Larchenwiesen (larch meadows) of the eastern Alps and Jura mountains (Italy, Switzerland, Norway)
	Boreal area	Lovangar (foliage meadows) of Sweden, reindeer husbandry in Finland (and other examples of wooded meadows from Sweden, Norway, Estonia)
	Mediterranean area	Dehesas in Spain, montados in Portugal, kouri (wood pastures) in Greece, pascoli arborati in Italy (and other examples from Korsica, Madeira, Slovenia, Croatia, Bulgaria Turkey, Albania)
	Pannonian area	Silvopastoral systems in lines in Hungary

practices is unknown. Yet as the legal land use classification does not recognize plots managed as mix of agriculture and forestry, parcel can be either one or other, we can assume that systems fitting the agroforestry systems would be small garden like and non-commercial.

We can find sparse general notes in agricultural history book from Beranová and Kubačák (2010) about usage of trees in animal diet and crop cultivation. According to the authors the feeding of tree leaves and branches was important in early medieval times, the forest pasture in Czech republic was a common practice which was banned through seventeenth and eighteenth century, while the collecting of acorns and other forest tree fruits and forest litter for bedding could locally be found until the Second World War. The authors mention these practices marginally and even present some of them as pure assumptions like the trees sheltering crops from frost or sunlight in prehistoric times (Beranová and Kubačák 2010). Alternatively, we can find notes about using trees as fodder and bedding before Second World War in Růžičková and Čeněk (2011). No greater attention has been given to Czech agroforestry since.

Consequentially, Czech Republic does not give information about the so-called "*other wooded areas*" (semi-natural or natural stands with scattered tree vegetation) noted for greater biodiversity than other contemporary forests for European comparative statistic (SoEF 2011). Similarly, there exist no official Czech data on other wooded areas which could be categorized as a natural cultural heritage or which are managed by historically valuable techniques (SoEF 2011).

11.2.1.1 Agroforestry in the Nineteenth Century Historical Records

While the current state may indicate that agroforestry ceased in the Czech Republic long time ago, the historical record has proven otherwise. During the previous analysis covering seventeen hundred of cadastral areas of Czech Franciscan cadastre (1824–45) it was found that there was an agroforestry tradition in Bohemia —the largest of Czech Habspurk provinces, at least untill the half of nineteenth century (Krčmářová and Jeleček in prep.).

Franciscan cadastre is one of the last taxing attempts of the Hapsburk monarchy. It covers its whole area providing accurate maps and figures about the land and its inhabitant's activity on it. Basic unit is a city or village with smaller settlements and its agricultural and forest surroundings. Every cadastral map is accompanied by tables summarizing areas covered by various land use categories and by texts evaluating the yields of individual categories and their classes.

In a table document stating the area of variously taxed land types (in original *Ausweis uber die Benutzung*) ten agroforestry categories were found. Namely fields, meadows, pastures and vineyards with several types of trees- in particular, fruit trees, olive trees and forest trees and slash-and-burn systems. Five of them could be found in Czech lands in substantial numbers (Krčmářová and Jeleček in prep.). These were fruit fields, meadows and pastures and wood meadows and pastures. The agroforestry allotments were also discernible in the maps (Fig. 11.1). As can be

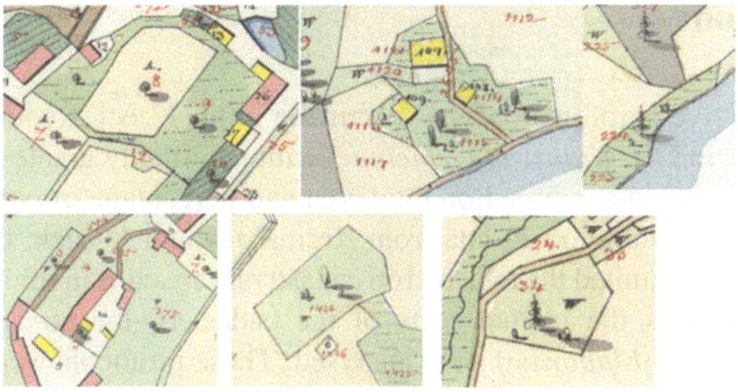

Fig. 11.1 Depiction of agroforestry on the maps (from *left above* to *right down*) **a** fruit field enveloped in fruit meadow; **b** and **c** wooded meadow; **d** fruit pasture; **e** and **f** pasture with wood usage

seen, where the tables discern only the fruit and non-fruit trees, the maps—or more precisely the map makers, seem to make a distinction also between coniferous and broadleaved trees. Moreover in the mentioned previous research it has shown that some of agroforestry categories were present in every one of the nearly seventeen hundred cadastral areas studied.

Briefly, there was a rich classification of agroforestry in the nineteenth century Hapsburg land tax system and the systems were well distributed though did not cover great areas. The quantitative analysis of the area and overall landscape and environmental context of the plots, however, does not reveal any details on features of these systems.

11.3 Aims

As the Franciscan cadaster recognizes agroforestry, the aim was to analyse its more narrative and textually descriptive materials in order to characterize the Bohemian agroforestry.

The following questions were leading the analysis:

(a) What picture of Bohemian agroforestry does the analysis of the Franciscan cadaster evaluative protocols (*Schatzung elaborate*) and yield protocols (*Roh Ertrages Protokol*) give us?
(b) In which parts of the protocols do we find information about agroforestry?
(c) Does the information coverage correspond with the area that agroforestry parcels cover in the villages or with the village's average soil fertility?

11.4 Methods

For each of the five agroforestry types occurring in the Czech Republic 30 villages (cadastral areas, in Ger. *Katastral Gemeinde*) with recordedly greatest occurrence of each category were chosen based on total area and percentage of cadastral area and percentage of agricultural area (data from Krčmářová and Jeleček, in prep.). When these three lists were united the total of 166 cadasters were chosen for further analysis.

For these villages the textual parts of the cadaster—the so-called evaluative protocols (*Schatzung Elaborate*) were analysed. These protocols were developed as accompanying materials during the revision of the original maps—making of the so-called cadaster duplicate. The elaborate is a pre-printed questionnaire consisting of fourteen paragraphs addressing both the environmental and cultural aspects of the village and its subsistence system.

The argumentation accompanying setting of the yield of the individual cultures and their classes is treated in separate document—*Roh Ertrages Protokol*, which is usually enclosed in the evaluative protocol. On several pages without a pre-printed structure the individual classes of the agricultural cultures are described. Also this document was analysed.

Data were collected in the National Archive from the stabile cadastre duplicate fond (*Národní archiv, stabilní katastr duplikát*) from November 2013 till February 2014.

The following data were collected and analysed:

(a) information coverage—the occurrence and amount of information on agroforestry in the various parts of the protocols,
(b) thematic information recorded.

The information coverage was further compared with village specific data on

(a) area covered by agroforestry in general and by its individual categories,
(b) soil fertility data.

The data about area of individual land use categories were transcribed from the formerly mentioned document *Ausweis űber die Benutzung* and analysed as percentage of the agricultural area (part of the dataset used in Krčmářová and Jeleček, in prep.).

To approximate the soil fertility of the historical cadastres the current data about the price of agricultural land was used. The prices are set by Ministry for Agriculture for every cadastral area based on the general soil fertility. They are updated and published as a supplement of Law nr. 151/1997 addressing property pricing.[1] The current cadastres were traced to historical cadastres with the help of archival inventories and current state administration system description and maps.

[1]Table "Základní ceny zemědělských pozemků" (Basic prices of agricultural plots) published as Amendment nr. 16 to Edict of Ministry of finances nr. 279/97 Sb.

The areas of agroforestry and the average soil fertility were compared between the villages where agroforestry was and was not mentioned. Welch t-test for comparing group means in groups with uneven variances was used in JMP 7.0 statistic package.

11.5 Results

The issue of agroforestry was addressed in 49 of the analysed 166 evaluative and yield protocols (*Schatzung elaborate* and the enclosed *Roh Ertrages protocol*) of Franciscan cadastre.

There are following 14 topics addressed in the evaluative protocol.

- Topography, geology and soil
- Climate
- Total area
- Settlements
- *Location of agricultural and forest cultures around the settlement*
- Location of nearest court of justice, administrative office, parsonage, school
- Historical landmarks
- Boarders
- Number of inhabitants and houses
- Usual food and drinks of inhabitants
- Number of domestic animals, their breed, *fodder* and usage
- Water bodies
- Ways
- Market
- *Products of various land use categories*
- *The state and management of individual cultures*
- Obstructions to cultivation on the general level and in particular culture types, obstructions in the breeding of domestic animals
- Machinery used
- Times of seeding and harvest
- Quality of products
- Ownership structure
- Houses and their state
- Industrial facilities

Agroforestry was mentioned in the parts emphasized by italic font—parts about localization of various cultures, animal diet, agricultural products and cultures management. If agroforestry was mentioned in the evaluative protocol, it was commonly treated also in the yield document.

The information coverage found and its correspondence with the actual area of agroforestry in the village is shown in Table 11.2. Whereas the fruit agroforestry coverage correlates positively with the area of the systems, opposite trend is found for the wood pastures. The records are, however, consistently more detailed in the

Table 11.2 Agroforestry information coverage in the evaluative protocol (*Schatzung elaborate*) and yields protocol (*Roh Ertrages Protokol*) and its relationship to agricultural land price and areas taxed as agroforestry

Evaluative protocol paragraph/yield protocol	Number of villages, where mentioned (out of 166)	Relation with agricultural land price	Relation with occurrence of *agroforestry areas merged*	Relation with occurrence of *fruit fields*	Relation with occurrence of *fruit pastures*	Relation with occurrence of *wood meadows*	Relation with occurrence of *wood pastures*
Location of agricultural and forest cultures around the settlement	21	|+|		|+|	|+|	|−|	
		$t = 2.35$		$t = 2.85$	$t = 3.25$	$t = 4.05$	
		$p = 0.03$		$p = 0.009$	$p = 0.003$	$p < 0.0001$	
Number of domestic animals, their breed, fodder and usage	11					|−|	
						$t = 4.43$	
						$p < 0.0001$	
Products of various land use categories	20	|+|	|+|	|+|	|+|		|−|
		$t = 2.5$	$t = 2.36$	$t = 3.18$	$t = 2.44$		$t = 2.98$
		$p = 0,02$	$p = 0.03$	$p = 0.005$	$p = 0.02$		$p = 0.004$
The state and management of individual cultures	18		|+|	|+|	|+|		|−|
			$t = 3.03$	$t = 3.45$	$t = 2.75$		$t = 2.67$
			$p = 0.006$	$p = 0.003$	$p = 0.01$		$p = 0.009$
Yields protocol	42	|+|	|+|	|+|	|+|	|−|	
		$t = 3.18$	$t = 2.12$	$t = 3.1$	$t = 4.13$	$t = 2.82$	
		$p = 0.003$	$p = 0.04$	$p = 0.003$	$p = 0.0001$	$p = 0.005$	
All parts	49	|+|		|+|	|+|	|−|	
		$t = 2.25$		$t = 2.78$	$t = 3.45$	$t = 3.12$	
		$p = 0.03$		$p = 0.007$	$p = 0.0009$	$p = 0.002$	

Results of Welch t-test. The test compared the agricultural land price and occurrence of individual agroforestry categories and agroforestry in general (agroforestry categories merged) in the cadastres where some information was found with those without any note about agroforestry provided

|+| where information coverage is connected with higher price/occurence in landscape |−| where opposite relationship found

Area of agroforestry categories was analysed as % of agricultural land. Only significant results shown (thus fruit meadows omitted altogether)

villages located in more fertile lands (approximated by the current agricultural land price).

The specific information about agroforestry found in the individual parts of the texts is summarized in Table 11.3 for evaluative protocol and Table 11.4 for yield protocol. We can learn where agroforestry plots were located, that the tree leaves or branches would serve as a fodder for animals, how the systems were managed, what was their vegetation structure and what products with what relative yield did they bring. Records document in vast majority agroforestry systems with fruit trees, wood pastures and meadows are with few exceptions not described.

11.6 Discussion

Analysis of space given to agroforestry in evaluative and yield protocols of Franciscan cadastre leads to at least two key findings. First it supports the assumption about existence of living agroforestry tradition up to at least the second half of nineteenth century in Bohemia. Second it enlightens the process of active forgetting of traditions during modernization.

11.6.1 Character of traditional Bohemian agroforestry

There were two major types of agroforestry—one with fruit trees and that for wood usage (in Ger. mit Obstbäumen, mit Obstnutzen and mit Holznutzen, in one case named mit Waldbäumen, with forest trees). Trees would be spread on fields, meadows and pastures.

The agroforestry systems are referred to as to mixed cultures (in Ger. gemischte Kuturen). The mixing is understood as sole occurrence of trees and agriculture on the same spot at the same time without any deeper reasoning about effects of cohabitation of the two elements. The products and management of the ground and tree component are always treated separately. There are no concrete practices which would enhance the "synergies of the tree and the crop" (Mosquera-Losada et al. 2008) documented neither any added general values of the arrangement noticed.

Agroforestry plots were economically relevant as they were included in the taxed land. Only the fruit tree agroforestry was, however, mentioned as commercially profitable as the fruit production brought extra yield to the plot. The trees on wood pastures were not commented on and the yield was computed only from the grass production. We could assume that the non-fruit trees were more relevant from the household economy view or of marginal economic relevance in general.

Systems with trees were mentioned to be located on distinctive plots like slopes, valleys or balks and smaller scattered plots between other cultures. In case of fruit systems the beneficial southern exposition was common, which would help the fruit to ripe. While the fruit systems could be located next to houses or right behind the

Table 11.3 Information coverage and location in various paragraphs of evaluating protocol (*Schatzung Elaborate*)

Paragraph	
Location of agricultural and forest cultures around the settlement[a]	Agroforestry plots mentioned **at the very end of the paragraph**
	Only fruit categories documented
	Fruit fields located on slopes or plateaus with southern exposition, close to the village, among or between the normal fields, protected/shielded locations, eastern slopes or close to buildings
	Fruit meadows mentioned to lay close to houses, on the river islands, scattered among other cultures, in higher places and strong slopes or protected plots on slight slopes, sometimes facing east
	Fruit pastures are described to be scattered, located on southern or western slopes, on smaller plots, being remote from the village, on the boundaries with fields, near houses or similarly located as other pastures
Number of domestic animals, their breed, fodder and usage[b]	Usage of trees mentioned in **section about animal fodder**
	In 12 cases the animal is **goat,** in 3 cases it is a **sheep**
	The fodder are **tree leaves or branches** (in ger. *Laubpauschen, Laub, Baumlaub, Laubreißig, Laubfutter, Laubblättern*) more commonly dried (in Ger. *trockene, getrockene, gedorrte*) served to animals in stalls in winter, yet also usage of fresh leaves is mentioned in the goat's diet in three cases during summer period
	The **species of tree** is not specified
	The animals are usually mentioned to be pastured throughout the spring, summer and autumn and the tree leaves are usually mentioned after the generally much more common hay and straw
Products of various land use categories[c]	Documented in a **separate note** following the pre-printed treeless categories sections
	Only fruit categories documented (17 times fruit fields, 10 times fruit pastures, 6 times fruit meadows)
	Ground and tree products mentioned separately
	- Ground products—hay and aftergrass from pastures and meadows, various "field products" in case of fields (in Ger. *Ackerproducte*)
	- Tree products described in 10 cases simply as fruit, more specifically as stone fruit and/or pome fruits (6 cases); or even classified with regard to genus (3 villages) mentioning in all these cases refined apples, pears and nuts on fields in one case accompanied by plums and cherries and apples and pears on meadows. Fruit trees were described to be in three cases as refined or in five cases of half-refined sorts
The state and management of individual cultures[d]	Described in a **separate section** following the pre-printed treeless categories (except 2 cases when treated by the treeless categories) **Only fruit categories** recorded
	The management of the ground and trees are treated separately- ground is always described to be managed as "pure" field or pasture and the tree as a tree in the garden (The field were usually ploughed, fertilized and harrowed, in the case of pastures ground was left without any kind of management. The trees in gardens were treated with care and given a proper management as including the spuding, fertilization, cleared from parasites and trimmed.)

[a]NA, SK-dupl signatures: Bidez 441; Bunzl 49, 331, 448, 458; Elbog 416; Kaurz 143, 319; Leitm 26, 31, 88, 114, 191, 221, 265, 275, 338, 345, 421, 598, 603

[b]NA, SK-dupl, signatures: Beraun 279; Budw 347, 366; Elbog 18; Kaurz 319,350; Koniggr 294, 483; Leitm 114,588; Saaz 96

[c]NA, Sk-dupl, signatures: Bunzl 17, 331; Czasl 421; Leitm 3, 26, 31, 88, 114, 191, 196, 221, 230, 265, 275, 315, 338, 345, 421, 598, 603

[d]NA, SK-dupl, signatures: Bunzl 49; Leitm 3, 26, 31, 88, 114, 191, 196, 221, 230, 265, 275, 315, 338, 345, 421, 598, 603

Table 11.4 Information about individual agroforestry types found in texts of yields protocol (*Roh Ertrage Protokoll*) of Franciscan cadastre

Fields with fruit trees[a]	One, two or three **classes** (11, 8 and 4 cases respectively) with distinctive plot character and yield
	Location, ground character and fertility, use, management and **yields** on average approximated to those of worse second class fields (some, however, even to first class fields)
	Located (in total of 8 records) on "*southern slopes*", first class fields on "*protected southern slopes*", second class in "*protected location ... with late frosts*" and those of third class in "*higher open locations…strong western or eastern winds at the time of flowering*"
	Trees planted over area from one-tenth to two-thirds, on average one-third
	Tree species in majority cases are a combination of apple, pear and plum trees, sometimes also mixed with cherries and nuts. They are in all cases half-refined or refined. The **health and care** given to the trees is described briefly as being good or poor
	The **monetary usage** of trees is in half of the cases mentioned together with a more or less precise description where these would be transferred to and sold
	The **yield** was separated into that of the ground (field of first, second or third class) and in that of the tree component (on average 1, 5 times the yield of first class fields)
Meadows with fruit trees[b]	One or two **classes** differing in the grass quality or plot character.
	Location described in one case as "*protected damp valley*", in second case as "*sandy and gravel flooded river Moldau islands*" in third only mentioned to be in "*favourable locations*"
	Trees on average planted over 1/3rd of the plots
	Tree **species** varied: mix of apples, pears, plums, cherries, nuts and mirrabells (2 cases), cherries (1 case of 2nd class meadow), plums (3 cases), plums and apples (2 cases)
	The **yield** was separated into that of the ground (as a meadow of certain class) and in that of the tree component (on average as the first class fields)
Pastures with fruit trees[c]	One **class** approximated with regard to location, soil fertility and yields to normal pastures
	Located on the field balks (4 cases), sunny southern or western slopes or plateaus (4 cases), cattle ways, small slopes or plots with stony character
	Usage usually not mentioned—once mentioned to be worse cattle or sheep pastures with mixed grass (sweet and sour grass), once as sheep pastures
	Trees on average planted over one-sixth of the plot
	Tree species described as a mix of apples, pears and plums (7 cases), plums (2 cases), cherries and plums (1 case), apples and pears (1 case) or walnuts (1 case)
	The **yield** was separated into that of the ground (yield of average pastures) and in that of the tree component (on average that of the first class fields)

(continued)

Table 11.4 (continued)

Wood pastures[d]	Only **four records**
	Located on scattered or enclosed plots on hills and slopes exposed to various cardinal signs
	Ground and soil character is sandy soils on sand with gravel in all cases
	Usage/products: cattle pasture with sweet grass, in one case distinguished also second class pastures with mixed or sour grass used mostly by sheep
	The **vegetation structure** is in three cases described as a pasture with birch trees planted one-thirds (2 cases) or one-fourth of the plot
	The trees are not taxed—the **yield** is derived from grass production approximated to fourth class meadows or 4/7 yield of third class meadow
	In one case the tree component is not specified yet there is—except the yield derived from the fodder, mentioned yield from wood of conifers

[a]NA, SK-dupl, signatures: Bidez 371; Bunzl 448, 458; Kaurz 319; Leitm 3, 141, 175, 191, 196, 221, 230, 250, 265, 315, 338, 345, 421, 603
[b]NA, SK-dupl, signatures: Kaurz 143, Leitm 15, 221, 250, 338
[c]NA, SK- dupl, signatures: Bidez 441; Bunzl 17, 116, 191, 224, 331, 458; Czasl 421, 508; Kaurz 319; Leitm 26, 88, 175, 191, 221, 230, 598; Prach 675, Rakon 317
[d]NA, SK-dupl, signatures: Budw 158, 347, 366; Bidez 376

gardens, the wood pastures were more associated with forests and declared to be remote. We can assume that the systems would be kept on places where other systems did not bring their full yield and the trees compensated this hindrance (case of fruit agroforestry) or on places which were not suited for anything else in case of wood pastures.

Such interpretation corresponds with the complexity of premodern land use system, in which a diversity of land uses mirrored the diversity of the landscape in respect to topography or microclimate. Usage of less favourable smaller, hilly, stony, remote or hard to reach plots shows the pressure to use all areas possible to produce at least something. It corresponds with the specific intensity of traditional systems noted by some scholars (e.g. Hakansson 2009). Also the labour intensive character of contemporary management is revealed as the more labour intensive land use systems like fields and orchards (and fruit fields, meadows and pastures) are located near the housing while pastures and forests (and wood pastures) on the outskirts.

In international perspective we can see several current ethnographic parallels and specifics of Bohemian agroforestry.

The fruit agroforestry could be paralleled with the German *Streuobs*t systems (Herzog 1998). Though mostly conceptualized in Germany this system is still found throughout Western, Central and Eastern Europe. It is defined as "*tall trees of different types and varieties of fruit, belonging to different age groups, which are dispersed on cropland, meadows and pastures in a rather irregular pattern*" (Lucke et al. 1992: 10). The tree density varies from 20 to 100 trees per hectare or more. *Streuobst* can also be organized in lines along roads as alleys (Herzog 1998).

The most common fruit types are apple (*Malus domestica* Borkh.), pear (*Pyrus communis* L.), plum (*Prunus domestica* L.) and mazard cherry (*Prunus avium* L.), others are sour cherry (*Prunus cerasus* L.), Persian walnut (*Juglans regia* L.) and other (Lucke et al. 1992).

While the resemblance of Streuobst with the Bohemian historic fruit fields, meadows and pastures with regard to species and tree coverage is striking, the wooded meadows in Germany are not a prioritized and restored agroforestry practice and cannot be compared. There are German traditional names assigned to wood meadows—*die Holzwiesen, die Laubwiesen*. This land use is considered traditional from the prehistoric perspective and is more of Baltic wood meadows nature (Leube 1992).

Nowadays most common example of agroforestry in European landscape—pasture in the forest or silvopasture (Mosquera-Losada 2008: 8) was not described in the studied parts of Franciscan cadastre. There is no indication of pasture in the forest in the classification or texts. No agroforestry category was assigned to the woodland. They would all be listed under agricultural lands and assigned either to fields, meadows or pastures.

However, the most common and abundant category on the landscape level—the category covering the greatest area, were wood pastures. From the maps we can also assume that there were either coniferous or broadleaved trees and bushes left on these meadows and pastures. As there is not any information given on the actual density of the tree cover, we cannot decline that these could be considered as pastured forests. However, we know that they would not be commercially logged—otherwise they would be appropriately taxed. Wood agroforestry use and management is not described nearly at all in the in the section about management of the evaluative protocols nor in the yields protocols. We can assume from the note in the forementioned summarizing table document—*Ausweis uber die Benutzung*, that these systems were pollarded (note "*Kopfholz*" behind the name of wood pastures category). Further from the four found records in the yield protocols we learn about one case, where the coniferous trees on the pasture were harvested as forest trees. In other three wooded pastures cases, for which some information was provided, there were birch trees mentioned on the cattle and sheep pasture. We can only assume, that birch leaves and branches could serve as fodder for goats or sheep (as mentioned for example by Bonfils 2009). In all other cases these categories were omitted from the protocols.

As a possible parallel to Bohemian wood pastures could be perceived Hungarian pastures with oak, wild pear, beech, hornbeam, ash and willow, which are thought to cover about 5500 ha there (Vityi et al. 2014) and being subjected to considerable scientific attention (e.g. Saláta et a. 2013; Varga et al. 2012). The fruit agroforestry is much less studied here. The record of birch trees ad conifers could also indicate similarity with pastures with birches, pines, willows and ashes grazed by sheep in Great Britain (Kirby et al. 1995), pastures/meadows with larches or spruce grazed by sheep and goats known from Swiss Alps (Gillet and Gallandat 1996) or other Alpine areas (Etienne 1996). From the observed range of the wood pasture/meadow systems we can, however, assume that these would have their specific character and variability in Bohemia.

The findings root Bohemian agroforestry in the Central European group differing form Mediterranean, Baltic and Scandinavian types with respect to the prevailing types and the main components.

11.6.2 Voicing Traditional Knowledge in Franciscan Cadastre

The 166 cadasters to start the analysis with were chosen based on the highest area covered by agroforestry from nearly 1700 villages studied. Only 49 of them actually documented the practice. How can we understand the limited amount of information recorded about agroforestry plots?

We can suggest, that the notes about agroforestry were not elaborated on because these plots were too normal, mundane, taken for granted as something which everybody knew. After all the plots were present in every village and the categorization was rich, which may indicate long tradition (Krčmářová and Jeleček, in prep.). Moreover there are parts of the evaluative protocols written like the reader presupposedly knows what the situation was like.

We can also assume that agroforestry was not mentioned due to the little area that it covered or small yields. However, it was not small spatial spread which caused omittance from the record. Wood pastures, for example covered area comparable with some other well documented categories, yet would be only rarely documented. On the other hand the fruit fields and meadows were far less common yet documented more properly. We may assume the greater informational coverage of a phenomenon in tax record was connected to the economical yield. This trivial conclusion could probably explain the difference between the fruit and wood agroforestry coverage.

However, the rationale behind the overall sparse documentation of agroforestry in evaluative protocol might be more complex. These protocols actually do not concern only economically relevant issues—they bring information about the monuments, people diet, industrial facilities, architecture and other. The information coverage of agroforestry theme is greater in villages with higher overall soil fertility or/and higher numbers of fruit agroforestry categories positively correlated with the fertility (Krčmářová and Jeleček, in prep.). The inclosure of the theme in the record was thus not determined only by the importance of the category itself but also by the village. In more agriculturally favourable areas the protocols were more detailed in every aspect. That brings us to a hypothesis that the evaluative protocols of the cadaster were besides the taxing purposes focused on the potential for industrialization and modernization of various areas. The fertile areas were in the focus—they were the ones "with the future" in modern agriculture, ones where the least investments of capital would bring the greatest possible profits (as seen in the land use changes, e.g. Bičík et al. 2001). Also the aforementioned coverage of modern specific themes (industrial equipment, transport networks, ethnographical details about the population diet, architecture or records about the monuments) in the protocol is supportive of this theory.

Agroforestry was probably not recorded because it did not fit into the set of upcoming modern innovations. The emphasis was on the monocultural land uses, simple specialized techniques applicable and transferable everywhere, machinery, chemistry, stability of yields and rationalization of the production. Trees would become an obstacle for machinery and new modern animal diet was concentrating more on use of industrial leftovers than natural products. Ability to be able to clearly define the yields and express these in monetary terms would be a priority as well (for more reasons for agroforestry abandonment see (Eichhorn et al. 2006). The most spatially common category—the wood pastures can serve as an example of opposite of all above mentioned. They were located on small, remote and scattered plots on unfavourable plots like slopes, balks, roadsides and others. The trees would not heighten the possible commercial yield and had their significant role probably only in household economics. Their yield was hard to calculate or generalize in the way used in the cadastre (amount per year, general standardize description). Moreover these parcels were—judging from the maps and four records more diverse than other agroforestry categories. We can come to a conclusion that at least some "mixed" systems of agroforestry were too multifunctional, diverse, local, with various non-monetary yields and resistant to generalization to be counted on in the new development scheme.

It is thought that one of the characteristics of modernization was the rejection of traditions—these were in the contemporary conceptualization of development considered unuseful or even harmful and these assumptions about traditions were vocalized in the educational or scientific literature (Szabo 2013). The disadvantage of traditional agricultural systems and the advantage of the modern innovations on the other hand were also discoursively supported. The fate of traditional practices was sealed when the features of the new conceptualization of agriculture were outlined and these would not fit in. In the Franciscan cadastre we can see the prevalent "innovators" discourse emphasizing the key ideas as the relevant topics while not recording and describing existing topics which would not adjust to new model. We can assume that this discourse demonstrated also in Franciscan cadastre texts helped to institucionalize the vision of industrial agriculture (e.g. Van Dijk 1989).

11.7 Conclusions

Agroforestry is a classical example of a traditional land use which has been locally smitten by the wave of modern innovations and is revived in the sustainability turn seen in European agricultural policy now.

It currently has a minor or nonexistence place in Czech agricultural history, theory or practices. However, Franciscan cadastre records on classification and areas of taxed lands as well as its accompanying texts draw a picture of widely distributed and diversified group of agroecosystems with trees—a living tradition just 150 years ago. The findings assign the found practice pattern to other nowadays Central European agroforestry systems—e.g. the Streuobst in Germany and elsewhere and wooded meadows and pastures of Hungary or Alpine and Atlantic areas.

The low amount of thematic information recorded in the textual parts of cadastre contradicts the wide distribution and rich classification of agroforestry. This finding is consistent with the contemporary modernization intensification and rationalization trends. It is argued that agroforestry could not fulfil the features of ideal modern land use and thus was not supported.

The fact that records were elaborated in greater detail in more fertile areas supports the theory that the evaluating protocol was developed—besides the taxing purposes, to asses the potential for industrialization and modernization of various areas. Focus of innovators was on the areas, where the modern methods would bring the greatest yields with the lowest investments. These areas were serving as case studies and templates for the development of other areas.

The treatment of agroforestry in texts of Franciscan cadaster can serve as an example of omitting the preindustrial traditions in the development schemes otherwise seen in agricultural political and educational discourse domains. While the systems had their part in contemporary landscape, mental horizon and economy they were considered no longer useful and were to a large extent intentionally or unintentionally ignored.

The legitimization of the new—modern ways of doing things was connected to more or less active forgetting of the old ways. Only recently we reach out for the slipping biocultural heritage uncovered in traditional agricultural knowledge and systems.

References

Alavalapati JRR, Nair PKR (2001) Socioeconomics and institutional perspectives of agroforestry. In: Palo M, Uusivuori J (eds) World forests, society and environment: markets and policies. Kluwer, Dordrecht

Association for Temperate Agroforestry, AFTA (1997) The status, opportunities and needs for agroforestry in the United States. AFTA, Columbia

Bennett B (1992) Plants and people of the Amazonian rain forests: the role of ethnobotany in sustainable development. Bioscience 42:599–607

Beranová M, Kubačák A (2010) Dějiny zemědělství v Čechách, na Moravě a ve Slezsku. KOSMAS, Prague

Bičík I, Jeleček L, Kupková L, Perlín R (2001) Long-term trends in the land use changes in Czechia and their societal driving forces in 1845–2000. Agric Econ 47(7):290–295

Bignal EM, McCracken DI (2000) The nature conservation value of European traditional farming systems. Environ Rev 8(3):149–171

Bonfils M (2009) Fodder trees in temperate climate. L'université populaire de permaculture. Available on the internet, http://www.permaculturefrance.org/en/content/article/fodder-trees-temperate-climate-marc-bonfils. Accessed on 10 Oct 2014

Bruchac MM (2014) Indigenous knowledge and traditional knowledge. In: Smith C (ed) Encyclopedia of global archaeology. Springer Science and Business Media, New York

Dahlstrom A, Rydin H, Borgegard SO (2009) Remnant habitats for grassland species in an abandoned Swedish agricultural landscape. Appl Veg Sci 13(3):305–314. doi:10.1111/j.1654-109X.2009.01068.x

Eichhorn MP, Paris P, Herzog F, Incoll LD, Liagre F, Mantzanas K, Mayus M, Moreno G, Papanastasis VP, Pilbeam DJ, Pisanelli A, Dupraz C (2006) Silvoarable systems in Europe—past, present and future prospects. Agrofor Syst 67:29–50

Eldredge N (1995) Dominion. Henry Holt and Co, New York

Etienne M (1996) Western European silvopastoral systems. INRA Editions, Science Update Series, Paris

FOREST EUROPE, UNECE and FAO (2011) State of Europe's Forests (SoEF). Status and trends in sustainable forest management in Europe

Geist H (2006) Agroforestry. In: Heist H (ed) Our earth's changing land: an Encyclopedia of land-use and land-cover change. A-K. Greenwood Publishing Group, Westport

Gillet F, Gallandat JD (1996) Wooded pastures of the Jura mountains. In: Etienne M (ed) Western European silvopastoral systems. INRA Editions, Science Update Series, Paris

Green, BH, Vos W (2001) Managing old landscapes and making new ones. In: Green BH, Vos W (eds) Threatened landscapes: conserving cultural landscapes. Spon Press, London. ISBN: 041925630X

Gustavsson R (2004) Exploring woodland design: designing with complexity and dynamics—woodland types, their dynamic architecture and establishment. In: Dunnet N, Hitchmough J (eds) The dynamic landscape: design, ecology and management of naturalistic urban planting. Spon Press, London

Hakansson NT (2009) A comparative survey of intensive cultivation systems in non-state social contexts. In: First world congress of environmental history, 4–8 Aug 2014, Copenhagen, Denmark

Herzog F (1998) Streuobst: a traditional agroforestry system as a model for agroforestry development in temperate Europe. Agrofor Syst 42:61–80

Hodder K, Buckland P, Kirby K, Bullock J (2009) Can the pre-neolithic provide suitable models for re-wilding the landscape in Britain? Br Wildl 20(5, suppl.): 4–15

ICSU (2002) Science and traditional knowledge. Report from the ICSU study group on science and traditional knowledge, March 2002, available on the internet. http://www.icsu.org/publications/reports-and-reviews/science-traditional-knowledge/Science-traditional-knowledge.pdf. Accessed on 9 Sept 2014

Jeleček L (1985) Zemědělství a půdní fond v Čechách ve 2.polovině 19.století. Academia, Prague

Krčmářová J, Jeleček L (in prep.) Agroforestry types, abundance and land-use context in 19th century Bohemia

Kirby KJ, Thomas RC, Key RS, McLean IFG (1995) Pasture-woodland and its conservation in Britain. Biol J Linn Soc, 56(Suppl.):135–153

Krech S, McNeill JR, Merchant C (2004) Traditional ecological knowledge in Encyclopedia of world environmental history O-Z. Routledge, New York

Leaf J (2003) Green revolution. In: Krech S, McNeill JR, Merchant C (eds) Encyclopedia of world environmental history, Vol 1–3. Routledge, New York

Leube A (1992) Studien zu Wirtschaft und Siedlung bei den germanischen Stämmen im nördlichen Mitteleuropa während des 1.–5./6. Jh. u. Z. Ethnographisch Archäologische Zeitschrift 33:130–146

Maffi L (2007) Biocultural diversity and sustainability. In: Pretty J, Ball A, Benton T, Guivant J, Lee D, Orr D, Pfeffer M, Ward H (eds) Sage handbook on environment and society. Sage Publications, London

Magyari EK, Chapman JC, Passmore DG, Allen JRM, Huntley JP, Huntley B (2010) Holocene persistence of wooded steppe in the Great Hungarian Plain. J Biogeogr 37(5):915–935. doi:10.1111/j.1365-2699.2009.02261.x

Malanima P (2009) Pre-modern European economy. One thousand years (10th–19th century). Brill, Leiden and Boston

McAdam JH, Burgess PJ, Graves AR, Riquero-Rodriguez A, Mosquera-Losada MR (2008) Classifications and functions of agroforestry systems in Europe. In: McAdam JH, Mosquera-Losada M, Riqueiro-Rodriquez A (eds) Agroforestry in Europe. Current status and future prospects. Springer Science and Business Media, Dodrecht

Mitlacher K, Poschlod P, Rosén E, Bakker JP (2009) Restoration of wooded meadows—a comparative analysis along a chronosequence on Öland (Sweden). Appl Veg Sci 5(1):63–73. doi:10.1111/j.1654-109X.2002.tb00536.x

Mitterauer M (1995) Peasant and non-peasant family forms in relation to the physical environment and the local economy. In: Rudolph RL (ed) The European peasant family and society. Liverpool University Press, Liverpool

Mosquera-Losada MR, McAdam JH, Romero-Franco R, Rigueiro-Rodríguez A (2008) Definitions and components of agroforestry practices in Europe. In: Rigueiro-Rodríguez A, McAdam J, Mosquera-Losada MR (eds) Agroforestry in Europe, vol 6. Springer, Dordrecht

Nair PKR (1993) An introduction to agroforestry. Kluwer, Dordrecht

Nair PKR (2008) Foreword. In: Rigueiro-Rodríguez A, McAdam J, Mosquera-Losada MR (eds) Agroforestry in Europe, vol 6. Springer, Dordrecht

Plieninger T, Höchtl F, Theo S (2006) Traditional land-use and nature conservation in European rural landscapes. Environ Sci Policy 9:317–321

Rapport DJ (2007) Healthy ecosystems: an evolving paradigm. In: Pretty J, Ball A, Benton T, Guivant J, Lee D, Orr D, Pfeffer M, Ward H (eds) Sage handbook on environment and society. Sage Publications, London

Richards JF (2003) The unending frontier: an environmental history of the early modern world. University of California Press, California

Rigueiro-Rodríguez A, Fernández-Núñez E, González-Hernández P, McAdam JH, and Mosquera-Losada MR (2008) Agroforestry systems in Europe: productive, ecological and social perspectives. In: Riqueiro-Rodriquez A, McAdam JH, Mosquera-Losada M (eds) Agroforestry in Europe. Current status and future prospects. Springer Science and Business Media, Dordrecht

Rois-Diaz M, Mosquera-Losada MR, Rigueiro-Rodríguez A (2006) Biodiversity indicators on silvopastoralism across Europe. EFI technical report 21, Joensuu, Finland. http://www.efi.fi/attacment/f5d80ba3c1b89242106f2f97ae8e3894/561a5d970cda8e0c77274808a4a8c16a/TR21.pdf. Accessed on 10 Feb 2007

Růžičková V, Čeněk M (2011) Historie chovatelství v Českých zemích z fotoarchivu Národního zemědělského muzea Praha. Profi Press, Prague

Saláta D, Geiger B, Pető Á, Horváth S, Kenéz Á, Malatinszky Á (2013) On the frontier of natural and historical sciences: an integrated multi-proxy approach to assess the landscape history and evolution og Kisgombos Wood Pasture, Hungary. In: Circulating natures: water—food—energy: abstract book of the 7th Conference of the European society for environmental history, 21–24 Aug, Munich, Germany

Sinclair F (1999) A general classification of agroforestry practice. Agroforestry systems 46:161–180

Skolimowski H (1994) The participatory mind: a new theory of knowledge and of the universe. Arkana, London

Szabo P (2013) They did not know what they were doing: the realities of the transition to "scientific" forestry in the Czech Lands. In: Abstract book of ESEH 2013—circulating natures: water—food—energy; seventh biennial conference of the European society for environmental history, 21–24 Aug 2013, Munich, Germany

Van Dijk TA (1989) Structures of discourse and structures of power. In: Anderson JA (ed) Communication yearbook 12. Sage, Newbury Park

Varga A, Molnár Z, Saláta D, Molnár K, Bölöni J (2012) Biocultural changes of wood pastures in the last 250 years (Hungary, Romania-Transylvania). In: Book of abstracts IUFRO 9 conference, Bosznia-Hercegovina, Szarajevó 9–11 May 2012

Vityi A, Marosvölgyi B, Szalai Z, Varga A (2014) Agroforestry research and development in Hungary. In: Book of abstracts of 2nd EURAF conference, 4–6 June 2014, Cottbus, Germany

Wulf M (2003) Forest policy in the EU and its influence on the plant diversity of woodlands. J Environ Manage 67:15–25

Chapter 12
Plant Toponyms as a Tool in Investigating Possible Links Between Cultural and Biological Diversity. The Case of Tuscany

Maria Adele Signorini, Bruno Foggi, Laura Cassi, Luca Ongaro and Federica Frondizi

Abstract Tuscan toponyms and their relations with forest vegetation were investigated, also with the aim of testing whether toponyms can be used as a source of information on vegetation changes. Main forest species growing in Tuscany were selected and referred to expressly defined *topo-species* (corresponding either to botanical species or to groups of species), according to their vernacular names. Tuscan toponyms were related to *topo-species* thanks to an expressly developed computer program. Over 2400 toponyms related with forest *topo-species* were identified, 1048 of which meaning plant communities. Each toponym was located by a GIS software. Distribution maps of toponyms were generated and compared with actual Tuscan forest vegetation. Only 249 cases (10.3 %) showed a perfect correspondence between *topo-species* and current vegetation; some of these inconsistencies are discussed.

Keywords Toponyms · Phyto-toponyms · Tuscany · Italy · Forest vegetation · *Topo-species* · Biocultural diversity

---

M.A. Signorini (✉) · B. Foggi
Department of BIO, University of Florence, Via G. La Pira 4, 50121 Florence, Italy
e-mail: mariaadele.signorini@unifi.it

B. Foggi
e-mail: bruno.foggi@unifi.it

L. Cassi
Department of SAGAS, University of Florence, Via San Gallo 10, 50129 Florence, Italy
e-mail: laura.cassi@unifi.it

L. Ongaro
IAO (Istituto Agronomico per l'Oltremare), Florence, Via A. Cocchi 4, 50131 Florence, Italy
e-mail: ongaro@iao.florence.it

F. Frondizi
Via Troghi 19, 50067 Rignano sull'Arno, Florence, Italy
e-mail: dede2803@tiscali.it

M. Agnoletti and F. Emanueli (eds.), *Biocultural Diversity in Europe*,
Environmental History 5, DOI 10.1007/978-3-319-26315-1_12

12.1 Introduction

Names are perhaps among the first products of human culture. Giving a name to a place is an important cultural act. It states that a place has a meaning which should be remembered. Place names often survive through changes in local landscape, in this way becoming a kind of historical documents of landscape dynamics. They reflect not only the main features of a place, but also those of people who lived there, in this way providing insights into different cultures and environmental perceptions.

A place name (*toponym*) can be considered as part of the process of verbal definition of an area, the product of cultural elaboration by a certain social group, thus being strictly linked to people who use it and to the meaning acquired over the years according to socio-economical changes; it is part of a cultural heritage shared by a local community (Stopani 1994).

Toponymy as a scientific branch of knowledge has been established only in recent times; its aims are mainly to examine place names related to cities, regions and geographical elements, evaluating their meaning, their historical changes and their connections to natural and cultural environment. Names of places are needed to form a spatial reference system to identify objects in a geographical area, and remained for a long time as the only instrument to define and distinguish land tenures, until the introduction of cadastral cartography (Cassi 2008).

Place names have been handed down for centuries and sometimes they have come to us so altered and complex that it is not simple to trace their meaning; as a matter of fact, this kind of analysis is mainly linguistic but is necessarily supported also by other disciplines, like geography, archaeology and anthropology. At the same time, toponymy can be considered a good instrument for investigations aimed to shed light on the processes of colonization of a certain area, on traditions and economic activities and also on landscape dynamics, especially by comparing the distribution of plant-related toponyms with that of botanical species and of plant communities, in this way also providing insights into possible past and recent climate changing (Pellegrini 1990). A previous work carried out in Tuscany (Italy) highlighted the link between place names related to beech (*Fagus sylvatica* L.) and holm oak (*Quercus ilex* L.) and the distribution of those species, considering possible inconsistencies (especially in the case of beech) as a clue of the existence of heterotopic stations of the species (Mori 1930). The research was later extended in the same region by Cassi (1973), who analyzed a higher number of species—including both woody plants and herbs, and wild or cultivated ones—and also took into consideration generic names of tree communities, such as those meaning 'forest' or 'wood'. In this way, 4804 toponyms related to plants were found and located on the map. This survey showed the importance of plant-related toponyms as means to understand the perception of the environment by people who live in a certain geographical area, and in retracing vegetation changes through the centuries, as well as the existence of different linguistic and dialectal areas. Toponyms from the above-cited surveys were recently used by Monacci et al. (2010) as a source of information in finding out some heterotopic stations of beech near Pisa (Tuscany, Italy), thus shedding light on ancient plants distribution, their indigenous status and their migration and/or colonization dynamics.

Toponomastic analyses are characterized by many uncertainties: we cannot overlook the long history of some toponyms and the conscious or unconscious transformations made by people who have been using them, or even by cartographers during their work; moreover, toponyms were originally handed down orally and were often originated and used only by a single family or by a little social group, making this kind of information ephemeral; most of them are also doomed to disappear due to socio-economical changes implying the abandonment of rural areas and the loss of many traditional activities linked to rural society.

It must also be stressed that the meaning of toponyms can often be ambiguous. Toponyms as linguistic signs are made up of two parts: signifier and signified (according to the linguistic model by Ferdinand de Saussure); but the correlation between the two can hardly be assumed (Pellegrini 1990). During the process of creating toponyms, in some atypical but quite recurring cases it is not possible to find the logic connection between the meaning of a place name and the corresponding geographical object (Gerola 1950). According to Ritchot (1989), toponyms normally stress what is perceived by local people as clashing with the surrounding environment, rather than describing what is usual, in this way expressing a will of contradiction. This can be of the highest importance in considering plant-related toponyms as a source of information in retracing past distribution of plants and vegetation in a certain area; the existence of both convergence and divergence between the distribution of plant-related toponyms and corresponding botanical species cannot lead to a defined conclusion (Pelé 1992). That is why studies concerning plant-related toponyms must involve scholars of different scientific and humanistic disciplines.

Dating toponyms with certainty also presents difficulties; in Tuscany most of them probably arose in the fourteenth century, with the spread of *mezzadria* (sharecropping); but it is hard to date them back more precisely.

Since toponyms as cultural products can be used as a source of information on vegetation and landscape biodiversity, the first aim of this research is to develop and test a method for searching plant-related place names, so that they can be subsequently investigated in their possible links with vegetation. All the toponyms that are supposed to originate from plant names will be referred to as *phytotoponyms*.

12.2 Plants and Toponyms in Tuscany

Plants are one of the most important sources for toponyms. In Tuscany, many place names are related to plants, and most of them to trees or woods (Cassi 1973). In our survey, only names of places related to trees have been analyzed, on the ground of the following considerations: no other group of plants has such an influence on human life; trees represent a source of building material, fuel, food and medications, but they also bear many symbolic meanings in different human cultures; thanks to their size, the length of their life cycle and their tendency to form big communities, they have a high impact on landscape. We excluded toponyms connected with generic words meaning ‘wood’, ‘forest’ or the like, because they give some non-specific or even ambiguous information.

The most relevant dominant forest tree species growing in pure or mixed woods in Tuscany were selected for this research. Some more sporadic species were also included, due either to their wide distribution or to their socio-economical importance. Among these, field maple (*Acer campestre* L.), which can be found in most Tuscan woods (Hofmann et al. 1998) and has been used for centuries in Tuscan agricultural practices; and the cork oak (*Quercus suber* L.), traditionally cultivated for centuries for cork production along the southern coast of Tuscany.

Vernacular, local and ancient names for each plant and plant community used in Tuscany and in the neighbouring regions (Liguria, Emilia Romagna, Marche, Umbria, Lazio) were obtained thanks to an in-depth research carried out on botanical dictionaries and other works reporting lists of plant and woodland names, such as forestry, linguistic and geographical papers (Bernetti and Mondino 1998; Cassi and Marcaccini 1998; Di Bérenger 1859; Gherardini 2006; Penzig 1924; Targioni Tozzetti 1809). A list of names of plant species and plant communities was subsequently set up, but many problems were encountered in establishing univocal correspondences between names and botanical species. For this reason we decided to refer the selected species to expressly defined categories named *topo-species*, that is, plant typologies identified on the ground of their vernacular names. A *topo-species* can exactly correspond to a single botanical species, as in the case of *castagno* (chestnut, *Castanea sativa* Miller); or to several species belonging to the same genus, as in the case of *aceri* (some maples, *Acer* sp. pl.); or even to species belonging to different genera, as in the case of *carpini* (hornbeam, *Carpinus betulus* L., and hop-hornbeam, *Ostrya carpinifolia* Scop.). This is because different sorts of plants are defined and identified on diverse bases by human communities and by botanists, as it is well known in ethnobotanical studies, where the concept of *ethno-species* (or *folk generic*) has already been established (Signorini et al. 2007; Berlin 1972, 1992). As a matter of fact, different species that show similar morphological and ecological characteristics and are used by people for the same purposes cannot be easily discernible for local communities, thus forming an elementary unit of folk taxonomy which does not exactly correspond to a botanical species; in addition to this—and in this way diverging from the concept of *ethno-species* or *folk generic*— two or more species can be actually identified and distinguished by local people because of their use or of different morphological or ecological characteristics and can consequently get different vernacular names, where a ‘specific’ epithet (usually an adjective) separates a species from another; but this kind of information can be lost in phyto-toponyms. Some examples related to the genus *Salix* can explain it: different species belonging to this genus are often distinguished by vernacular names, mainly on the ground of the colour of their branches, but generally phyto-toponyms only refer to the generic local name used for all the species of the genus *Salix*. In addition to this, the vernacular name 'vinco'—from the Latin verb *vincĭo*, meaning 'to tie'—is referred to those species of *Salix* that were commonly used for traditional agricultural practices (especially to tie vegetables or grapevines to their support), not to a well defined botanical species. On the contrary, some species are so recognizable for local people that vernacular names are not ambiguous, as in the case of 'faggio' (beech, *F. sylvatica*), 'castagno' (chestnut, *C. sativa*) and 'cipresso' (cypress, *Cupressus sempervirens* L.); in these cases, phyto-toponyms can be clearly correlated to a single plant species. Some toponomastic surveys have also been used in this study as a source of information (Cassi 1973; Del Giudice 1988; Millemaci 1999; Pellegrini 1990; Pellegrini et al. 1990; Pieri 1898, 1919, 1969; Rossi 1985; Stopani 1994), in order to detect plant-related place names with a somehow ambiguous origin. As a matter of fact, many phyto-toponyms derive from Latin terms combined together and in resulting words the original meaning could be hidden: an example can be the phyto-toponym 'Fatimondo', deriving from the two Latin words *fagetu mundu*, meaning 'tidy beech forest' (Pieri 1919). Some toponyms have gone through such transformations that they can be interpreted only with the aid of linguistic studies. This is the case, for example, of the toponym 'Impruneta', that comes from the Latin expression *in poneta* (in pine forest), and is consequently related not to the word *prunus* (brier), but to the word *pinus* (pine): the linguistic root of the phyto-toponym has been so altered, that the true etymology is currently not known even by local people (Pieri 1919).

For this research, a total of 21 *topo-species* were considered, corresponding to the most relevant forest tree species in Tuscany (Table 12.1).

12.3 Methodological Approach

The Istituto Geografico Militare (IGM) maps of Tuscany at the scale 1:25,000, 'Serie 25', were used as a source for toponyms. They were chosen because the number of toponyms reported is higher than in other even more recent and detailed maps. A digital list of all the Tuscan toponyms recorded on these maps has been obtained, in which every place name is linked to the UTM coordinates of the bottom left corner of the corresponding map text location.

An ad hoc computer program was expressly created to compare vernacular names of tree plants with Tuscan place names drawn from IGM maps. The program has been written in Python, using the Levenshtein extension module,[1] which provides a library of functions for fast computations of text string similarity. The software automatically checks the map names against a reference file containing an array of ancient and vernacular common names for each of the selected *topo-species*, measuring their similarity according to Jaro-Winkler distance (Winkler 1990). A stop words file is also provided as input, containing a list of articles, prepositions, most

[1]https://github.com/ztane/python-Levenshtein.

Table 12.1 *Topo-species* selected and analyzed in the present study

Topo-species	Botanical species	English common names
Abeti	*Abies alba* Mill.	Fir and spruce
	Picea abies (L.) H. Karst.	
Aceri	*Acer platanoides* L.	Norway and sycamore maples
	Acer pseudoplatanus L.	
Betulla	*Betula pendula* Roth	Birch
Carpini	*Carpinus betulus* L.	Hornbeam and hop-hornbeam
	Ostrya carpinifolia Scop.	
Castagno	*Castanea sativa* Mill.	Chestnut
Cerro	*Quercus cerris* L.	Turkey oak
Cipresso	*Cupressus sempervirens* L.	Cypress
Faggio	*Fagus sylvatica* L.	Beech
Frassini	*Fraxinus angustifolia* Vahl s.l.	Common and narrow-leaved ashes
	Fraxinus excelsior L.	
Leccio	*Quercus ilex* L.	Holm oak
Olmi	*Ulmus glabra* Huds.	Elms
	Ulmus minor Mill.	
Ontani	*Alnus cordata* (Loisel.) Loisel.	Alders
	Alnus glutinosa (L.) Gaertn.	
	Alnus incana (L.) Moench	
	Alnus viridis (Chaix) DC.	
Oppi	*Acer campestre* L.	Field, Montpellier and Italian maples
	Acer monspessulanum L.	
	Acer opalus Mill. s.l.	
Orniello	*Fraxinus ornus* L.	Manna ash
Pini	*Pinus halepensis* Mill.	Mediterranean pines
	Pinus pinaster Ait.	
	Pinus pinea L.	
Pioppi	*Populus alba* L.	Poplars (white and black poplar)
	Populus nigra L.	
Querce	*Quercus petraea* (Matt.) Liebl.	Deciduous oaks (except Turkey oak)
	Quercus pubescens Willd. s. l.	
	Quercus robur L.	
Robinia	*Robinia pseudoacacia* L.	Black locust
Salici	*Salix alba* L.	Willows
	Salix caprea L.	
	Salix cinerea L.	
	Salix eleagnos Scop.	
	Salix pentandra L.	
	Salix purpurea L.	
	Salix triandra L.	
	Salix viminalis L.	

(continued)

Table 12.1 (continued)

Topo-species	Botanical species	English common names
Sughera	*Quercus suber* L.	Cork oak
	(*Quercus crenata* Lam.)	
Tremolo	*Populus tremula* L.	Aspen

common words such as 'casa', (house), 'monte' (mount) etc., to be filtered before processing. Considering the value 1 as the exact correspondence between two strings and the value 0 as two completely different strings, a threshold value for each *topo-species* was empirically chosen after many test runs of the program, to be the minimum limit of similarity that could be considered. For all the *topo-species* the final similarity threshold was set at about 0.96. With this value, even phyto-toponyms, quite different from vernacular names but clearly linked to forest species, could be detected; at the same time, the number of false positives (see below) proved to be passably low. The computer program was also very efficient in overcoming some common typos in the original maps, mostly linked with the inhomogeneous use of accents and apostrophes, which are characteristic of Italian language.

Thanks to this program, 62,720 Tuscan toponyms could be analyzed in a very short time (in the order of magnitude of few seconds on a standard desktop computer) and a digital table was automatically generated, including 2910 possible phyto-toponyms related with selected *topo-species*.

Every toponym was subsequently manually checked in order to delete false positives. A list had been previously created of those words that should be excluded from automatic comparison, being quite similar to plant names but clearly not related to them. For example, the word 'colmo' (meaning either 'peak' or 'overflowing', speaking of the sedimentation process in a river) contains the word 'olmo', that is the Italian name for elms (*Ulmus* sp. pl.), but it has a completely different meaning and toponyms originated from it are not at all related with elms. In order to clarify specific problems concerning phyto-toponyms with ambiguous meanings, many linguistic and toponomastic works were consulted (Ambrosini 2006; Del Giudice 1988; Ferro 1964; Finamore 1980; Millemaci 1999; Nocentini 2010; Pellegrini 1990; Pellegrini et al. 1990). Names remaining still not clear after these analyses were definitely excluded. Many toponyms were also removed as probably linked to family names and not to plants, especially when plant names follow words such as 'casa' (house), 'podere' (holding, plot of land), etc., because it was hard to ascertain whether they had been originated by a plant name or instead by the name of the owner of a plot of land or a building. This is the case, for example, of the name 'Villa Pini': this toponym is probably linked to a Pini family, even if a connection with the existence of a group of pines (in Italian: 'pini') cannot be excluded. Following up a critical examination of all the 2910 toponyms automatically detected by the program, 487 of them (16.7 %) proved to be false positives and were manually deleted. In the final table, 2423 phyto-toponyms related

Table 12.2 Table of phyto-toponyms (part)

Topo-species	Phyto-toponym	UTM *x*	UTM *y*
Faggio	La Faggionaia	686,072	4,756,080
Faggio	Monte di Fò	682,709	4,883,180
Faggio	Monte Faggio all'Ombrellino	687,137	4,882,420
Faggio	Monte Faggiolo	724,923	4,854,990
Faggio	Monte Faieto	593,872	4,892,490
Faggio	Montefatucchio	736615	4,847,310
Faggio	Poggi di Faeta	700,378	4,739,150
Faggio	Poggio del Faggio	763,714	4,850,380
Faggio	Rio di Faggeto	714,635	4,882,360
Faggio	Sorgente della Faggeta	668,066	4,884,970
Faggio	Villa Fattucchia	68,488	4,845,160
Faggio	Ville di Faggeta	742,838	4,816,820
Faggio	Zona Faggio al Respiro	715,320	4,835,520
Frassini	Casa Fràssine	717,595	4,866,670
Frassini	Casa Frassinello	637,886	4,806,740
Frassini	Casa Frasso	645,892	4,856,980
Frassini	Fosso dei Frassini	722,554	4,757,890
Frassini	Frassanello	715,833	4,881,030
Frassini	Frasseta	655,883	4,780,420
Frassini	Monterifrassine	696,852	4,854,570

Each toponym is associated to a *topo-species* and to geographical coordinates

with the selected forest topo-species were reported; each single toponym being associated both to a topo-species and to geographical coordinates (Table 12.2).

Phyto-toponyms were later split into three groups: toponyms indicating a single plant, a group of plants, or a plant community. These were usually detected thanks to suffixes like '-eto/a', '-aio/a', '-ino/a', but also with the aid of toponomastic works in the case of ambiguous forms. Phyto-toponyms related to plant communities (1048 toponyms, that is, 43 % of the total) were then used for further comparisons with actual Tuscan forest vegetation (see below).

Using a GIS software, every geo-referenced phyto-toponym was plotted on a digital map of Tuscany, and different distribution maps were generated: a general map showing the distribution of all the phyto-toponyms (Fig. 12.1a) and specific maps with the distribution of one (or in a few cases of two) *topo-species* (Fig. 12.1b). Distribution maps of phyto-toponyms were later analyzed and compared with the most significant available documents on actual Tuscan forest vegetation (Arrigoni 2001; Bernetti and Mondino 1998; Hofmann et al. 1998) and with the Corine Land Cover map of Tuscany,[2] in order to investigate possible

[2]http://www502.regione.toscana.it/geoscopio/servizi/wms/USO_E_COPERTURA_DEL_SUOLO.htm.

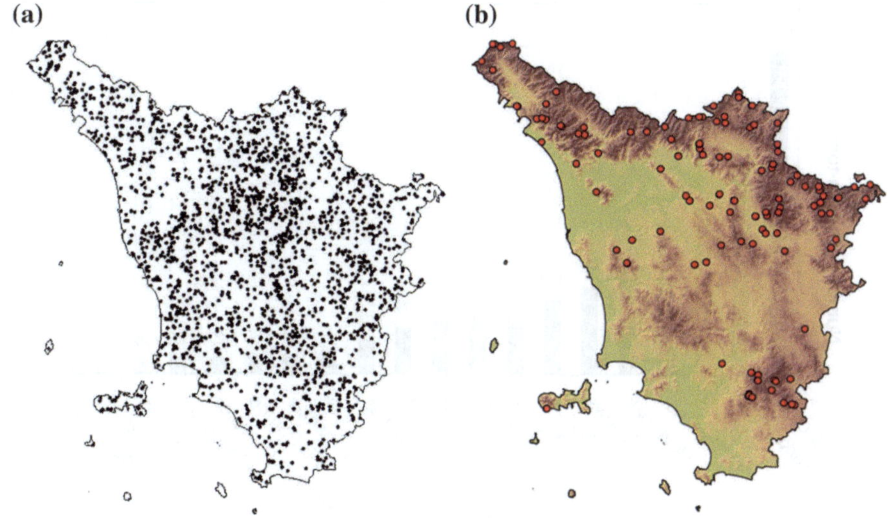

Fig. 12.1 Distribution maps of phyto-toponyms. **a** General map showing the distribution of phyto-toponyms related to all the selected *topo-species*. **b** Distribution map of toponyms related to the *topo-species faggio* (beech, *Fagus sylvatica* L.)

relationships between phyto-toponyms and vegetation. In discussing the results, it must be considered that coordinates and elevation of each phyto-toponym on the maps were assumed to be those of the bottom left corner of the corresponding map text location on IGM maps, and consequently it is possible that they do not exactly correspond to those of the place identified by that toponym.

12.4 Results and Discussion

Generally speaking, *topo-species* associated with the highest number of toponyms resulted to be the most common forest species in Tuscany today (Fig. 12.2): deciduous oaks (*Quercus* sp. pl.), chestnut (*C. sativa*), beech (*F. sylvatica*), hornbeams (*O. carpinifolia*, *C. betulus*). When comparing the distribution of actual Tuscan forests (Hofmann et al. 1998) with that of phyto-toponyms linked to forest communities, a quite good correlation was found (Fig. 12.3). The percentage of toponyms related to chestnut and holm oak communities turned out to be lower than the one related to corresponding woods, but it must be considered that the number of toponyms would increase if some place names, not considered because generic or ambiguous, were also taken into account. Toponyms such as 'selva' (wood) and 'palaia' or 'palina' (coppice) are about 300 and most of them are probably to be referred to chestnut woods; similarly, the names 'forteto' and 'macchia' can be at least partly related to Mediterranean woods and scrubs dominated by *Q. ilex*. Yet in

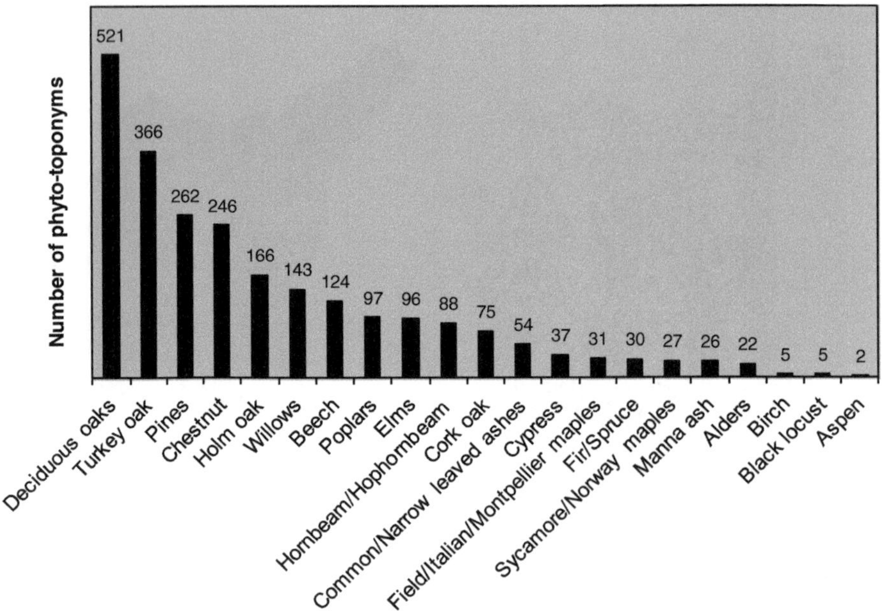

Fig. 12.2 Distribution of Tuscan phyto-toponyms per *topo-species*. 'Deciduous oaks' include *Quercus petraea* (Matt.) Liebl., *Q. pubescens* Willd. s. l., *Q. robur* L

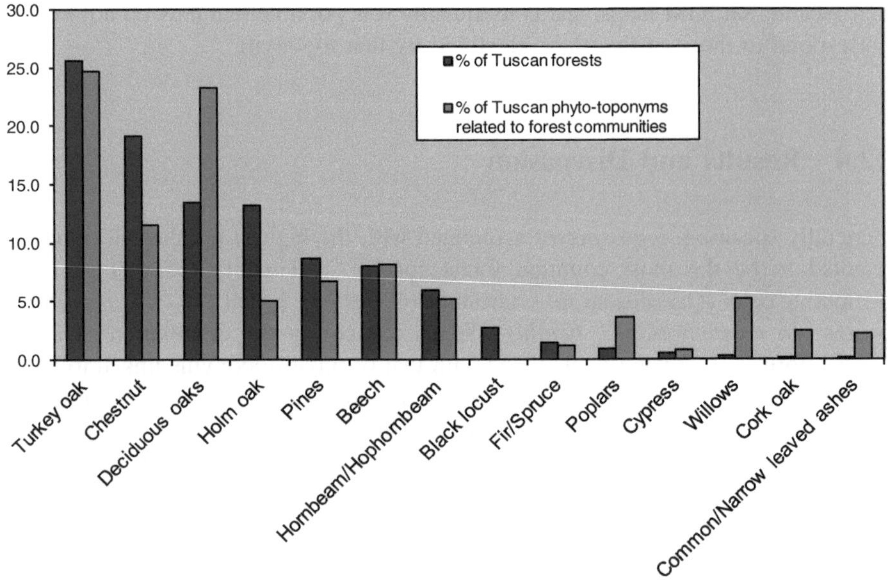

Fig. 12.3 Correlation between the distribution of actual Tuscan forests according to Hofmann et al. (1998) and the distribution of phyto-toponyms linked to forest communities

other instances different explanations can be found. Black locust (*Robinia pseudoacacia* L.) woods are more common than corresponding phyto-toponyms, most likely because the diffusion of this species in Tuscany dates from relatively recent times, so that there was not enough time for toponyms to be established. On the other hand, toponyms related to some sporadic species like cork oak (*Q. suber*) and ashes (*Fraxinus* sp. pl.) are more frequent than corresponding woods. Explanations for these inconsistencies can be found in the small extent of woods, in the possible presence of extra-zonal communities, in the cultural and economic importance of some tree species that can lead to an over-representation in toponyms (in the case of *Q. suber*); but in some cases vegetation changes can also be involved, such as the recent vanishing of some floodplain forest in the case of ashes. Only a detailed study on each phyto-toponym can succeed in explaining its relationship with forest vegetation, perceived as something common or, on the contrary, as something unusual.

Surprisingly, when comparing tree-linked toponyms with the forest vegetation map of Tuscany (Arrigoni 2001), only 877 phyto-toponyms fall within forest areas (36.2 %) and only 249 cases (10.3 %) show a perfect correspondence between *topo-species* and dominant species in current vegetation. These inconsistencies can be partly related to methodological aspects of the map, where small woods, vegetation growing on forest borders, riparian communities and locally dominant species have probably been underestimated. Yet in some cases, landscape changes can also be involved.

The importance of methodological aspects becomes evident when comparing plants-linked toponyms with Corine Land Cover map. In this case, 1818 toponyms (75 %) fall within woods. This higher percentage can be partially explained with the different definition of woods and the more detailed scale of this map. Nevertheless, some general considerations on tendencies emerging from these comparisons can be made. For example, in the case of toponyms linked to the *topo-species olmi* (elms, *Ulmus* sp. pl.), toponyms falling within woods are 23 % according to forest vegetation map of Tuscany and 57 % according to the Corine Land Cover map. Regardless of the two different values, in both cases the percentage is quite low. This can be related to the fact that elms used to be a component of rural landscapes rather than forests, as they were commonly planted to be used in agricultural practices, while being only species of secondary importance in woods. Similar considerations can be done for other *topo-species*, such as *cipresso* (cypress, *C. sempervirens*), *oppi* (field maples, *Acer* sp. pl.) and *salici* (willows, *Salix* sp. pl.). On the contrary, in the case of typical forest trees, the percentage of phyto-toponyms falling within woods is generally quite high, with some exceptions for *querce* (deciduous oaks) and *castagno* (chestnut). Yet as already observed above, this can be partially due to the exclusion of some generic or ambiguous names related to forest communities.

Anyway, in many cases the relatively low number of phyto-toponyms falling within current woods can also be due to landscape changes. This is the case, among the others, of the Tuscan village Castagneto Carducci, whose name is clearly connected with the word 'castagneto' (chestnut grove). Chestnut groves for fruit production used once to be common near Castagneto Carducci, but in the past 150 years they have been progressively abandoned as a consequence of socio-economic changes, as the population mostly turned from farming, breeding and forest activities to olive, cereals and grape cultivation and other economic activities (Agnoletti 2009). At present times, chestnut groves are definitely not common in the surroundings of Castagneto Carducci, but the toponym still bears memory of this past land use.

When considering the toponyms elevation, a high correlation between the distribution of toponyms and that of the corresponding *topo-species* was found for some trees, such as holm oak (*Q. ilex*) and cypress (*C. sempervirens*), possibly because in these cases phyto-toponyms have a clear and unambiguous botanical meaning and each *topo-species* exactly corresponds to a single botanical species, with a more or less limited ecological range. On the other hand, the altitudinal range proved to be very wide in the case of other species. This can be due to the fact that some *topo-species* include species with different auto-ecology or with a wide ecological range, as it is probably the case with the *topo-species aceri* (i.e. some maples, *Acer* sp. pl.); but environmental and/or cultural reasons may also be involved. For example, in the case of phyto-toponyms related to the *topo-species abete* (*A. alba* Miller and possibly also *Picea abies* (L.) Karst.), the wide elevation range could be related to the plantations of firs (*A. alba*), created and maintained near monasteries or churches located at different altitudes. On the other hand, the wide elevation range for toponyms linked to beech (*F. sylvatica*) can be partially explained with the existence of extra-zonal beech woods (Monacci et al. 2010). On this aspect, phyto-toponyms as a record of vegetation history can represent a testimony of ancient climate changes. Considering the distribution of the 66 toponyms linked to beech falling within woods, only 25 of them (37.9 %) are located in woods currently dominated by beech: most of them (41, i.e. 62.1 %) fall within different kinds of woods and only eight of these (12.1 % of the total) are located near already described extra-zonal beech forests (Monacci et al. 2010), that are regarded as important evidences of ancient climatic events. Perhaps phyto-toponyms could be used as a clue in further researching extra-zonal communities.

12.5 Conclusions

Comparing our results with previous studies on the same subject (Cassi 1973; Mori 1930), the advantages of a computerized method are evident, especially the outstanding rapidity and precision in detecting plant-related names among toponyms. Automatic comparison between toponyms and common plant names can allow researchers to process a lot of data in a very short time, also with the possibility of investigating toponyms concerning wide areas. Moreover, human errors in detecting toponyms and in manually calculating geographical coordinates and elevations can be eliminated, even if some data can be less accurate, as they are referred to the bottom left corner of the map text location and not to the exact location of the place. Anyway, on the average elevation values for each *topo-species* proved to be essentially comparable. Even if the number of phyto-toponyms found with this automated method cannot be directly compared with those obtained in previous studies because of some different linguistic and methodological choices, the adopted method proved to be both quick and effective (Table 12.3).

Problems connected with the peculiar multi-disciplinary kind of research of phyto-toponomastic are many. The time of formation of most toponyms is hardly known with certainty; some place names can be ambiguous because they have been altered through centuries and in many cases today their meaning can hardly be understood; they have been modified or omitted by cartographers and/or forgotten by local communities together with rural abandonment. In addition to this, in many cases it is difficult to set a well defined connection between a phyto-toponym and a botanical species. Nevertheless, as phyto-toponyms mostly reflect both local vegetation and socio-cultural features of local communities, research on place names can give an important contribution in understanding links between cultural and environmental heritage of an area. Any inconsistency should be carefully analyzed,

Table 12.3 Tuscan phyto-toponyms related to forest species: results of the present study and of a previous research (Cassi 1973), in which phyto-toponyms had been manually detected

Topo-species	Number of phyto-toponyms			
	Total		Related to plant communities	
	Present study	Cassi (1973)	Present study	Cassi (1973)
Abeti	30	23	12	6
Aceri	27	18	17	–
Betulla	5	4	4	3
Carpini	88	72	51	40
Castagno	246	197	113	>70
Cerro	366	328	243	203
Cipresso	37	33	8	5
Faggio	124	105	80	42
Frassini	54	40	22	15
Leccio	166	148	49	–
Olmi	96	89	13	5
Ontani	22	23	11	–
Oppi	31	21	1	–
Orniello	26	25	17	8
Pini	262	240	66	39
Pioppi	97	84	35	–
Querce	521	414	229	146
Robinia	5	–	0	–
Salici	143	90	51	>25
Sughere	75	71	24	18
Tremolo	2	3	2	–

but many appearing contradictions can be explained with the double meaning of toponyms, linked to something that can be either very common in an area, or exceptional. In other instances, they can be related to methodological aspects of vegetation studies and derived maps, but the history and evolution of woods and landscape must also be considered. Forests have been modified in their extent and specific composition by past and recent climate changing, fires and plant diseases and social transformations; and these changes are still ongoing. Through toponyms, we have the opportunity to process a great amount of different data which bear valuable information about both cultural and environmental factors in the past. If we can associate each phyto-toponym with the approximate time of its formation, we can shed some light on the history of forest vegetation and also on landscape changes, but always critically considering every single case and watching out for generalizations.

References

Agnoletti M (2009) Il paesaggio come risorsa: Castagneto negli ultimi due secoli. ETS, Pisa

Ambrosini R (2006) Lucca e il suo territorio: toponomastica, dialettologia, critica linguistica. Maria Pacini Fazzi, Lucca

Arrigoni PV (2001) A forest vegetation map for Tuscany. An ever-changing document. Parlatorea V:119–127

Berlin B (1972) Speculations on the growth of ethnobotanical nomenclature. Lang Soc 1:51–86

Berlin B (1992) Ethnobiological classification: principles of categorization of plants and animals in traditional societies. Princeton University Press, Princeton

Bernetti G, Mondino G (1998) I tipi forestali. Boschi e macchie di Toscana. Edizioni Regione Toscana, Firenze

Cassi L (1973) Distribuzione geografica dei toponimi derivati dalla vegetazione in Toscana. Riv Geogr Ital LXXX(4):389–432

Cassi L (2008) Il bosco nei nomi di luogo della Toscana. Regione Toscana. http://www.arsia.toscana.it/antichimestieri/percorsi/universita/bosco/bosco.htm. Accessed 3 Nov 2011

Cassi L, Marcaccini P (1998) Toponomastica, beni culturali e ambientali. Gli "indicatori geografici" per un loro censimento. Società Geografica Italiana, Roma

Del Giudice CA (1988) La toponomastica del Comune di San Vitale (Massa di Lunigiana). Aedes Muratoriana, Modena

Di Bérenger A (1859–1863) Dell'antica storia e giurisprudenza forestale in Italia. Longo, Treviso e Venezia

Ferro G (1964) La toponomastica ligure di interesse geografico. Università degli studi di Trieste, Facoltà di magistero, Pubblicazioni n. 5, Trieste

Finamore E (1980) I nomi locali italiani. Origine e storia. Edizioni Nuovo Frontespizio, Rimini

Gerola B (1950) Sul rapporto logico tra etimo e toponimo. Archivio per l'Alto Adige LXVI (1950):429–462

Gherardini R (2006) Termini rustici nel territorio di Pontassieve e di Bagno a Ripoli. Società Editrice Fiorentina, Firenze

Hofmann A, Goretti D, Merendi GA, Tabacchi G, Vignoli M, Bernetti G (1998) L'inventario forestale. Boschi e macchie di Toscana. Giunta Regionale, Regione Toscana

Millemaci G (1999) Dizionario toponomastico delle Valli della Bure. Società Pistoiese di Storia Patria, Pistoia

Monacci F, Focardi D, Tomei PE (2010) Stazioni eterotopiche di faggio e cartografia storica. Il caso della Toscana Centrale. In Atti 14ª Conferenza Nazionale ASITA, 1363–1368. Brescia. 9–12 novembre 2010

Mori E (1930) Faggio e leccio in Toscana. Saggio di ricerche fitotoponomastiche. Riv. Geogr. Ital. 37(4–6):123–128

Nocentini A (2010) L'etimologico. Vocabolario della lingua italiana. Le Monnier

Pelé I (1992) Utilisation géographique du fichier Rivoli. Acta Geographica 90: 27-37

Pellegrini GB (1990) Toponomastica italiana. Hoepli, Milano

Pellegrini GB, Gasca Queirazza G, Marcato C (1990) Dizionario di toponomastica. Storia e significato dei nomi geografici italiani. UTET, Torino

Penzig O (1924) Flora popolare italiana, vol. 2. Ed. a cura dell'autore, Orto Botanico della R. Università, Genova

Pieri S (1898) Toponomastica delle valli dei Serchio e della Lima. Arch. Glottol. Ital., Suppl. Period. V, Firenze (reprint 1936 in Atti Accad. Lucchese Sci. Lett. Arti, II

Pieri S (1919) Toponomastica della valle dell'Arno. Accademia dei Lincei, Roma

Pieri S (1969) Toponomastica della Toscana meridionale e dell'arcipelago toscano. Accademia Senese degli Intronati, Siena

Ritchot G (1989) Géographie structurale et toponymie contradictoire. Cah Géogr Qué 33(88):67–71

Rossi F (1985) Stemmi e toponimi dei Comuni della Provincia di Firenze. All'Insegna del Giglio, Firenze

Signorini MA, Lombardini C, Bruschi P, Vivona L (2007) Conoscenze etnobotaniche e saperi tradizionali nel territorio di San Miniato (Pisa). Atti Soc Tosc Sci Nat Mem Ser B 114:65–83

Stopani A (1994) La toponomastica del comune di Radda in Chanti. Centro di studi chiantigiani "CLANTE", Radda in Chianti

Targioni Tozzetti O (1809) Dizionario Botanico Italiano. Parte seconda. Guglielmo Piatti, Firenze

Winkler WE (1990) String comparator metrics and enhanced decision rules in the Fellegi-Sunter model of record linkage. In: Proceedings of the section on survey research methods, ASA 1990: 354–359

Chapter 13
Traditional Agricultural Practices, Land Cover Diversity and Biodiversity in the Southern Podpolanie Region

Martina Slámová, Peter Jančura, Vladimíra Fabriciusová, Boris Beláček, Katarína Zrníková and Zlatica Prídavková

Abstract Studied area is located in the vicinity of the Poľana Protected Landscape Area, where "Man and the Biosphere Programme" is implemented under the auspices of the UNESCO organization. Region was colonized during the 17th century by pastorals and later by peasants. Relatively, late colonization process was caused by harsh climatic, soil, and terrain conditions, which were not very favourable for agriculture. Region was sparsely populated. Later, after deforestation, specific form of scattered settlements developed and together with small agricultural plots have persisted till today in the landscape. Region has never undergone land consolidation reforms. Traditionally, beef cattle is bred. Cereals, potatoes, and cabbage are planted on slightly fertile sandy loam Cambisols. Summer seasons are usually dry and it was the reason for the development of irrigation channel system. We identified traditional agricultural plots, historical irrigation channels, and biotopes in the field (2013). We evaluated diversity and heterogeneity of land cover and we assessed significance of biotopes. Data are evaluated in a transect (from the valley flat to the watershed). Traditional agricultural plots with irrigation channels positively affected land cover diversity and biodiversity. We suggest preservation of heterogeneous land cover, which is managed by traditional agricultural practices by

M. Slámová (✉) · P. Jančura · K. Zrníková · Z. Prídavková
Technical University in Zvolen, T.G. Masaryka 24, 960 53 Zvolen, Slovakia
e-mail: mslamova@hotmail.com

P. Jančura
e-mail: jancura@tuzvo.sk

K. Zrníková
e-mail: katarina.zrnikova85@gmail.com

Z. Prídavková
e-mail: pridavkova.z@gmail.com

V. Fabriciusová · B. Beláček
Landscape Protected Area Poľana, J. M. Hurbana 20, 960 01 Zvolen, Slovakia
e-mail: vladimira.fabriciusova@sopsr.sk

B. Beláček
e-mail: borisoff211@yahoo.com

M. Agnoletti and F. Emanueli (eds.), *Biocultural Diversity in Europe*,
Environmental History 5, DOI 10.1007/978-3-319-26315-1_13

local inhabitants, regarding principles rising from the European Landscape Convention (CoE 2000).

Keywords Traditional agricultural plots · Historical irrigation channels · Biotopes · Land cover diversity · Landscape value · Biocultural diversity

13.1 Introduction

13.1.1 Protection of Landscape Values And Risks of the Vanishing

Traditional agricultural landscapes constitute conditions for high level of the landscape biodiversity (Lindborg and Eriksson 2004) and traditional agro-ecosystems are characterized by a high diversity (Kadoya and Washimati 2010). Presence of the rural landscape, but mainly traditional cultivation of land that supports the heterogeneity of cultural landscape, add attractiveness to landscapes (Gobster et al. 2007; Daugstad 2008). Natural and cultural values of these cultural landscapes are generated by historical landscape structures. Their value rises primarily from high level of biodiversity and diversity of land cover, preserved traditional agrarian forms of land cultivation and technologies, socially and culturally significant historical buildings (Stefunkova et al. 2011). Land accounts made by the EEA (1990–2000) in many European nations, with the exception of, e.g. Spain, France and Greece, show that afforestation trends prevail, and the extension of forests often creates homogeneous land cover, which reduces landscape diversity (Agnoletti 2008). Slovakia occupies 49,036 km^2. Agricultural landscape covers 44.5 % of the country (Izakovicova and Bezak 2010) and traditional rural landscapes occupy the territory 2369 km^2; that means 4.8 % of the country (enumerated according to GIS layer available by WMS; GIS layer is a part of the work: Miklos and Izakovicova (2006) (Fig. 22.1a). Traditional agricultural landscapes are located in Slovakia predominantly in the submountain and mountain regions (EEA 2010). For the last 50 years this cultural landscape type has become rare and it is also the consequence of land consolidation reforms which underwent agricultural landscapes and which were interconnected with the withdrawal of local inhabitants (Chrastina 2005). Traditional agricultural landscapes disappeared from many submountain and mountain regions of Slovakia (Petrovic 2006a, b; Olah 2009). Land consolidation reforms also caused the intensification of agriculture tending to decline or extinction of the traditional agrarian forms, reducing the diversity of landscape and biodiversity (Bezak and Halada 2010). The intensification of agriculture negatively influences ecosystems (Billeter et al. 2008), however, also land abandonment is the reason of negative phenomenon, as landscape fragmentation or successive processes (Bielsa et al. 2005; Pinto-Correia and Vos 2004; Poyatos et al.

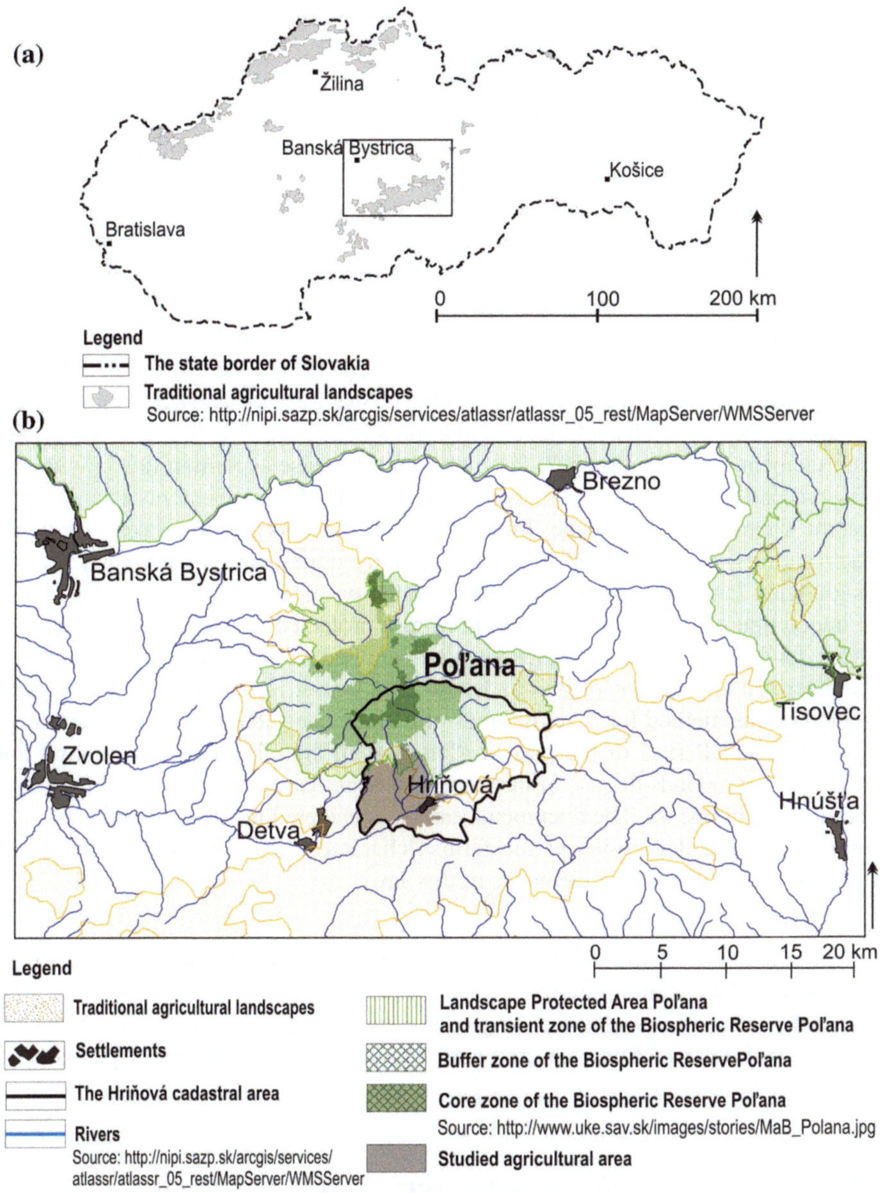

Fig. 13.1 Traditional agricultural landscapes in Slovakia (**a**). Overlapp of the Polana Biosphere Reserve, traditional landscapes and the study area (**b**)

2003). Erosion processes increased due to the abandonment of terraced fields in the southeastern Spain. The overall pattern displayed a shift from traditional and dispersed agriculture to large-scale and intensive practices (Faulkner et al. 2003).

Similar situation of land abandonment is evident in many regions of Slovakia (Danis 2008; Gallay and Gallayova 2011; Midriak and Zauskova 2011). Vanishing of cultural and historical values of traditional landscapes has an accelerating trend (Eetvelde and Antrop 2004). Re-naturation and re-occurence of nature regulative processes, or an intensification of erosion processes cause a financial loss which we can exactly enumerate (Bezak and Halada 2010). Toward the aforementioned argument, enumeration of financial loss of the vanished agricultural landscapes is not so easy. Since these areas are not a subject of special protection at present and since there are no specific policies intended to preserve or manage cultural value in agricultural landscapes, the landscape in general is an object of several national policy documents. Some documents are partially or more focused on historical agricultural landscapes (Spulerova and Petrovic 2011). The Act of National Council of the Slovak republic no. 543/2002 Coll., on nature and landscape preservation introduces the term "characteristic landscape's appearance" in § 1-1., and "significant element of the landscape". It is important, because according this law we can preserve parts of cultural landscapes. Significant landscape elements contribute to specific character of landscape: forests, bogs, riparian vegetation, lakes, wetlands, water courses, cliffs, ravines, stone fields, sandy dunes, parks, alleys and hedges. Asignificant landscape element that has a function of a biocenter, a biocorridor or an interaction element can be pronounced as a protected landscape element (§ 25). These landscape elements represent also land cover structures contributing to characteristic appearance of the landscape (Jancura et al. 2006). As authors commented on the method for "Programmes of maintenance about Protected Landscape Areas/National Parks" in the part of "Passport of significant parts of natura and landscape", it extends the use of the term significant landscape elements in the term traditional settlements. The documents can be considered insufficient and too vague when presenting definitions of the terms "characteristic landscape's appearance" and "significant landscape element" in this law.

Aims of the contribution are

- An evaluation of the diversity of the traditional agricultural landscape in the southern Podpol'anie region
- Identification of the historical irrigation system in agrarian plots with relatives to the distribution of European and nationally significant biotopes of plants.

13.1.2 Genesis of the Specific Agricultural Landscape in the Podpolanie Region

Cadastral area of Hrinova is located mainly in the Zvolenska kotlina basin and partially in the Polana Mt.. It belongs to the "Podpolanie" region and characteristic is the presence of scattered settlements which are regionally called "kopanice". The area is a part of the traditional landscape with scattered settlements of the Slovenske

Stredohorie Mts. and Krupinska planina Mts. The Polana Mt. represents an individual sub-unit in extensive area of traditional landscapes (Ira et al. 2008). Slovakia has five regions of traditional landscapes with scattered settlements (Petrovic 2006a, b). Study area underwent Wallachian colonization processes relatively late in the comparison with similar areas in Slovakia. Human activities have begun more intensively influencing the landscape in the sixteenth and eighteenth centuries (Ira et al. 2008). Scattered settlements formed on the boarder of the past ecumene or anecumene. Harsh climatic, soil and terrain conditions were not very favourable for agriculture in the studied region. However, on the other hand, they contributed to the preservation of the traditional agricultural landscape (Omasta 2011).

The specificity of the Hrinova cadastral area is based on persisting of the form of dispersed settlements with relatively original landscape structure. Maintenance of the traditional form of farming and unique landscape is due to the fact that this area was unaffected by agricultural collectivization following the World War II. (Mojses and Petrovic 2013). Later, in the nineteenth and at the beginning of the twentieth century was established glass and wood processing industry. Heavy industry has started since twentieth century and it has caused a population growth (Omasta 2011). Farmers were organized in the association of the self-employed farmers before and during the communist era and the management of agricultural landscape has continued till the present (Stefunkova et al. 2013). Grasslands in surroundings of scattered settlements are cultivated traditionally, a triple field system is applied and arable land is fertilized by organic farmyard manure. Meadows are mowed twice a year and at the end of the growing season they are grazed by cattle. We can divide grasslands into four categories: wetlands, fens and springs; waterlogged meadows and associations of high herbs; mesophylic meadows; associations of pastures and successive phases on pastures (Ujhazy and Kistin 1998). Traditionally, a beef cattle and sheep are bread. Cereals (barley, oat and wheat), potatoes, cabbage and technical crops (flax, hemp) are planted (Jancura et al. 2011).

Agricultural land is located mostly on little fertile sandy loam Cambisols (VUPOP 2013). Annual rainfall is within the range 700–900 mm (Fasko and Stastny 2002). However, based on a microclimatic measuring is shown, that the lack of precipitation is in the period of the vegetation growth (Fric et al. 2011). Bedrock consists mainly of granodiorite rocks which have a low water permeability and prevail less intensive springs (0.1– 0.4 $l\ s^{-1}$) (Malik et al. 2002). Water absorbs rapidly and flows over the surface of granodiorite rocks to the lower horizons. A similar process appears in the sandy loam weathering mantle that has locally a depth of 2 m. We suppose, that combination of soil conditions with local meso-clima were reasons why irrigation channels—*catchworks* were developed in the past on the slopes of the agricultural landscape. *Catchwork* is a term used in England. "*Irrigation schemes along narrow brook valleys were implemented through diversion structures connected to a system of contour-parallel ditches*" (Leinbundkut and Kohn 2014). As authors note, the well-documented tradition of irrigation is found all over the alpine piedmont landscapes as well as in England and other European countries.

The position of the study area in Slovakia is displayed in Fig. 13.1a. The Polana Biosphere Reserve (BR) is a part of the Hrinova cadastral area (20,360 ha) and occupies 1324 ha (Pilarikova and Gallayova 2012) (Fig. 13.1b). The main specific feature of the BR is a co-existence of the man and nature, it appears as extraordinary natural diversity which complete exceptional cultural diversity and landscape character which are manifested in folk architecture (Ira et al. 2008), concerning the wealth, material and non-material culture (Huba 2009). The Polana BR is covered mostly by forests (85 %). This area belongs to the least urbanised areas in Slovakia. There are only three settlements (Iviny, Snohy a Vrchslatina) with about 400 permanent inhabitants (Fabriciusova 2012).

13.2 Methodology

13.2.1 Land Cover Diversity Indicators

The evaluation is based on the results of the doctorate thesis which elaborated Zrnikova (2014). Indicators were calculated and evaluated in a vector map of land cover. It was derived from orthophotomaps by ditization of different land cover categories. We identified the following structures: meadows and pastures, arable land, settlements, water courses and non-forest wooded vegetation formations. We distinguished sub-structures in the last category because it contains successive areas and further this category has a variety of structures. We differed these structures: patches, lines, clusters, groups and solitaires. Orthophotomaps with a resolution 0.5 m (2013) were provided by TU—Zvolen in frame of the project CEX ITMS 26220120069. Land cover diversity was evaluated in ArcGIS 10 ™ using the module Patch Analyst Statistic 5.1™ (Rempel et al. 2012). We evaluated the following indicators: absolute class area (CA) [ha]; relative class area (CA) [%]; absolute total area (TA) [ha]; relative total area (TA) [%]; number of patches (NP); mean patch size (MPS) [ha]; total edge TE [m]; edge density (ED) [m/ha]; average amount of edge per a patch (MPE) [m]; ratio of patch perimeter to area (MPAR) [m/ha].

Specific landscape character and extraordinary and valuable parts of the cadastral area were published by Jancura (2000). Author divided area into squares and we used one representative square (SQ 100 ha) for purpose of this work. The square represents features of landscape character. We compared the diversity of land cover in the square with diversity of land cover in the agricultural landscape (AL) of the Hrinova cadastral area. Correlations were evaluated by Pearson correlation coefficient r which was calculated in MS Office Excel 2010™:

$$\text{Correl}\,(X,Y) = \frac{\sum(x-\bar{x})(y-\bar{y})}{\sqrt{\sum(x-\bar{x})^2\sum(y-\bar{y})^2}}$$

Correlation coefficients (in absolute values) which are less than 0.35, are generally considered to represent low or weak correlations; 0.36–0.67 modest or moderate correlations; 0.68–1.0 strong or high correlations; and with *r* coefficients more than 0.90 very high correlations.

13.2.2 Mapping and Classification of Biotopes

Biotopes of grassland were excerpted from GIS database of DAPHNE and biotopes are characterized in the Catalogue of biotopes in Slovakia (Stanova and Valachovic 2002).

Each unit is briefly described according to its physiognomic and ecological characteristics which are used for easy and exact identification in the field. If one of the units contains a different syn-morphological and ecological type (soil and clime) it is mentioned in its characteristic. Species abundances are displayed in one block of information and it contains scientific names of plants in the alphabetic order and with differentiated dominant and diagnostic species in each unit (by the font type). Diagnostic species are characteristic and differential for each unit. If possible, taxons of the Red book of endangered species were included into units (Cerovsky et al. 1999), mainly from localities with higher abundancies, in order to support protection of their biotopes. Individually are signed biotopes of endemites of the Carpathian–Pannonian flora (Kliment 1999). Emphasis is given to the Natura 2000 species which are numbered by an international code [n]. Localities of the national significance are indicated in the decree of the Ministry of the Environment SR no. 24/2003 Z.z., which is the appendix no. 1 of the law no. 543/2002 Z.z.. Plant nomenclature is united according to the latest list of plants (Marhold and Hindak 1998). Biotopes are listed in maps and indicated by abbreviations of biotope data and by the percentage abundance of each biotope in a particular polygon. In case of biotopes LK5, Lk6 should be noted, that they can not be clearly distinguished, and they were mapped together and therefore are marked as Lk5, 6 in maps.

13.2.3 Evaluation of Biotopes in Catchworks

Catchworks were identified in the field by GNSS Leica GS05, in the autumn 2013 (8 and 10 October, 17, 27 and 28 November). Catchworks were considered to be terrain depressions, or planed surfaces that are oriented in the direction of contours on slopes in the Riecka catchment. Total length of mapped catchworks was 3736.68 m. We used digital terrain model for the calculation of the relief parameters of catchworks that are: altitude above sea level [m] and slope [°]. Raster grid had a resolution 10 m and it was provided by "Urad geodezie, kartografie a katastra Slovenskej republiky" under the licence contract no. 877-11-5725/2014. Categories of the slope are classified after 50 m (500–550 m; 550–600 m; 600–650 m; 650–700 m; 700–750 m) and slopes are classified according to criteria which are applied in the soil ecological units (Linkes et al. 1996): 0°–1°; 1°–3°; 3°–7°; 7°–12°; 12°–17°; 17°–25°. Further we evaluated the functionality of catchworks according to the presence of water or according to plants which indicated groundwater and we divided them into the following categories: functional, partially functional and dysfunctional.

Biotopes were verified in the field in catchworks. Detail mapping of land cover and biotopes are presented in transect which was constructed from the valley flat to the watershed.

13.3 Results

13.3.1 Land Cover Diversity

Cadastral area spreads in 12,631.52 ha and studied agricultural area occupies 2827.97 ha (22.38 %). The prevailing part of the specific historical agricultural area spreads into the submountain region (420–1050 ASL); an altitude difference is 630 m. We evaluated the area of the following land cover categories: meadows and pastures (938.93 ha; 33.20 %), arable land (925.31 ha; 32.71 %), wooded vegetation formations (799.54 ha; 28.27 %), settlements (120.23 ha; 4.25 %), roads (37.98 ha; 1.34 %), streams (5.98 ha; 0.21 %).

Wooded vegetation formations include areas with wooded non-forest successive vegetation. These areas indicate a lack of anthropic pressure and insufficient maintenance about the cultural landscape. Succession on meadows represents 231.28 ha (28.9 %) of wooded vegetation formations and 8.18 % of the studied agricultural area. On the other hand, wooded vegetation formations represent very important land cover category contributing significantly to diversity of the traditional agricultural landscape. This category consists of gardens with fruit trees (156.18 ha); ecologically valuable riparian vegetation (82.52 ha) and several types of wooded vegetation formations on hedgerows of land parcels and on agrarian terraces (234.97 ha). The area of all of the formations is shown in Table 13.1.

Landscape matrix of the studied landscape is represented by meadows, pastures and arable land. These categories prevail over the all ones. Meadows and pastures spread in large homogeneous areas, mostly along streams and nearby slopes. Land cover indexes show that wooded non-forest vegetation formations and plots of arable land strongly contribute to land cover diversity. The representative square documents the arrangement of land cover categories in a map (Fig. 13.2a). Based on the evaluation of the correlation coefficient of the selected land cover categories we consider this square as an area representing a hight degree of land cover diversity in the studied agricultural landscape (Table 13.2).

Table 13.1 Classes area of land cover in the agricultural landscape of Hrinova (2013)

CA (structure/substructure of land cover)	Absolute area (ha)	Relative area (%)
Meadows and pastures	938.93	33.20
Arable land	925.31	32.71
Wooded vegetation formations (WVF)	799.54	28.27
WVF: diffused all over the area-meso-structures	*99.68*	*3.52*
WVF: diffused all over the area-micro-structures	*21.71*	*0.77*
WVF: sparsely diffused-meso-structures	*90.67*	*3.21*
WVF: sparsely diffused-micro-structures	*19.22*	*0.68*
WVF: gardens	156.18	5.52
WVF: forests	48.30	1.71
WVF: intravillan-gardens, parks	46.28	1.63
WVF: areal-compact, meso-structures	46.72	1.65
WVF: areal-compact, micro-structures	31.01	1.09
WVF: areal-lines, meso-structures	6.19	0.22
WVF: areal-lines, micro-structures	4.21	0.15
WVF: predominantly lines	15.52	0.55
WVF: small patches (100–500 m^2)	11.48	0.41
WVF: riparian vegetation	82.52	2.92
WVF: lines	39.70	1.40
WVF: plates	34.70	1.23
WVF: fans	39.08	1.38
WVF: solitaires	3.61	0.13
WVF: clusters	2.75	0.10
Settlements	120.23	4.25
Roads	37.98	1.34
Streams	5.98	0.21
TA	2827.97	100

Classes with areas of successive plants are italicized in the table

A significant correlation between indexes documenting land cover diversity exists among the land cover categories: wooded vegetation formations and arable land. Evaluating the selected indexes (CA, MPS, TE, ED, MPAR) we calculated the correlation as very strong in case of wooded vegetation formations (1) and strong correlation (0.79) in case of arable land. Evaluating all of the indexes (CA, NP, MPS, TE, ED, MPE, MPAR) we observed that these values declared less significant correlation; Moderate correlation (0.56) is case of wooded vegetation formations and strong is in case of arable land (0.80). Formations of wooded vegetation are relative small patches in the country and they significantly contribute to land cover diversity in the landscape. Besides, a high number of patches formations have variable forms as we document in Table 13.1 and in Fig. 13.2b. Overall number of the different formations is 19. Very interesting visual appearance in the landscape have formations representing hedges of agrarian plots. Visually, they highlight

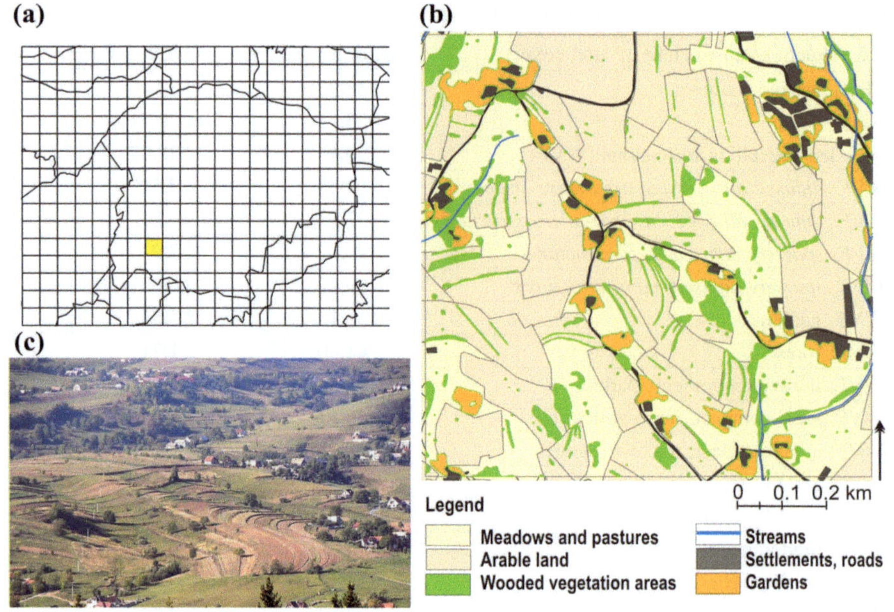

Fig. 13.2 Position of the representative *square* in the cadastral area (**a**), arrangement of land cover in rugged terrain (**b**), photo of the *square* (**c**)

Table 13.2 Evaluation of land cover diversity in the agricultural landscape and in the representative square

Indexes of the diversity	Wooded vegetation forms[a]	Arable land
LA		
CA [ha]	*548.86*[b]	*925.31*
NP	*6194.00*	1149.00
MPS [ha]	*0.09*	*0.81*
TE [m]	*911,731.85*	*611,401.77*
ED [m/ha]	*322.39*	*216.20*
MPE [m]	147.20	*532.12*
MPAR [m/ha]	35,781.97	*54,453.18*
SQ 100 ha		
CA [ha]	*6.83*	*51.49*
NP	*271.0*	78.0
MPS [ha]	*0.03*	*0.66*
TE [m]	*23,511.11*	*33,636.70*
ED [m/ha]	*235.07*	*336.31*
MPE [m]	86.76	*431.24*
MPAR [m/ha]	30,295.35	*26,712.96*

[a]Excluded are: gardens, parks in intravillan, forests
[b]Indexes documenting a high degree of the diversity are italicized

boarders of plots and formations visually appear as fans, lines and plates. They appear as a regular geometrical pattern that is visually very attractive for tourists. The square representing landscape pattern as a visual feature of landscape character of the southern Podpolanie region is displayed in Fig. 13.2c. On the other side, a relatively high value of MPE index was evaluated in maps by vectorisation of ortophotomaps. That means a shape complexity and it is an opposite case to the simple geometrical pattern which we can perceive visually in the landscape. Moreover, arable land has more complex patches than wooded vegetation formations. We suppose that it is caused by terrain, soil quality and by plots fragmentation into very small patches (several tens of square metres to hectare). We could identify that complexity in maps but it is not visible in real landscape because the complexity of parcels exhibited only in great detail. In case of agrarian land, we did not document individual plots, because private ownership was not detectable in orthofotomaps. Plots are displayed in cadastral maps, but in case we identify individual plots, indexes of land cover diversity could refer to higher degree of the diversity. Anyway, both categories contribute significantly to diversity of land cover.

13.3.2 Contribution of Historical Catchworks to Biodiversity

Catchworks were situated mainly on the slopes with meadows and pastures. Prevailing part of catchworks was situated in an altitude within the range 650–700 m ASL (37.88 %), as well as we identified them in the following altitudes: 500–550 (20.76 %), 550–600 (22.65 %) and 600–650 (18.1 %). Only few catchworks were located within the range 700–750 m ASL (0.56 %) because the agricultural landscape is limited mainly up to 650 m ASL (Fig. 13.3a). Catchworks were predominantly located on the slopes within the range 7°–12° (32.45 %) and 12°–17° (26.83 %). Less are found in these categories: 0°–3° (0.35 %), 3°–7° (12.38 %), 17°–25° (22.37 %), 25°–30° (5.61 %). Functional catchworks had a length of 1600.22 m (42.82 %), partially functional had a length of 85.19 m (2.28 %) and dysfunctional represented 2002.58 m (53.59 %).

Catchworks corresponded with the European significant NATURA 2000 biotopes. Lowland hay meadows [6510] (Lk1) were the most frequent in catchworks, abundance was within the range 100–60 % and they appeared in 8 of 6 localities. Very frequent were biotopes of hydrophilous tall herb fringe communities of plains and of montane to alpine levels [6430] (LK5) and nationally significant biotope of waterlogged mountain meadows (LK6). They occured in the northern part of the Riecka catchment (85 %) and in other parts of the catchment these biotopes had abundance only within the range 30–2 %. Nationally significant biotope, mesophilic grazed pastures and meadows (Lk3) had abundance within the range 38–20 % and occured in two localities. Transition mires and quaking bogs [7140] occured in one locality and its abundance was only 6 % (Fig. 13.3b).

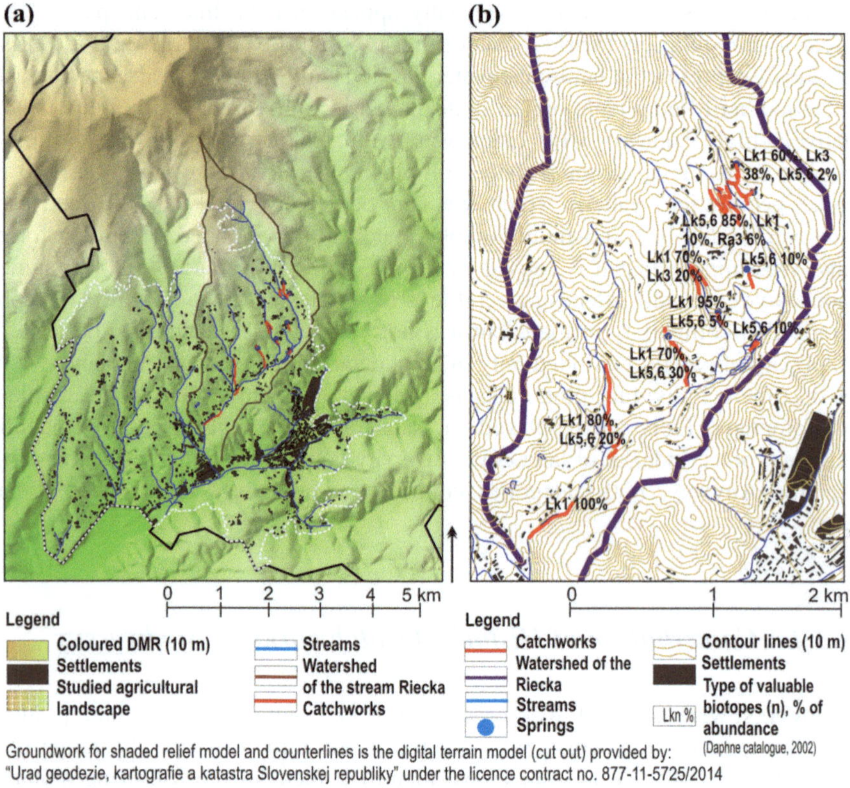

Fig. 13.3 Watershed of the Riecka stream and scattered settlements in Hrinova (**a**), Location of catchworks in the Riecka catchment and biotopes (**b**)

13.3.3 Evaluation of Biotopes in the Transect

The presented transect had a length of 258 m and elevation from the valley flat to the watershed was 39 m (min. 673 m ASL to max. 712 m ALS) (Fig. 13.4a). We recorded 10 changes in land cover (a length is in brackets) and six different land cover categories (Fig. 13.4b): alluvial meadows with valuable biotopes Lk5, 6 (15 m), meadows/pastures with valuable biotopes Lk1 (150 m), ploughland with corn (82 m), road (2 m), ploughland with corn and segetal vegetation (5 m), hedge with segetal vegetation and *Rosa* sp. (4 m) (Fig. 13.4c). That means, prevailing part of the transect (64 %), was represented by land cover categories comprising valuable biotopes on meadows and pastures.

We identified two biotopes of the Euroepan significance.

The biotope Lk1 [6510] occupies fertilized, one or two times per year mowed meadows with a predominance of tallgrass and valuable grass for feeding (*Arrhenatherum elatius* (L.) P. Beauv. ex J. Presl et C. Presl, *Alopecurus pratensis*

(a)

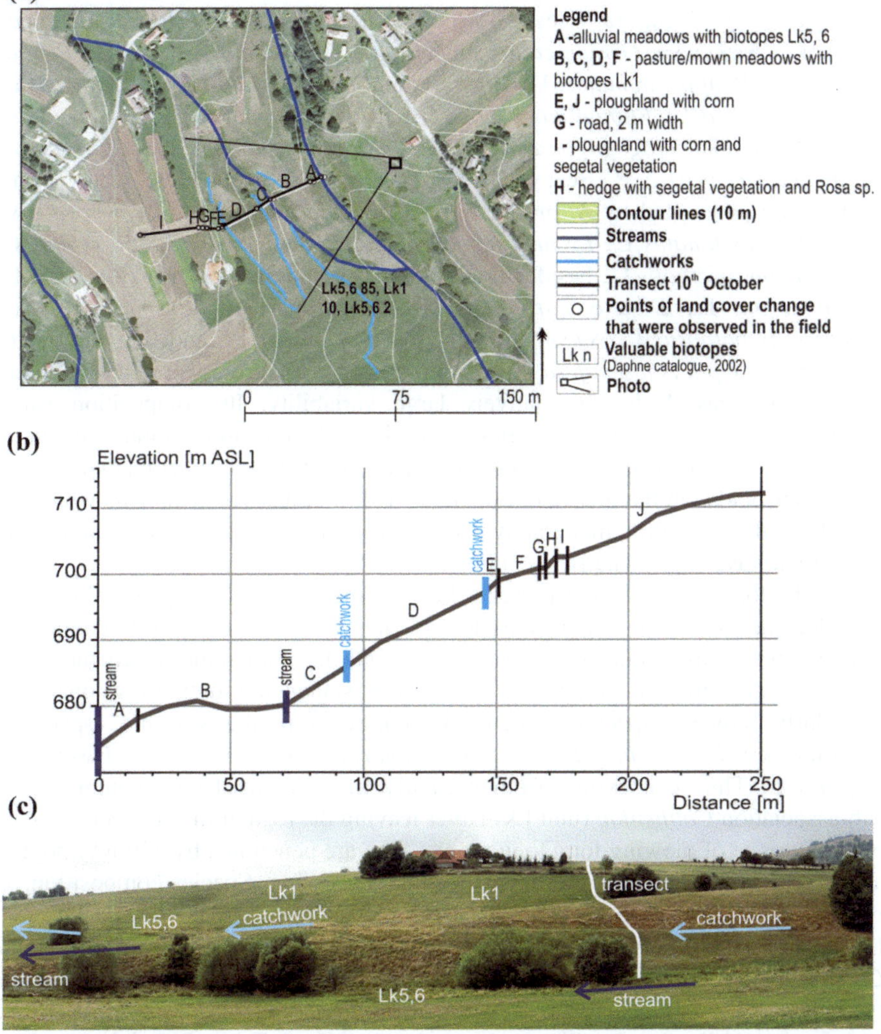

(b)

(c)

Fig. 13.4 Land cover with catchworks and biotopes (**a**), graph of the transect (**b**), valuable hydrophilic (Lk 5,6) and mesophylic biotopes (Lk1) in photo (**c**)

L., *Trisetum flavescens* (L.) P. Beauv., *Anthoxanthum odoratum* L., *Festuca rubra* L.). Herbs are represented by: *Achillea millefolium* L., *Agrostis capillaris* L., *A. elatius* (L.) P. Beauv. ex J. Presl et C. Presl, *Avenula pubescens* (Huds.) Dumort., *Bromus hordeaceus* L., *Campanula patula* L., *Carum carvi* L., *Cerastium holosteoides* Fr., *Crepis biennis* L., *Dactylis glomerata* L., *Dactylorhiza sambucina* (L.) Soo, *F. rubra* L., *Galium mollugo* L., *Geranium pratense* L., *Heracleum sphondylium* L., *Jacea pratensis* Lam., *J. pseudophrygia* (C. A. Mey.) Holub, *Knautia arvensis* (L.) Coult., *Leontodon hispidus* L., *Lotus corniculatus* L., *O. morio* L., *O. mascula subsp. Signifera* (Vest) Soo, *O. ustulata* subsp. *Aestivalis* (Kümpel) Kümpel et Mrkvicka, *Ornithogalum umbellatum* L., *Pastinaca sativa* L., *Phleum pratense* L., *Pimpinella major* (L.) Huds., *Plantago lanceolata* L., *Poa pratensis* L., *Ranunculus acris* L., *Rhinanthus minor* L., *Saxifraga granulata* L., *Silene vulgaris* (Moench) Garcke, *Tragopogon orientalis* L., *Trifolium dubium* Sibth., T. *pratense* L., *T. flavescens* (L.) P. Beauv., *Veronica chamaedrys* L.. More watterlogged habitats are penetrated by *A. pratensis* L., *Colchicum autumnale* L., *F. pratensis* Huds., *Holcus lanatus* L., *Lychnis flos-cuculi* L., *Potentilla alba* L., *R. repens* L., *Sanguisorba officinalis* L., besides drier and warmer habitats occupy *B. erectus* L., *C. glomerata* L., *Daucus carota* L., *F. rupicola,* Heuff. *Primula veris* L., *Ranunculus bulbosus* L., *Salvia pratensis,* L. *a S. minor* L., and frequent are *O. militaris* L. a *Ophrys insectifera* L. This biotope occupies from wet to drier habitats in warmer areas. It has a relatively large variability. Its composition varies according to ecological characteristics of the site and the method of management. It is rich in species. It occurs in alluvium of rivers, on the slopes and embankments, on the former fields, on the grassy fallow fields and in orchards—with soils which are slightly acidic to neutral, moderately deep to deep, moderately moist to slightly dry and which have a good nutrients supply.

The biotope Lk5 [6430] represents flowery meadows with a predominance of broadleaf herbs in year-round moist to wet sites in alluviums of watercourses, in terrain depressions and in spring areas. Herb stands have often mosaic character and species composition is highly variable. Herb stands are only occasionally or irregularly mowed. Species particularly significant in the seasonal aspects of blooming are: *Filipendula ulmaria* subsp. *ulmaria* L., *G. palustre* L. a *Lysimachia vulgaris* L. They can evolve from regularly mowed grassland communities of sub-association *Calthenion* (unit Lk7) after leaving the regular management. In case of the absence of mowing for a longer time, they are penetrated by willows, poplars and alders, which indicate the line of further succession. Species composition: *A. pratensis* L., *Aegopodium podagraria* L., *Angelica sylvestris* L., *Caltha palustris* L., *Carduus personata* (L.) Jacq., *Cirsium oleraceum, C. paludosa* (L.) Scop., *Chaerophyllum hirsutum* L., *Epilobium hirsutum* L., *F. ulmaria* subsp. *Ulmaria* L., *G. palustre* L., *Iris sibirica* L., *Lysimachia vulgaris* L., *Lythrum salicaria* L., *Mentha longifolia* (L.) L., *Phragmites australis* (Cav.) Trin., *Pseudolysimachion longifolium* (L.) Opiz.

We identified one biotope of the national significance.

The biotope Lk6 was mowed regularly in the past and at the present is less used. It is represented by one-time of two times yearly mowed wet meadows on the waterlogged alluviums of watercourses, in surroundings of the slope and the sub-slope spring areas, and spreads also in littoral zones of water reservoirs behind the zone of reed stands. Optimal conditions for that biotope are in the mountain and submountain areas, eventually occur in mosaics with other types of wet meadows, or occupy smaller areas in terrain depressions with mesophilic biotopes. This habitat is characterized by a persistently elevated groundwater. Species composition: *A. pratensis* L., *Angelica sylvestris* L., *Caltha palustris* L., *Carex cespitosa L.,*

Carex nigra (L.) Reichard, *Carex panicea* L., *Cirsium canum* (L.) All, *Cirsium heterophyllum* (L.) Hill, *Cisrium oleraceum* (L.) Scop., *Cirsium palustre* (L.) Scop., *Cirsium rivulare* (Jacq.) All., *C. paludosa* (L.) Moench, *D. majalis* (Rchb.) P. F. Hunt et Summerh., *Deschampsia cespitosa* (L.) P. Beauv., *Equisetum palustre* L., *F. rubra* L., *F. ulmaria* (L.) Maxim., *G. palustre* L., *Geum rivale* L., *H. lanatus* L., *C. hirsutum* L., *Juncus acutiflorus* Ehrh. ex Hoffm., *Juncus conglomeratus* L., *Juncus filiformis* L., *Lysimachia vulgaris* L., *Lythrum salicaria* L.,*Myosotis scorpioides* L., *P. trivialis* L., *Ranunculus repens* L., *Scirpus sylvaticus* L., *Senecio erraticus* Bertol., *Tephroseris crispa* (Jacq.) Rchb., *Trollius altissimus* Crantz, *Valeriana officinalis* L. and in drier sites occur also *Gladiolus imbricatus* L., *Scorzonera humili* L., *Serratula tinctoria* L., *Succisa pratensis* Moench.

13.4 Discussion

Our findings about highly heterogeneous land cover correspond with results of other authors. Olah et al. (2008) found out that among the five landscapes in the Stiavnicke vrchy Mts., Kremnicke vrchy Mts., Cerova vrchovina Mts., Javorie Mts. and Polana Mts., which were compared in terms of land cover diversity, the most heterogeneous is the landscape under the Polana Mt..

Localities with catchworks correspond with valuable biotopes of European and national significance, as we observed in the transect. Biotopes which we used in this work were usually mapped as complexes of biotopes which have no precise location in land cover polygons and this is the reason why there is defined only a relative abundance within the range of percents. In this phase of the investigation we cannot say that the presence of biotopes is caused by catchworks. Rather the reason is the persisting traditional management of meadows and pastures. However, in the future, it would be appropriate to make a detailed field survey focused on the occurrence of biotopes in surroundings of catchworks. As we observed in the field, catchworks provide conditions (soil moisture) for biotopes evolution (Lk5, Lk6), however, locally we observed these biotopes only near springs or water sources, but did not appear in the whole length of catchworks. Biotope Lk1 is maintained by mowing and fertilizing.

Successive processes are present only in the peripheral zones of the studied agricultural landscape and represent an area 8.18 %, what is relatively small part of the investigated landscape. Warning, however, could be a radical change in land use of the Polana BR. Grasslands occupied 3 256.80 ha (16 %) in the 1950s and only one 790.31 ha (8.79 %) in 2003 of the total area of the BR Polana (Gallayova 2007). An extensive farming closely correlates to the land abandonment; in Hrinova it means 17 % of the previous agricultural land (Stefunkova et al. 2013). This trend causes decrease of valuable habitats of birds (Kristin 2010). A negative trend proves also our findings that dysfunctional catchworks represent 53.59 % of the total length of catchworks. These facts draw attention to the unfavourable state of the biotopes' management.

Natural conditions are a handicap for the agricultural production in the areas with traditional submountain and mountain regions in Slovakia (Spulerova and Petrovic 2011). Therefore, in Slovakia, agri-environmental payments primarily affect the sustainability of extensive land management. The degree of association between the natural conditions of agricultural land management, which are expressed by the range of soil price categories, and the company entering into an agri-environmental commitment is high and statistically verifiable. The measures are increasingly allocated to companies that operate on less productive soil categories, which exhibit worse natural conditions (Blaas 2011). Currently, traditional agricultural landscapes require financial support for preservation of their character. Regions with naturally disadvantaged production are significantly supported by agri-environmental payments in Slovakia. We do not consider this manner of allocating subsidies to agriculture to be sustainable; however, agri-environmental payments are useful in the initial phases of agricultural revival. Based on the aforementioned text, we have to highlight a new approach in the management of the Polana BR. Its main goal is providing and supporting a positive attitude of stakeholders and residents to the existence of the BR. The BR Polana was recently criticized by the UNESCO headquarter, due to a lack of population and insufficient human influence on nature. Therefore, under the project "Development of Conservation and Protected areas in the Slovak Carpathians", was prepared a scheme of viable activities for local farmers to preserve traditional agriculture. Furthermore, the Civic Association PRONATUR developed the project "Nature for people—people nature" which opens up opportunities for the development of business activities, respectively, to the conservation of biodiversity. Activities are coordinated by the BR's management and by local inhabitants. The City Council agreed with the extension of the Polana BR to a transitional zone. It includes traditional agricultural landscape. An inclusion of the traditional agricultural landscape in the BR should be a guarantee for the preservation of the valuable landscape character and it could be beneficial in terms of easier and more effective applications for finances via grant schemas.

Fleischer and Tsur (2000) exactly proved in case of Israel, that agricultural landscape could generate a higher financial potential from tourism activities than the only income of the traditional agriculture. Extensively used traditional rural landscape with its natural character is very attractive for recreation and tourism for citizens (Walker et al. 2003). An attractive cultural landscape provides the financial potential for farms which can be implemented into the practise by agro-tourism activities (Daugstad et al. 2006). Thus, Lovell and Johnston (2009) pointed, that concept of diversified, multifunctional agricultural landscape, respecting biodiversity of biotopes is topical in many countries over the world.

13.5 Conclusions

The southern Podpolanie region has been known in Slovakia mainly due to preserved folk traditions. Recently, more discussed is the protection of the unique character of landscape. Additionally, we observed in the field, historical technical objects—catchworks that have not yet been documented at any locality in Slovakia. The investigated landscape has heterogeneous land cover with persisting traditional agriculture. A large representation of woody vegetation formations and their variability supports the ecosystem stability of the territory. It also enriches the variability of land cover patches which is so characteristic for this landscape. A number of European and nationally significant biotopes of plants is present in the territory. This work directly does not prove the diversification of biotopes by catchworks. But based on the functionality of catchworks we show the need for their adequate maintenance. This step should be prevention before the loss of biotopes due to successional processes.

Acknowledgements The paper was financially supported by the Cultural and Education Grant Agency—contract no. 011TUZ-4/2012 (90 %). The work is supported by the European Regional Development Fund, Research and Development Operational Programme, Grant No. ITMS 26220120069—Centre of Excellence for Decision Support in Forest and Country.

References

Agnoletti M (2008) European agricultural policies and landscape: effects strategies and perspectives. In: Proceedings/actes of the seventh meeting of the workshops of the Council of Europe for the Implementation of the European Landscape Convention, Piestany, pp 41–60

Bezak P, Halada L (2010) Sustainable management recommendations to reduce the loss of agricultural biodiversity in the mountain regions of NE Slovakia. Mt Res Dev 30:192–204

Bielsa I, Pons X, Bunce B (2005) Agricultural abandonment in the north eastern Iberian Peninsula: the use of basic landscape metrics to support planning. J Environ Plan Man 48(1):85–102. doi:10.1080/0964056042000308166

Billeter R, Liira J, Bailey D et al (2008) Indicators for biodiversity in agricultural landscapes: a pan-European study. J Appl Ecol 45(1):141–150. doi:10.1111/j.1365-2664.2007.01393.x

Blaas G (2011) Economic farm level impact of agri-environmental payments in Slovakia. In: Proceedings of Soil Science and Conservation Research Institute, vol 33, VUPOP, Bratislava, pp 5–21 (in Slovak)

Cerovsky J, Ferakova V, Holub J, Maglocky S, Prochazka F (1999) Red data book of endangered and rare plant and animal species of Czech Republic. Vascular Plant Species, Priroda, Bratislava (in Slovak)

Chrastina P (2005) Landscape evolution as the environmental history phenomenon (example of Trencianska hollow basin and its mountain edge. Historicka geografie 33:9–19 (in Slovak)

Council of Europe (CoE) (2000) The European landscape convention—Firenze, 20.X.2000 (ETS No.176). Official Text in English and Explanatory Report. Council of Europe, Strasbourg. http://www.coe.int/t/dg4/cultureheritage/heritage/Landscape/default_en.asp. Accessed 8 May 2011

Danis D (2008) Prognosis of succession processes dynamics and its influence on Pieniny National Park Landscape. Partner, Poniky (in Slovak)

Daugstad K, Rønningen K, Skar B (2006) Agriculture as an upholder of cultural heritage? Conceptualizations and value judgements—A Norwegian perspective in international context. J Rural Stud 22(1):67–81. doi:10.1016/j.jrurstud.2005.06.002

Daugstad K (2008) Negotiating landscape in rural tourism. Ann Tourism Res 35(2):402–426. doi:10.1016/j.annals.2007.10.001

Euroepan Environmental Agency (EEA) (2010) Foresight analysis of rural areas of Europe, 2007–2010, EEA report, 2010. http://www.eea.europa.eu/publications. Accessed 10 June 2012

Eetvelde VV, Antrop M (2004) Analyzing structural and functional changes of traditional landscape—two examples from Southern France. Landsacpe Urban Plan 67:79–95. doi:10.1016/S0169-2046(03)00030-6

Fabriciusova V (2012) Natural values in the area of the CHKO-BR Polana. In: Proceedings of the project acting together for the preservation of natural values of the Polana Area and the Borzsony HUSK/1101/2.2.1/0352, Lesy SR, štátny podnik, IPOLY ERDO Zartkoruen Mukodo Reszvenytarsasag, Detva, pp 25–26. http://huskforwildlife.eu/assets/docs/Zbornik_referatov.pdf. Accessed 25 July 2014 (in Slovak)

Fasko P, Stastny P (2002) The annual average of rainfall (1961–1990). In: Miklos L, Hrnciarova T (ed) Landscape atlas of the Slovak republic. MZP SR, Bratislava, SAZP, Banska Bystrica. http://globus.sazp.sk/atlassr/. Accessed 25 July 2014 (in Slovak)

Faulkner H, Ruizb J, Zukowskyj P, Downward S (2003) Erosion risk associated with rapid and extensive agricultural clearances on dispersive materials in Southeast Spain. Environ Sci Policy 6:115–127. doi:10.1016/S1462-9011(02)00126-0

Fleischer A, Tsur Y (2000) Measuring the recreational value of agricultural landscape. Eur Rev Agric Econ 27(3):385–39. doi:10.1093/erae/27.3.385

Fric M, Orena M, Strelcova K, Jezik M, Sitkova Z (2011) The changes of stem circumference and microclimatic parameters of spruce forests in BR Poľana during the vegetation season of 2009. In: Stredova H, Roznovsky J, Litschmann T (ed) Microclima and mesoclima of landscape structures and anthropogenic environment. Proceedings, Skalni mlyn, Cesky hydrometeorologicky ustav, Praha (in Slovak)

Gallay I, Gallayova Z (2011) Overgrowing of non-forestry area in Slovak's protected areas during last 15 years (on the base of corine land cover). Naturae Tutela 15(2):125–135 (in Slovak)

Gallayova Z (2007) A quantitative comparison of permanent grasslands abundance in the year 1956 and 2003 in the Biosphere reserve Polana. In: Dreslerova J, Grohmanová, L (ed) Rural landscape 2007. Proceedings of the 5th annual international interdisciplinary conference, Ekologický institut Veronica, Hostetin, Regionální organizace IALE ČR, Brno, pp 24–28 (in Slovak)

Gobster PH, Nassauer JI, Daniel TC, Fry G (2007) The shared landscape: what does aesthetics have to do with ecology? Landscape Ecol 22:959–972. doi:10.1007/s10980-007-9110-x

Huba M (2009) Transformation of a Region with mountain farmers (exemplified by the Polana Mountains and the Podpolanie Region). Slovak Ethnology 3:312–330

Ira V, Huba M, Podolak P (2008) Quality of life of population living in a protected area and its sustainable development (examples of the protected landscape area Polana). Geographia Slovaca 25:97–122 (in Slovak)

Izakovicova Z, Bezak P (2010) Development scenarios of representative landscape ecosystems in the Slovak Republic considering global changes: final report of the project SAV-FM-EHP-2008–03-09, Ústav krajinnej ekológie SAV, Bratislava. http://www.uke.sav.sk/phocadownload/ZS%20SCERES%20-%20zaverecna%20sprava.pdf. Accessed 12 Mar 2014 (in Slovak)

Jancura P (2000) Identification of visual landscape and landscape character on the example of the subregion Detva-Hrinova. Acta Facultatis Ecologiae:127–141(in Slovak)

Jancura P, Bohalova I, Surovcova S (2006) Human activities and the landscape of the Pieniny. Pieniny - Przyroda i Człowiek 9:133–142 (in Slovak)

Jancura P, Cihlarova R, Jakubec B (2011) Detva, values of landscape. Partner, Poniky. http://web.detva.sk/Detva_hodnoty_krajiny.pdf. Accessed 3 Feb 2014 (in Slovak)

Kadoya T, Washitani I (2010) The Satoyama index: A biodiversity indicator for agricultural landscapes. Agric Ecosyst Environ 140:20–26. doi:10.1016/j.agee.2010.11.007

Kliment J (1999) Commented overview of the vascular plants of the Slovakian flora, listed as as endemic taxons in the literature. Slovak Botanical Society, Bratislava (in Slovak)

Kristin A (ed) (2010) Birds of the Polana protected landscape area. Slovak ornithological Society—BirdLife Slovensko, Bratislava (in Slovak)

Leibundgut CH, Kohn I (2014) European traditional irrigation in transition part I: Irrigation in times past—a historic land use practice across Europe. Irrig Drain 63(3):273–293. doi:10.1002/ird.1826

Lindborg R, Eriksson O (2004) Historical landscape connectivity affects present plant species diversity. Ecology 85(7):1840–1845. doi:10.3410/f.1023334.269679

Linkes V, Pestun V, Dzatko M (1996) Handbook for using maps of land-evaluation and soil-ecological units. VUPU, Bratislava (in Slovak)

Lovell ST, Johnston DM (2009) Creating multifunctional landscapes: How can the field of ecology inform the design of the landscape? Front Ecol Environ 7:212–220. doi:10.1890/070178

Malik P, Svasta J, Jetel J, Hanzel V, Gedeon M, Scherer S, Fendek M (2002) Hydrogeological conditions. In: Miklos L, Hrnciarova T (ed) Landscape atlas of the Slovak Republic. MZP SR, Bratislava, SAZP, Banska Bystrica. Available at http://globus.sazp.sk/atlassr/. Online access 25 July 2014 (in Slovak)

Marhold K, Hindak F (ed) (1998) Checklist of non-vascular and vascular plants of Slovakia. Veda, VSAV, Bratislava (in Slovak)

Midriak R, Zauskova L (2011) Land abandonment and abandonment of landscape in Slovakia. Geographia Cassoviensis V 1:79–83 (in Slovak)

Miklos L, Izakovicova Z (ed) (2006) Atlas of the representative geo-ecosystems of Slovakia. UKE SAV, MZP SR, MS SR, Bratislava. http://nipi.sazp.sk/arcgis/services/atlassr/atlassr_05_rest/MapServer/WMSServer. Accessed 30 June 2014 (in Slovak)

Mojses M, Petrovic F (2013) Land use changes of historical structures in the agricultural landscape at the local level—Hriňová case study. Ekologia (Bratislava) 32(1):1–12. doi:10.2478/eko-2013-0001

National Council of the Slovak Republic (2002) The Act of National Council of the Slovak republic no. 543/2002 Coll., on nature and landscape preservation. In: Collection of laws of the Slovak Republic. http://zbierka.sk/zz/predpisy/default.aspx?PredpisID=16747&FileName=02-z543&Rocnik=2002. Accessed 6 Sept 2013

Olah B, Gallay I, Oravcova M (2008) Evaluation of secondary landscape structure using landscape indexes. In Novaková M, Svicek M (ed) Proceedings of the scientific seminar: environmental aspects of the analysis and evaluation of the landscape: identification and determination of indicators (and indexes) based on landscape survey and remote sensing data, VUPOP, Bratislava (in Slovak)

Olah B (2009) Transformation of the Slovak cultural landscape since the 18th century and its recent trends. J Landscape Ecol 2:41–55

Omasta S (2011) Dispersed rural settlement—socioeconomic and historical aspects. Environ 45 (1):43–47 (in Slovak)

Petrovic F (2006a) Relationship between dispersed settlement and morphometric-position attribute of the relief. Acta Environmentalica Universitatis Comenianae (Bratislava) (in Slovak) 14 (1):105–113

Petrovic F (2006b) The changes of the landscape with dispersed settlement. Ekologia (Bratislava) 25:65–89

Pilarikova M, Gallayova Z (2012) Changes of grasslands overgrown by woodland in the part of the Hrinova cadastral since 1969 to 2010. Acta Facultatis Ecologiae 27:89–98 (in Slovak)

Pinto-Correia T, Vos W (2004) Multifunctionality in mediterranean landscapes—past and future. In: Jongman RHG (ed) The new dimensions of the European landscape, Springer, Berlin, pp 135–165

Poyatos R, Latrone J, Llorens P (2003) Land use and land cover change after agricultural abandonment. The case of a mediterranean mountain area (Catalan Pre-Pyrenees). Mt Res Dev 23:362–368. doi:10.1659/0276-4741(2003)023[0362:LUALCC]2.0.CO;2
Rempel RS, Kaukinen D, Carr AP (2012) Patch analyst and patch grid. Queens Press, Ontario Ministry of Natural Resources, Ontario. http://www.cnfer.on.ca/SEP/patchanalyst/index.html. Accessed 30 June 2013
Spulerova J, Petrovic F (2011) Historical agricultural landscape as a subject of landscape ecological research. Hrvatski Geografski Glasnik 73(2):155–163
Stanova V, Valachovic M (ed) (2002) Catalogue of biotopes in Slovakia. DAPHNE—Institute of Applied Ecology, Bratislava (in Slovak)
Stefunkova D, et al (2011) Attractiveness of the Carpathian landscape, focusing on the historical agrarian structures and biodiversity. UKE SAV Bratislava, KARTPRINT (in Slovak)
Stefunkova D, Spulerova J, Dobrovodska M, Mojses M, Petrovic F (2013) Traditional agricultural Landscapes—a model of detailed land use mapping Tájökológiai Lapok 11(1):1–21
Ujhazy K, Kristin A. (1998) Biodiversity of functional areas. In: Slavikova D, Krajcovic V, et al (eds) Protection of the biodiversity and management of permanent grasslands in the CHKO—BR Poľana 2, IUCN Slovakia, Bratislava (in Slovak)
VUPOP (2013) Typology of Production Categories. Soil Portal, Information Service VUPOP, Bratislava. http://www.podnemapy.sk/portal/verejnost/tpk_pp/tpk.aspx. Accessed 20 Feb 2013
Walker PA, Marvin SJ, Fortmann L (2003) Landscape changes in Nevada County reflect social and ecological transitions. Calif Agric 57(4):115–121. doi:10.3733/ca.v057n04p115
Zrnikova K (2014) Biodiversity protection in the pelation with arrangement of historical landscape structures. Unpublished PhD. thesis, TU in Zvolen, Zvolen (in Slovak)

Author Biography

Martina Slámová Martina Slámová has a PhD degree in ecology. She is an assistant professor of the Department of Landscape Planning and Design at the Technical University in Zvolen. The teaching activity is focused on landscape planning and design. She is engaged in research about techniques of landscape visualization. Currently, she is a leader of the cultural and educational project about historical structures and cultural landscapes in Slovakia.

Chapter 14
Ecological and Cultural Diversity of Traditional Mountain Agricultural Landscape: A Case Study from Slovakia

Jana Špulerová, Marta Dobrovodská, Stanislav Dávid, Ľuboš Halada and Peter Gajdoš

Abstract The traditional agricultural landscape in Slovakia reflects the history and long-term mutual relationship between the landscape and generations of farmers, resulting in the occurrence of specific agricultural features. The ecological and cultural diversity of traditional agricultural landscape is evaluated using the example of the area around Liptovská Teplička. This area constitutes a Carpathian montane landscape of steep slopes with small-scale arable fields, grasslands and unproductive elements of plots: balks such as terrace slopes, step bounds, heaps or mounds. Due to heterogeneous natural conditions on the one hand, and specific cultural-historic conditions on the other, there has arisen an agrarian landscape of high landscape diversity and biodiversity. It represents mosaics of unique islands of species-rich plant and animal communities, originated by continuous succession over centuries and dependent on traditional cultivation. We have studied the connections between development of cultural-historic conditions (settlement, land law, social and political environment, methods of agricultural cultivation) and the biodiversity of individual historical landscape elements (plots of arable land, grassland, balks). Our ecological assessment was centred on evaluation of species richness, habitat diversity and vulnerability of species based on vegetation and zoological surveys and existing ecological conditions. The cultural and historical value was

J. Špulerová (✉) · M. Dobrovodská
Institute of Landscape Ecology, SAS, Štefánikova 3, 814 99 Bratislava, Slovak Republic
e-mail: jana.spulerova@savba.sk

M. Dobrovodská
e-mail: Marta.Dobrovodska@savba.sk

S. Dávid · Ľ. Halada · P. Gajdoš
Institute of Landscape Ecology, SAS, Akademická 2, 949 01 Nitra, Slovak Republic
e-mail: stanislav.david@savba.sk

Ľ. Halada
e-mail: lubos.halada@savba.sk

P. Gajdoš
e-mail: p.gajdos@savba.sk

M. Agnoletti and F. Emanueli (eds.), *Biocultural Diversity in Europe*,
Environmental History 5, DOI 10.1007/978-3-319-26315-1_14

determined by the extent of preservation of traditional cultivation techniques and original land terracing with preserved forms of anthropogenic relief, as well as by the presence of small architecture elements.

Keywords Biodiversity · Bounds · Traditional agricultural landscape · Cultural value · Biocultural diversity

14.1 Traditional Agricultural Landscape as an Expression of Agrarian Culture

Agrarian culture and the peasant mindset is one of the roots of European culture. Agriculture was very important for the population living on the territory of Slovakia as a principal occupation and a crucial part of the productive forces. Currently, agricultural land occupies almost half of the area of Slovakia. The level of agrarian culture is a reflection of historical, natural and socio-economic conditions (Freyfogle 2001). The development of agrarian culture in Slovakia is the result of the efforts of local residents to meet their growing material needs and at the same time it is also evidence of a degree of cultural maturity in how they humanized the natural environment in which they lived. The peasant life was strongly influenced by material, social and spiritual expressions of their cultures, relationship to nature, community and family structure and value orientations (Slavkovský 1998), which developed in Central Europe from the intersection of many cultures. One of the tangible expressions of agrarian culture is the traditional agricultural landscape which carries certain natural and human–social information written into the landscape over time (Štefunková and Dobrovodská 2009) by way of features such as plot division, shape and size of fields, land use, presence of balks, regional and local differences as the result of the interaction of natural conditions, geographical location, cultural-historical and economic development. Traditional agricultural landscapes in Slovakia comprise a mosaic of small-scale arable fields and permanent agricultural cultivations, including grasslands, vineyards and high-trunk orchards originating from continuous succession over centuries (Štefunková and Dobrovodská 1997). They were not directly affected by the collectivization of agriculture in the second half of the twentieth century and at present they cover less than one percentage of the entire area of Slovakia, for which reason special attention should be paid to them (Špulerová et al. 2011). The agricultural landscape of the Carpathian region has for a long time preserved its archaic forms of life, since its natural conditions are less favourable for agriculture than the lowland of the Pannonia-Potisc area (Nováková and Slavkovský 2010). Traditional agricultural fields were more likely to persist on steep terrain, less fertile soils and locations that were closer to the settlements but more isolated from the regional capital cities (Lieskovský et al. 2014). Humans by their activities and transformation of forest to agricultural landscapes created new sources of biodiversity, in particular meadows, pastures, bounds as forms of anthropogenic relief (terrace slopes, steep bounds, mounds and walls, etc.) and field edges. Bounds are an important element of traditional agricultural landscape, due to not only their ecological, but also their cultural and historical significance. They reflect the ways in which the land was settled and then divided as a result of inheritance. The presence of very small plots and the associated high proportion of bounds of different character was a typical feature of Slovak territory until the middle of the twentieth century.

The aim of our research is to examine the mutual relationship between the biological and the cultural diversity of traditional agricultural landscapes. Ecological and cultural diversity was studied using the example of the cadastral territory of the Liptovská Teplička village. Liptovská Teplička is one of three study areas where detailed biodiversity research has been conducted. The village is among the highest altitude agricultural settlements in Slovakia with preserved traditional agricultural landscape of arable land and grassland.

14.2 Study Area and Methods

The Liptovská Teplička study area is part of Prešov region and Poprad district. It lies in the basin of the Black Vah. The area is part of the Low Tatras National Park (NAPANT) and its buffer zone. The total area is 9869 ha, of which 81 % is a forest, 17 % of the agricultural land is represented mostly by grassland, nearly 1 % is built-up areas, and less than 1 % is water surface. Typical elements of the agricultural land are narrow-strip plots of arable land and grassland with various forms of anthropogenic relief, which we regard as traditional agricultural landscape. The altitude ranges from 846 to 1429 m a.s.l. Natural conditions are characterized by a broad spectrum of geological bedrocks (predominantly limestone and dolomites) and soils associated with them, mostly steep (12°–17°) or moderate slopes, relief segmentation and varied microclimatic conditions.

Plots of seven different types of traditional agricultural landscape, varying with regards to different natural conditions, structural characteristics, land use, presence and type of bounds and/or intensity of utilization, were selected for study (Table 14.1, Fig. 14.1).

The cultural-historical value expresses the degree of preservation of structural features including: original plot division and land terracing with bounds, the land's shape and size, the type of bounds and land use and traditional field and bound cultivation techniques. These latter include ploughing, mowing, grazing and stone removal. The cultural and historical value of the area was increased by the presence of small architectural elements and historical farm buildings that we observed in the study area. Identification of the traditional agricultural character was derived from historical documents, field research and interviews with local people. The study areas were assigned three grades of cultural-historical value: (1) low; (2) medium and (3) high.

Table 14.1 Characteristics of study plots in Liptovská Teplička

	Type of bounds	Land use before collectivisation	Present management	Altitude (m a.s.l)
(1) Extensively used grasslands on silicaceous substrate	Terrace slopes	Terraced grassland, over-grassed before collectivization	Occasionally grazed and irregularly mown	924–956
(2) Extensively used grasslands on calcareous substrate	Mounds	Mosaic landscape—rotation of arable land, meadow and fallow land	Occasionally grazed and irregularly mown	1045–1081
(3) Extensively used grasslands on calcareous substrate	Mounds	Mosaic landscape—rotation of arable land, meadow and fallow land	Regularly mown and occasionally grazed	970–978
(4) Extensively used grasslands on silicaceous substrate	Mounds	Mosaic landscape—rotation of arable land, meadow and fallow land	Occasionally grazed and irregularly mown	1048
(5) Extensively used mosaic of grasslands and arable fields	Terrace slopes	Mosaic landscape—rotation of arable land, meadow and fallow land	Regularly mown and occasionally grazed	922–934
(6) Extensively used mosaic of grasslands and arable fields	Mounds	Mosaic landscape—rotation of arable land, meadow and fallow land	Regularly mown and occasionally grazed	957–958
(7) Extensively used grasslands on calcareous substrate	Terraced slopes	Mosaic landscape—rotation of arable land, meadow and fallow land	Occasionally grazed and irregularly mown	960–961

The ecological evaluation of mountain traditional agricultural landscape was based on a detailed inventory of vegetation and selected bioindication animal groups: Molluscs (Mollusca), Millipedes (Diplopoda), Spiders (Araneae), Grasshoppers (Orthoptera), Beetles (Coleoptera) and Birds (Aves). Special attention was given to forms of vegetation connected to the balks (forms of anthropogenic relief), which are considered as refuges for many plant and animal species in the agricultural landscape.

The final ecological evaluation was expressed as the sum of biodiversity and conservation assessments. Biodiversity assessment was centred on (1) species richness, (2) the Shannon biodiversity index and (3) index of equitability. Conservation value assessment was performed based on the presence of protected and vulnerable species included in the National Red Species List of protected species declared by national law (decree No 23/2004) and European Conservation Directives (Annexes of Birds Directive and Habitats Directive). The final evaluation scale was divided into five categories with equal intervals: sites of (1) very low

Fig. 14.1 Liptovská Teplička study area and delineation of study plots

significance (1–10 points), (2) low significance (11–20 points), (3) medium significance (21–30 points), (4) high significance (31–40 points) and (5) very high significance (41–50 points) (Halada 2000). Hemeroby, synanthropy and other vegetation characteristics were not considered in this assessment.

14.3 Results

14.3.1 Cultural-Historical Value

The landscape of the Liptovská Teplička cadastral area is characterized by its strip-like structure of small-scale plots, which beautifully frame the forested landscape of the Low Tatras Mountains. The village, which still preserves typical wooden architecture, was colonized in the seventeenth century by Goral settlers.

Specific forms of anthropogenic relief were created during the period of terrain modification for agricultural production.

Based on the inventory of original forms of anthropogenic relief, which are one of the characteristic features of traditional agricultural landscape, four types of bounds were distinguished in the Liptovská Teplička study areas (Špulerová and Dobrovodská 2010): (1) terraces slopes, (2) step bounds, (3) mounds and (4) slope mounds and heaps on terrace slopes (Figs. 14.2 and 14.3). The bounds had varying soil skeleton contents: muddy, muddy-rocky, rocky-loamificated and rocky.

The origination of bounds and their structure was associated with a high relief slope of agricultural production land, with a high soil skeleton content, and high fragmentation of the land into narrow and long ploughed plots. Terraces and step bounds are the result of improvement in relief soil quality, and they continued to be cultivated after their initial creation. The mounds are the result of soil skeleton removal and usually they were not further cultivated after construction. Slope mounds and heaps on terrace slopes are the result of both of the aforementioned methods of soil cultivation.

Based on the preservation of structural features of traditional agricultural landscape, cultivation techniques and historical and present land use, the following three degrees of cultural-historical values were assigned to the study areas:

1. **Low cultural-historical value** areas were constituted by traditional agricultural landscape without regular management, with signs of gradual abandonment indicated by shrub and tree overgrowth. These areas were frequently observed in the forest contact zone and in highly terraced areas.

Fig. 14.2 Terrace slopes

Fig. 14.3 Mounds. Photo made by Špulerová in 2009

2. **Medium cultural-historical value** areas comprised regularly mown or extensively grazed former grassy fields or former mosaics of arable fields and grassland. Most former arable fields covered by grass in the last 50 years are now used as extensive meadows or pastures. Meanwhile, most grassland in Liptovská Teplička is managed by the Agricultural Cooperative and subsidized by agri-environmental schemes.
3. **High cultural-historical value** was designated for mosaics of arable fields and permanent grassland with almost completely preserved forms of traditional use. These traditional landscapes were created by arable crop rotation, traditional hay-making and use of draught animals, and they were mainly located in the most accessible areas near dwellings. Steep slopes and long narrow fields are not accessible to heavy machines and therefore traditional management has been preserved there.

   No cultural-historical value was reported for reclaimed intensively used grassland and arable land.

History and agrarian culture is also represented by preserved rural architecture such as the wooden houses in the village, and the wooden barn and about 500 earth cellars located on the slope, behind the urban areas of the municipality. Two-meters-deep cellars are still used to store the harvest of potatoes, vegetables and flowers as they keep a stable temperature throughout the year (Fig. 14.4). An inseparable part of traditional agricultural landscape are the elements of small architecture—such as crucifixes, crosses, statues, haylofts or spring shelters situated on the field edges or crossroads—which sometimes become dominant elements in the landscape (Štefunková and Cebecauer 2006).

Fig. 14.4 Preserved earth cellars in Liptovská Teplička (Špulerová 2010)

14.3.2 Ecological Evaluation

Our final ecological evaluation was expressed as a combination of the biodiversity assessment and conservation assessment (Table 14.2). Species richness expressed by diversity indices and the number of species points to very high species diversity of habitats. More than 190 species of vascular plants, 55 species of birds (Aves), 23 species of mollusc (Mollusca), 83 species of orthopterans (Orthoptera), 15 species of Diplopoda and 146 spiders (Araneae) were recorded on the study plots, including several rare, vulnerable or protected species.

Table 14.2 Overall ecological assessment of studied traditional agricultural landscape plots in Liptovská Teplička

	Araneae		Aves		Diplopoda		Mollusca		Orthoptera		Vegetation		Overall assessment	
	BD	CA	BD	CA	BD	CA	BD	CA	BD	CA	BD	CA	Σ	Category
LT1	5	5	2	2	5	0	4	0	1	3	5	0	32	4
LT2	4	1	5	4	4	0	1	5	5	5	4	2	40	4
LT3	4	5	1	4	4	0	5	5	2	4	5	3	42	5
LT4	2	0	1	1	4	0	4	0	5	4	3	0	24	3
LT5	5	2	3	1	5	3	5	0	3	3	3	0	33	4
LT6	4	1	1	3	1	0	4	5	5	4	3	1	32	4
LT7	1	3	5	5	3	0	5	5	1	5	5	5	43	5

BD biodiversity, *CA* conservation value assessment

The conservation value assessment confirmed our hypothesis that traditional agricultural landscape with bounds forms islands of biodiversity, including the presence of many rare, vulnerable and threatened species, particularly from animal groups. The highest conservation value was reported for orthopterans (Orthoptera), birds (Aves) and molluscs (Mollusca).

The overall ecological value assessment is focused on comparison of and highlighting differences between types of traditional agricultural landscapes that are present in our study areas.

14.3.2.1 Extensively Used Terraced Grasslands on Silicaceous Substrate (LT1)

The former arable field on terraces and terraced slopes of muddy-rocky soil skeleton content were created by gradual ploughing and later changed to grassland. Today they are extensively used and occasionally grazed. Terraced slopes constitute habitats with the highest species richness of spiders, three of them vulnerable (Dankaninová and Gajdoš 2012). A high species richness of molluscs and occurrence of endangered species from Orthoptera was also recorded on grassland terraces, leading to this landscape structure being assessed as having a high overall ecological value.

14.3.2.2 Extensively Used Grasslands with Mounds on Calcareous Substrate (LT2, LT3)

Extensively used mosaics of grassland and mounds arose from former arable fields being overgrown with grass and mounds which arose mostly on shallow stony soils as a result of soil skeleton removal and gradual deposition of stones into mounds. Over time, gradual succession created species-rich grasslands close to semi-natural habitats, which are characterized by high species richness with the occurrence of rare and endangered plant species. Mounds on calcareous substrate represent the habitat with the highest diversity of spiders, and high diversity and occurrence of endangered species of Orthoptera, mollusc and birds; the overall assessment of animals and plants showed this to be a very significant habitat. The biodiversity of adjacent grassland was slightly lower, but the overall ecological assessment was high.

Within the study plot LT3 semi-natural meadow was situated. They were managed as grassland for a long time and were not affected by the reclamation of agricultural landscape. Their ecological value was higher than the previously discussed grassed arable fields, and based on the overall ecological assessment they belong to sites of very high ecological significance.

14.3.2.3 Extensively Used Grasslands with Mounds on Silicaceous Substrate (LT4)

Species richness of extensively used grassland on shallow soils on silicaceous substrate, which are grazed or mowed, is lower compared to the communities on calcareous substrate. Nevertheless, they are considered regionally rare communities, because they occurred only on this site. These habitats are particularly important for Orthoptera, whose species richness makes this a close-to-natural habitat. Our overall ecological assessment of this site is one of medium ecological significance.

14.3.2.4 Extensively Used Terraced Mosaic of Arable Fields and Grasslands (LT5)

Terrace slopes in combination with small-block fields and grassland document traditional management of the area. Species richness was higher if terrace slopes were mowed and decreased in the vicinity of arable land and fallow land. The overall ecological assessment of these former arable fields is as sites of higher ecological significance, noting also the positive development of habitat succession to semi-natural grassland. Our overall assessment of biodiversity was relatively high due to the regular mowing of meadows but slightly declined due to only occasional mowing of mounds in the vicinity of arable land.

14.3.2.5 Extensively Used Mosaic of Arable Fields and Grasslands with Mounds (LT6)

The extensively used mosaic of grasslands and arable fields with mounds were characterized by a similar overall ecological assessment as terraced mosaics which ranged from medium to high ecological significance. A lower biodiversity value was reported on mounds in the vicinity of arable fields, the highest value on mounds next to grassland.

14.3.2.6 Extensively Used Terraced Grasslands on Calcareous Substrate (LT7)

Extensively used terraced grasslands represent a mosaic of former arable fields that were changed to grasslands. From the selected fauna biodiversity point of view, the terraced slopes were evaluated as sites of high significance, but the overall ecological assessment is as a site of very high ecological significance.

Intensively used reclaimed grassland was not subject of our interest, but two sites with such habitats were selected as control plots in the vicinity of study plots

LT2 and LT6. Our ecological assessment confirmed our assumption that overall ecological value of this habitat is low.

Intensively used large-block fields of arable land represent another control plot, which was established within the LT2 study plot for comparison of biodiversity assessment in relation to intensity and style of land management. Partial results of biodiversity and conservation value assessment of vascular plants and selected groups of animals pointed to this site being of very low ecological significance.

The highest ecological value was recorded for extensively used grassland with different types of bounds (LT3, LT7) or semi-natural meadow without bounds, indicating a character of semi-natural habitats. Medium values of biodiversity were recorded for mosaics of arable land and grassland and the lowest biodiversity was confirmed to be on intensively used areas (reclaimed meadows, large-block arable fields).

14.4 Connecting Ecological and Cultural-Historical Diversity

The assessment of the ecological and cultural-historical value of the Liptovská Teplička study area confirmed that this area represents a region with a specific combination of natural and cultural diversity, including high visual quality of the landscape. Maybe due to its isolated location and the physical proximity of dwellings, the local community still retains traditional values in matters both material and spiritual.

The significance of the cultural-historical value lies especially in fact that they are repositories of knowledge regarding former ways of agrarian management. The term 'cultural landscape' has been connected with environmental management since 1960 (Jones 2003). Based on structural features, such as shape and size of land parcels, type of bounds, land use and extant way of land cultivation (including ploughing, mowing, grazing), we can retrospectively construct a model of the development of the agricultural activities and the historical landscape which the preserved traditional agricultural landscapes constitute.

The results of ecological and cultural-historical evaluation are sometimes contrary to each other. Sites of very high ecological significance are not always at the same time very significant from a cultural-historical point of view. We can assume that the ecological value of traditional agricultural landscapes was much lower in the heyday of small-scale farming dominated by arable fields. The relicts of such farming occur very rarely today as most of the former arable fields were changed to grassland due to low production and therefore have been as sites of very high cultural-historical value.

Comparing the biodiversity of types of grassland, the highest species richness was recorded for semi-natural grassland, lower species richness for extensively used grassland (including grassed former arable fields) and the lowest for intensively used grassland. The bounds proved to be islands of high species richness, including habitats for selected fauna groups. The very high species richness also owes something to the heterogeneity of the geological substrate, as carbonate and siliceous substrates meet in this area. Even if nitrophilous species occur in mosaics with arable fields, their proportion was not very high, reflecting the semi-natural character of these traditional agricultural structures developed over the centuries. Other studies found that traditional extensive maintenance of the agricultural landscape help to maintain biological and genetic diversity in these ecosystems (Altieri and Merrick 1987) and preserve optimal environmental characteristics of soils (Krnáčová et al. 2013).

Land use changes of the last decades caused creation of such new habitats as fallow land, extensively used or partially abandoned pastures, and small patches of forested habitats, all of which increase landscape diversity, offer refuges for other species in an intensively or extensively utilized agricultural landscape, and at the same time increase overall stability of soil ecosystems. The positive result of this development is that there are no visible signs of abandonment or significant degradation of traditional agricultural landscape that could threaten these valuable landscapes as has happened in other regions of Slovakia or within Europe (Mojses and Petrovič 2013; Agnoletti 2007). Some agricultural land is still regularly managed by small farmers but the biggest local user of agricultural land is the Agricultural Cooperative in Liptovská Teplička. The Agricultural Cooperative has carried out ecological farming on more than 1200 ha of land leased from land owners since 1996. Based on our research and the current state of land management, we consider the present state of the traditional agricultural landscape to be favourable. To preserve this valuable landscape, it is necessary to continue the ecological farming which is in operation today thanks to agro-environmental subsidies. The biological or cultural-historical values of these landscapes are increased also by the historical uniqueness of this area, the visual aspect of the landscape and other cultural and regulating ecosystem services that they provide, including the impact of bounds on the physical properties of soil (Fazekašová et al. 2012). Even if the traditional agricultural landscape belongs to those historical landscape types which are very vulnerable, proper legislation for their protection and management is currently absent (Hrnčiarová 2004).

Acknowledgments This research was conducted within the grant project of the Ministry of Education of the Slovak Republic and the Slovak Academy of Sciences Grant No. 2/0158/14 "Diversity of agricultural landscape and its ecosystem services" and the project of the Slovak Research and Development Agency No. APVV-0669-11 "Atlas of landscape archetypes of Slovakia".

References

Agnoletti M (2007) The degradation of traditional landscape in a mountain area of Tuscany during the 19th and 20th centuries: implications for biodiversity and sustainable management. For Ecol Manage 249(1):5–17

Altieri MA, Merrick L (1987) In situ conservation of crop genetic resources through maintenance of traditional farming systems. Econ Bot 41(1):86–96

Dankaninová L, Gajdoš P (2012) Epigeické pavúky historických štruktúr poľnohospodárskej krajiny (vinohradnícka krajina Svätý Jur). Folia faunistica Slovaca 17(3):275–290

Fazekašová D, Kotorová D, Boburská L (2012) Development of biochemical parameters of soil in conditions of sustainable use of soil. Ekológia (Bratislava) 31(2):238–247

Freyfogle ET (ed) (2001) The new agrarianism: land, culture, and the community of life. Island Press, Washington

Halada Ľ (2000) Landscape-ecological evaluation of vegetation as a part of the landscape-ecological planning. Ekológia (Bratislava) 19(2):99–106

Hrnčiarová T (2004) Natural and cultural aspects of landscape and its potential. Životné prostredie 38(2):61–65

Jones M (2003) The concept of cultural landscape: discourse and narratives. Landscape interfaces. Springer, The Netherlands, pp 21–51

Krnáčová Z, Hreško J, Kanka R, Boltižiar M (2013) The evaluation of ecological factors affecting environmental functions of the soils in area of traditional agrarian structures. Ekológia (Bratislava) 32(2):248–261

Lieskovský J, Kenderessy P, Špulerová J, Lieskovský T, Koleda P, Kienast F, Gimmi U (2014) Factors affecting the persistence of traditional agricultural landscapes in Slovakia during the collectivization of agriculture. Landscape Ecol 29:867–877

Mojses M, Petrovič F (2013) Land use changes of historical structures in the agricultural landscape at the local level—Hriňová case study. Ekologia 32(1):1–12

Nováková K, Slavkovský P (2010) Traditional agrarian culture in Slovakia—part of European agrarian heritage. HISTORIE—OTÁZKY—PROBLÉMY. 2. http://ucd.ff.cuni.cz/en/veda-a-vyzkum/historie-otazky-problemy/hop-archiv-cisel/historie-otazky-problemy-2-2010-c-2/tradicna-agrarna-kultura-slovenska-sucast-europskej-rolnickej-civilizacie/. Accessed 6 Feb 2015

Slavkovský P (1998) Tradičná agrárna kultúra Slovenska. Veda, Bratislava

Špulerová J, Dobrovodská M (2010) Significance of landscape diversity of historical agricultural landscape structures on the example of Liptovská Teplička study area. In: Nováček P, Huba M (eds) Udržitelný rozvoj—stav a perspektivy v roce 2010: sborník z konference. Univerzita Palackého v Olomouci, Olomouc, pp 241–251

Špulerová J, Dobrovodská M, Lieskovský J, Bača A, Halabuk A, Kohút F, Mojses M, Kenderessy P, Piscová V, Barančok P, Gerhártová K, Krajčí J, Boltižiar M (2011) Inventory and classification of historical structures of the agricultural landscape in Slovakia. Ekológia (Bratislava) 30(2):157–170

Štefunková D, Cebecauer T (2006) Visibility analysis as a part of landscape visual quality assessment. Ekológia (Bratislava) 25(1):229–239

Štefunková D, Dobrovodská M (1997) Historické poľnohospodárske formy využitia zeme—ich úloha v trvalo udržateľnom rozvoji. Acta Environmentalica UC, 149–153

Štefunková D, Dobrovodská M (2009) Preserved European cultural heritage in agrarian landscape of Slovakia. Tájökológiai Lapok 7(2):283–290

Chapter 15
Religion and the Management of the Commons. The Sacred Forests of Epirus

Kalliopi Stara, Rigas Tsiakiris, Vasilis Nitsiakos and John M. Halley

Abstract Sacred natural sites (SNS), and especially forests, constitute almost certainly the world's oldest conservation systems. The reasons for their maintenance are related very often with concrete ways of managing local resources and ecosystems, through religious rules. In Zagori and Konitsa, NW Greece sacred forests exist in most villages. Their vegetation and forest structure variety along with cultural elements, such as identities of the communities who had established them, the purpose of their establishing, the different rituals implemented for their transformation from profane to sacred, associated taboos, and their particular history create their unique character. Accepted uses in sacred forests are depended to the purpose of their establishment. More often hunting, grazing, collection of plants, mushrooms, and dead branches are allowed, while taboos are mainly connected with the trees themselves. Sacred forests display nowadays a newly emerged value for biodiversity conservation and they can serve as a locally adapted exemplar of successful historical conservation systems.

Keywords Sacred Natural Sites · Old growth forests · Northern Pindos National Park · Greece · Mediterranean mountains · Biocultural diversity

---

K. Stara (✉) · J.M. Halley
Department of Biological Applications and Technology, University of Ioannina, University Campus, 45110 Ioannina, Greece
e-mail: kallio21@hotmail.com; kstara@uoi.cc.gr

J.M. Halley
e-mail: jhalley@cc.uoi.gr

R. Tsiakiris
Forestry Service of Ioannina, M. Kotopouli 62, 45445 Ioannina, Greece
e-mail: rigastsiakiris@gmail.com

V. Nitsiakos
Department of History and Archeology, University of Ioannina, University Campus, 45110 Ioannina, Greece
e-mail: bnitsiak@cc.uoi.gr

M. Agnoletti and F. Emanueli (eds.), *Biocultural Diversity in Europe*, Environmental History 5, DOI 10.1007/978-3-319-26315-1_15

15.1 Introduction

Sacred Natural Sites (SNS) have been defined as areas with special significance to people and communities (Oviedo and Jeanrenaud 2007). Sacred sites can be the home of gods, places related with spiritually significant historical events or people, or places of worship, expressing a unique sense or spirit of place (Antrop 2005; Dudley and Higgins-Zogib 2012). They are often associated with temples, shrines, mosques, churches, pilgrimage trails, or epiphanies of the divine in natural features (Verschuuren et al. 2010). Their sacredness is often related with a foreboding expressed as a taboo that prevents their exploitation for private use. Because sacred sites have been managed as protected areas they have been recognized as the world's oldest conservation systems (Dudley et al. 2009).

SNS exist all over the world. In spite of the common Christian view of sacred nature as idolatry (Dudley and Higgins-Zogib 2012; Frascaroli 2013), in Greece there are well-known examples of Christian Orthodox SNS such as the rock pillars of Meteora (Lyratzaki 2007) and the monastic community of Mt Athos (Papayannis 2007). However, spirituality is also expressed on a smaller scale by outlying churches and their emblematic trees or groves that magically encircle or mark village boundaries with the power of the cross (Stara et al. 2012). Such sacred sites often specify liminal or dangerous places, boundaries, significant routes, important resources, or drinkable water (Nixon 2006). SNS also denote concrete ways of managing local resources through religious rules (i.e., Anderson et al. 2005; Chandran and Hughes 2000; Sharma et al. 1999; Virtanen 2002; Watley and Colfer 2004). In historical contexts where religious institutions were closely connected with political authorities, the term "sacred" functioned as a synonym for a community-based management of common resources though social taboos (Byers et al. 2001; Rutte 2011). Religion can attach sacred values to specific resources regulating their use by means of threats of punishments of a supernatural character for the trespassers (Colding and Folke 2001; Rutte 2011).

In this paper, we present the natural and cultural characteristics of 16 representative sacred forests in the mountains of Epirus in North West Greece that serve as exemplars for managing natural resources and conserving ecosystems through religious rules.

15.2 Materials and Methods

15.2.1 Description of the Study Area

Our study areas include the municipalities of Zagori and Konitsa, in the Pindos range of NW Greece. These areas flourished under Ottoman occupation (1479–1912) but nowadays are among the least populated areas of Greece. Mt Grammos (2520 m) defines the northern limit, while Mt Smolikas (2637 m), the second highest mountain of Greece, is located to the east. The *Katára* (literally "accursed") pass, near Métsovo, and newly constructed Egnatia highway demarcate the southern boundary, while the long ridge of Mt Mitsikeli completes the square to the west.

Our research area encloses a large diversity of mountains, including rough terrain with steep rocky slopes, scree-slopes, escarpments, and ravines of the hard limestone or erodible flysch and ophiolite that develop into badlands. The deep gorges and steep valleys of the Vikos, Voidomatis, and Aoos rivers created suitable shelter during the ice ages with flourishing microhabitats that still host a great biodiversity of flora and fauna that feature many endemic or rare species of Europe. Descending from the alpine and subalpine anthropogenic meadows above tree limit, Bosnian and black-pine forests (*Pinus heldreichii and P. nigra*) in the form of open wood pastures give way to dense fir, beech, and oak forests (*Abies x borisii-regis*, *Fagus sylvatica*, *Quercus* spp) mixed with several other broadleaved species. Pastoral woodlands and open scrublands created a diverse cultural landscape dominated by rangelands until recently but now, due to land abandonment, the landscape is mainly dominated by young mixed broadleaved forest and dense schlerophyllus scrublands. The recent homogenization of the vegetation has also hidden the mosaic of cultivated terraces, shredded oakeries, orchards, rivulet gardens, and extended overgrazed bushland near most mountain villages.

The local identities of the various peoples who came to this area are imprinted on the landscape through the long interaction between people and Nature. Descending again from the mountain peaks to the lowlands in Zagori, but focusing on people this time; we first see the summer pastures above the tree line corresponding to pastoral transhumant Sarakatsani, who started to use the land from the eighteenth century. After the Second World War (WWII) many of these people settled permanently in the villages of the area (Campbell 1964). In the oak vegetation zone of the central and west parts of Zagori, where most of the villages are located, the landscape presents relics of the characteristic mosaic of permanent mountain settlements. The inhabitants of that area were initially settled agriculturalists. From the seventeenth century men started to immigrate to Europe, Africa, and America. That, in combination with the privileges that Zagorians enjoyed during the period of the Ottoman occupation, has resulted in an impressive prosperity that is still visible in the settlements and landscape. From the thirteenth century south and south-east Zagori (*Vlachozágoro*) was occupied by the linguistically distinct Vlach transhumant pastoral community. Today the forested areas of south-east Zagori are inhabited by settled Vlachs, still known as "woodcutters."

As far as the area of Konitsa is concerned, there is also a quite clear correspondence between ecological zones and cultural unities. The high mountain pastures are occupied by Vlach transhumant pastoralists who spend their summer time in the villages of Aetomilitsa (on Grammos Mt) and Fourka (on Smolikas Mt) and move to the plains of Thessaly and Macedonia for the winter time. What mostly characterizes these communities from an ecological point of view are the customary practices of managing collective resources on the basis of a balance between human population, animals, and the available pastures, which belong to the whole community. Sedentary Vlachs reside also in the Aoos valley in a cluster of settlements lined up along the river, which developed historically their own distinctive local culture based on forestry and agriculture, with complementary domestic animal husbandry. The lowland in the zone between the settlements and the river is cultivated; in the past the zone of the cultivated land used to extend to the north of the settlements. Above the settlements there are also zones of preserved woodland and thickets, as well as pastures, and further away forests, which constitute an important factor of the local economy. In the past this population specialized also in crafts and professions related to forests such as sawyers and tar-dealers.

As to the Sarantaporos valley, it is a unity which specialized historically in crafts (stone masonry, painting-hagiography, wood carving, carpentry), that is why all these villages are called *mastorohória* (literally "craftsmanvillages"). Occupying the oak zone, these communities were initially agropastoral. At some historical moment their male population turned to technical specialization (eighteenth–nineteenth century) due to demographic and economical hardship. This phenomenon started to decline at the end of the nineteenth century and disappeared in the second-half of the twentieth century.

Another unit is formed by the villages of the Konitsa plain, which combine agriculture with animal husbandry. These villages used to belong to local *beys* in Ottoman times, but after their liberation they became free and developed a local economy based on small holdings plus a small-scale domestic animal husbandry. People from the town of Konitsa itself own also land in the plain, which was distributed to them by the state (a large part of them are Asia Minor refugees that settled in Konitsa in 1923). The town of Konitsa itself constitutes the administrative, commercial, and cultural center of the municipality. Konitsa has always been a center and a crossing. It has been a place of cultural input, where various cultural elements met, and has functioned as a channel for the osmosis among different people and cultures. To all this the well-known bazaar played a central role (Nitsiakos 2008).

National laws and presidential or ministerial edicts protect part of the cultural heritage of the area, mainly Byzantine and post-Byzantine monuments, private, or public buildings or entire settlements. Infrastructure related to the network of transportation (bridges, stone-paved lanes) or with pre-industrial agriculture (water mills, threshing floors) that date back as far as the fifteenth century also constitute characteristic examples of local architecture. Moreover, the area is well known for its high ecological value and has a great conservation importance. Therefore, it is protected by several Greek and European laws and part of it has been designated as the National Forest of Pindus (1966), the National Forest of Vikos-Aoos (1973), the biogenetic reserve of Pindos (1976), the Northern Pindos National Park (2005), the Vikos-Aoos UNESCO Geopark (2010), and includes eleven NATURA 2000 sites: Special Protected Areas (SPAs) for bird conservation and Sites of Community Importance (SCIs) for habitat and species protection.

15.2.2 The Study on the Sacred Natural Sites of Epirus

The study of the SNS of Epirus started in 2000 and was initially involved 23 villages in Zagori (Stara et al. 2015a). Work resumed in 2012 through an inter-disciplinary research programme based at the University of Ioannina (THALIS-SAGE 2012–2015 "Conservation through Religion: the Sacred Forests of Epirus"). Our aim was the study of sacred forests as an effective conservation system and their bio-cultural value, focusing not only in their cultural and spiritual value, but also in their biodiversity importance using as criteria groups of organisms depending on mature trees such as lichens, birds, bats, fungi, saproxylic insects and soil biodiversity (Read 2000; Rackham 2006).

Through extensive archival and ethnographic research and fieldwork, we located the SNS of every village, but we orientated our study on forests excluding isolated mature trees or small groves in church yards. We asked people to tell us about the reasons of maintenance of their sacred forests, their history, ritual activities, supernatural guardians, accepted and non-acceptable uses, taboos, and stories relating to trespassing in these forests. We confirmed people's narratives in community archives and we visited the forests accompanied with locals where possible.

In order to create common research areas for different scientific disciplines, including colleagues who studied biodiversity in detail, we chose a subset of the sacred forests of the area using as first priority criteria their classification in different geographic and cultural units (Fig. 15.1). This covered most of the diversity existing in the area, inasmuch as we chosed at least one representative forest from each geographic-cultural unit (Fig. 15.2; Table 15.1).

We selected forests that were bigger than 4.5 ha applying also a rough criterion of 70 % tree cover, excluding those that were open woodland pastures in the past, with an exception if this vegetation type was only represented in this form. In addition we chose representatives of at least one of all vegetation types existing in the area. Good preservation status was also considered, as some sacred forests have been partly destroyed. Sacred forest boundaries based on ethnographical and archival research were later identified and mapped by the use of orthorectified aerial photographs of the year 1945, which is the oldest complete set of aerial photographs in the area.

During our visit to each forest we recorded the geology and bedrock type, checked for the presence of landslides, fallen rocks, water, and springs and noted also vegetation type, forest structure, past, and recent uses, existing buildings or other artifacts. We defined their recent and older borders in as much detail as possible. A rapid biodiversity survey was focused on easily identifiable large mammals and raptors. All data were entered into a GIS environment for further analysis (topography, inclination, etc.).

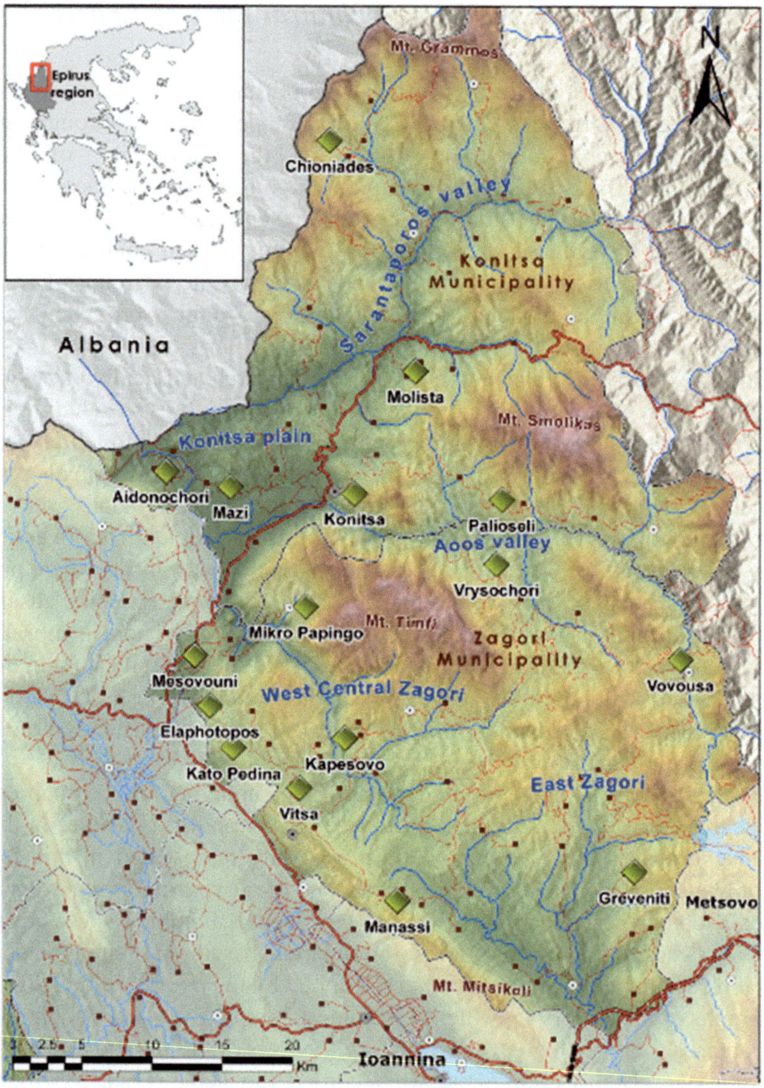

Fig. 15.1 Location of the research area and the selected 16 characteristic sacred forests in Konitsa and Zagori municipalities in Epirus, NW Greece

15.3 Results and Discussion

15.3.1 The Sacred Commons

All different cultural units in the area preserve sacred forests and used religion as a strategy to control management during the Ottoman occupation. In that period these

Table 15.1 Physical characteristics of the 16 sacred forests of our study

	Forest name (in Greek)	Forest name (in English)	Town/Village	Cultural/Geographic unit	Vegetation type	Area (ha)	Altitude (m)
1	Kourí	Protected forest	Kónitsa	Konitsa plain	Black pine	115.70	760–1600
2	Kourí	Protected forest	Mázi	Konitsa plain	Broadleaved oak	10.40	560–645
4	K(ou)rí stin Panagiá	Virgin's forest	Aidonochóri	Konitsa plain	Mixed broadleaved	19.80	715–960
3	Kourí	Protected forest	Chioniádes	Konitsa, Sarantaporos valley	Mixed broadleaved	41.30	1120–1300
5	K(ou)rí tis Panagiás	Virgin's forest	Elaphótopos	Central-West (CW) Zagori	Prickly oak	29.10	660–890
6	Livádi	Meadow	Manassí	CW Zagori	Mixed broadleaved	53.7	1000–1400
7	Livádi	Meadow	Vrysochóri	East Zagori, Aoos valley	Broadleaved oak	10.40	930–1150
8	Livadákia	Small meadows	Vítsa	CW Zagori	Broadleaved oak	4.90	917
9	Mereáo	Common land	Palioséli	Konitsa, Aoos valley	Black pine	22.40	1100–1340
10	Agía Paraskeví	Santa Paraskevi	Vovoússa	Konitsa, Aoos valley	Black pine	6.80	980–1070
11	Agios Charálampos	Saint Charalampos	Mesovoúni	CW Zagori	Broadleaved oak	17.02	530–600
12	Tráfos	Ditch	Mólista	Konitsa, Sarantaporos valley	Black pine	43.29	880–1220
13	Plái	Slope	Mikro Pápingo	CW Zagori	Stinking, Grecian junipers	28.23	1200–1550
14	Gradísta	Fortress	Kapésovo	CW Zagori	Broadleaved oak	23.6	950–1170
15	Anília	Non sunny	Káto Pediná	CW Zagori	Prickly oak	10.3	848–963
16	Toúfa (the core area)	Thicket	Greveníti	East Zagori	Beech, mixed broadleaved	117.2	1060–1380

mountain communities enjoyed privileges and a kind of "autonomy" from the Ottomans, while the Church found herself in the special position to substitute political power and judicature (Mihailaris 2004). Local authorities comprehending the importance of forests, especially in cases where these could be used as protective wood belts against natural hazards, decided on their protection with the support of religion as supernatural guardian. Supernatural fears, based on pre-Christian beliefs of nature-spirits (Stara 2012a), coalesced in the idea of a dangerous and powerful sacred, which aided the political power to enforce, among the members of a closed society, conformity of moral respect toward common important resources, such as forests. As conceptualized elements sacred forests functioned also as cultural codes (Colding and Folke 2001) that could protect social order, secure the prosperity of the community, delineate admissible relations between its members and define categories such as good and bad, inside and outside, safe, and dangerous (Douglas 1966).

Our archival work revealed communal agreements, which prove that some sacred forests used to be private land that belonged to different villagers. Narratives enforce such decisions as people often refer to toponyms that indicate property rights of certain villagers inside the sacred forests. Moreover, existence of private trees inside the sacred forests, such as sweet chestnuts that ex-owners of the land are allowed to harvest, are indirect indicators that such areas were once private land.

15.3.2 The Names

Sacred forests are refereed to with the same term *ierá* in Greek literature, but in spoken language the preferred general term is *klisiast(i)ká*, which are literally those belonging to the Church. Another synonym is *vakoúfika*, from the Turkish *vakuf*, which signifies a bequest from which the income should be used for public benefit (Moutafchieva 1988).

Many sacred forests are named *kourí* or with syncope *kri* with the meaning of a piece of forest kept as protected with the potential to be used as a reserve for community needs or Church income. Other sacred forests are called *livádia* (in plural). *Livádi* literally means meadow, but our field research revealed that these areas do not have the typical appearance of an open grassland. On the contrary they are woodlands, which function as protective wood belts above villages. Therefore, we relate the etymology of the world with the ancient Greek *livás* (literally "water drop") (Stara and Tsiakiris 2010). Moreover, the Palioseli village sacred forest is named *mereáo* (*miriye* in Turkish, *meriás* in Greek) and means the public land (Moutafchieva 1988).

When the forest is dedicated to a church that is found in its vicinity, the protector saint gives his or her name not only to the sacred forest, but also to the terrain around it. This expresses the domestication of the land, the protection by the saint, and the subjugation of the forces of nature to the spiritual power of the holy personage to which the site is dedicated (Nixon 2006; Stewart 1991). Other sacred forests take their names from the salient features of the landform, following the classical rules of landscape onomatopoieia (Martin 1995).

15.3.3 The Physical Characteristics of the Sacred Forests

Most sacred forests are located above villages and with a size of no more than 50 ha (Table 15.1). Most mountain villages of the area are located in the altitudinal zone where oak forests grow. Accordingly, most selected sacred forests are dominated by mixed broadleaved woodlands consisted mainly by broadleaved oaks (*Q. pubescens*, *Q. frainetto*, *Q. trojana*, *Q. cerris*), while mixed broadleaves have also a big percentage, with maples (*Acer monspesulanum*, *A. obtusatum*), horbeams (*Carpinus orientalis*, *Ostrya carpinifolia*) being the commonest trees reaching the crown of these forests. Monumental trees of beeches, black pines, and junipers (*Juniperus foetidissima*, *J. excelsa*) can be found in respective forests. Thus, the biggest prickly oaks (*Q. coccifera*) in the wider area are to be found only in SNS in the vicinity of the villages in the lower altitudes. Dendrochronology is underway for the oldest trees in the above forests, but many old trees are aging more than 200 years, while there are examples of trees (junipers, oaks, black pines) older than 350 years (Aris Kyparissis and Valentino Marini Govigli, *personal comm.;* Sarris 2008).

Two sacred forests are exceptionally large: the beech forest above Greveniti village (117.2 ha) and the black-pine forest above Konitsa town (115.7 ha) according to their borders shown in the aerial photographs of year 1945. While most of the 16 selected sacred forests are located above 1000 m this of Konitsa and the unique juniper forest of Mikro Papingo reach the highest altitudes (1600 and 1550 m, respectively). Konitsa town sacred forest has the widest range of all, starting from 760 m altitude just above the last houses and reaching the top of the steep slope that hangs above the town. By contrast, the abandoned village of Mesovouni has the lowest sacred forest (of broadleaved oak) starting at just 530 m altitude.

Inclination of the bedrock, which was mainly limestone or sometimes flysch, was in most cases very steep and our field survey revealed that rockfalls, landslides, and even snow avalanches are not rare phenomena and are imprinted in the forest history. In certain cases, as in Manassi village, rockfalls destroyed recently a church inside the forest, while in Palioseli village rolling rocks have been found to have stricken the standing trees and these were later stabilized by local people. Landslides even today frequently destroy roads, bridges, and infrastructure mainly in the Eastern Zagori, such as the destruction of buildings in Greveniti village (2013), where the most extensive sacred beech forest is found. It was not a surprise to find that most sacred forests also host springs, rivulets, or even water reservoirs.

Our detailed biodiversity survey has not finished yet, but our visits to the field so far reveal that all sacred forests serve as breeding sites of raptorial species, especially sort toed eagle (*Circaetus gallicus*) and honey buzzard (*Pernis apivorus*) that seem to prefer those forests, while evidence of the presence of tawny owl (*Stix aluco*) was found in most cases. This was expected, as raptors in general show site fidelity and those forest patches were probably the only breeding sites containing enough big trees and having less disturbance in the near past. Bio-indicators confirmed the presence of the brown bear (*Ursus arctos*) also in all the sacred forests we visited.

15.3.4 The Ritual and Other Praxes

Dedication to a church and most often to the church of the village patron, even if it lies in village's center or in the vicinity of the forest, was the commonest praxis to divide sacred from secular land. Interestingly *aphorismós* (excommunication) was also often used as a sanctification practice. Excommunication in the first Christian centuries constituted a pure ecclesiastical punishment visited on people. However, during the fourteenth–eighteenth century the practice was extended to applications in private issues of economic or social nature. In that context it could be used as an abstract threat, capable of stigmatizing someone (e.g., a trespasser on a sacred forest) with exclusion from the Church, its mysteries and social events related to religious life. Moreover, it could impose to trespassers and their generation a cursed live and damn their souls to the heritage of eternal hell (Mihailaris 2004). Locals recall an atmosphere of intense religiosity and fear. The ceremony was implemented ad loc, at the edge of the forest, singing certain curse psalms of David (e.g., Ps 58, Ps 109), ringing the bell, holding black candles, using an upside down cauldron, and generally reversing the regular order of things. Use might even be made of the number of priests that announced the excommunication. Numbers three and seven especially strengthened its force and this is the reason why some excommunicated forests are called *eftapápada* (literally excommunicated by seven priests) (Stara 2012b).

Dedication and excommunication were not mutually exclusive. On the contrary, excommunication, and in some cases also communal agreements, could be used to enforce dedication. After the establishment of the modern Greek state, excommunication gave way to new forms of Modern European Justice and communal agreements were addressed to the Forestry Service, which published special forestry regulations for the management of the sacred forests. However, locals continued to relate regulations to taboos associated with the supernatural guardians of the sacred forests.

15.3.5 The Supernatural Guardians of the Sacred Forests

The Virgin Mary (*Panagía*), Saint Nicholas (*Ai Nikólas*), Santa Paraskevi (*Agía Paraskeví* or *Stavínere* in Vlach), and Saint Charalampos (*Ai Charálampos*) act in the sacred forests as the intermediaries between the God and people. St. Paraskevi is one of the dragon-killer saints, the protector of fountains in local culture, thus we expect her appearance in sacred forests, where aquifer protection is likely to be among their services. St. Nikolas is protector of travelers in folk religion and thus beloved to Epirus where every family had migrant sons (Stara et al. 2015as). However his spring celebration [May 20th, relics translation to Bari-Italy, which is one of the most important pilgrimage centers of St. Nikolas in Europe (Ševčenko 1983)], also coincides with the period that transhumant shepherd families return from winter to summer pastures and this is accompanied with passage rites such as food offerings. In the sacred forest of St. Nicholas in Vitsa village this date used to regulate the opening of the wood pasture for the village livestock, allowing it to graze the sacred forest just after Service and celebrations in situ until the end of the season.

In cases of trespassing, saints anger or supernatural power of curses in the form of punishments could range from warnings to severe consequences. An instant freezing of the hand that goes to cut the sacred wood, a snake appearing or an accident during cutting are common saint's warnings. After the sacred contagion, saints can appear in trespassers dreams and oblige them to return what they have taken back to the forest where it belongs, or cause them delayed maladies, bring about severe accidents or even sudden death. This could be to wrong-doer himself or to innocent members of his family or his animals (Stara et al. 2012).

In the case of excommunication, negative examples characters who had not feared the anger of God were invoked: the lepers of Giezi (2 Kings 5: 26–27), the gibbet of Judas (Acts 1: 18), the tragic end of Dathan and his family (Old Testament, Book of Numbers 16:31). Apart from material calamities those coming into contact with excommunicated forest might also be denied care and prayers for their dead body increasing possibilities of an unfavorable fate at the Day of Judgement. As with all Orthodox punishments, excommunication aimed at the moral improvement of trespassers and their reintroduction at the Church community. Thus, there was no formal fix imposition, in order to do not discourage the sinner from breaking away from the flock, eternally excommunicated from the Orthodox faith (Mihailaris 2004).

15.3.6 Reasons of Protection and Maintenance of the Sacred Forests

Pindos geology and the inherent vulnerability of the landscape (Bailey et al. 1998), in combination with the high precipitation of West Greece and the long harsh winters, makes protection of settlements compulsory. Frequently landslides and torrents and sometimes avalanches cause severe damage to buildings and roads even today. It is well known that forests have a protective character against natural hazards due to the barrier effect of standing and fallen trees (Volkwein et al. 2011). Our ethnographic and field research confirmed this protective function of sacred forests against natural hazards. Additionally, these protective sacred forests often hosted rivulets and springs, regulating runoff precipitation, and replenishing village aquifers (see also Chandran and Hughes 2000).

Other sacred forests were protected as reserves in case of need. Evergreen prickly oaks especially, could serve as winter fodder preventing livestock starvation during a long harsh winter. In such cases the community controlled harvesting though communal agreements with a certain fee being paid to the Church. Sacred Forests could also provide an income for the Church and serve the community's everyday requirements, such as firewood for or the school, the church during the service or the priest when he had no property in the village.

Sacred forests might exceptionally even give their timber for the benefit of the community. In such a case, the rule of protection could change easily particularly when the reasons were purely religious, as seen in other societies (Chandran and Hughes 2000). An oral tradition from village Megalo Papingo says that the protective juniper forest of this village gave its timber for the construction of the central church. As this forest was just above the village reasons of security against erosion required its replantation with black pines by the Forestry Department in 1959 (Stara 2000). But religious use was not the only exception. For example, part of the forest of Gradista in Kapesovo village was clear-cut when benefactors brothers Paschali decided to finance the construction of the famous *Pascháleios School* in Kapesovo (1864). Gradista was dedicated to village patron St Nicholas, excommunicated and guarded with communal agreements. Local authorities decided to use its wood to co-finance the project indirectly for exporting timber and lime, necessary for the completion of the works. Similarly in 1925 the community council of Elaphótopos village decided the establishment of a limekiln in the village's sacred forest dedicated to the Virgin, in order to repair the council hall and the schools of the village. However, the lime kiln ultimately never worked, due to a series of tragic events that befell the promoter of the project and his family, which contributed decisively to its cancellation. To this day, the inhabitants of Elaphótopos link this tragedy with the wrath of the Virgin, falling mercilessly on those who went messing with her forest (Stara 2009).

Other practical reasons for protection have been also been mentioned by locals, such as the use of the bigger trees as *stálos* (literally livestock shelter from sun in heat of the summer). Locals refer also to aesthetics relating to visual aspects and the beauty that a forest provides in the landscape and annual celebrations associated with forests that surround outlying churches. Moreover historical reasons were offered, such as exemption from Ottoman taxes or the potential of the forest to act as a shelter, given recent memories from the WWII.

Lastly, many outlying churches related to sacred forests, constitute formerly central churches of abandoned now settlements. Such sacred forests attach in many cultures taboos that control access and behavior (Chouin 2002). The more the place is used as a sacred place, like, e.g., St Nicholas sacred forest in Vitsa's abandoned village where a local bequest finances annual celebrations, the more protected is the sacred forest. By contrast, when a church is abandoned its forest is gradually neglected.

15.3.7 Conservation Practices—Accepted and not Accepted Uses

Thus, the effective conservation practice is case dependent and is related to the reasons of establishment and maintenance for every sacred forest in Epirus. They can vary from strict protection to controlled management (Table 15.2).

When a church foundation and sacred forest foundation is rationalized as epiphany of the divine, which is a common occurrence in our study area (Stara et al. 2015as), protection is strict. This is the case of the sacred forest of St. Paraskevi in Vovoússa, preserved for ceremonial reasons according to local beliefs. St. Paraskevi is the miraculous village patron who guards and protects the village and its

Table 15.2 Cultural characteristics of the 16 sacred forests of our study

	Forest name (in Greek)	Town/Village	Ritual praxis	Protector saint	Protection reasons	Protection practices
1	Kourí	Kónitsa	Communal agreements. Only the name refers to religious practices	–	Protective wood belt, aesthetics	**Protection of mature trees**. Cutting of fuel wood during WWII is reported as trespassing
2	Kourí	Mázi	Dedication, communal agreements	Virgin Mary, old village's patron	Protective wood belt, protection of water aquifer, livestock shelter from sun in heat of the summer (stálos), aesthetics	**Protection of mature trees**. Grazing and dead wood collection was accepted. Cutting for poles is reported as trespassing. The report of a lime kiln in the forest edge indicates heavy management in the past. Today the forest is fenced and grazing is illegal
3	K(ou)rí stin Panagiá	Aidonochóri	Dedication	Virgin Mary	Forest of an abandoned monastery according to the tradition. Annual celebration is alive	**Protection of mature trees** in the view shed of the church. Abandonment of the place is obvious
4	Kourí	Chioniádes	Dedication	St. Paraskevi	Protective wood belt, protection of water aquifer, worship, aesthetics	**Strict protection**. Shredding is reported as trespassing. Mature trees around the church have been cut recently for safety reasons, the income saved for the church
5	K(ou)rí tis Panagiás	Elaphótopos	Dedication, communal agreements	Virgin Mary	Reserve, Ottoman taxes exclusion management for Church and Community income	**Controlled management**. Protection of mature trees in the view shed of the icon stand. Controlled grazing, acorn collection, shredding, cutting for poles or fuel wood, charcoal or lime kilns are reported
6	Livádi	Manassí	Unknown, in the interviews it is referred as vakoúfiko	–	Protective wood belt, protection of water aquifer, livestock shelter from sun	**Protection of mature trees**. Grazing and dead wood collection was accepted. Shredding is reported as trespassing

(continued)

Table 15.2 (continued)

	Forest name (in Greek)	Town/Village	Ritual praxis	Protector saint	Protection reasons	Protection practices
7	Livádi	Vrysochóri	Dedication	St. Charalampos, village's central church	Protective wood belt	**Strict protection**. Collection of dead wood and shrubs for bon fires was accepted. Grassing is reported as trespassing
8	Livadákia	Vítsa	Dedication, communal agreements	St. Nicholas, old village's patron	Forest of the abandoned village according to the tradition, annual celebration is alive	**Protection of mature trees**. Collection of dead wood and controlled grazing after the 20th of May and until the end of the season. The place is well preserved
9	Mereáo	Palioséli	Excommunication, communal agreements	–	Protective wood belt, protection of water aquifer. Narratives tell that the oaks inside the forest have been planted to support the forest's protective character	**Protection of mature trees**. Occasional grazing and cutting for fuel wood are reported as trespassing. Timber cutting for the construction of a bridge during the WWII is reported as mandatory
10	Agía Paraskeví	Vovoússa	Dedication	St. Paraskevi, village's patron	Worship. The location of the church in a hill by the river was chosen through epiphany by the saint's icon herself. Possibly the forested hill is associated with the flow of Aoos river on hill foot St Pararaskevi's fountain, and the protection of bridges and other infrastructures related to the river	**Strict protection**. Even hunting and wild honey collections are not accepted. For recent path repairing and safety reasons selected trees cut, but trunks remain on site

(continued)

Table 15.2 (continued)

	Forest name (in Greek)	Town/Village	Ritual praxis	Protector saint	Protection reasons	Protection practices
11	Agios Charálampos	Mesovoúni	Dedication	St. Charálampos, old village's patron	Forest of the abandoned village according to the tradition, annual celebration is alive	**Protection of mature trees** in the view shed of the church and in the fenced churchyard. Abandonment of the place is notable
12	Tráfos	Mólista	Excommunication	–	Protective wood belt, protection of water aquifer, Ottoman taxes exclusion	**Strict protection**. Occasional grazing and shredding are reported as trespassing. Villagers used the forest as a shelter during the WWII
13	Plái	Pápingo	Excommunication	–	Protective wood belt, livestock shelter from sun	**Protection of mature trees**. Grazing was accepted, occasional cut for fuel wood or poles is reported as trespassing
14	Gradísta	Kapésovo	Dedication, excommunication, communal agreements (1842, 1957)	St. Nicholas, village's patron	Reserve for community public works and later management for Church's income, Ottoman taxes exclusion, village protection from the north wind, livestock shelter from sun	**Protection of mature trees**. Grazing was accepted. Part of the oaks cut for the construction of the village school in 1864. After that controlled collection of dead wood and shrubs was accepted. Frequent shredding and cutting for fuel wood is reported as trespassing
15	Anília	Káto Pediná	Dedication, communal agreements	St. Nicholas	Protected wood belt, reserve for winter fodder, livestock shelter from sun, aesthetics	**Protection of mature trees**. Collection of dead wood and shrubs for bon fires, cutting for poles and grassing were accepted. Shredding, is reported as trespassing
16	Toúfa	Greveníti	Excommunication	–	Protective wood belt above village, protection of water aquifer. Villagers used the forest as a shelter during the WWII when Nazi burned their village	**Strict protection**. Occasional collection of dead wood and shrubs is reported as trespassing

Fig. 15.2 Management of sacred forests can vary from strict protection to controlled intensive management: **a**. Vovoússa village, black pine forest, strictly protected for ritual reasons, **b**. Greveniti village, beech forest strictly protected, excommunicated as protective wood belt **c**. Mikro Pápingo village, juniper wood pasture excommunicated as protective wood belt, **d**. Elaphótopos village, prickly oak forest dedicated to the Virgin and protected as village's reserve and for Church's income.

inhabitants. Many deaths are associated with trespassing in her sacred forest; therefore even harvests, such as hunting, collecting of bees or the cultivation of the land in the forest foothills, are avoided. Permitted actions are limited to the renovation of the church and the maintenance of the trail, which has been done with minimum interventions aimed at ensuring safe access for visitors.

In most sacred forests, hunting and collection of dead wood, acorns, mushrooms and other non-timber wood products (NTWPs) is allowed. Grazing in most cases is allowed freely or for certain time periods, expect from protective forests in very steep slopes. Shredding is not allowed, but it is reported occasionally as trespassing: in evergreen prickly oak forests it is done for winter emergency fodder, while mixed broadleaved forests trespassers during spring and summer are seeking fresh fodder for kids and lambs of their flocks.

In all forests, prohibitions are related with mature trees. However, as we have already mentioned, controlled management could permit even timber cutting for necessary public works or intense management such as in the extreme case of Elaphótopos village where even cutting for the establishment of a lime kiln to contribute to Church's income might be allowed. However any intervention was avoided within sight of the small icon stand. As a result, today difference between the spiritual center and the borders of the sacred forest are obvious in forest structure.

Regarding the breaking of these rules, some tolerance was extended to the members of lower social strata who when trespassing were asking for God's mercy, Church's help and community's charity. As a consequence a looser morality was reserved for them, who as an internally distinct social unit, were expected to pay with misfortune their failings (Stara 2009).

15.4 Sacred Natural Forests in Modern Times

Since the WWII, changing patterns of land use and population decline have had a dramatic effect on the social structure, management practices, and cultural landscapes of the area, as happened generally in most mountains of Greece. The same forces that threaten the cultural landscape of mountainous Greece, threaten also its sacred elements. Abandonment, collapse of local management systems and the indifference of state authorities has caused many SNS derelict, forgotten and left to be swallowed up by the natural succession that, right across the Mediterranean is

Fig. 15.3 Outlying icon stands and crosses, here in Aoos valley villages in Konitsa, mark paths to churches, or remind people borders of sacred areas. Often they consist of directed devotions (támata) that are offered in return of gratitude to saints

transforming the diverse cultural mosaic into impenetrable scrubland, prone to fire (Koutsias et al. 2012). Old beliefs and taboos are often neglected, not only by locals hungry for "progress," but also by many Church authorities who stigmatize them as superstitions. Despite this, considerable respect for sacred areas endures in many local communities and constitutes an important part of people's sense of identity (Fig. 15.3).

15.5 Conclusions

Today Sacred Forests are recognized as the world's oldest form of nature protection. Their cultural value and their contribution to ecosystem services are well understood. Some sacred forests, especially those with a clear and current protective character against natural hazards, remain stable over time, but others are in danger of losing their meaning because of land abandonment and changing community practices. The most characteristic such examples are the prickly oak sacred forests. Although these forests still protect soil and have aesthetic value, they are losing their value as livestock shelter or last-resources fodder for hard winters, because of the reduced goat numbers and replacement of prickly oak foliage by hay (Stara et al. 2015).

The rise of environmental concerns and ecological awareness in the last two decades have recast SNS as key and secure areas of biodiversity conservation (Dudley 2009). This brings into play a new dimension in the values of SNS all over the world. However, in the mountainous regions of Epirus as in other parts of the Mediterranean, modern ideas about ecological value are not easily accepted by local communities or local authorities. Thus, we could say that these SNS contribute to a large, unrecognized "shadow" conservation network. Biodiversity conservation adds a new value for many forgotten sacred forests and could potentially enforce their acceptance in modernity. However, it is important for local communities to be able to tell their story of protection of old growth forests as successful local means of protection that though established in the distant past preserve their sacred character and continue to hold the sense of spirituality in modern times.

Acknowledgments This research has been co-financed by the European Union (European Social Fund—ESF) and Greek national funds through the Operational Program "Education and Lifelong Learning" of the National Strategic Reference Framework (NSRF)—Research Funding Program: THALIS. Investing in knowledge society through the European Social Fund. We would like to thank our colleagues Alkis Betsis for the creation of the map and GIS work, Vaggelis Roumpas, Dr Giorgos Korakis, Dr Vasilis Dalkavoukis, Grigoris Manopoulos, Aris Kyparissis and Valentino Marini Govigli for data exchange, Dr Jenny Wong and professor Oliver Rackham for their support and professor Mauro Agnoletti for giving us the opportunity to contribute to this volume. Moreover, we would like to thank the Forestry Services of Ioannina, Konitsa, and Metsovo for the permission to study their archives. Lastly, we would like to thank all local participants to the study for time and hospitality during field work. Without them this research will be never been accomplished and the sacred forests will be never protected.

References

Anderson DN, Salick J, Moseley RK, Xiaokun O (2005) Conserving the sacred medicine mountains: a vegetation analysis of Tibetan sacred sites in Northwest Yunnan. Biodivers Conserv 14:3065–3091. doi:10.1007/s10531-004-0316-9

Antrop M (2005) Why landscapes of the past are important for the future. Landsc Urban Plan 70:21–34. doi:10.1016/j.landurbplan.2003.10.002

Bailey GN, Chr Ioakim, King GPC, Turner G, Sanchez-Goni MF, Sturdy D, Winder NP (1998) Northwestern Rpirus in the Palaeolithic. In: Leeuw SE (ed) Understanding the natural and anthropogenic causes of land degradation and desertification in the Mediterranean basin. The Archaemedes project, Office for Official Publications of the European Communities, Luxembourg

Byers BA, Cunliffe RN, Hudak AT (2001) Linking the conservation of culture and nature: a case study of sacred forests in Zimbabwe. Hum Ecol 29(2):187–218. doi:10.1023/A:1011012014240

Campbell JK (1964) Honour, family and patronage. A study of institutions and moral values in a Greek Mountain Community. Clarendon Press, Oxford

Chandran MDS, Hughes JD (2000) Sacred groves and conservation: the comparative history of traditional reserves in the mediterranean area and in South India. Environ Hist 6:169–186

Chouin G (2002) Sacred Groves in History: Pathways to the Social Shaping of Forest Landscapes in Coastal Ghana. IDS Bulletin 33 (1):39–46

Colding J, Folke C (2001) Social taboos: "invisible" systems of local resource management and biological conservation. Ecol Apl 11(2):584–600. doi:10.1890/1051-0761(2001)011[0584:STISOL]2.0.CO;2

Douglas M (1966) Purity and danger: an analysis of concepts of pollution and taboo. Routledge and Kegan Paul, London, New York

Dudley N, Higgins-Zogib L (2012) Protected areas and sacred nature: a convergence of beliefs. In: Pungetti G, Oviedo G, Hooke D (eds) Sacred species and sites. Advances in biocultural conservation. Cambridge University Press, UK

Dudley N, Higgins-Zogib L, Mansourian S (2009) The links between protected areas, faiths and sacred natural sites. Conserv Biol 23(3):568–577. doi:10.1111/j.1523-1739.2009.01201.x

Frascaroli F (2013) Catholicism and conservation: the potential of sacred natural sites for biodiversity management in central Italy. Hum Ecol 41:587–601. doi:10.1007/s10745-013-9598-4

Koutsias N, Arianoutsou M, Kallimanis AS, Mallinis G, Halley JM, Dimopoulos P (2012) Where did the fires burn in Peloponnisos, Greece the summer of 2007? Evidence for a synergy of fuel and weather. Agric Fort Meteorol 156:41–53. doi:10.1016/j.agrformet.2011.12.006

Lyratzaki I (2007) Meteora world heritage site. Thessaly Greece. In: Mallarach J, Papayannis T (eds) Protected areas and spirituality. Proceedings of the first workshop of the Delos initiative, Montserrat 2006. IUCN, Gland, Switzerland

Martin GJ (1995) Ethnobotany. Charman & Hall, London

Mihailaris PD (2004) Aforismos: I prosarmogi mias poinis stis anagkaiotites tis Tourkokratias (Excommunication: adaptation of a punishment to the needs of the Ottoman Occupation), 2nd edn. National Research Foundation—Center of Modern Greek (In Greek)

Moutafchieva V (1988) Agrarian relations in the Ottoman Empire in the fifteenth and sixteenth centuries. Columbia University Press, New York

Nitsiakos V (2008) I Konitsa ke ta horia tis. Politismou anatomi (Konitsa and its villages. Culture's dissection). Prefectural government of Ioannina, Epirus AE, Ioannina (In Greek)

Nixon L (2006) Making a landscape sacred. Outying churches and icon stands in Sphakia, Southwestern Crete. Oxbow Books, Oxford

Oviedo G, Ieanrenaud S (2007). Protecting sacred natural sites of indigenous and traditional people. In: Mallarach J, Papayannis T (eds) Protected areas and spirituality. Proceedings of the first workshop of the Delos initiative, Montserrat 2006. IUCN, Gland, Switzerland

Papayannis T (2007) Mount Athos Greece. In: Mallarach J, Papayannis T (eds) Protected areas and spirituality. Proceedings of the first workshop of the Delos initiative, Montserrat 2006. IUCN, Gland, Switzerland

Rackham O (2006) Woodlands. Collins, London

Read H (2000) Veteran trees: a guide to good management. Veteran trees initiative. http://publications.naturalengland.org.uk/publication/75035. Accessed 15 Oct 2014

Rutte C (2011) The sacred commons: conflicts and solutions of resource management in sacred natural sites. Biol Conserv 144:2387–2394. doi:10.1016/j.biocon.2011.06.017

Sarris D (2008) Kateuthintiries grammes parakolouthisis typon oikotopon (Guideless for different habitat types monitoring) Management Agency of Vikos - Aoos and Pindos National Parks, unpublished technical report

Ševčenko N (1983) The life of Saint Nicholas in Byzantine art. Bottega d' Erasmo, Torino

Sharma S, Rikhari HC, Palni LMS (1999) Conservation of natural resources through religion: a case study from Central Himalaya. Soc Nat Resour 12(6):599–612. doi:10.1080/089419299279461

Stara K (2000) The impact of perceived land-use values on biodiversity conservation in the Vikos-Aoos National Park, Papigo, Greece. Dissertation, M.Sc., University of Wales, Bangor

Stara K (2009) Sacred woods and groves in the Vikos-Aoos National Park, NW Greece. Locally adapted management systems, perceptions and values of local societies for the conservation of their natural environment. Dissertation, Ph.D., University of Ioannina, Greece (in Greek)

Stara K (2012a) Trees and the sacred in modern Greece. Langscape 11(2):60–63. http://www.terralingua.org/blog/2012/09/06/terralingua-langscape-volume-2-issue-11 Accessed 16 Oct 2014

Stara K (2012b). Northern Pindos National Park excommunicated forests. In: Mallarach JM (ed) Spiritual values of protected areas of Europe Workshop proceedings. Federal Agency for Nature Conservation, Germany. http://www.bfn.de/fileadmin/MDB/documents/service/Skript322.pdf. Accessed 16 Oct 2014

Stara K, Tsiakiris R (2010) The sacred woods called "meadows" in Zagori, Pindos mountains, Greece In: Sidiropoulou A, Mantzanas K, Ispikoudis I (eds) Proceedings of the 7th Panhellenic Rangeland Congress in Xanthi, Thessaloniki (in Greek with english abstract), 14–16 Oct 2010. http://www.elet.gr/pages/wp-content/uploads/09-Pages-from-7o-praktika.pdf. Accessed 16 Oct 2014

Stara K, Tsiakiris R, Wong JLG (2012) Sacred trees and groves in Zagori, Northern Pindos National Park, Greece. In: Gl Pungetti, Oviedo G, Hooke D (eds) Sacred species and sites. Advances in biocultural conservation. Cambridge University Press, UK

Stara K, Tsiakiris R, Wong JLG (2015) Valuing trees in a changing landscape: a case study from Northwestern Greece. Hum Ecol 43(1): 153–167.doi:10.1007/s10745-014-9706-0

Stara K, Tsiakiris R, Wong JLG (2015a) The trees of the sacred natural sites of Zagori, NW Greece. Landsc Res 40(7): 884–904. doi:10.1080/01426397.2014.911266

Stewart Ch (1991) Demons and the devil. Moral imagination in modern Greek culture. Princeton University Press, New Jersey

Verschuuren B, Wild R, McNeely JA, Oviedo G (2010) Sacred natural sites. Conserving nature and culture earthscan, London

Virtanen P (2002) The role of customary institutions in the conservation of biodiversity: sacred forests in Mozambique. Environ Values 11:227–241

Volkwein A, Schellenberg K, Labiouse V, Agliardi F, Berger F, Bourrier F, Dorren LKA, Gerber W, Jaboyedoff M (2011) Rockfall characterisation and structural protection—a review. Nat Hazards Earth Syst Sci 11:2617–2651. doi:10.5194/nhess-11-2617-2011

Watley RL, Pierce Colfer CJ (2004) Sacred forest, hunting, and conservation in West Kalimantan, Indonesia. Hum Ecol 32(3):313–338. doi:10.1023/B:HUEC.0000028084.30742.d0

Chapter 16
Historical Development of Forest Patterns in Former Slash and Burn Sites in Southern Estonia

Pille Tomson, Robert Gerald Henry Bunce and Kalev Sepp

Abstract Slash and burn is one of the oldest methods of clearing and fertilizing land for growing crops. The practice continued in Estonia until the beginning of the twentieth century. The historical background to the use of fire in Estonia is first described, including the legal controls that were set up. After cultivation the slash and burn fields left as fallow, before trees were allowed to regenerate for up to 20 years, when the cycle started again. The term *buschland* from the local Baltic German dialect was used on documents to designate the land that had been used for regular burning—a land use category that no longer exists. The literature on the environmental impact of slash and burn is discussed but there is no agreement about the effects of the practice on soil. The effects on biodiversity have also not been widely studied. The study area was the Karula National Park in Southern Estonia. Maps from the end of nineteenth century were compared with contemporary digital map and databases using Mapinfo. In the nineteenth century the *buschlands* occupied 34 % of farmland but the analyses showed that now 77 % had become forest, with the remainder being grasslands or arable. These changes were because the sites were on hills and the slopes were too steep for modern agriculture. The dominant tree species, forest types and soils are also described and are associated with infertile conditions.

Keywords Slash and burn · Land use history · Forest biodiversity · Biocultural diversity

P. Tomson (✉) · R.G.H. Bunce · K. Sepp
Institute of Agricultural and Environmental Sciences, Estonian University of Life Sciences, Kreutzwaldi 5, 51014 Tartu, Estonia
e-mail: pille.tomson@emu.ee

R.G.H. Bunce
e-mail: bob.bunce@emu.ee

K. Sepp
e-mail: kalev.sepp@emu.ee

M. Agnoletti and F. Emanueli (eds.), *Biocultural Diversity in Europe*,
Environmental History 5, DOI 10.1007/978-3-319-26315-1_16

16.1 Introduction

Slash and burn is one of the oldest methods of agriculture for clearing and fertilizing land to grow crops. The practice consists of a cycle of burning trees, growing crops for some years, leaving the land fallow and then allowing regeneration of trees before starting again. During the Neolithic period in Europe slash and burn was widely used, as described by Pyne et al. (1996). In the modern world the practice is usually associated with the tropics.

In Northern Europe slash and burn cultivation lasted until the twentieth century. In Finland the use of slash and burn is well known, as Heikinheimo (1915) compiled a comprehensive overview of this method when it was still being used. In the eastern part of Finland—in Savo and Karjala—fire cultivation continued until the 1930s, as described by Voionmaa (1987). In Southern Sweden slash and burn was also used in Smaland. The forest Finns were responsible for the introduction of the practice to central and western areas of Sweden, such as Värmland and Dalecarlia (Hamilton 1997; Wedin 2003). Forest Finns also introduced fire cultivation into Norway (Wedin 2003). In the forests of North-Western Russia slash and burn was used even until the 1960s (Bobrovskii 2010).

Often the impacts of fire and fire cultivation on ecosystems are examined together, as described by Pyne et al. (1996) and Goldhammer and Furyaev (1996). In a different context, Grant et al. (2012) also emphasized, in a review of the effects of rotational burning of moorlands, that, even today, the full environmental impacts of the use of fire are still not understood. Some relevant literature is not widely available because it is written in languages such as Swedish, Finnish and Russian.

In Estonia it is widely understood that slash and burn cultivation belonged to prehistoric times. It has been suggested that in slash and burn regions, dry oligotrophic pine forests underwent soil erosion following burning (Laasimer 1958). Some species-poor spruce forests are also thought to be the result of prehistoric slash and burn cultivation (Rõuk 1995; Paal 1997).

At the same time, the Estonian historians have described the wide use of fire in agriculture between the seventeenth and nineteenth centuries (Ligi 1963; Kahk 1992; Tarkiainen 2014). Slash and burn cultivation was a characteristic feature of pre-modern agriculture before the changes that took place at the end of the nineteenth century. The recent literature shows that, according to the records of ethnographical inventories, slash and burn cultivation persisted until the end of nineteenth century and in some cases even in the twentieth century (Jääts et al. 2010). Anecdotal evidence comes from an 86-year-old lady who recalled hearing in childhood from her grandfather about slash and burn at the end of the nineteenth century, as recorded in summer 2014 by the senior author in Rõuge parish, Southern Estonia.

The aim of present study is to provide the first description of current forests on former slash and burn sites, using existing databases and maps from the nineteenth century and then to analyse the factors underlying the formation of these forests.

16.2 Fire Cultivation in Estonia

Ligi (1963) described slash and burn as part of the agricultural land use system typical of the Southern Estonia. In the eighteenth century approximately half of the annual crop yield was harvested from slash and burn fields (Ligi 1963). In Northern and Western Estonia the limestone bedrock meant that the ash from burnt wood did not have the same fertilizing effect as in Southern Estonia, where the soils are acid because they are on moraines derived mainly from the Devonian sandstone bedrock, although other materials are also present. Forest regeneration on abandoned slash and burn fields on shallow limestone soils is also relatively slow in comparison with Southern Estonia (Ligi 1963).

The contrasting history of slash and burn management in the regions of Estonia is also reflected in legislation. Southern Estonia belonged in the nineteenth century to the Livonian Covernatore, whereas Northern Estonia was in the Estonian Covernatore. In the Livonian Peasant Law from 1819 (Lihwlandi-ma tallorahva… 1820) mainly targeted to the liberation of the peasants, there are strict rules for slash and burn management. The rules are that the break between the burnings must be 24 years and that no more than three crops can be taken from the same field during this cycle. In contrast the concurrent Estonian Peasant Law (Eestima Tallorahwa… 1816) provides only the rules for liberation and the peasant's self-government.

In the nineteenth century slash and burn was only used in young forests. The regular burning of the old forests was stopped in Estonia already in the seventeenth century due to the lack of timber (Etverk 1974).

In young forests the regular burning cycle took place every 30 years in the seventeenth century but the interval between burnings was decreased until 15–20 years in the nineteenth century (Meikar and Uri 2000). After burning the land was ploughed, harrowed and seeded or the seeds were put into soil with a type of harrow. Rye, barley and turnip were the most common crops in slash and burn fields. Fields were cultivated for 2-5 years after burning and after that were left as fallow and used for grazing. Tree regeneration then commenced and after 10–15 years the *buschland* was ready for the next burning event (Meikar and Uri 2000).

The special land category, termed *buschland* in the local Baltic German dialect, was used on maps and documents to designate the land that had been used for regular slash and burn cultivation. (Meikar and Uri 2000). The local south Estonian dialect did not separate the forests used for timber and the *buschlands*, as both were named "*mõts*". The term "*mõts*" was also used even if there was no tree cover on the *buschland* at that particular time. Even today old people in the southern Estonia use the term "*mõts*" if they are talking about herding cattle to pasture. This name may be connected with the point that at an earlier date slash and burn cultivation was also used in mature forests. However, the use of the term also shows that slash and burn cultivation was an integral part of many wooded areas. This peculiarity of terminology might be caused some misunderstandings of ethnographical records, because in the official Estonian the word "*mets*" is used only in the meaning of forests.

There have been different views as to the fate of *buschlands* after the decline of slash and burn management. They have been regarded as grasslands (Liitoja-Tarkiainen 2006) or as pastures or forests with wood production becoming increasingly important (Ratt 1985; Meikar and Uri 2000). *Buschlands* were also considered to be the main land resource to increase arable land (Ligi 1963; Koppel 2005; Tarkainen 2014).

16.3 Biodiversity Connected with Slash and Burn Cultivation

There are different opinions about the tree species composition of *buschlands*. Many authors state that birch was the dominant species (Ligi 1963; Öpik 1992; Viires 2000). However, mixed forests with birch and spruce are also mentioned (Öpik 1992; Meikar and Uri 2000). In Finland grey alder is also common in former slash and burn sites (Heikinheimo 1915). Probably, in the *buschlands* birch or grey alder were the main species and in the old forests slash and burn cultivation were carried on in mixed stands.

Heikinheimo (1915) described the impacts of fire cultivation on the forests in Finland according to different zones around the settlements. The nearest and most regularly used areas were dominated by grey alder. In the second less used areas birch was the major species and in the outermost zone pine dominated. Spruce was dominant only outside the third zone which had not been used for slash and burn. Also, aspen was present in slash and burn areas.

Pine seeds are more resistant to fire than those of spruce. In addition, pine trees were often available for seed dispersal as they were common in uncultivated sites such as rocky outcrops, bogs and sandy places inside the slash and burn zones (Heikinheimo 1915). The special method "huuhta" was used in mature coniferous forests in the eastern part of Finland (Tarkiainen 2014).

Many authors have pointed out that slash and burn has caused soil degradation and thus has led to changes in vegetation (Laasimer 1958; Rõuk 1995; Paal 1997). Laasimer (1958) in particular considers that the dry oligotrophic pine forests in South-Eastern Estonia are a result of soil erosion after slash and burn cultivation. Rõuk (1995) and Paal (1997) consider that species-poor spruce forests are the result of slash and burn cultivation and Paal (1997) also includes species-poor spruce, aspen and birch forests.

Historical data about soil quality in slash and burn sites are not consistent. There are opinions that the soils in *buschlands* were poorer (Ligi 1963) or more fertile (Koppel 2005) than in permanent arable fields.

The impacts of fire on soils mainly depends on the intensity of the heat, which is related to the amount of organic material, which in slash and burn will be correlated with the length of time since the last burning but also the weather conditions (Delgado-Matas 2004). All authors agree that the pH of soils rises immediately after burning but the organic matter decreases, the amount of cations increase and mineral nitrogen suitable for plant growth also increases after burning (Viro 1974; Reintam and Moora 1983; Pyne et al. 1996; Delgado-Matas 2004), but there are few data indicating the persistence of these changes. According to Viro (1974) some changes may persist until 50 years.

The same authors describe erosion after slash and burn management, but this is the result of opening the soil surface and is not specific to the use of burning. Some authors argue that slash and burn cultivation promotes podsolization (Reintam and Moora 1983). Ligi (1963) on the other hand states that regular slash and burn cultivation slowed down the rate of podsolization.

16.3.1 Materials and Methods

On the base of historical data, it is clear that slash and burn was a significant factor in Southern Estonian rural landscapes. In order to investigate impacts of this factor, a case study was carried out in the Karula National Park (Fig. 16.1). The Park (12,300 ha) is an area with a complex relief and variable mosaic landscapes located in the uplands of South-Eastern Estonia. The conservation aims of the Park are to preserve the natural biotopes and also to maintain semi-natural habitats as well as

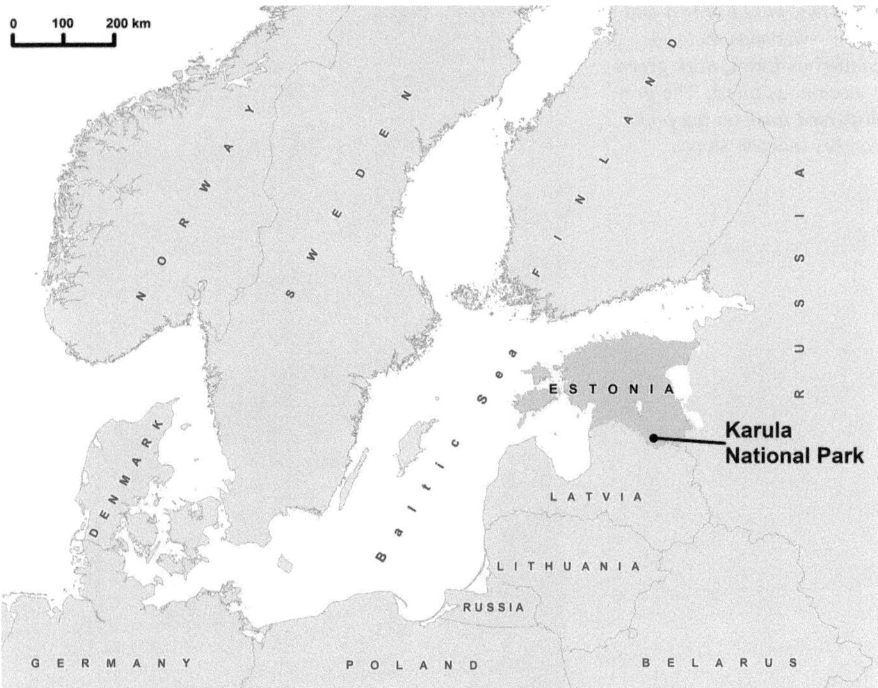

Fig. 16.1 Location of study area

traditional rural landscapes and their associated cultural heritage. Karula National Park does not correspond exactly to Category II of the International Union for the Conservation of Nature (IUCN), which national parks consisting of natural or near-natural habitats (Dudley 2008). Agricultural land covers 12 % of the Park, whereas forests and wetlands cover 83 %, water bodies 3 % and other categories 2 % (Karula rahvuspargi kaitsekorralduskava 2007).

In order to characterize pre-modern land use and locate the former *buschlands* 51 maps (Estonian National Archives, record EAA 2469) from 1870 to 1880 were geo-referenced and digitized (Fig. 16.2). The maps (scale 1:4200) cover a 3755 ha in study area. These maps were drawn up for land purchase at the end of the nineteenth century and the distribution of different land cover types is shown. The maps cover account for 62 % of the historical farmlands located on the territory of present-time Karula National Park so the results of study are representative of the

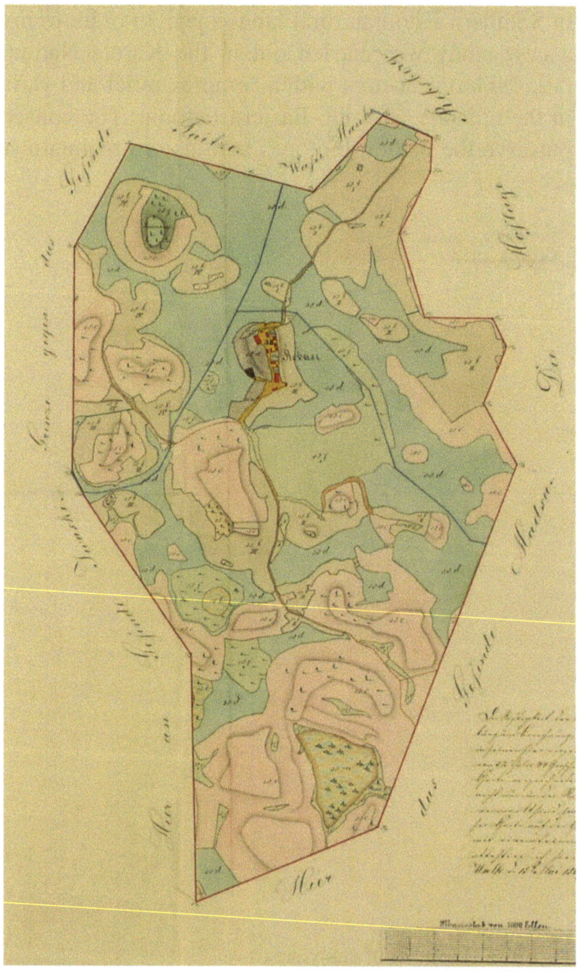

Fig. 16.2 The nineteenth century map of Rebase farm. *Pink areas* indicate the *buschlands*, *yellow with red dots*—farmyard, *dark grey*—vegetable garden, *grey*—permanent arable fields, *olive green*—pastures, *green*—hay meadows, *mixed brown and green*—wetlands, *beige*—coniferous forest, *dark green*—deciduous forest. The *grey dispersed lines* on the *pink patches* indicate slopes

whole rural landscape of Karula. On these maps, there are marked permanent arable fields, *buschlands*, hay meadows, pastures and other land use patches present in the nineteenth century, making it possible to examine their historical allocation.

To identify the character of the former slash and burn lands today comparisons were made with contemporary maps and databases. The current Estonian Basic Map, which is digitally available, was used to identify the contemporary situation.

The Estonian Digital soil map was used to identify the soil conditions. This database consists the data about soil types according to Estonian soil classification (Vabariigi digitaalse… 2001). In this paper the names of soils are presented according to the World Reference Base (WRB), adapted by Astover et al. (2012).

The digital forest management dataset was used to establish the main forest characteristics. This dataset consists of information on each forest stand. The age, dominant species and types of forest were used from this database. In the forest management dataset the forest type site classification of Lõhmus (2004) was used. In the present paper the similar forest classification of Paal (1997) is used to describe the forest site types and provide names. For the forest site type classes the names are used according to Paal (2007), because Lõhmus (2004) gives no names in English. The "age" and "dominant tree species" refer to the upper tree layer.

MapInfo Professional 10.0 programme was used for comparing the maps. Digitized maps were used to calculate the areas of land use types. In the Park borders the *buschland* and other land use patches layers were overlaid with the other databases and results were analysed in Microsoft Excel 2010. Although two farms were partly outside the Park, the forest and soil analyses were only made inside the boundary.

The Latin names of the species quoted in the text are as follows: birch (*Betula pendula*), spruce (*Picea abies*), pine (*Pinus sylvestris*), grey alder (*Alnus incana*) and aspen (*Populus tremula*).

16.3.2 The Characteristics of Former Slash and Burn Sites

The analyses of nineteenth century maps show that *buschlands* form the main part of farmlands at that time, and cover 1267 ha (34 %). The arable fields covered 825 ha (22 %), hay meadows 655 ha (17 %), pastures 382 ha (10 %), woodlands 195 ha (5 %), wetlands 276 ha (7 %), water bodies 48 ha (<2 %), vegetable gardens 25 ha (<1 %), roads and yards 23 ha (<1 %) and areas unidentifiable on the maps cover 57 ha (<2 %).

The comparison of the nineteenth century map with Estonian Basic Map shows that at the present time the landscapes have changed and the former slash and burn lands are now mainly covered by forests: 952 ha (77 %) are forest, 135 ha (11 %) are mapped as grasslands, 124 ha (10 %) as arable fields, 16 ha (1 %) as other open areas and 8.6 ha (<1 %) as all other classes (water, wetlands and scrub). Secondary forest is therefore the main land cover containing biodiversity in the former slash and burn parcels.

As elsewhere in Estonia as well as in Karula National Park much former agricultural land has been covered by regenerated forest during the second half of the twentieth century. Among these secondary forests (1567 ha in study area) forests on former *buschlands* cover the largest area 911 ha (58 %). Forests have colonized grasslands cover 472 ha (30 %) and those growing on former permanent arable fields 184 ha (12 %).

The soil types on *buschlands* are mainly Albeluvisols 660 ha (53 %), which are soils typical of the southern margin of the boreal forest that have been affected by peri-glacial activity and are somewhat acidic. In study area the most common soil unit of Albeluvisols is Haplic. Regosols in study area are the eroded soils with shallow humus layer and they cover the 346 ha (28 %). Less widespread are Luvisols, brown soils of the cool temperate zone, 94 ha (8 %) and Podzols 26 ha (2 %). Other soil types cover 108 ha (9 %). In the secondary forests on the former *buschlands* the proportions of soils are a little different: Albeluvisols cover 572 ha (61 %), Regosols 189 ha (20 %), Luvisols 56 ha and (6 %), Podzols 26 (2 %) and other soil types cover 108 ha (9 %).

The main species in the forests formed on the *buschlands* are pine 353 ha (41 %), birch 267 ha (31 %), grey alder 116 ha (13 %), spruce 114 ha (13 %) and aspen 17 ha (2 %).

The forests of former slash and burn sites according to the forest database are mostly the *Oxalis* site type mesotrophic boreal forests, 447 ha. The *Hepatica* site type (172 ha) forests are more fertile in their character and occur the site type class eutrophic boreo-nemoral forests according to Paal (1997). The more fertile *Aegopodium* site type occurs less widely (55 ha). The more dry and poor soil site types are represented by the *Oxalis-Vaccinium vitis idea* (80 ha) and *Oxalis-Vaccinium myrtillus* (16 ha) site types. All the other less presented site types cover 60 ha. The *Oxalis* forests are typical to the former *buschlands*, 69 % of this type forests on farmlands grow on *buschlands*, 8 % of *Oxalis* type forests on farmlands grow on the former arable fields, 7 % on the former pastures, 6 % on the former hay meadows and 10 % were forests and wetlands in nineteenth century.

The age of these secondary forests is approximately between 10 and 170 years, with the most common age group being between the ages of 60 and 70 years, as shown in Fig. 16.3. During the period after the Second World War, forest also developed on much agricultural especially on hay meadows. In the former *buschlands* the regeneration of forests started much earlier and continued until the early twentieth century.

16.3.3 Discussion: Factors Shaping the Biodiversity of Former Slash and Burn Sites

In Karula farmlands the *buschlands* were the most common land use at the end of nineteenth century, with other types being less common. On the old maps they were clearly linked with hills as visible in Fig. 16.2. The slopes and tops of these small

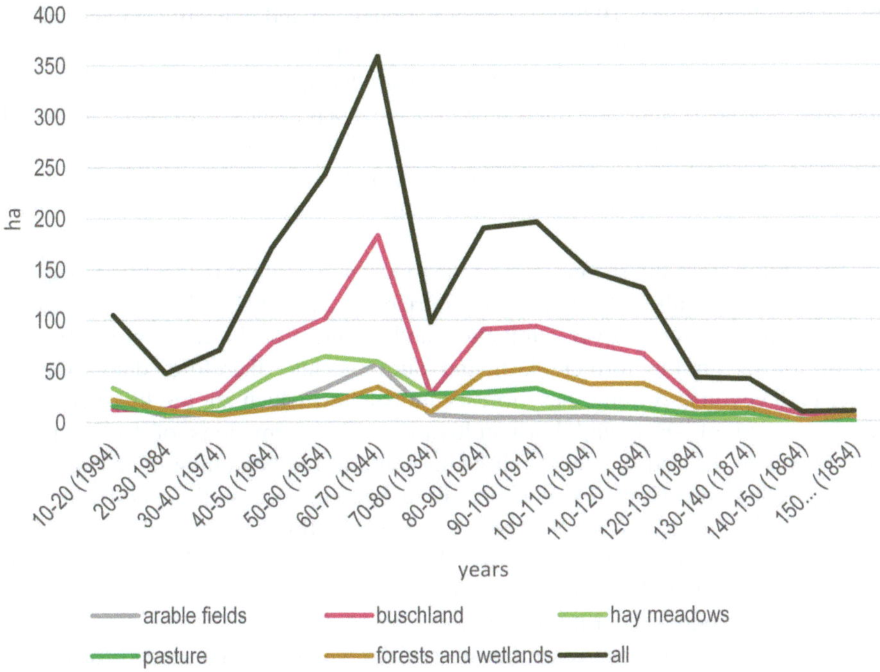

Fig. 16.3 The area of age classes of agricultural patches in the study area

hills were not easy to cultivate as permanent arable fields because of difficulties in transporting the dung up to the slopes for fertilization, as well as problems of ploughing the steeper slopes. Slash and burn cultivation evolved as an effective way of growing crops in those areas and, as shown below, they were also linked with acidic soils. Relief was therefore probably the one of the factors which determined the extent and persistence of fire cultivation on hills in Karula throughout the nineteenth century. Tarkiainen (2014) argues that according to the agricultural books from the seventeenth century the slopes were most suitable for slash and burn, because of the dry soils. Open slopes were probably preferred, because the cut wood needed to dry before burning.

Even although there are no ethnographical records from the Karula parish about slash and burn, the large area of *buschlands* suggests that the fire cultivation was as important here as elsewhere in Southern Estonia.

The former *buschlands* lost their previous function by the end of nineteenth century and were mainly converted into fields, grasslands or forests (Meikar and Uri 2000; Koppel 2005; Liitoja-Tarkiainen 2006; Tarkiainen 2014, Tomson et al. 2015). However, by the end of twentieth century most of the former *buschlands* had become forests.

The main reason for the land abandonment during the second half of the twentieth century has been mechanization. In the former Soviet countries the impacts of collectivization need to be considered (Nikodemus et al. 2005).

In Estonia during 1961–1989, 21.2 % of nationalized and collectivized agricultural land was transferred into state forest land (Kasepalu 1991).

The same trends can be also observed in Karula (Fig. 16.3). The abandonment of the wet grasslands and cultivated areas on the steep slopes is evident. Some differences are present notably the rapid increase of forest regeneration in the period following the end of the Second Wold War and the slow increase in secondary forest on the former of slash and burn sites during the nineteenth and early twentieth centuries.

The period from the second part of the nineteenth century is identified as a time of change in agriculture by Estonian historians (Kahk 1992). There was introduction of crop rotation, clover cultivation and an increase in cattle breeding. The oldest forests on the former slash and burn patches, therefore, date from the nineteenth century. Among these forests are patches, which are the first generation after the end of slash and burn cultivation. The formations of forest on the *buschlands* at the end of nineteenth century and at the beginning of the twentieth century reflect the changed understanding of farmers because the woodlands were progressively valued as a resource to supply the timber for modernizing farm complexes and living houses. The development of these new forests was a slow process, which lasted for decades. Other types of agricultural land were not remarkably abandoned during that period.

The need for timber increased because of the new buildings needed for modern cattle breeding and the wood was more valued, so some farmers decided to encourage forest regeneration at the beginning of the twentieth century. Before the twentieth century forest management was important in the forests belonging to Manor houses but not to those linked to farmland (Meikar 2014). The plantation of coniferous became more common on farmlands in 1930s (Etverk 1974). Also, some forest patches on farms showed re-growth following clear cuts at the beginning of the twentieth century, especially in the category "forests and wetlands" of Fig. 16.3. Increases in land drainage also encouraged forest regeneration on wetlands.

The major part of the former cultivated land, on the previous *buschland*, turned into forest before the Second World War. However, the peak of forest regeneration came between 1944 and 1954 which is not possible to explain with mechanization because the tractors were introduced into Karula in the middle of 1950s so must have been due to social factors discussed below. Some forests also started to colonize former permanent arable fields and wet meadows.

In 1941 and 1949, the Soviet authorities deported a large number of people from Karula parish (Merila-Lattik 2005) which could have led to the decrease in cultivation in subsequent years. The years during and just after the Second World War were hard for rural people and fields became abandoned causing the observed increase in forest cover. Tree regeneration started not only in abandoned fields, but also in forest land after clear cutting, but it is unlikely that the latter was widespread during the war, when the economy was under pressure and the reduced manpower available was needed to ensure survival.

The subsequent land abandonment, caused mainly by mechanization, lasted to the 1970s. This process is similar for that which took part throughout Estonia

(Kasepalu 1991). Probably, the regeneration of clear cuttings is more important in this period.

Among the secondary forests in Karula National Park the forests in the former slash and burn cultivation sites cover the biggest proportion first, because *buschlands* were most common in farmlands and second, more of the former *buschlands* were abandoned than permanent arable fields. The second major groups are the wet forests on the former grasslands. These forests are highly visible elements in the landscape because of their location on the hills.

These forests can be characterized by two age groups: the forests that started to grow before the Second World War and those that started to grow after, because the factors that led to their formation were different. The different age groups on the former *buschlands* are characterized also by different dominant tree species (Fig. 16.4). Birch and alder grow in the younger patches which started to generate after the Second World War, when Estonia became a part of Soviet Union. The large area of fast-growing deciduous trees such as birch and grey alder is specific to abandoned arable land. Spruce and pine were also planted and afforestation on agricultural land was also common in the Soviet period (Kasepalu 1991). The older *buschland* forests are dominated by pine. Fire cultivation can also encourage pine (Heikinhemo 1915).

Albeluvisols were appropriate for fire cultivation because of their relatively high acidity which could be improved by the wood ash and also because of the location of these kinds of soils in the moraine hills (Astover et al. 2012). There are some differences in the soils of these two age groups of forests in former *buschland*

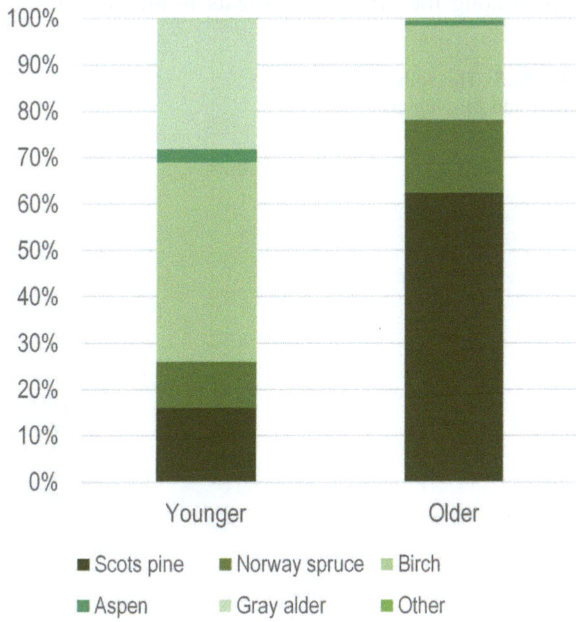

Fig. 16.4 The dominant tree species of the forests in former *buschland*

(Fig. 16.5). The eroded soils (Regosols) are more common in the younger groups and the infertile Podzols are more widely present in the older forests.

The major portion of eroded soils in the areas of younger forests reflect the longer period in cultivation and probably also the impact of mechanized cultivation. Probably, the patches in better soils were cultivated longer and were more affected by erosion. Also, the fertilization and tillage during the management of the fields has probably influenced later forest growth, but how much is not clear.

The smaller proportion of eroded soils among the soils of older forest patches reflects the lack of mechanized tillage and shorter cultivation period with a longer time for soil recovery. The occurrence of Podzols among the old *buschland* forest soils shows that in the period of modernization of agriculture at the end of nineteenth century, the areas with the poorest soils were abandoned first thus repeating a pattern that is well known elsewhere in Europe.

In low-intensity agriculture, the soil properties do not change very much, specially the subsoils (Köster and Kõlli 2013). Some impacts of fire may persist for 50 years or more (Viro 1974), but this do not change the soil type. The podzolization must be depressed on the *buschlands* in comparison with other forests on the same soils. The herb cover in the pasture stage of fallow and the subsequent cover of deciduous trees do not promote leaching.

The results do not confirm that slash and burn cultivation in the Nordic region has caused permanent soil degradation because of erosion, as argued before by Laasimer (1958).

Soil types are connected with forest site type (Lõhmus 2004). The typical forest site on Albeluvisols according to Paal (1997) is the *Oxalis-Vaccinium myrtillus* site type, which is less common on former *buschlands* in Karula. The most common site type among the *buschland* forests is the *Oxalis* site type, which is also the most

Fig. 16.5 The soils of afforested *buschlands*

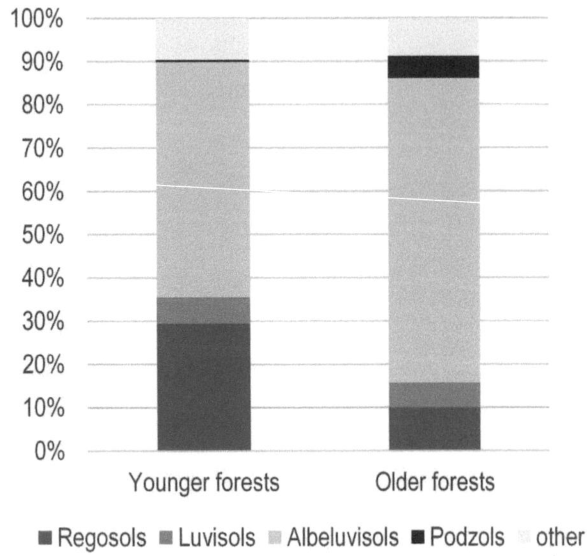

common type in the Southern Estonia (Paal 1997). The *Oxalis* site type is common in forests on former farmlands but less widespread on state forest land, which originally in the nineteenth century belonged to the owners of manor houses. More fertile areas suitable for cultivation were colonized by forests in historical times and Albeluvisols, which today is considered largely as not suitable for agriculture (Astover et al. 2012) were often used for slash and burn cultivation (Ligi 1963).

Oxalis site type of forest is also common in the former *buschlands* with eroded soils. The eutrophic boreo-nemoral forests cover relatively larger area amongst the eroded soils compared with those in other forest types. The more fertile soils were more impacted by agriculture.

The results do not confirm, or reject, the hypothesis that the dry oligotrophic pine forests were formed by cultivation practices in the former slash and burn areas, as argued by (Laasimer 1958). These results do, however, fit the opinion, where species-poor spruce forests are connected with slash and burn cultivation in many areas (Paal 1997; Rõuk 1995), because usually the dominant tree species in the *Oxalis* site type is spruce (Paal 1997). This is probably because of the selection of this particular soil type for slash and burn agriculture due to its relatively low fertility and location.

The long-term impact of slash and burn on soil quality has not been finally determined. It is difficult to see that any further progress can be made without intensive studies of soil characteristics in slash and burn sites in comparison with control areas on the same soil type. The same applies to impacts of the practice on biodiversity although a study is in progress to assess differences in vegetation between slash and burn sites and other forests.

16.4 Conclusions

Fire cultivation is a land use management practice, whose impact has been insufficiently studied in Estonia. In Southern Estonia such cultivation was still used even in the nineteenth century. Comparisons between old and modern maps enabled the characteristics of the former slash and burn sites to be determined. Slash and burn covered about 34 % of farmland in Karula at the end of the nineteenth century. The changes in agricultural practices, which started at the end of that century, created new forest patterns connected with the intensification of farming and the associated decline in fire cultivation. In the second half of the twentieth century, the forest cover in the former slash and burn sites increased again due to their location on slopes and tops of hills, not suitable for mechanized cultivation. Former slash and burn parcels have therefore had the most changeable land use in Karula. The forests on sites formerly used for slash and burn cultivation are therefore of relatively recent origin in farmland landscapes.

Forests in former slash and burn patches are mostly of the mesotrophic boreal forest, *Oxalis* site type. This forest site type is not so common on forests on other former agricultural land. The younger forests on formerly slash and burn lands also have impacts following the cessation of agriculture. Whether the forest type is the result of fire cultivation, or if fire cultivation was practiced on land suitable for the Oxalis forest site type, is not clear.

The history of slash and burn cultivation offers a good example of how inherent natural conditions have led to the formation of the current pattern of land use. Land use history needs to be understood in order to interpret the contemporary biodiversity patterns. The extent and strength of the impacts of fire cultivation on biodiversity, particularly vegetation, needs future study.

Acknowledgments This research has been financed by the institutional research funding IUT21-1 of the Estonian Ministry of Education and Research and by the European Union supported Environmental Conservation and Environmental Technology R&D programme project ÖKOMAA.

References

Astover A, Kõlli R, Roostalu H, Reintam E, Leedu E (2012) Mullateadus. Õpik kõrgkoolidele (Soil science. Textbook for Highschools). Eesti Maaülikool, Tartu

Bobrovskii MV (2010) Effect of the historical land use on the structure of forest soils in European Russia. Eur J Soil Sci 43(13):1458–1466. doi:10.1134/S1064229310130053

Delgado-Matas C (2004) Dynamics of nutrients in slash and burn agroforestry in Koli National Park. Dynamics of nutrients in slash and burn agroforestry in Koli National Park. In: Lovén L (ed) Tutkittu ja tuntematon Koli. Metsäntutkimuslaitoksen tiedonantoja, vol 915, pp 29–46

Dudley N (ed) (2008) Guidelines for applying protected area management categories. IUCN, Gland. http://cmsdata.iucn.org/downloads/guidelines_for_applying_protected_area_management_categories.pdf. Accessed 26 Oct 2014

Eestima Tallorahwa Seädmissed (1816) Estonian Peasant Law. http://digar.nlib.ee/digar/show/?id=100923. Accessed 26 Oct 2014

Etverk I (1974) Metsa õpitakse tundma ja kasutama (Forest had been learned to know and use). In: Valk U, Eilart J (eds) Eesti metsad (Estonian forests). Valgus, Tallinn

Goldammer JG, Furyaev VV (eds) (1996) Fire in ecosystems of boreal Eurasia. Kluwer Academic Publisher, Dordrecht

Grant MC, Mallord J, Stephen L, Thompson PS (2012) The costs and benefits 565 of grouse moor management to biodiversity and aspects of the wider 566 environment: a review. RSPB, Bedfordshire

Hamilton H (1997) Slash and burn in the history of Swedish forests. Rural Development Forestry Network Paper, 21f, pp 19–24. http://www.odi.org/sites/odi.org.uk/files/odi-assets/publications-opinion-files/1162.pdf. Accessed 26 Oct 2014

Heikinheimo O (1915) Kaskiviljelyksen vaikutus Suomen metsin (The impacts of slash and burn cultivation in the Finnish Forests). Metsähallituksen julkaisuja II, Keisarillisen senatiin kirjapainossa, Helsinki

Jääts L, Kihno K, Tomson P, Konsa M (2010) Tracing fire cultivation in Estonia. For Stud 53:53–65. doi:10.2478/v10132-011-0089-3

Kahk J (1992) Uuenduste sissetung 18. sajandi lõpust 19. sajandi keskpaigani (Innovations invasion in the late 18th century to mid-19th century). In: Kahk J (ed) Eesti talurahva ajalugu (History of Estonian peasants). Olion, Tallinn

Karula rahvuspargi kaitsekorralduskava 2008–2018 (The management plan of Karula National Park 2008–2018) (2007). http://www.keskkonnaamet.ee/public/Karula_rahvuspargi_kodulehe_materjalid/Karula_RP_KKK_2008-2018.pdf. Accessed 10 Oct 2014

Kasepalu A (1991) Mis peremees jätab, selle mets võtab (That which the master leaves, the woods will take). Valgus, Tallinn

Koppel K (2005) Maakasutuse uurimise metoodika Kasaritsa uurimisala (Rõuge kihelkond) 17.-19. sajandi külamaastike kujunemise näitel (The research methods of historical land use as exemplified by the formation of rural landscapes on Kasaritsa study area (Rõuge parish) in 17th–19th centuries). Retrieved from http://dspace.utlib.ee/dspace/handle/10062/1263

Köster T, Kõlli R (2013) Interrelationships between soil cover and plant cover depending on land use. Est J Earth Sci 62(2):93–112. doi:10.3176/earth.2013.08

Laasimer L (1958) Nõmmealade geobotaaniline iseloomustus ja areng Eestis. Nõmmealade taasmetsastamise ja nõmmemetsade majandamise küsimusi (The geobotanical characterization and development of oligotrophic sites in Estonian. The reforestation oligotrophic sites and management of oligotrophic forests). ZBI, Tartu, pp 29–43

Ligi H (1963) Põllumajanduslik maakasutus Eestis XVI–XVII sajandil (Agricultural land use in Estonia in XVI–XVII century). Tallinn

Lihwlandi-ma tallorahva Seädus (1820) Essimenne-Kolmas jaggu (Agrarian Law of the Livonian Governorate. First-Third Section). Tartu

Liitoja-Tarkiainen Ü (2006) Talukõlvikud Vooremaal 17.-19. sajandil (Farmland's land use patches in Vooremaa in 17th–19th century). In: Tamberg T (ed) Vene aeg Eestis. Uurimusi 16. sajandi keskpaigast kuni 20.sajandi alguseni (Russian time in Estonia. Studies from the mid-16th century until the early 20th century). Eesti Ajalooarhiivi Toimetised, vol 14(21). Tartu, pp 229–254

Lõhmus E (2004) Eesti metsa kasvukohatüübid (Estonian forest site types), 2nd edn. Eesti Loodusfoto, Tartu

Meikar T (2014) Riik ja talumetsad Eesti Vabariigi eel (State and forests of farms before Estonian Republic). Eesti Mets 2:44–50

Meikar T, Uri V (2000) Võsamaade majandamisest Eestis. Eesti metsad ja metsandus aastatuhande vahetusel (Of the management of bushland in Estonia). In: Meikar T, Etverk I (eds) Akadeemilise Metsaseltsi Toimetised, vol XI. Tartu, pp 103–120

Merila-Lattik H (2005) Karm ja kaunis Karula. Elust, olust ja inimestest muinasajast tänapäevani (Rough and beautiful Karula. The peoples and way of life from ancient times to the present day). Eesti Folkloori Instituut, Tartu

Nikodemus O, Bell S, Grīne I, Liepinš I (2005) The impact of economic, social and political factors on the landscape structure of the Vidzeme Uplands in Latvia. Landscape Urban Plan 70:57–67. doi: 10.1016/j.landurbplan.2003.10.005

Öpik E (1992) Maaviljelus (Agriculture). In: Kahk J (ed) Eesti talurahva ajalugu 1 (History of Estonian peasants, vol I). Olion, Tallinn

Paal J (1997) Eesti taimkatte kasvukohatüüpide klassifikatsioon (Estonian vegetation site type classification). Keskkonnaministeerium, Tallinn

Paal J (2007) Loodusdirektiivi elupaigatüüpide käsiraamat (Handbook of habitat types of habitat directive). Digimap OÜ, Tallinn

Pyne SJ, Andrews PL, Laven RD (1996) Introduction to wildland fire. Wiley, New York

Ratt A (1985) Mõnda maaviljeluse arengust Eestis läbi aegade (About agricultural development in Estonia). Valgus, Tallinn

Reintam LJ, Moora TH (1983) O vljanii podsečnogo zemledelija na svojstva počv (Impact of slash farming on properties of soils). In: Čelovek rastitelnost i počva (Man, vegetation and soil). Academy of Science, Tartu

Rõuk A-M (1995) Inimtegevus ja selle mõju loodusele (Human activities and their impacts to the nature). In: Raukas A (ed) Eesti Loodus (Estonian nature). Valgus and Eesti Entsüklopeediakirjastus, Tallinn

Tarkiainen Ü (2014) Aletamine maastikukujunduse- ja kultuuritegurina Põhja-Euroopas ja Eestis (Slash and burn cultivation as factor of landscape design and culture in Northern Europe and in Estonia). Tuna. Ajalookultuuri ajakiri 1:26–41

Tomson P, Bunce RGH, Sepp K (2015) The role of slash and burn cultivation in the formation of southern Estonian landscapes and implications for nature conservation. Landscape and Urban Planning 137:54–63. doi: 10.1016/j.landurbplan.2014.12.015

Vabariigi digitaalse suuremõõtkavalise mullastiku kaardi seletuskiri (2001) (Headnotes of digital large scale soil map of Estonia). Maa-amet, Tallinn. http://www.maaamet.ee/docs/kaardid/mullakaardi_seletuskiri.pdf. Accessed 25 Oct 2014

Viires A (2000) Puud ja inimesed (Trees and people). Ilmamaa, Tartu

Viro PJ (1974) Effects of forest fire on soil. In: Kozlowski TT, Ahlgren CE (eds) Fire and ecosystems. Academic Press, New York

Voionmaa V (1987) History of the Karelian Tribe in Finland-Extracts. Suomen Antropol 4:207–212

Wedin M (2003) Forest finns vs. Swedish commons. In: Berge E, Carlsson L (eds) Proceedings from a workshop on "commons: old and new", Oslo, 11–13 March 2003. Norwegian University of Science and Technology, Trondheim, pp 59–60

Part II
Management and Conservation Approaches

Chapter 17
The Garfagnana Model for Exploiting Agrarian and Cultural Biodiversity: The White Garfagnina Sheep Breed, a Case Study of Sustainable Local Development

Francesca Camilli and Sandro Pieroni

Abstract This work shows and discusses ten years of activities of the Union of Municipalities of Garfagnana to support the protection of local agrarian biodiversity in a sustainable rural development perspective. Such activities were based on the involvement of local community as the main protagonist of the process of development together with the coordination role of local institutions. The key elements of the success of the Garfagnana model was the exploitation of the multifunctionality of local resources and the diversification of local economies, taking into account both local knowledge on agricultural traditions and the contribution of research, as in the case of the white Garfagnina sheep breed attests.

Keywords Biodversity · Landscape · Rural development · Multifunctionality · Diversification · Biocultural diversity

17.1 Introduction

17.1.1 The Protection of Vegetal and Animal Agrarian Biodiversity

In every territory, special productions can be recognized as elements of the local environment which is meant as the result of a system of interactions, occurring during time, between material and immaterial resources. Such productions are part

F. Camilli (✉)
Institute of Biometeorology, National Research Council, IBIMET-CNR, Via Caproni 8, 50145 Florence, Italy
e-mail: f.camilli@ibimet.cnr.it

S. Pieroni
Unione dei comuni della Garfagnana, Via Vittorio Emanuele, 9, 55032 Castelnuovo di Garfagnana, Lucca, Italy
e-mail: sandro.pieroni@ucgarfagnana.lu.it

M. Agnoletti and F. Emanueli (eds.), *Biocultural Diversity in Europe*, Environmental History 5, DOI 10.1007/978-3-319-26315-1_17

of the local identity and are often represented by animal and vegetal resources that follow the evolution of the local environment intended as the ensemble of natural, social and cultural assets. Local productions, thus, represent a huge set of values both as a genetic reservoir and a pool of elements of the local culture that cannot be lost, although they do not always seem to be attractive as they do not provide immediate relevant benefits especially in terms of local economic growth.

In Garfagnana, the *equilibrium* between the natural and anthropological components can still be observed because during the centuries, the local population has been safeguarding local nature. As a consequence, a unique genetic reservoir, enclosing important resources for the conservation of genetic variability, has been locally maintained. Indeed, biodiversity in Garfagnana does not only supply genetic and agrarian variability but also attests the cultural peculiarities of the local community. One of the most important elements of this genetic-cultural interaction is local knowledge that has been built by "cultivating" continuous relationships between the cultural and the natural environment. In fact, indigenous people who defend for a long time the variety of environmental local resources, get more interested in conserving biodiversity and this is crucial in generating the ecological services and natural resources on which people depend (Gadgil et al. 1993).

In 1992, the Convention on biological diversity approved at Rio de Janeiro recognized, for the first time at international level, the linkage between local rural communities and biological resources and the need to protect and preserve traditions and cultural heritage belonging to local communities (art. 8, subparagraph J; art. 10, subparagraph C). Cultural factors strongly contribute to biodiversity and also affect agrarian biodiversity, so that domestication has been defined as "the basis of biodiversity of agrarian cultures" (Frankel et al. 1995). Local knowledge, as an outcome of model-making about the functioning of the natural world, is a value to manage ecosystem and biodiversity in a sustainable way (Gadgil et al. 1993) and, together with the community experience, forms a sort of "capital" that, if ignored, can lead to resource exhaustion and environmental degradation (Gadgil et al. 1993). Furthermore, local knowledge can be considered complementary to science because it can support the analytical approach of a situation, from local to global (Reid et al. 2006). The involvement of the community elicits a direct linkage between conservation and local benefits being a direct incentive for local people to protect biodiversity in the long term (Salafsky and Wollenberg 2000) and a basis for capacity building which is critical in generating success across all outcomes (Brooks et al. 2013).

In Garfagnana, the linkage between local community and the environment has been very tight and steady during time. Still, nowadays, it is possible to exploit local biodiversity as a real resource, effectively providing benefits just because of the natural characteristics of the environment and the knowledge in managing it, developed by the local community. Biodiversity is an important commons not only for humanity but also for the development of regional and local benefits such as ecotourism and other services for human well-being at the community level (Capistrano et al. 2005). To be aware of the value of local vegetal and animal biodiversity, as well as food derived from it, means to be aware of our own roots and traditions. This is a cultural process itself that supports the local community in retrieving the sense of belonging to the territory. Thus, the exploitation of material (natural and agrarian biodiversity) and immaterial resources (local culture, knowledge, and traditions), and the relationship between them are fundamental to foster local identity and need to be considered as a means to improve local economy and support rural development. At the same time, this means to preserve the landscape features that are a driving force for tourism activities, while increasing the quality of life for local populations, supporting their recreational, emotional and spiritual needs, and their sense of identity.

However, the integration of biodiversity conservation and livelihood-needs requires building the capacity to deal with multiple objectives, the use of deliberative processes, learning from commons research and, in general, developing a complexity approach for commons governance (Berkes 2007). In the past ten years, many local resources in Garfagnana have been exploited through a process of implementation of actionresearch that has been underpinning local development within the framework of several projects. The latter have tracked a route which has required steady coordination on behalf of local governmental institutions and a growing awareness of the community on both cultural values provided by local resources and the importance of sharing efforts in working together to achieve common goals. This has been the basis for the development of a rural area, where integrated strategies engaged public institutions, the production system, and local administrations into the recovery, preservation, and maintenance of the linkages between cultural and biological diversity. In fact, this local complexity has to be managed through the involvement of networks and partnerships at various levels of government, private sector, and civil society (Brown et al. 2005). Many elements of local identity can provide a rural area with an added value that enables that area to become more competitive in a sustainable way as it can define its own identity and make it unique. In order to do this, the involvement of local communities is required in conservation acts as a means to mitigate possible local resistance. At the same time, protecting the productivity of a resource may be used as a means to enhance local livelihoods and development (Berkes 2007) and, if very little exits in literature about biological conservation and rural development (Berkes 2007), the case of the exploitment of white Garfagnina sheep breed can represent an important example of conservation aimed at local rural development.

17.1.2 Description of Garfagnana Area

Garfagnana is an area of the north western Tuscan Apennine, in the province of Lucca. The characteristics of this land are typical of mountain areas (Fig. 17.1). The peculiar pattern of Garfagnana climate, together with the anthropogenic influences, has determined the present landscape showing a “mosaic” of agrarian and natural ecosystems such as, forests of deciduous and coniferous trees, chestnut trees, pastures and arable lands. From a pedological point of view, the soil of Garfagnana

Fig. 17.1 Landscape of Garfagnana Mountains

is shallow, mostly acid or subacid, with a prevalent sandy texture, the major part of it showing low fertility (Benvenuti et al. 2008). The elevation of the agro-ecosystems ranges between 300 and 800 m above the sea level and the local climate is characterized by high rainfall (1100–1300 mm per year) and wide seasonal and daily temperature variations: average summer and winter temperatures range, respectively, between 30–15 °C and 10–5 °C. Agriculture in Garfagnana is a marginal economic activity, if it is compared with agricultural activities of areas managing more intensive productions. This is due to several geographical and socio-economic features such as land terrain, climate, land fragmentation, structural weakness of the economy, and infrastructure facilities. In fact, local agriculture is mainly characterized by the interaction of two different factors: (1) the mosaic management of the agro-ecosystem based on natural woods, chestnut woods, arable land, and pastures; (2) the support of environmentally sustainable agronomic techniques aimed at maintaining traditional agronomic practices (no use of pesticides and synthetic fertilizers, use of manure rotations with perennial forage crops). The preservation of traditional agronomic practices of this agro-ecological system is coherent with organic management practices (Altieri 1995) and has allowed the maintenance and improvement of species and varieties that otherwise would have suffered strong genetic erosion or extinction as shown, among others, by the cases in Italy of saraceno grain and Garfagnana spelt, two endangered genetic resources that were basic for the survival of communities in marginal areas (Scialabba 2003).

In spite of the geographical and economic marginality of local economies of Garfagnana, satellite activities, upstream and downstream of the agricultural production supply chain must be considered as positively affecting other important aspects, such as the environment and the tourism sector.

The physiographic patterns of Garfagnana have influenced the settlement system which is characterized by the spread of pastures at high altitudes, where once most part of the summer livestock activities were conducted. Nowadays, livestock farming is still characterized by small size farms which, in spite their low productivity, play an important role in maintaining the landscape features, but also in developing strategies that can preserve the environment and biodiversity. The latter become, thus, a sort of capital whose values can be transferred to locally made, typical products. This has been shown, for example, by the quality of typical cheese which is the result of the interaction among specific pedoclimatic characteristics, autochthonous genetic variations, as well as anthropic components and local traditions which are difficult to be reproduced elsewhere (Turchi et al. 2011). Thus, the combination of different factors has a fundamental role in protecting not only the land but also the complex structure of the *terroir.*

In this chapter, the descriptions of the recovery and reintroduction of the white Garfagnina sheep breed (Fig. 17.2) is reported as a case study of local development. Different steps, occurred during ten years of activities related to the exploitation of this sheep breed, will be described. Furthermore, the factors which have been responsible for the successful exploitation of the sheep breed will be elicited and discussed. These aspects will regard: (i) the test of the economic feasibility of the exploitation of the native sheep; (ii) the development of the concept of multifunctionality of a local resource; and (iii) the exploitation of the social component of Garfagnana to safeguard social cohesion and support the process of local development.

Fig. 17.2 Garfagnina sheep breed

17.1.3 Sheep Breeding in Garfagnana

Sheep farming has been traditionally one of the main livestock activities in this area and has influenced the environment characterized by large pastures, generally at 1300–1500 m of elevation. In 1950, in Garfagnana, sheep livestock counted for more than 50,000 heads; in 1999–2007 sheep decreased to about 3000 heads (Table 17.1). Settlements called "Alpe" (alp) or "Capanne" (huts) were spread all over the land of Garfagnana. They are the special buildings that once were used as pastures during summer time and nowadays are evidences of the important activities carried on by shepherds in the past.

The decrease of sheep breeding activities led to a progressive abandonment not only of pastures but also of the small settlement of houses that could guarantee the preservation and the control of the precarious environmental *equilibrium* of this land. Economic incentives provided to local farmers and aimed at exclusively maintaining local activities, without any strategy directed to enhance the value of the final products and territorial environmental resources, have not been successful. The progressive abandonment of high altitudes areas used for sheep breeding has been causing the loss of morphological features linked to the traditional landscape. Pastures have gradually come back to a natural state and have been deteriorated; terraces, where once typical mountain products, such as potatoes and rye were cultivated, have been abandoned. As a consequence, the aesthetical and economic value of the landscape toward a new tourism strategy of local development, is

Table 17.1 Data on sheep in Garfagnana 1999–2007, by municipality (*Source* Local Health Authority)

Municipality	No. of sheep in 1999	No. of sheep in 2007
Camporgiano	93	65
Careggine	254	140
Castelnuovo Garf.na	356	132
Castiglione Garf.na	508	191
Fosciandora	355	151
Gallicano	27	13
Giuncugnano	252	93
Minucciano	1215	510
Molazzana	318	165
Piazza al Serchio	493	294
Pieve Fosciana	705	529
San Romano Garf.na	76	0
Sillano	112	223
Vagli sotto	60	0
Vergemoli	103	0
Villa Collemandina	808	407
Total	**5735**	**2913**

Table 17.2 Livestock—number of farms in the period 1999–2007 by class dimension

No. of sheep	No. of farms in 1999	No. of farms in 2007
<30	106	47
From 31 to 50	20	17
From 51 to 80	25	5
>80	23	11
Total	**174**	**80**

Source of data Local Health Authority
Elaboration of data Union of Municipalities of Garfagnana

decreased and has been only partly compensated by the naturalistic evolution that represents, in an area of progressive abandonment, the prevailing spread of woods onto the cultivated areas.

Data related to the number of sheep breeding farms by dimension (Table 17.2) showed that the farms with the highest number of heads per herd are a few ones, whereas the majority manages herds with less than 30 heads. The small and micro sheep farms cannot even be acknowledged as conventional farms. Since the early 1970s, the sector showed low dynamicity and competitiveness even though, the wide distribution was indicating an unexpressed potential of development. This was also due to the lack of a generational turnover of breeders whose approach to innovation has become more and more rigid over time.

17.1.3.1 Sheep Farming and Traditional Sheep Breeding

According to Carlo De Stefani, who wrote a detailed report on agriculture of Garfagnana at the end of nineteenth century (De Stefani 1878), livestock with more than 50,000 heads of sheep could be found in the area, mostly flocks of about 100 sheep. Sheep were led to graze on the pastures of the Apennines in summer time but, in winter, the transhumance to Maremma or to the lands around Pisa and Livorno was common. During the transhumance the whole family of the sheep breeder (men, women and children) often moved to the bottom of the valley while, sometimes elderly remained at home. At the end of September, real "caravans" moved along the main road called "Via Nazionale" toward Lucca; the caravan was led by the shepherd (the "vergaro") followed by other men, the boys and a procession of mules and horses together with all the family. By the beginning of May, the shepherds moved back again to the highlands. The timing for animals to be conducted to pastures was regulated by specific rules of the municipalities: for example, land owners used to produce the hay at the same time, so to allow the next pasture. When chestnuts started falling, the flock could not graze longer in the chestnut woods until the harvesting of the fruits wad ended. At the end of 1800, De Stefani estimated a production of 40,000 lambs (minus the part used for breeding) that were sold to external markets during the period between March and May. The overall gain from sheep breeding was primarily obtained from the production of milk, the meat of lambs and wool. The income obtained from selling the manure and the sheep from comeback share was also economically appreciated.

17.1.3.2 The White Garfagnina Sheep Breed

The white Garfagnina sheep breed probably derives from the Apennine sheep breed strain. Originally, it was considered a threefold attitude sheep (milk, wool, and meat) which was well adapted to the hard environment of the poor Apennine pastures (Fig. 17.3). This white medium-sized sheep breed was typical of the Tuscan Apennines, from Lucca up to the borders with Umbria. After the 1950s, the decrease of livestock was followed by the progressive replacement of the white Garfagnina sheep with the Massese sheep breed which, although considered more productive, did not adapt to the Appennine environment showing to be much more delicate and prone to get diseases. In 2003, only 70 heads of the white Garfagnina sheep were left in Garfagnana together with a pool of sheep bred by the Forest Service, in Abruzzo, at the Center for Sheep Breeding in Castel del Monte (L'Aquila). The exploitation of white Garfagnina sheep aimed not only at protecting biodiversity, but also, at strengthening the economic value of products related to this resource.

Fig. 17.3 Grazing sheep in chestnut woods

17.2 Ten Years of Projects Actions

Starting from 2004, different projects were performed in order to support the exploitation of the white Garfagnina sheep and a series of actions were directed toward the development of local economies linked to this resource. The overall aims of such projects, promoted and coordinated by the Union of Municipalities of Garfagnana (former Mountain Community) were

- to exploit and preserve the native germplasm of white Garfagnina sheep breed;
- to develop a sustainable economic and environmental model to exploit local mountain resources;
- to study, develop and support sheep breeding, an important agricultural activity embodying strong cultural and agricultural features of local identity in order to characterize the *terroir*.

In 2004, the project "Terre Alte" started the activities directed at exploiting a traditional sheep breeding based, in the past, on the pasture system or "alpeggio" diffused along the Apennines. The project name "Terre Alte" (highlands in English) indicates those pasture areas, mostly at higher elevations of Garfagnana mountains, that are currently unused or under-utilized but, nevertheless, still showing interesting potentials as regards to the quality and the productivity of grazing. Because of their location at over 1200 m of elevation, shepherds whose farms were located at lower elevations (between 600 and 1000 m), used to go to the highlands during summer for vertical transhumance. Some of these pastures are currently private properties, but the majority of them are public, being owned by the regional government. In the latter case, pastures are often allocated to the few shepherds still working in the area, through a credit system called "fida pascolo." The exploitation of the areas located at high altitudes may represent a model for the correct use of these lands, as an ideal source of pastures for both cattle and sheep breeding, with positive effects on the quality of the final products.

However, it must be underlined that "Terre Alte" has been important for it has been able to exploit areas which are peculiar for the local environmental *equilibrium*: in facts, the correct use of these lands implies the protection of the environment. Finally, it is not secondary to consider the positive feedback "Terre Alte" provided to the development of tourism in this area: in this respect, the several public properties (a recovery plan of these properties aimed at developing trekking and hospitality structures was promoted by the former Comunità Montana) play an important role as they have been able to provide added value to the whole "Garfagnana system."

As a matter of fact, the public properties could preserve the traditional activities such as grazing animals, in the wide prairies of the Apennines and, thus, contribute to characterize the landscape of the area making it more attractive for tourism as it is shown by the increased level of attractiveness toward the excursion tourism. In facts, this land is characterized by two main components. The first one is the combination of two ranges of mountains showing completely different morphological elements and confined into a quite small area. One range of mountains is formed of the Apuans Alps, characterized by a rocky soil and a typical alpine morphology: high peaks and steep mountain walls that separate Garfagnana from the Tyrrhenian Sea. The other range of mountain is composed of the Northern Apennines showing more rounded peaks and characterized by wide clearings at high elevations. The second component of Garfagnana is provided by the population distribution which is detectable in the urbanization of hundreds of small villages and in the care of the land use as it is indicated by cultivated terraces or small parcels of land at 300–1000 mt of elevation. Both the components provide the tourists with the feeling to visit a place that, despite being hard to live, is nevertheless able to provide stillness and calm.

Garfagnana is the place where, in 1985, the former Comunità Montana arranged one of the first Italian trekking routes, made of nine steps allowing the whole tour of the valley between the Apennines and the Apuans. These steps refer to a well defined and organized system of trekking trails, shelters, rest stops but, above all, they rely on an efficient and accurate information and hospitality system coordinated by the Center for Rural Tourism that, over the years, became the Tourist Information Bureau of Garfagnana. Unused or underused public facilities belonging to the agricultural heritage of the Forest Region of Tuscany were involved, together with some private actors, in building up the overall system. For this reason, the highlands of Garfagnana could become a real container of all those projects that aim at providing value to the lands and the agriculture activities even at highest altitudes, by integrating those features strictly related to agriculture production with those of tourism and local culture and traditions.

Every year, the environment of Garfagnana can attract thousands of tourists mostly interested in hiking and outdoor sports, but also significantly representing an important market for local products. In this way, the specificity of the products, the landscape and the environment have turned into the sustainable competitive asset for the Garfagnana valley (Fig. 17.4).

In this framework, in 2004, the need to increase the identity of Garfagnana area through the exploitation of traditional productions, together with the need to preserve the landscape connected to cultural and agrarian biodiversity supported the project aimed at reintroducing the white Garfagnina sheep breed. At this time, different actions started supporting also the development of the "Cerasa Farm" (Fig. 17.5). This farm belongs to the Regional Forest Heritage Trust and is administrated by the Union of Municipalities of Garfagnana. It is run by a family who has got a long experience in traditional sheep breeding. The abilities and competences of this family were fundamental to grant the success of the reintroduction of the white Garfagnina sheep breed and test the benefits and the economic sustainability of its exploitation. Human capital owing the necessary know-how to validate the actions toward the exploitation of the white Garfagnina sheep breed was, thus, crucial: no exploitation of local resources would have been possible without evaluating if the use of traditional breeding standards (from feeding to animal health and production) could guarantee a real economic viability for the farm.

Fig. 17.4 Landscape of Garfagnana valley

Fig. 17.5 The "Cerasa Farm"

Thanks to the collaboration with MIPAAF,[1] at the Center for Sheep Breeding in Castel del Monte, L'Aquila, 20 Garfagnina sheep were reintroduced at the "Cerasa Farm." A pilot center to safeguard the white Garfagnina sheep breed was built up, as a basis for research activities to be conducted in collaboration with research institutions. In the same period, the project for the preservation of agricultural biodiversity through the research of ancient cultivated varieties was launched. Home gardens in small villages of the valley started cultivating an extraordinary variety of vegetables, fruits, grains that could represent the history and the culture of the territory.

This was possible by exploiting the local genetic resources through a detailed investigation in the area, firstly, involving local schools. In a second phase, scientific analysis and characterization of germplasm (varieties characterization, conservation techniques) were performed in collaboration with research institutes and universities. The Region of Tuscany also played an important role, particularly through the role of ARSIA, the former Regional Agency for Development and Innovation in the Agricultural sector, which conducted specific studies on the germplasm of the white Garfagnina breed. This was possible thanks to the project "Bank of the germplasm" promoted in 2005 by the Region of Tuscany through the regional law, L.R. 64/04 that sustained the exploitation of 200 horticultural and fruit varieties and the establishment of 32 "coltivatori custodi" or guardian farmers in Garfagnana. As regards to the animal resources, the white Garfagnina sheep could be exploited at the "Cerasa Farm." The University of Pisa carried on the scientific actions aimed at analyzing the qualitative characteristics of the products derived from Garfagnina sheep. The positive results provided by this approach made other breeders become very interested toward the project. The Union of Municipalities also took in charge the organization of regular meetings among farmers, using focus groups methodology, in order to detect problems and involve farmers in confronting themselves and sharing experiences. The scientific validation, together with the operational actions of all the breeders involved, was indispensable to define protocols and standards at the basis of the production phases. The implementation of the "Garfagnana model" continued by defining production protocols and standards which became the study object of the project "Vagal." Vagal is a cross-border IT-FR Maritime project aimed at the characterization of germplasm resources of rural Tuscan lands. Beyond studying milk and meat production by Garfagnina sheep, in 2007 Vagal started studying also the production and processing of wool. The first steps of the project showed that wool could represent a form of income integration for the local breeders and artisans. Wool produced by the white Garfagnina show average diameter and length fiber values of, respectively, 18.1 μm and 85 mm (Bacci et al. 2013), a pale yellow color and, even if it is produced from a sheep breed whose main production attitude is milk, it shows to be suitable for spinning and easy to be woven. The activities aimed at exploiting wool

[1]Ministero delle politiche agricole alimentari e forestali, Italian Ministery of agriculture, food and forestry.

have been conducted through three steps: (i) the definition of standards parameters for processing the raw material by identifying proper manufacturing structures for scouring and spinning small amounts of greasy wool; (ii) the training of local artisans to properly process the wool (natural dying, hand weaving, use of the waste produced by the first phases of the textile production to make felt); iii) the collaboration with other regional and national projects related to the exploitation of natural fibers.

A center for educational and information activities on rural development was also established at the "Cerasa Farm" during the "Vagal" project. The center provides the access to the Garfagnana model of exploiting local resources and support local production supply chains. Furthermore, the development of production rules and protocols were implemented at the center. In 2008–2011, the Union of Municipalities of Garfagnana participated to the project "Percorsi di Orientamento" (Vocational Guidance Paths),[2] collaborating to a research on the development of female entrepreneurship in the agriculture and the textile craft sectors and the possible rebuilding of a short textile processing chain in Tuscany. In this project, a group of women from Tuscany working in the agriculture sector (sheep breeding) and the textile manufacturing sector (hand weavers, tailors, felters, and embroiderers) was involved in activities related to the creation and development of entrepreneurship. Some women from Garfagnana were included in this group and took part to a study on the assessment of competences aimed at career development (Di Fabio et al. 2011).

The group of women (entrepreneurs and women interested to become entrepreneurs) attended also workshops on enterprise management, organization and product marketing. Such activities led to the creation of a network interested into the production of textile craft products by using local resources, their craft abilities and knowledge of local textile cultures. The collaborative working relationships among these women were tested in 2011 during the last year project within a particular initiative called DNA "Design, Natura, Artigianato" (designnaturaartigianato.wordpress.com). The initiative consisted in one workshop involving women in agriculture and handicraft, together with young designers, in the design and production of textile prototypes made of Garfagnina wool, as a resource of this rural territory. The workshop's aim was to show how wool could be introduced into a local supply chain, potentially being transformed from an undervalued waste material to a valuable textile product, capable to transfer the meaning of material agricultural resources and craft culture, as an expression of local landscape.

[2]The collaboration between the Union of Municipalities of Grafagnana (former Mountain Community) and the Institute of Biometeorology of the National Research Council, IBIMET-CNR, was based on the convergence of objectives toward the development of female enterprises and entrepreneurship through the sustainbale exploitation of tangible and intangible resources of rural areas, in particular, those related to the possible development of "short" textile supply chains. This was the main objective of the project "Percorsi di Orientamento" (2008–2011), financed by the Ministry of Labour and Social Affairs and coordinated by CNR-Ibimet in four Italian regions: Tuscany, Emilia Romagna, Sardinia and Campania.

17.3 Results

Almost ten years of activities on local development designed and coordinated by the Union of Municipalities of Garfagnana together with the engagement of the community within local and cross-border projects allowed the reintroduction of the white Garfagnina sheep breed in Garfagnana. The first result of this process was the conservation of local biodiversity and the "creation" of a unique local resource. This important result was necessary but not sufficient to justify the exploitation of the native sheep if further steps had not been accomplished, as shown by the following outcomes. The recovery of the native livestock proved to be quantitatively significant as the number of white Garfagnina sheep breed increased from a few heads in 2004 to over 800 in 2013 (almost 100 new heads per year). This sustainable and replicable model of good practices, aimed at local development, could be built up as it was implemented during the years as the result of a process involving the whole territory through the exploitation of local expertise and know-how in agriculture and manufacturing together with the contribution of scientific research.

17.3.1 The Community

One of the results of Garfagnana development process was the retrievement of the local culture linked to traditional forms of farming and, more generally, the sense of belonging of the farmers to their land. In facts, the exploitation of an ancient local breed has relied on the acceptance of local farmers to feel as part of Garfagnana and actors of the process of development. In this respect, the older farmers represented the driving force of the younger ones. It must be underlined that the attention of the media (magazines and TV programs) toward the experience of the reintroduction of the white Garfagnina sheep breed, built up a very positive message of Garfagnana and its local resources outside the territory. This, in return, generated a feeling of pride and belonging in local breeders, beyond providing an important added value to the whole territory, either directly or indirectly, through the dissemination and promotion of the project activities and the creation of forms of sustainable competitive advantages. Another important factor influencing the positive path of the project's activities has been the strengthening of the relationships among the community membres (in particular, the farmers). This developmental model could not be regardless of engaging farmers in sharing their experiences. A system of thematic focus groups was brought off and showed to be positive in encouraging the new generations of shepherds in confronting themselves with the elder ones. The closer relationships among sheep breeders could be further strengthened by their sense of belonging to a common and highly valued cultural experience and, thus, could define people behavioral attitudes that showed to be particularly positive for the success of the developmental model. The reference role of the Union of Municipalities in coordinating such actions has proved to be essential for this process.

17.3.2 The Local Economies

The economic viability of the exploitation of the white Garfagnina sheep breed was a necessary step to follow. Indeed, the highest value gained by products, such as cheese and meat, along with the decrease of running costs, was one of the key element of the success of the rapid reintroduction of the sheep. Without a clear economic validation the exploitation of the white Garfagnina Sheep breed would have failed. The increase of the demand of cheese and meat and the relative increase in the price provided farmers with confidence in their businesses. The decrease of the running costs—in particular, those related to animal feeding and care—due to the strong capacity of this native breed to well adapt to the local environmental conditions, contributed to the economic sustainability of the breeding activities. The results of the characterization analysis of milk and meat (by ARSIA[3] and the University of Pisa) are currently available and represent a good basis for marketing the products. An identification brand also contributed to the increase of the economic value of the Garfagnina sheep meat. The farmers who produce lambs could verify the increase of the price of the lamb meat shifting from 3.5 eu/kg in 2004 to 5.5–6.0 eu/kg in 2013. Today, the income from the sale of lamb produced in Garfagnana is estimated about € 65.000 € per year while the sale of the animals for breeding is estimated about € 10.000 € per year.

As regards to the cheese made according to local traditional methods, all the production is almost entirely locally sold, especially to farm visitors and tourists. Currently, the annual production of cheese is estimated about 10 tons with a value of about € 150.000. Sheep breeders who need technical support for the management of the white Garfagnina sheep breed or breeders eager to start a new business can get support at the Cerasa farm that was acknowledged as a pilot center for breeding the Garfagnina sheep. Here, farmers are involved in joint activities (summer pasture recovery system called "guardiania unica",[4] seminars and workshops on key issues such as livestock feeding, animal health, cheese production). In this perspective, the "Cerasa Farm" is becoming a place where it is possible to conduct training and dissemination activities about the local sheep breed and, more generally, about the strategies for the sustainable rural development of the area.

[3]Former "Agenzia Regionale Sviluppo e l'Innovazione nel settore Agricolo" (regional agency for the development and innovation in the agricultural sector).

[4]"Guardiania unica" is an ancient practice used by many breeders who leave their animals to a sheperd entrasted to looking after the herd during summer. Sometimes the sheperd looking after the livestock can be one of the breeders confering the sheep. The shepherd is rewarded with a sum of money shared by all the breeders according to the number of animals confered. Once this practice was used also for cattle (summer pastures) and pigs (oak woods) for short periods.

17.3.3 Multifunctionaly: The Role of Wool

From an economic point of view, wool is the minor part of the sheep breeding production. In comparison with meat and cheese, the wool supply chain is composed of a larger number of processing phases, some of which (wool scouring and spinning) need industrial structures placed outside the territory. The income from the sale of spun Garfagnina wool has been estimated about € 5.000 per year. Despite the low monetary impact, due also to the costs of managing wool—from the collection phase, at farms level, until the spinning phase, at the spinning manufacture—the exploitation of Garfagnina wool has enabled the reactivation of a short textile supply chain which was at risk of disappearance. Wool could pass from being a no value by-product that is difficult to discharge[5] into a valuable product providing economic benefits and a kind of extra income for breeders. The exploitation of wool has to be evaluated also from a social perspective as it was linked to the starting up of training courses involving some women of Garfagnana in natural dyeing and textile traditional activities such as, hand weaving and felting that, since the beginning of the '900, was a quite spread craft activity in this area. Furthermore, the textile handworks, made of Garfagnina wool and produced by local artisans have been proven to function as a vehicle of attention and interest toward the territory when they are shown and sold at local and regional craft fairs and farmers' markets. These activities can provide artisans with some income but are also a means to promote local products and landscape through a material, the wool, which would have been otherwise wasted. "Percorsi di Orientamento" provided the possibility to technically test the use of Garfagnina wool and wool threads to create formally and functionally innovative textile products. As the applied design criteria were chosen considering the peculiarities of the rural land of Garfagnana and, more generally, of the Tuscan landscape, the handcraft prototype products could be possibly recognized by sustainable tourism markets and by rural and quality tourism, as well as.

17.3.4 The Landscape and Tourism

The increase of the number of animals raised the demand for areas to be allocated to pasture, especially in summer. The Union of Municipalities supported this demand through the recovery of high altitude pasture areas that regained their primary function and the role of key element providing the landscape with new value, as the landscape itself could be used as an instrument for tourism attractiveness. In this

[5]Commission Regulation (EU) No 142/2011 of 25 February 2011 implementing Regulation (EC) No 1069/2009 of the European Parliament and of the Council laying down health rules as regards animal by-products and derived products not intended for human consumption and implementing Council Directive 97/78/EC.

respect, the renewed presence of herds of grazing animals has been receiving a positive response on behalf of tourists. The exploitation of the Garfagnina Sheep breed has been a process through which also the landscape features, especially those related to sheep breeding such as, pastures and constructions linked to them, have been either recovered or improved with positive consequences on the landscape. The increase of the tourism offer was evident not only in the landscape improvement but also in the increase of local agrifood productions. This trend was connected to the raise in the demand for local food that was supported by inhabitants and local restaurants (tourism demand for local food). Tourists are very much attracted by products holding a strong territorial identity. Many farms carrying out breeding activities are also agritourisms. The diversity of activities has enabled the farms to maximize the added value of products and spread the image of local products.

17.4 Conclusions

The average very small dimension of farms in Garfagnana and the marginality of the land do not allow the yield of competitive agrofood products on a large-scale market. Therefore, the remarkable heritage linked to traditional knowledge and biodiversity can be the essential element to strongly connote small amounts of high quality productions. Thus, the strong identity of this rural area can provide local productions with a high added value, making them culturally relevant and more competitive and profitable. The awareness of local identity on behalf of the local community has, then, radically changed the production approach and the structure and organization of farms that have increased and improved typical productions deriving the white Garfagnina sheep breed. This was possible by diversifying the farm activities that, in return, influenced the landscape, thus, meeting also the demand of rural tourism and agritourism visitors searching for special destinations. As regards to the Garfagnina wool, different considerations have to be made because of the more complex supply chain if compared to agrifood productions. Nevertheless, among local productions, Garfagnina wool has shown to play an important role for possible economic and employment opportunities as it appears by the increased quantities of wool produced year by year. Furthemore, Garfaganina wool initiatives have allowed exchanges with similar projects outside Garfagnana making it possible to start evaluations of the wool processability and textile prototyping. Through the natural dyeing courses, the exploitation of wool has also given way to the integration of different supply chains. Future actions to improve the promotion of wool textiles from Garfagnana should also be made within the framework of the promotion of the different lands of Tuscany, a region which is able to hit foreigners' imaginary being possibly useful to successfully market also this type of local craft productions. Further steps need to be undertaken, but the multifaceted results so far obtained by the community of Garfagnana, by exploiting the multifunctionality of one of its more relevant local resources, outline promising perspectives toward the sustainable development of this rural area.

References

Altieri MA (1995) Agroecology: the science of sustainable agriculture. In: Agroecology: the science of sustainable agriculture. Westview Press, Boulder

Bacci L, Camilli F, Di Lonardo, Duce P, Vagnoni E, Mauro (2013) Neglected wools: fundamental steps to counteract the loss of potentially valuable materials derived from native sheep breeds. Hindawi Publishing Corporation conference papers in materials science, vol 2013, Article ID 402372, p 7

Benvenuti S, Loddo D, Macchia M (2008) Biodiversità della "seedbank" di fitocenosi spontanee nella coltura del farro (*Triticum dicoccum* (Schrank) Schübler) in agroecosistemi della Garfagnana (Lucca) Ital J Agron/Riv Agron 1(Suppl):47–57

Berkes F (2007) Community-based conservation in a globalized world. PNAS 104(39):15188–15193

Brooks J, Waylen K, Borgerhoff Mulder M (2013) Assessing community-based conservation project: a systematic review and multilevel analysis of attitudinal, behavioral, ecological and economic outcomes. Environ Evid 2:2

Brown K, Mackensen J, Rosendo S, Viswanathan K, Cimarrusti L, Fernando K, Morsello C, Muchagata M, Siason IM, Singh S, et al (2005) Ecosystems and human well-being: policy responses. In: Millennium Ecosystem Assessment. Island Press, Washington, DC, pp 425–465

Capistrano D, Samper C, Lee MJ, Raudsepp-Hearne C (2005) Ecosystems and human well-being, vol 4. In: Millennium Ecosystem Assessment. Island Press, Washington

De Stefani C (1878) Delle proprietà comuni e dei limiti alle proprietà private in alcune parti dell'Appennino. In: Storia dei Comuni di Garfagnana. Riv. europea, IX, Modena, pp 115–126

Di Fabio A, Palazzeschi L, Camilli F, Raschi A (2011) Interventi psicologici a sostegno dell'imprenditoria femminile: evidenze empiriche di efficacia nel contesto italiano (Psychological interventions to support entrepreneurship of women: Empirical evidence of effectiveness in the Italian context). Risorsa uomo 4:455–470

Frankel OH, Brown AHD, Burdon JJ (1995) The conservation of plant biodiversity. Cambridge University Press, Cambridge

Gadgil M, Berkes F, Folke C (1993) Indigenous knowledge for biodiversity conservation. Ambio, Biodiv Ecol Econ Policy 22(2–3):151–156

Reid WV, Berkes F, Wilbanks T, Capistrano D (eds) (2006) Bridging scales and knowledge systems. In: Millennium Ecosystem Assessment. Island Press, Washington, DC

Salafsky N, Wollenberg E (2000) Linking livelihoods and conservation: a conceptual framework and scale for assessing the integration of human needs and biodiversity. World Dev 28:1421–1438

Scialabba NE-H (2003) Organic agriculture: the challenge of sustaining food production while enhancing biodiversity. Secretary Priority Area for Inter-Disciplinary Action on Organic Agriculture, Food and Agriculture Organization of the United Nations, Rome, Italy. United Nations Thematic Group, Sub-Group Meeting on Wildlife, Biodiversity and Organic Agriculture, Ankara, Turkey

Turchi B, Nuvoloni R, Fratini F, Pedonese F, Ebani VV, Cerri D (2011) Caciotta della Garfagnana cheese: selection and evaluation of autochthonous mesophilic lactic acid bacteria as starter cultures. J Anim Sci

Chapter 18
Resilient Agrarian Landscapes in Face of Changes: The Coevolutive Approach to Understand the Links Between Communities and Environmental Characters

Catherine Dezio, Aurora Cavallo and Davide Marino

Abstract The paper focuses on a conceptual framework through analyses of the relationship that links and communities to their places and determines the capacity for resilience of those landscapes in the face of changes. The coevolutionary approach has been recognized as a key framework for understanding change in complex social–ecological systems and as a foundational concept for ecological economics (Costanza et al. in Introduction to ecological economics. St Lucie Press, Florida, 1997). The coevolutive approach describes as a dynamic relationship between environmental systems and social systems. Coevolution is different than mere co-dynamic change, in that at least one—social or environmental—system is evolving or changing through variation, selection and inheritance. This leads to the necessity of identifying the set of anthropic—or cultural—and natural—biological—relationships that influence change within these relationships, determining their destiny or, in other words, whether these landscapes are conserved or lost. The need to protect local resources, and conserve the functions that come from natural capital, determines processes that can be guaranteed in a more efficient and sustainable way by local communities. The paper discusses the conceptual hypothesis on a case study of the UNESCO Vineyard Landscapes, based on the Switzerland case of Lavaux. The research concludes with a view of adaptive management, aimed to the preservation of landscape values, while respecting the natural dynamics of the landscape evolution.

C. Dezio
University La Sapienza, Rome, Italy
e-mail: catherine.dezio@gmail.com

A. Cavallo (✉) · D. Marino
DiBT University of Molise, Campobasso, Italy
e-mail: auroracavallo@libero.it

D. Marino
e-mail: dmarino@unimol.it

M. Agnoletti and F. Emanueli (eds.), *Biocultural Diversity in Europe*,
Environmental History 5, DOI 10.1007/978-3-319-26315-1_18

Keywords Coevolutive approach · Rural landscapes · Resilience · UNESCO vineyard landscapes · Biocultural diversity

18.1 Introduction

The framework in which Cultural Landscape started to be discussed is the nineteenth century cultural geography, product of the encounter between cultural anthropology and geography. Later, the subject has been explored in landscape aesthetics, in semiotics, in landscape geomorphology and many other areas. What is included in all these aforementioned disciplines is the temporal factor, for which the Cultural Landscape is a humanized landscape, man-made, shaped by culture—which is its driving force (Sauer 1931)—, a result from the interaction between natural space and human influence in history.

If we talk about the Agrarian Cultural Landscape, we will discuss about systems that contribute to both aesthetic aspects and ecological integrity, "derived from familiar methods and/or subsistence, where the dominant features are the result of a traditional approach or locally adapted to management" (Institute for European Environmental Policy 2007). They represent areas organically in a continuous evolution and with an active role in current society, they are ideal contexts in which it becomes necessary understanding how these sets of resources can play an active role in territorial policies and how conservation could be declined in a dynamic project, which seeks a dialectic and harmonic agreement between growth and protection, and between development and conservation. In this context, the coevolutionary perspective of Norgaard (1984) provides a link between ecological and economic paradigms: business decisions, market dynamics, interactions with the productivity structure are all elements that define the processes of use of natural resources by humans. Again, Norgaard defines how human activities can alter the ecosystem and how responses given by the ecosystem could provide an explanation of new social actions (Norgaard 1984). The approach to which our analysis belongs is linked to the coevolutionary model, which connects the changes brought about by man's actions to the territory, and as results to the landscape.

Coevolution has been recognized as a key framework for understanding change in complex, socioeconomic and ecological systems (Costanza et al. 1997; Folke et al. 2005; Kallis and Norgaard 2010). This leads to the necessity of identifying the set of anthropic and natural relationships that influence change within these relationships, determining their destiny or, in other words, whether these territories are conserved or lost. These dynamics can be due to natural causes, such as adaptation and mitigation management in response to climate change and the associated risks, but they can also be caused by social and economic factors, such as the question of depopulation. This environmental and economic crisis underlines how some parts of a territory, for example its internal areas—with their wealth of environmental resources, knowledge, manufacturing and potential uses—are reservoirs of resilience that can be called into play in future relationships with less resilient areas. The need to protect local resources, and conserve the functions that come from natural capital, determines processes that can be guaranteed in a more efficient and sustainable way by local communities.

How, therefore, can care for territorial resources be organized? Which initiatives best describe the relationships linking local communities to landscapes, and so determine their capacity for resilience? In this paper, we focus on these questions, also discussing a case study on UNESCO Cultural Agricultural Landscapes of Lavaux in Switzerland.

The paper is organized as follow: Sect. 18.2 focus on the conceptual framework, Sect. 18.3 analyses the role played by communities in driving the resilience process of a landscape, while Sect. 18.4 discuss the case study. Finally, some brief conclusions are provided.

18.2 Understanding Resilience in Agrarian Cultural Landscapes: The Role of the Coevolutionary Perspective

The agrarian cultural landscapes, as result of a long history of land usage and layering and overlapping of human interventions in the past, represent a rural heritage, both tangible and intangible, nowadays bearer of sustainable development.

The Institute for European Environmental Policy defines agrarian cultural landscapes as "(those) derived from domestic and/or subsistence procedures, where the dominant features are the result of a traditional approach—or one adopted locally by the management". Given this definition, we can already realize that it is about systems in which the aesthetic characteristics,—but also the ecological ones-, are actually the product of a community—managed approach to the territory. In order to study this relationship and the changes that occur in the evolutionary dynamics of a region, we use the coevolutionary approach. Coevolution refers to the joint evolution of more species belonging to the same environment, interacting among themselves so to influence one another; in the case of the agrarian landscape, business decisions, market dynamics, interactions with production, the economic system and the tradition of the local community are all elements that interact with natural resources, influencing its evolution and image overall (Norgaard 1984). Thus, the landscape could also be defined as "frutto della coevoluzione in una condizione di reciprocità del sistema naturale ed antropico" ("the result of coevolution in a condition of the reciprocity of natural and anthropic system"); a distinctive example of this balanced relationship is the agrarian cultural landscape, as evident result of layering and overlapping human interventions in the past. Hence, within the term "cultural", which distinguishes these landscapes from other landscapes, relies not only the temporality of historicity, which is missing in spatiality of the single word "landscape", but even that peculiar balance, and system of relationships that allows them to be preserved.

What we currently define as "cultural landscape" is a *modus dicendi* of a society, a labyrinth of relationships, mediations and dynamics, an union between environment and historicity, spatiality and temporality and horizons of naturalness and anthropocentrism that give way each other.

To understand this, knowing how the constituents of the environment are morphologically structured is not enough, nor it is knowing how the physiology of perception; works one needs to know the cultural, social and historical determinations, that is the exogenous factor which disturbs, in an incisive but contextually harmonic way, the natural ecosystems.

Thus, what characterizes the agrarian cultural landscapes could be read as an expression of the coevolution of the territory among objective characteristics (environmental constraints, such as climate) and subjective characteristics (business goals—and thus, human factors); hence, such cultural landscape could be analyzed as a complex matrix of different factors and historical layers through the structures composing it, or through economic, social and natural capital (Marino and Cavallo 2012).

This coevolutionary approach leads us to highlight how factors characterizing these landscapes—recently identified by Solymosi (2011) with isolation, marginalization; morphological and climate constraints, sustainable and traditional use of the land and its resources, small-scale, local residents who differ from the surrounding majority—are actually factors of a more complex evolution matrix. These, and other identifying features, are indicators of definition, but at the same time they are the conditions necessary to ensure the continuity of the landscape. Therefore, it is interesting to observe that the traditional economy and the sustainable management start from the constraints: the isolation, the small-scale, the difficult morphology and climate; therefore such a management actually is the result of adaptations to constraints and transformations. Hence, we can affirm these sites are actually perfect prototypes of resilient territories, as they are formed through cycles of adjustments, moving and keeping their balance.

The landscape is itself a product of the man on the natural environment and protecting it does not mean to deny the advancing artificiality, or to eliminate the disturbing factors. Rather, it is the cycle of upheaval, adaptation and (re) stability that shapes culture, beliefs and values of the community, which are real responsibles for evolutionary dynamics—and it is when this cycle is concluded that development occurs. We could say then that the potential of the cultural landscape essentially depends on the ability to close these cycles while maintaining the balance between man and environment, despite the changes (and thus of their resilience). The answer to the various pressures that may destabilize a landscape is measured by this resiliency, which introduces the understanding of how the community and its territories may face and build scenarios of development, in a framework influenced by interferences and changes. These process, within the coevolutive perspective is described in Fig. 18.1.

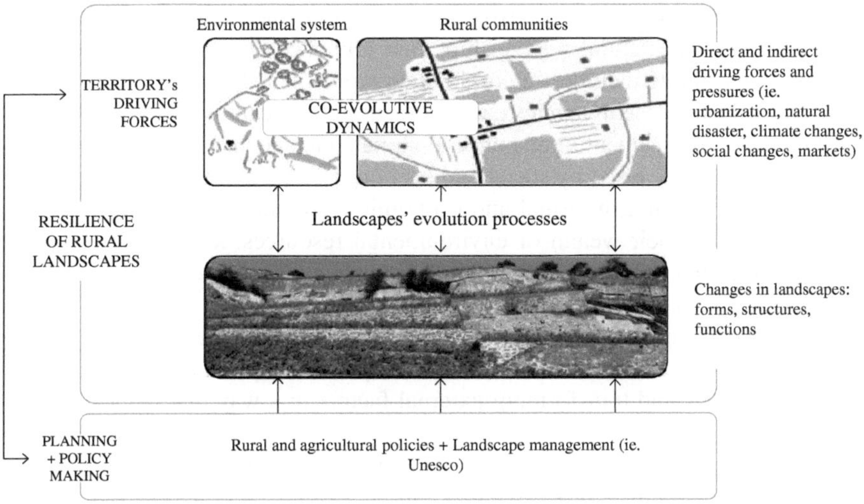

Fig. 18.1 The Coevolutionary approach as a framework to analyse the relationship between communities and their landscape

In order to study the resilience, Holling (1973) takes into account the SES systems (Social Ecological System): complex systems which intertwine social and ecological components, which inherently show adaptive cycles—of collapse-adaptation-balancing—and interact on different scales. The defining characteristics of SES relate to diversity (both biological and cultural components), non-homogeneity, complexity and connection. While using a coevolutionary approach, the use of ecological principles to analyze social dynamics becomes, in this context, an effective methodology, according to which the limits of adaptation are explained through ecological and social limits.

18.3 Conserved or Lost? The Role of Communities in Landscape Resilience

The aim of this section is to outline a conceptual framework through which we can analyze the relationship between man's social project and landscapes. In other words, we wish to understand the relationships that bind local communities to places and study the properties of the various initiatives for conserving landscapes and those of the territory in general (socioeconomics and environmental aspects), that contribute towards determining the resilience of these territories to change. The approach to which our analysis belongs is linked to the coevolutionary model, which connects the changes brought about by man's actions to the territory.

Coevolution has been recognized as a key framework for understanding change in complex, socioeconomic and ecological systems (Costanza et al. 1997; Folke et al. 2005; Kallis and Norgaard 2010). This leads to the necessity of identifying the set of anthropic and natural relationships that influence change within these relationships, determining their destiny or, in other words, whether these landscapes are conserved or lost. These dynamics can be due to natural causes, such as adaptation and mitigation management in response to climate change and the associated risks, but they can also be caused by social and economic factors, such as the question of depopulation. This environmental and economic crisis underlines how some parts of a places—with their wealth of environmental resources, knowledge, manufacturing and potential uses—are reservoirs of resilience that can be called into play in future relationships with less resilient areas.

We define a resilient rural landscape is a landscape where the community is able to mitigate, adapt and recover from external pressures promoting collective (or individual) actions and transforming external factors in a way that reduces chronic vulnerability.

The need to protect local landscapes, and conserve the functions that come from natural capital, determines processes that can be guaranteed in a more efficient and sustainable way by local communities. Protecting the territory means defining the optional value generated, involuntarily, by the processes of underutilization of natural resources. However, safeguarding must be understood in a wider sense, as relating not only to natural capital, but also to the processes and functions deriving from it and, therefore, to the ecosystem services provided, connecting the investment in natural capital to the services it can provide (Cavallo and Marino 2013).

The term safeguarding, along with the concept of protection, has been widely used with a restricting meaning, rather than in the sense of care for territorial resources, where, for example, the intervention processes require actions of a commanding or controlling nature rather than seeking the involvement and participation of local communities.

An aspect, strictly linked to the agricultural landscapes, is linked to care for territorial resources is that of maintaining natural capital, that is, the natural resources within the territory and the processes that generate them. These processes can either be natural or caused by human action—and are often a combination of both—proving that there is a tight interconnection between safeguarding natural capital and the conservation of material culture, local knowledge and wisdom, and the identity of places. Identity in this sense is linked to both social capital and anthropic capital, especially for its aspects relating to settlement and infrastructure.

In this framework, there is a tight connection between maintenance—linked to agricultural use of an area—and prevention, that is, awareness of possible external pressures. By this we mean the prevention of disorders, or damages, connected to the natural world—hydro-geological instability, fires, loss of biodiversity—or to socioeconomic factors such as changes linked to various kinds of developments, that may be political normative, infrastructural or even technological.

As part of the wide theme of safeguarding a landscape, its safety is a further precondition—together with basic services—to reviewing settlement models. We can consider these models as the set of relationships that tie an area to the identity of the people who live there and change it, to the relationships that take place there, to the history that over time has taken root there.

The second, central, question is how to structure in an innovative way the relationship between the management of the landscapes and its resources, and the production of goods and services? How can we go from safeguarding a territory to creating production processes—for goods and services—that have a significant impact on the local economy? In other words, what are the economic organizations, whereby the natural and social capitals of the territory become the input for production chains with added value, especially at local level? How can we develop a plan to achieve the effective integration of social, economic and environmental processes, and so reduce the polarization between production systems and natural systems? Which forms of multi-level collaboration seem necessary?

Engraved within the human territorial space is what Augé (1993) defines the rules of residence. They represent a content that is, at the same time, both spatial and social, so that the quality of the space depends on the relationships that take place within it.

Social capital established in that relational space creates a favourable environment for the rules of residence to be learnt and circulated, thereby, through processes of exchange and sharing, encouraging new knowledge and contributing towards increasing the social capital itself. This process of acquiring knowledge is helped by the diversity of the players' fields of interest and their cross fertilization, by means of their openness to new relationships, knowledge and practices, setting the bases for new innovation processes at an operative and strategic level. Within these processes, the actors involved negotiate principles and objectives, they mobilize and share resources, stimulating a virtuous mechanism that is transmitted at territorial level to all the stakeholders, to civil society, to the economic actors and to the institutional framework.

Among the dimensions that we have identified, relocation covers a key role: active, efficient protection means local management, access to resources and self-government (Ostrom 1990). Cases of abandoned landscapes are determined, not only by demographic or social aspects, or those relating to settlement issues, but also by the lack of a local resource management, leaving room to external management, as is the case, to give an example, of access to land or the management of water resources. Relocation management allows local inhabitants to choose the intended use of their resources and make the decisions, such as whether the primary sector should produce food or energy.

Local resource management is more efficient. It carries, for example, a lower cost, connected to knowledge of the territory, local wisdom, cultural capital and the optimisation of benefits.

In this framework, a key role is covered by the creation of new collective knowledge by organizing intense knowledge brokering between the specialized expertise domains that people and services have developed over the years and which today, on the contrary, need to be integrated more closely in order to face and solve the emerging issues.

Another dimension that can explain the function of local communities within the conservation of local resources is linked to the coproduction of innovative services (Ostrom 1990; Murray et al. 2010), where the integration between public and private players has the aim of promoting a better mobilization of resources available locally and co-designing practices that are more in line with the trends in changes taking place and with the needs of the different kinds of local players.

Strongly connected to the topics covered up to this point is the necessity of a different governance to manage local resources, capable of promoting positive effects in economic terms, through new procedures or technologies, by transforming positive external factors (mainly involuntary) into services (voluntary operations), in particular, services tied to the landscapes. The whole process is described in Fig. 18.2.

In this perspective, there is a role for new models of allocation involving subjects already within the territory, for example farms, and for public functions relating to the landscape. On top of this, there is the role played by new forms of management and private–public governance, which ensures that work rules are shared, that government actions are put to best use, that there is complementarity between resources and public and/or private behaviour in order to achieve the contextual production of public goods (landscape itself, or health, environment, knowledge) and private goods (value creation, access to food and free choice, including

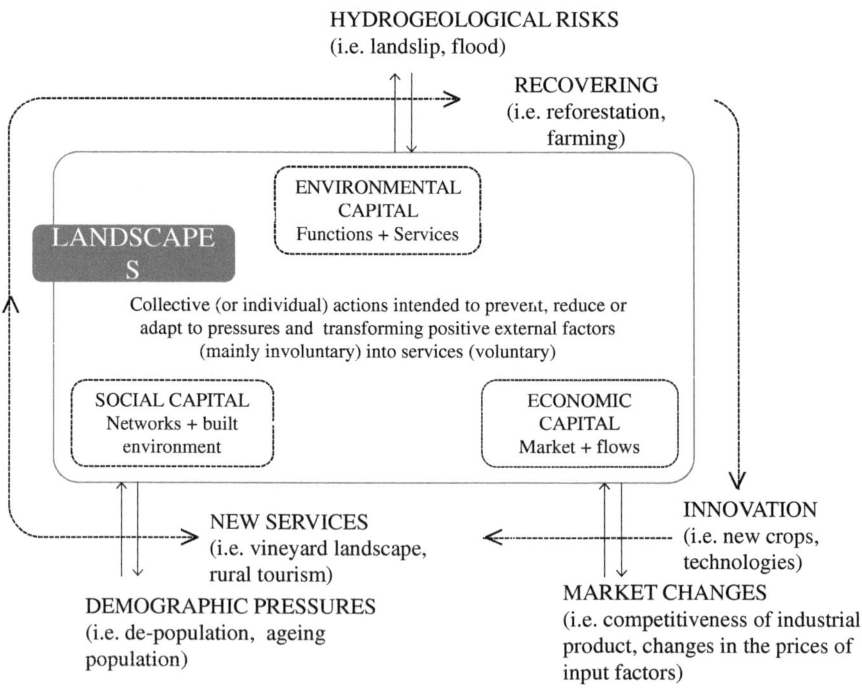

Fig. 18.2 Understanding the role of rural communities in driving the resilience process

hedonistic choices, on how to behave). Such mechanisms can be started through reallocation processes, using appropriate methods at local level that involve quota mechanisms with exchange markets at territorial level or between different territories, or through compensation mechanisms strictly linked to changing landscapes.

18.4 The Case Study: The Agricultural Cultural Landscape of Lavaux

A Cultural landscape has been defined by the World Heritage Committee (WHC) as "cultural properties (that) represents the combined works of nature and of man" (1992). This definition has been adopted by United Nations Educational, Scientific and Cultural Organization (UNESCO), in 1992, to introduce the link between biological and cultural diversity, and between tangible and intangible heritage. The WHC (1992) identifies three categories of cultural landscape: a "landscape designed and created intentionally by man" (such as historic parks and gardens), an "organically evolved landscape" (which may be a "relict or fossil landscape" or a "continuing landscape") and an "associative cultural landscape" (with a "religious, artistic or cultural associations of the natural element").

UNESCO Agrarian Cultural landscapes in particular, belonging to the category of organically evolved landscapes, testify the creative genius of man and social development: they are part of the collective identity, they illustrate the evolution of human society, they are the daily work of populations and they support the biodiversity of the area.

In order to be included in the World Heritage List—as regulated by the Convention Concerning the Protection of the World Heritage of 1972—the site has to possess the Outstanding Universal Value (OUV), where for Outstanding we mean "the representative of the best", while for "Universal" we mean a "value recognized in the world" and for which the entire humanity is responsible.

The Ministry prepares the tentative list with those sites that they decide to subject to registration in the following ten years; however, the actual application starts from the site itself: the local community prepares a dossier to sustain its candidacy and sends it to the State Party, which revises it and eventually sends it to the World Heritage Centre. If the file is complete, it is forwarded to the Advisory Body to be evaluated. This application form must show several points linked to the site, as its OUV, its authenticity, its integrity, protection granted by national and local regulations; also, the site must meet at least one of ten UNESCO criteria (such as representing a masterpiece of human creative genius, bearing a unique or at least exceptional testimony to a cultural tradition).

It is important to note that the registration to the UNESCO list is not in itself an act of protection, nor it constitutes a "freeze"; it represents a recognition of quality already achieved over the centuries and it is an opportunity for international exposure. Being a local initiative, the application process becomes a resilient project: it is an opportunity to enhance the uniqueness of the landscape with local projects.

One of the Agrarian Cultural Landscape enrolled in the UNESCO list in 2007 is the Swiss wine-growing landscape of Lavaux, a vineyard terraced landscape, which has remained intact since the eleventh century, when the monks gave the first impulse to shape the area through traditions handed down from generation to generation. In June 2013 we conducted an interview with the managers of the site, with some residents and as well some tourists of Lavaux, in order to investigate the role of the UNESCO application process in the context of the historical evolution of the landscape.

The site operators reported that the initiative of the application process was created to highlight the exceptionality of Lavaux. The desire to preserve intact the landscape existed before the UNESCO enrolment. The law for the protection of the site "Loi sur le plan de protection de **Lavaux**" (LPPL) was written in 1979, launched by the initiative "Sauver Lavaux" (Save Lavaux) and by a referendum, and it will become "LLavaux" in 2012.

The initiative of the application was started by a member of the Waldensian Great Council, who, visiting Asia and noting the terraced rice fields, made an interpellation to the Grand Council, asking the registration of Lavaux to the UNESCO list. In 2005, the local community instituted an association for preparing the application; at the end of 2005, the practice had been delivered, and in the summer of 2007, after some corrections, Lavaux was enrolled.

After enrolment, during the working at the management plan, five major guidelines were pursued: planning, economy, tourism, research and culture; and the landscape is the holistic picture of individual components. What is interesting the most to note is the community participation. The administration followed in fact a "bottom-up" methodology, meaning that the management of the site depends on the local community; even the drafting of the management plan of the site has followed this approach with a consultation among all stakeholders. In the interview, the managers of the site pointed out how important it is for the landscape protection to be in the hands of its inhabitants, and as well the support for small producers by institutions: as long as there will be wine-growing activities, the landscape will remain intact. Despite Lavaux has remained intact over the centuries, in fact, the local wine-growing economy remains a cause for concern; while erosion was slowed and accessibility improved, the risk of an economic crisis due to competition brought by large multinationals is still alive, with a consequent risk to the landscape. For this reason, the participation of all stakeholders in the governance of the territory was essential.

In June 2013, the "Lavaux Association Patrimoine Mondial" was created, composed of actors of different skills, and whose activities revolve around the wine-growing economy: for example, the development of an online website, www.lavaux-vignerons.ch, which provides information about the open cellars in the region. With a long-term view, there are planned strategies to raise awareness of landscape and/or oriented towards its protection over time, as the climate problem. The Association currently operates about forty initiatives, such as the creation of guided tours, the training of tourist guides, the project of a World Heritage centre, and the project "Lavaux en Scènes"; the latter being the reason of a phase of scientific research on Lavaux, made by the University of Lausanne, followed by a disclosure of the results to the public.

The interviews with residents and small producers of Lavaux (8, aged between 50 and 70) confirmed unanimous will and desire to maintain and enhance the landscape. Interviewed tourists (7, mostly from Asian countries) considered vineyards as the main characteristic of the landscape, and they expressed a desire to see Lavaux even more pristine than its current conditions, without cars or public transport.

Simplifying the evolution of a landscape, you can observe two distinct possibilities of reaction to human activities within or outside the site: development through adaptation to change, or collapse. Both options depend on the landscape management by its local community, intended as institutions, associations, residents, producers. The community assumes, in fact, a dual role in determining the evolution: it has modelled the landscape, it can enhance it and, eventually, it bears the risk of losing it.

In our case, Lavaux has embarked an adaptation path to development, while respecting the landscape values; the process has been summarized in Fig. 18.3.

The referendum initiative and the LPPL law of 1979 had as its objective the landscape protection; the subsequent projects—the application process and the Association Lavaux Patrimoine mondial—were no more intended as a protective

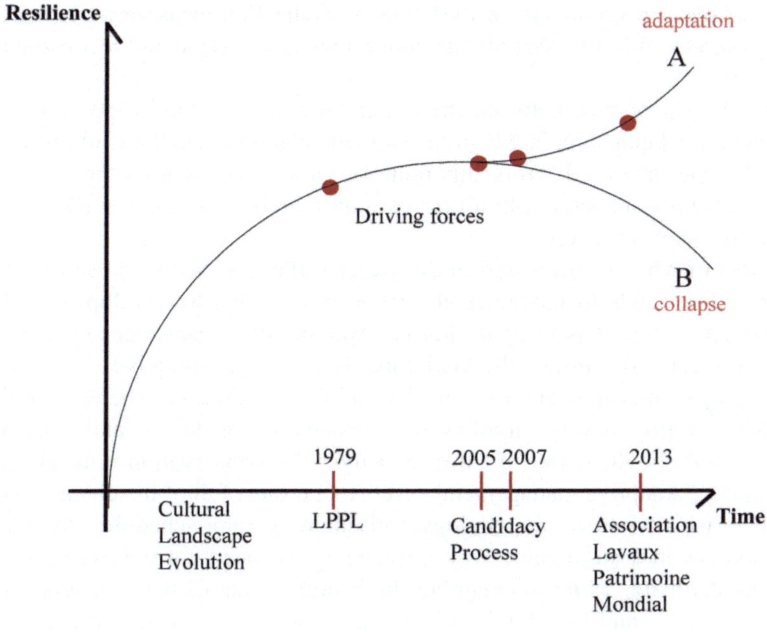

Fig. 18.3 The evolution of the cultural agricultural landscape of Lavaux

actions, but rather as resilient projects of adaptation to a growing development and of strengthening of a fragile economy.

The UNESCO application process is proposed as a quality response to development, globalization and international competition, which would otherwise deteriorate the landscape; moreover, starting from the local community, it embodies a participatory process and an opportunity to (re-) construction of territorial identity, which is the basis for the resilience.

Lavaux represents a *place* where the *community* was able to mitigate, adapt and recover from external pressures, promoting collective actions—although started by an individual initiative—in a way that reduces chronic vulnerability and facilitates inclusive growth within the candidature process, which has catalyzed the recognition process between the community and its territory.

18.5 Final Remarks

Talking about cultural landscape implies studying a network of relationships, not just an object; talking about resilient project means focusing on how to approach and react, rather than purely planning. Therefore, it can be said that, in order to protect cultural landscapes, it is necessary to apply a unique—yet ambivalent—perspective, both ecological and aesthetic at the same time, which is rooted primarily in its community and in the social system. This approach, the cornerstone of the project, should assume the role of regulator of relationships; it should include protective measures, conversion measures, revitalization measures, a balanced use of its resources and to be screened at continuing construction and reconstruction of its identity.

The interdependence between the community and the landscape is a complex and dynamic relationship, in which the strategic planning and the land management should be both analyzed. From this point of view, it is essential the role of community as significant actor both for the evolution of the landscape and for its care—connected to conservation.

Symbol of what we discussed in this paper is the UNESCO application process: this procedure tends to recognize the value (OUV) and to develop according to conservation, while it is itself a virtuous example of project management, which aims to protect and promote the landscape through local initiative.

The project management for agrarian cultural landscapes—being it linked to UNESCO or not—always implies the capacity of resilience and adaptability; therefore, we should define it through a dynamic conservation and an adaptive governance. "Adaptive management" refers to a project full of the characteristics of resilience, which follows the ecology and environmental schedules, by using the soil resources in a sustainable way, privileging the needs of development and of territorial diversity, while maintaining the historic value of the site, with its characterizing factors and the identity of the local community as overall pivots.

References

Augé M (1993) Nonlieux, Edition du Seuil

Cavallo A, Marino D (2013) Building resilient territories in the face of changes. In: Rural resilience and vulnerability:the rural as locus of solidarity and conflict in times of crisis. Proceedings of XXVth Congress of the European Society for Rural Sociology, Florence, 29 July–1 Aug 2013

Costanza R, Cumberland J, Daly H, Goodland R, Norgaard RB (1997) Introduction to ecological economics. St Lucie Press, Florida

Folke C, Hahn T, Olsson P, Norberg J (2005) Adaptive governance of social-ecological systems. Ann Rev Environ Resour 30:441–473

Holling CS (1973) Resilience and stability of ecological systems. Ann Rev Ecol Syst 4:1–23

Kallis G, Norgaard RB (2010) Coevolutionary ecological economics. Ecol Econ 69:690–699

Marino D, Cavallo A (2012) L'analisi della dimensione territoriale dell'agricoltura: una proposta di lettura. Agriregione Europa

Murray R, Caulier-Grice J, Mulgan G (2010) The open book of social innovation. Nesta, London

Norgaard R (1984) Coevolutionary agricultural development. Economic development and cultural change. The University of Chicago Press, Chicago

Ostrom E (1990) Governing the commons: the evolution of Institutions for Collective Action. Cambridge University Press, Cambridge

Sauer C (1931) The morphology of Landscape. Univ Calif Publ Geogr 2–2:19–54

Solymosi K (2011) Indicators for the identification of cultural landscape hotspots in Europe. Landscape Res 36(1):3–18

Chapter 19
Chestnut Management Practice as Tool for Natural and Cultural Landscaping

Giacomo Tagliaferri and Sara Di Lonardo

Abstract Chestnut (*Castanea sativa* Mill.) is one of the most economically multipurpose species of the Mediterranean region, important not only for the production of fruit and timber, but also for its contribution to landscape and environment and its socio-economic and cultural value as eco-cultural niche. For these reasons, chestnut management is today discussed since certain practices could affect soil properties and also be associated with losses of biological and cultural diversity, included provision of socio-cultural and environmental services. The objective of this work is to evaluate management options in chestnut coppices and of how much they could represent a choice from both plant ecophysiology and cultural point of views. Soil and plant ecophysiology measurements were performed in Tuscan Apennine areas before and after two cutting methods. Data showed that pollarded tree physiology was similar to pruning tree one's when it was wet but it was statistically different in the summer seasons affecting also soil water retention characteristics. These results demonstrated the importance of these measures in describing and indicating tree functional status. Moreover, the silvicultural practices modified not only the ecosystem attributes and the canopy cover characteristics but they also transformed the forest perception, an important factor to consider if expecting to engage local people in making a contribution for active landscaping.

Keywords *Castanea sativa* Mill. · Plant ecophysiology · Soil measurements · Eco-cultural niche · Silvicultural practices · People perception of forest · Biocultural diversity

---

G. Tagliaferri · S. Di Lonardo (✉)
Institute of Biometeorology-National Research Council (IBIMET-CNR),
Via G. Caproni 8, 50145 Florence, Italy
e-mail: s.dilonardo@ibimet.cnr.it

G. Tagliaferri
e-mail: g.tagliaferri@ibimet.cnr.it

M. Agnoletti and F. Emanueli (eds.), *Biocultural Diversity in Europe*,
Environmental History 5, DOI 10.1007/978-3-319-26315-1_19

19.1 Introduction

Sweet chestnut (*Castanea sativa* Mill.) is a long-lived plant (e.g., high presence of monumental trees all over the world) of ancient origins that has been intensively cultivated for centuries as a monoculture (coppices and orchards). Since Roman times, vast areas of Southern and Western Europe have been covered by groves and coppices of sweet chestnut trees. Having replaced the original broadleaved forest, they used to play a vital role in traditional agriculture also demonstrated by the numerous citations of poets and, in general, intellectuals (e.g., G. Pascoli, G. Carducci, G. Parini, G. Boccaccio as well as G. Orwell, J.-P. Sartre and H. Hesse) and in many folk legends. Chestnut cultivation was even more important in terms of producing a substitute for cereals ("bread of poors") than for the production of timber but it has also been used as medicinal plant for its antiseptic properties (Okuda 2005). With changing economic conditions and diseases (e.g., the onset of chestnut blight, ink diseases, or galls by Asian wasps) in the twentieth century, in European countries chestnut cultivation lost almost all of its economic and cultural significance; as a result, both coppices and orchards were abandoned and gave way to a more natural forest development (Arnaud et al. 1997; Conedera et al. 2001). In the past 20 years, there has been a renewed interest in preserving and replanting chestnuts that includes marketing strategies, landscape conservation and cultural history aspects (Eriksson et al. 2005; Bender 2010) since today chestnut forest ecosystems represent an important landscape component in the mountainous regions around the European Mediterranean basin and in the Southern Alps, covering more than 2.2 million ha (Conedera et al. 2004). Presently, the coexistence of large areas, both in abandonment and under traditional coppice system raises a lot of questions about management and forest policy of chestnut woods. On one hand, it has been highlighted that both, intensive exploitation and complete abandonment produce dysfunction at different levels such as bio-geochemical and water cycles, biological diversity and landscape patterns (Romane and Houssard 1995). Moreover, according to the IPCC (2007), the vulnerability of Mediterranean systems to water scarcity is predicted to increase in the near future as a consequence of larger inter-annual rainfall variability and higher frequency and intensity of extreme events such as droughts and heat waves. In this context, the identification and implementation of adaptation measures aimed at enhancing the resilience of the agroecosystems to water scarcity is a key priority to maintain both the quality and quantity of crop productions and protect water resources. On the other hand, besides the consolidated interests for chestnut coppice assortments, the demand for high-quality wood increased consistently in the last few years, bringing about an increased interest towards chestnut stands as well. But high-quality wood production cannot be achieved with the traditional short-rotation coppice, aimed at the production of poles and posts, or with the complete abandonment of silviculture. All these aspects have led to reconsider the management of chestnut woods: new management systems as the adoption of long rotation periods or the conversion into high forest in order to produce sawlogs have recently been adopted evermore frequently, beside the traditional coppice system with short rotation periods. They imply the choice of long rotation periods (30–50 years) necessary to produce sawlogs, and periodical thinnings, which, as a consequence of the ecological characteristics of the species and of technical and economical aspects, should be mainly low and heavy. This last element discriminates the system from the others proposed in the past and is characterized by frequent and low, light thinning. Although the interest for these new management options was emphasized even by the recent guidelines on forest policy, which suggest reconsidering management systems based on an over-exploitation of forest natural resources and adopting systems which take into account not only the natural ecological processes but also provision of socio-cultural and environmental services, the topic has not been investigated in detail. The aim of this chapter is to contribute to an evaluation of management options in chestnut coppices and of how much they could represent a choice from both plant ecophysiology and recovered soil water content and cultural point of views in term of forest perception.

19.2 Materials and Methods

19.2.1 Experimentation on Chestnut Trees

19.2.1.1 Study Area and Meteorological Data

Physiological responses of European chestnut to different management options such as pruning and pollarding were studied at three closed sites located within a chestnut coppice area (Luco del Mugello [44° 00′N and 11° 23′E; 360 m a.s.l.], about 3 ha). This area has a continental Mediterranean climate with a mean annual rainfall of about 996.2 mm; summertime is characterized by a high daily temperature range: the average difference between minimum and maximum temperatures is 15–17 °C. The meteorological station closest to the sites is placed in Borgo San Lorenzo (43° 57′ 31.72″N, 11° 23′ 29.5″E; 196 m a.s.l.); the meteorological station is part of the Tuscan Hydrological Service network and has been recording data almost continuously since 1933. Meteorological data used in this work are from the year 2002 to 2003 and the long-term series 1961–1990.

19.2.1.2 Field Experiment Protocol and Water Relations Analysis

Field data were collected from May to September in 2002 and 2003, when the coppice were about sixty years old. The experimental protocol included control trees (release of the standing crop and removal of dried up stems on the ground, only), pruning trees (light thinning, removal of second and third branches, poorly shaped, both withered and green) and pollarded trees (heavy thinning, total removal of the canopy with a directly cut over the shoot). Every tree cutting was done in correspondence of the vegetative rest, when all chestnut trees where without leaves (October–November 2001). Replicates of each plot (ten for each thesis) were located close to one another, to facilitate sampling and minimize microhabitat-induced variability in measured variables. Investigations at these sites focused on the responses to different type of tree management at the chestnut leaf (gas exchanges, such as transpiration and leaf water potential, temperature) and soil level (soil humidity at different depth and temperature). Replicates for each site were sampled monthly throughout the growing season, from May to September in both 2002 and 2003.

Leaf water potential and leaf gas exchange were measured in each experimental plot in cloud free days. Leaf water potential was measured on current-year shoots according to the procedures of Padgett-Johnson et al. (2000) at predawn (05.00–05.30 am and ended before sunrise) and midday (01.00–02.00 pm), using a pressure chamber (PMS, Instrumentation Co. Corvallis, OR, USA). The time between leaf excision and chamber pressurization was generally less than 15 s. Stomatal conduttance (g_s) and leaf transpiration (Tr) were determined on two leaves per sample tree with diffusion porometer (Steady State Porometer Li-1600, LI-COR Inc, Lincoln, Nebraska, US) within a 4-h period around solar noon (about 01.00 pm local time). Leaf temperature and relative humidity were also recorded from the Li-1600 output.

Soil moisture and temperature were measured directly with 5TM soil probes (Decagon Devices, USA). Measurements were taken at the base of each tree, at depths of about 20, 40 and 60 cm.

Correlations among all parameters were examined with Pearson's correlation analysis ($p \leq 0.05$) and t test using the statistical programme Statistica 6.0 (StatSoft 2003). Mean separation was obtained by means of the Tukey HSD test.

19.2.2 Survey About Forest Perception

A survey (questionnaire) was conducted among 204 people participating at the Chestnut Festival in Marradi (Italy) in October 2013. The main instrument of data collection was a structure questionnaire and administered interpersonally. The questionnaire was composed of 25 items grouped in five blocks: Identity (Q1–Q6), Knowledge and attitudes about utility of pruning (Q7–Q12), Knowledge of tree care practices (Q13–Q16), Tree appearance (Q17–Q23) [pictures with different chestnut tree management], Functions of forest trees (Q24–Q25). The questionnaire has been translated in English to be available for this article and it is freely accessible at web link: http://goo.gl/forms/Pg1x5wsKSE.

19.3 Results and Discussions

19.3.1 Field Experiment

Abundant precipitation was recorded from August to October 2002 (340 mm above the period averages) and the lowest temperatures in the same year in correspondence of July, August and September (−2.0, −2.3 and −3.6 °C lower that the monthly averages) (Table 19.1).

For this reason, soil water content remained quite high through the whole summer in all the three tree management areas (Fig. 19.1). Predawn water potential provided estimates of both the water status of accessible soil water and the degree of plant water stress (Slatyer 1967; Pallardy et al. 1991) by reflecting abundant soil water during the wet 2002 growing season (Fig. 19.2). In all treatments, high predawn leaf water potentials (−0.45 to −0.90 MPa) were observed during the wet 2002 seasons, showing no critical conditions for plant physiology, which is ascribed to water availability in soil layers. In the absence of drought, stomatal closure at midday was not observed and variations in g_s were primarily correlated with leaf temperature (Table 19.2a).

Results suggest that tree cutting management was the most critical parameter over the 2003 dry study period. The summer/autumn months were drier in 2003, with about 230 mm of rain below the period average recorded from May to September. Drought conditions were aggravated by high air temperatures in 2003 since mean temperatures in June, July and August were 2.8, 1.2 and 3.0 °C, respectively, higher than the monthly averages (Table 19.1). In the extremely 2003 year, soil moisture was low (Fig. 19.3) and water potentials decreased and varied from −1.46 to −1.72 MPa in late summer (Fig. 19.4). Predawn water potential values showed a small variation during the 2003 drought season, following the pattern reported by Cubera and Moreno (2007) and Moreno et al. (2007) for *Quercus ilex* trees growing in a drier site (precipitation: 500–600 mm). Midday water potentials did not decline noticeably during either drought in 2003, reflecting the isohydric nature of this species (Figs. 19.2 and 19.4). The maximal Tr and the minimum g_s were measured during August, which can be positive in terms of leaf cooling and association with soil drying but there were differences among the three tree managements since pollarded tree showed a different pattern since no g_s regulation was recorderd (Fig. 19.4). Stomatal conductance and midday leaf water potential were significantly correlated: the correlation was negative in pruning trees, indicating that midday leaf water potential was not limiting g_s while in control and pollarded trees it was highly positive (Table 19.2b) but the diversity in stomatal conductance could not be explained by the differences among soil water content at different depth for control, pruning and pollarded sampled areas (Table 19.3).

Table 19.1 Difference between meteorological data (monthly minimum, mean and maximum temperatures and cumulated precipitation) of the year 2002 and 2003 and climatic series values (1961–1990) from Borgo San Lorenzo meteorological station

	Meteorological series from 1961–1990				Differences between year 2002 and historical series				Differences between year 2003 and historical series			
Month	Temperatures (°C)			Cum. precip. (mm)	Temperatures (°C)			Cum. precip. (mm)	Temperatures (°C)			Cum. precip. (mm)
	Min	Mean	Max		Min	Mean	Max		Min	Mean	Max	
1	−0.5	4.3	9.1	96	−0.7	−1.0	2.7	−78	2.0	−0.5	0.5	−46
2	0.2	5.4	10.6	82	3.2	1.6	3.7	48	−2.5	−3.6	−1.9	−29
3	2.2	8.1	14.1	98	1.9	0.9	4.2	−96	0.2	0.2	3.8	−34
4	5.7	12.4	19.1	84	0.9	−1.4	−0.8	29	0.1	−1.9	−0.5	−3
5	8.8	16.2	23.7	70	2.2	−0.9	0.0	43	2.5	1.3	4.9	−60
6	12.4	20.2	27.1	56	2.6	−0.1	2.7	−22	4.0	2.8	7.2	−41
7	14.2	22.7	31.2	29	1.8	−2.0	−2.0	44	3.0	1.2	1.7	−13
8	14.0	22.6	31.2	73	1.9	−2.3	−1.8	125	4.9	3.0	3.5	−63
9	12.0	19.3	26.7	81	0.5	−3.6	−2.3	78	0.4	−1.3	−0.2	−53
10	8.2	14.4	20.6	94	1.9	−0.5	0.3	101	0.4	−2.3	−2.6	51
11	4.7	9.4	14.1	131	4.3	2.2	1.7	46	2.4	0.3	1.0	74
12	1.3	5.6	9.9	103	3.5	−0.2	−0.4	66	0.6	−0.3	1.3	−28
				996 (annual)				385 (annual difference from 1961–1990)				−244 (annual difference from 1961–1990)

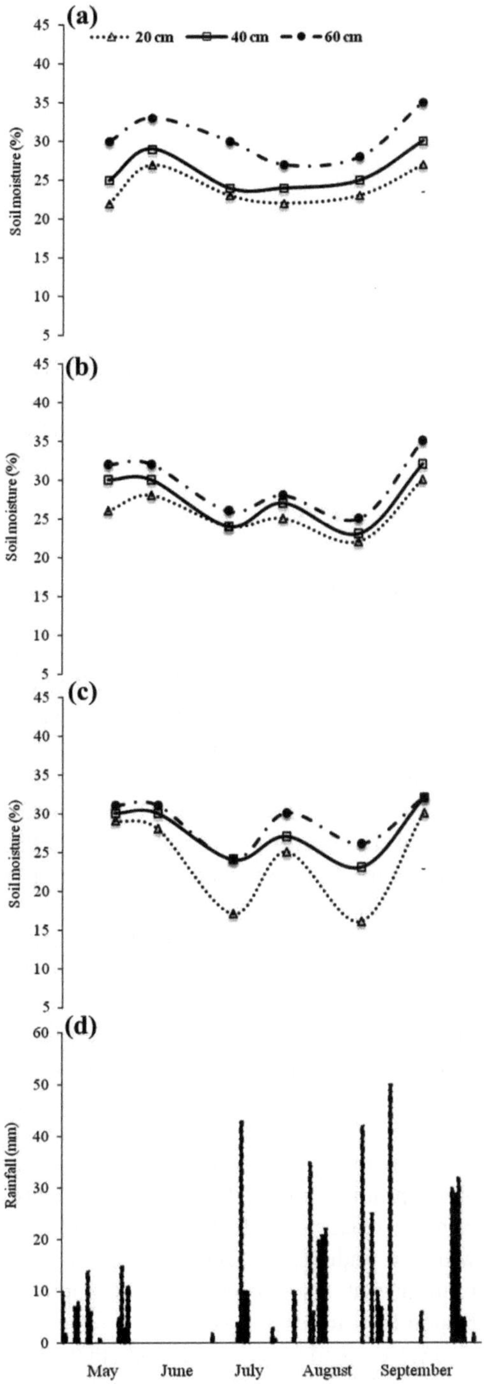

Fig. 19.1 Soil water content (%) at different depth **a** control, **b** pruning, **c** pollarded trees and rainfall **d** from May to September 2002, during the growing seasons. Error bars were little so they were not represented on the plot. Each point is a mean of five values

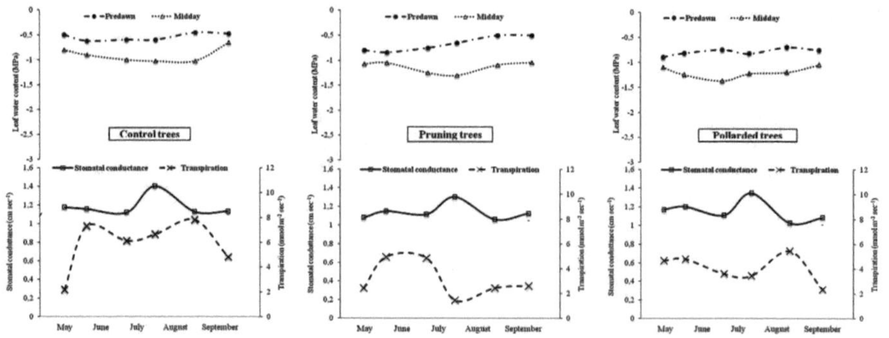

Fig. 19.2 Seasonal trends in predawn and midday leaf water potential and stomatal conductance and transpiration rate of control, pruning and pollarded trees during the 2002 growing season. Error bars were little so they were not represented on the plot. Each point is a mean of ten values

19.3.2 Survey

The socio-economic characteristics of the 204 respondents are summarized in Table 19.4.

Out of all the sampled respondents, 118 (58 %) of them were males while the remaining 86 (42 %) were females. The age group ranged from 18 to 75 years old and are equally distributed in all age classes. The level of formal education among the respondents was quite high ranging from primary to tertiary education (77 %). 130 respondents were self employed (34 %) and public workers (30 %) and the majority was from cities (68 %) surrounding the Festival area. The attitudes provided were generally favourable toward the importance and benefits of trees to the community. People placed great importance on trees; 80 % rating importance high or very high and only 7 % rating it low or very low. Tree importance was significantly affected by education, age and gender with trees generally rated more important as people' education increased. Most interviewed people (175; 86 %) thought that trees could improve the appearance of a community and maintain a healthy community environment. Respondents thought pruning can help tree health and so labours of maintenance are generally important but sometimes it would be better tree substitutions. The third item concerned "knowledge of tree care practices". Knowledge on tree care practices did not vary significantly with education but knowledge of urban trees and tree care was highly associated with thoughts about pruning utility. People preferred pruning trees but not pollarded ones. The primary reason people react negatively to pollarding was aesthetics going along with the findings presented earlier, in which respondents agreed about the utility pruning but also aesthetics was important. To explore this further, we asked respondents to react to seven pictures of trees pruned, pollarded or not cutted. They also were asked to choose the one that looks worst and the ones most like the utility pruning. When they were asked about the functions of trees, aesthetic and ecological functions were the most chosen answers. In Fig. 19.5 are summarized the

Table 19.2 Statistically significant ($p \leq 0.05$) correlation coefficients for stomatal conductance (g_s), predawn (Ψ_p) and midday (Ψ_m) water potential, transpiration (Tr) and leaf temperature (LT) of control, pruning and pollarded trees during the 2002 (**a**) and 2003 (**b**) growing seasons. ns = not significant

		g_s	Ψ_p	(Ψ_m)	Tr	LT
a						
Control	Stomatal conductance	–	ns	ns	ns	−0.55
	Predawn water potential	ns	–	ns	ns	−0.60
	Midday water potential	ns	ns	–	0.61	−0.75
	Tr	ns	ns	0.61	–	−0.33
	Leaf temperature	−0.55	−0.60	−0.75	−0.33	–
Pruning	Stomatal conductance	–	ns	ns	ns	−0.58
	Predawn water potential	ns	–	−0.09	−0.45	0.09
	Midday water potential	ns	−0.09	–	−0.44	0.08
	Tr	ns	−0.45	−0.44	–	−0.43
	Leaf temperature	−0.58	0.09	0.08	−0.43	–
Pollarded	Stomatal conductance	–	ns	ns	ns	−0.84
	Predawn water potential	ns	–	ns	0.27	−0.60
	Midday water potential	ns	ns	–	0.10	−0.73
	Tr	ns	0.27	0.10	–	ns
	Leaf temperature	−0.84	−0.60	−0.73	ns	–
b						
Control	Stomatal conductance	–	0.27	0.38	0.80	0.79
	Predawn water potential	0.27	–	−0.45	−0.03	0.33
	Midday water potential	0.38	−0.45	–	0.24	0.38
	Tr	0.80	−0.03	0.24	–	ns
	Leaf temperature	0.79	0.33	0.38	ns	–
Pruning	Stomatal conductance	–	0.24	−0.63	0.91	0.88
	Predawn water potential	0.24	–	−0.78	0.48	−0.20
	Midday water potential	−0.63	−0.78	–	−0.79	−0.21
	Tr	0.91	0.48	−0.79	–	ns
	Leaf temperature	0.88	−0.20	−0.21	ns	–
Pollarded	Stomatal conductance	–	0.65	−0.49	ns	0.91
	Predawn water potential	0.65	–	ns	ns	0.64
	Midday water potential	−0.49	ns	–	ns	−0.16
	Tr	ns	ns	ns	–	ns
	Leaf temperature	0.91	0.64	−0.16	ns	–

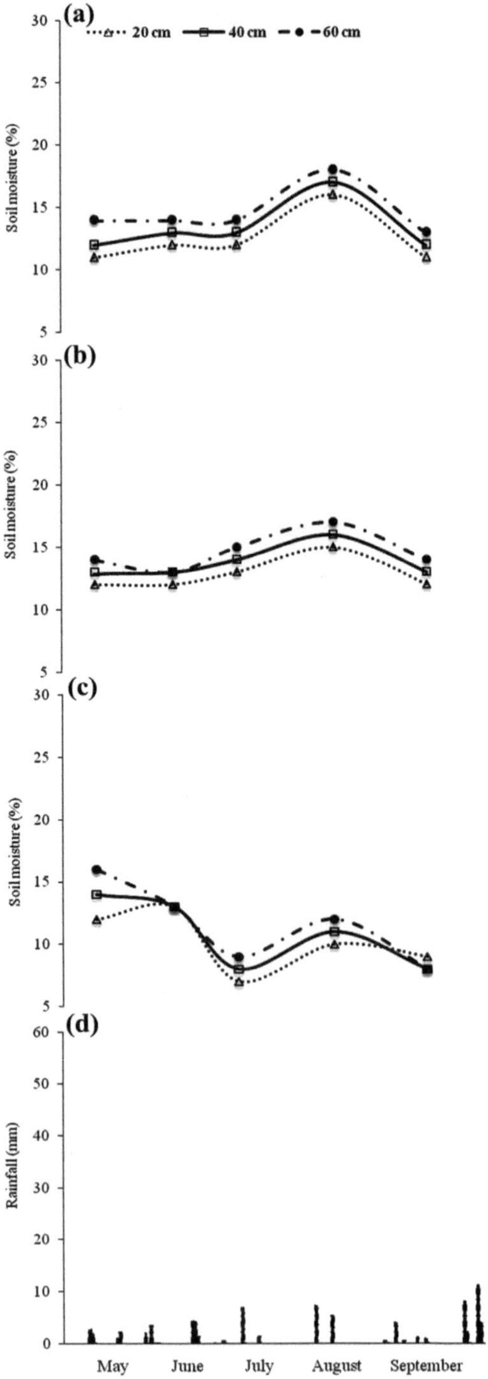

Fig. 19.3 Soil water content (%) at different depth. **a** control, **b** pruning, **c** pollarded trees and rainfall **d** from May to September 2003, during the growing seasons. Error bars were little so they were not represented on the plot. Each point is a mean of five values

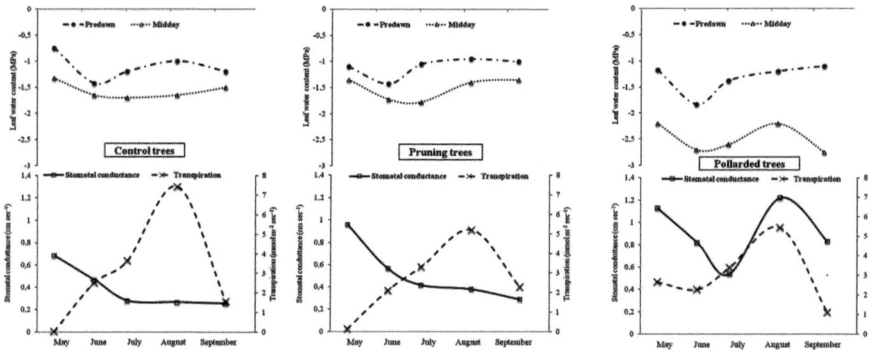

Fig. 19.4 Seasonal trends in predawn and midday leaf water potential and stomatal conductance (g_s) of control, pruning and pollarded trees during the 2003 growing season. Error bars were little so they were not represented on the plot. Each point is a mean of ten values

Table 19.3 Correlation coefficients between predawn water potentials of control, pruning and pollarded trees and soil water content at three different depth during the dry 2003 growing season

	Soil depth (cm)	*r*	*P*
Control	20	0.378	0.0053
	40	0.340	0.0021
	60	0.309	0.0001
Pruning	20	0.424	0.0034
	40	0.424	0.0026
	60	0.566	0.0012
Pollarded	20	0.031	0.0044
	40	0.298	0.0022
	60	0.444	0.0006

perceptions on the tree service functions. For respondents Trees were mostly important for the provision of nontimber forest products (e.g., mushroom, fruits and leaves) (35.5 %); according to the table, they are made up of recreation (22.1 %) and wild animal products (12.0 %). Regulation of watershed, provision of medicinal active molecules were also the mentioned by the respondents; they accounted for 6.5 % and 5.5 %, respectively.

Table 19.4 Socio-economic characteristics of respondents

Variable	Frequency	%
Age		
18–25	32	16
26–35	30	15
34–45	31	15
46–55	38	19
56–65	34	17
66 and above	39	19
Total	204	100
Gender		
Male	118	58
Female	86	42
Total	204	100
Educational qualification		
No education	2	1
Primary school certificate	45	22
Secondary school certificate	43	21
High school diploma	72	35
Degree in humanities	24	12
Degree in sciences	18	9
Total	204	100
Occupation		
Public worker	61	30
Private worker	69	34
No job	26	13
Retired	48	24
Total	204	100
Address		
City	138	68
Country	66	32
Total	204	100
Nationality		
Italian	189	93
Other	15	7

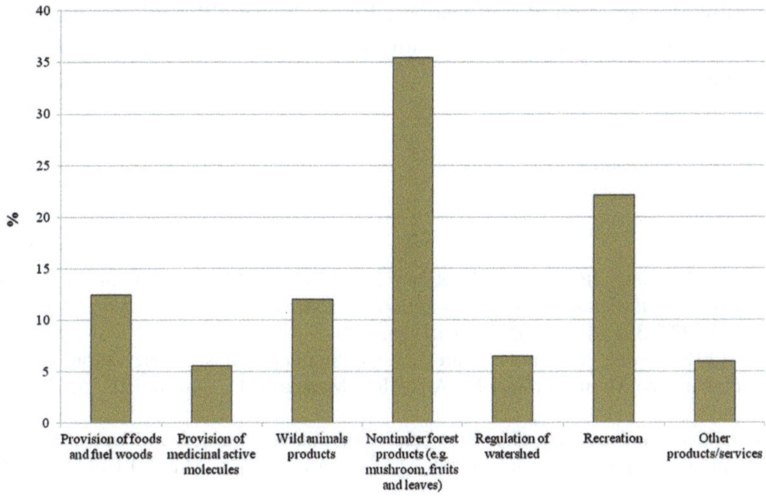

Fig. 19.5 Products and services derived from trees

19.4 Conclusions

This study provided links between plant ecophysiology-soil research and landscape social research and results can be resumed as follows:

(1) Correlations between soil and plant variables showed that the different chestnut management practices had also a strong effect on soil function conditions. In general, pruning increases soil moisture availability comparing with pollarding and reduces plant water stress by reducing stand transpiration as already stated by Bréda et al. (1995) but it can lead to higher soil evaporation as soil temperatures increase and increased understorey leaf areas and hence transpiration (Macfarlane et al. 2010; Martins et al. 2010). These changes may improve soil moisture availability and the water status of the retained leaves, as indicated by little increases in stomatal conductance in dry season. These leaf-level compensatory responses to pruning are often transitory, disappearing as the leaf area is rebuilt (Pinkard et al. 1998; Quentin et al. 2011). Pruning can immediately reduce transpiration by removing part of the trees' leaf canopy.

(2) The silvicultural practices modified not only the ecosystem attributes and the canopy cover characteristics but they also transformed the tree perception, an important factor to consider if expecting to engage local people in making a contribution for active landscaping, accordingly with a survey submitted in an important Chestnut Festival held in Italy in October 2013. Outreach and education about pruning techniques and the importance of maintaining clearance is very important and has to be taken into account also for regional development of tourism (Bender 2010). Landscape protection involves the

sustainable valuation of tree resource by considering the role of various actors, administrative planning, conservation authorities and organization as well as population.

References

Arnaud MT, Chassany JP, Dejean R, Ribart J, Queno L (1997) Economic and ecological consequences of the disappearance of traditional practices related to chestnut groves. J Environ Manage 49:373–391. doi:10.1006/jema.1995.0120

Bender O (2010) Sweet chestnut cultures in the Southern Alps: conservation and regional development. Eco Mont-J Protect Mt Areas Res Manage 2(1):5–14. doi:10.1553/eco.mont-2-1s5

Bréda N, Granier A, Aussenac G (1995) Effects of thinning on soil and tree water relations, transpiration and growth in an oak forest (*Quercus petraea* (Matt.) Liebl.). Tree Physiol 15:295–306

Conedera M, Stanga P, Oester B, Bachmann P (2001) Different post-culture dynamics in abandoned chestnut orchards and coppices. For Snow Landsc Res 76(3):487–492

Conedera M, Krebs P, Tinner W, Pradella M, Torriani D (2004) The cultivation of *Castanea sativa* (Mill.) in Europe, from its origin to its diffusion on a continental scale. Veg Hist Archaeobot 13:161–179. doi:10.1007/s00334-004-0038-7

Cubera E, Moreno G (2007) Effect of single *Quercus ilex* trees upon spatial and seasonal changes in soil water content in dehesas of central Western Spain. Ann Sci 64:355–364

Eriksson G, Pliura A, Fernandez-Lopez J, Zas R, Silva RB, Villani F, Bucci G, Casasoli M, Cherubini M, Lauteri M, Mattioni C, Monteverdi C, Sansotta A, Garrod G, Mavrogiannis M, Scarpa R, Spalato F, Aravanoupoulos P, Alizoti E, Drouzas A (2005) Management of genetic resources of the multi-purpose tree species *Castanea sativa* mill. Acta Hortic 693:373–386

IPCC (2007) Summary for policymakers. In: Climate Change, IPCC

Macfarlane C, Lardner T, Patterson K, Grigg AH (2010) A new model for predicting understorey leaf area from biomass in eucalypt forest to test the ecohydrological equilibrium theory. Methods Ecol Evol 1:371–379

Martins A, Raimundo F, Borges O, Linhares I, Sousa V, Coutinho JP, Gomes-Laranjo J, Madeira M (2010) Effects of soil management practices and irrigation on plant water relations and productivity of chestnut stands under Mediterranean conditions. Plant Soil 327(1–2):57–70

Moreno G, Obrador JJ, García E, Cubera E, Montero MJ, Pulido F, Dupraz C (2007) Driving competitive and facilitative interactions in oak dehesas through management practices. Agrofor Syst 70:25–40

Okuda T (2005) Systematics and health effects of chemically distinct tannins in medicinal plants. Phytochemistry 66:2012–2031. doi:10.1016/j.phytochem.2005.04.023

Padgett-Johnson M, Williams LE, Walker MA (2000) The influence of Vitis riparia rootstock on water relations gas exchange of *Vitis vinifera* cv. Carignane scion under non-irrigated conditions. Am J Enol Viticult 51:137–143

Pallardy SG, Pereira JS, Parker WC (1991) Measuring the status of water in forest trees. In: Hinckley TM, Lassoie JP (eds) Techniques and approaches in forest tree ecophysiology. CRC Press, Boca Raton, pp 27–76

Pinkard EA, Beadle CL, Davidson NJ, Battaglia M (1998) Photosynthetic responses of *Eucalyptus nitens* (Deane and Maiden) Maiden to green pruning. Trees Struct Funct 12:119–129

Quentin AAPOG, Beadle CL, Pinkard EL, Worldege D (2011) Responses of transpiration and canopy conductance to partial defoliation of Eucalyptus globulus trees. Agric For Meteorol 151:356–364

Romane F, Houssard C (1995) The sustainability of chestnut ecosystem. In: Romane F (ed) Sustainability of Mediterranean ecosystems. Case study of the chestnut forest, Report EUR 15727 EN, European Commission, DG XII, Brussels, Ecosystems Research, Report 19, pp 173–176. http://bookshop.europa.eu/en/sustainability-of-mediterranean-ecosystems.-ecosystem-research-report-19-pbCGNA15727/. Accessed 1 Nov 2014
Slatyer RO (1967) Plant-water relationships. Academic Press, London, p 366

Chapter 20
Does the Conservation Status of a Caledonian Forest also Indicate Cultural Ecosystem Value?

David Edwards, Tim Collins and Reiko Goto

Abstract The Black Wood of Rannoch is one of the largest remnants of ancient, semi-natural Caledonian pine forest in Scotland. It is culturally important as a location for aesthetic and spiritual experience, artistic inspiration, biocultural heritage and a sense of identity and belonging by local communities and urban visitors. Despite this, the values that inform its management are almost exclusively those associated with biodiversity conservation. This focus has proved very effective in protecting the forest from destructive economic interests. However, it could have two negative effects. First, it could continue to downplay the significance given to the full range of cultural benefits, and constrain their recognition and expression to those that are realised by a policy of conservation. Second, it could be used as a 'precautionary principle' to justify exclusion of visitors. This paper reports on a series of workshops, discussions, events and residencies, led by two environmental artists, to prompt a debate that rethinks existing narratives of the value and management of the Black Wood and the Caledonian forest more broadly. We analyse the range of cultural ecosystem services associated with the forest, and discuss how these might be realised through six alternative management scenarios. In conclusion, we suggest that the new concept of biocultural diversity offers a unifying principle with which to guide a new vision of social and ecological restoration.

Keywords Caledonian forest · Scots pine · Cultural ecosystem services · Aesthetics · Woodland history · Biocultural diversity

D. Edwards (✉)
Social and Economic Research Group, Forest Research,
Northern Research Station, Roslin EH25 9SY, Scotland, UK
e-mail: david.edwards@forestry.gsi.gov.uk

T. Collins · R. Goto
Collins and Goto Studio, 1m Whisky Bond Building 2 Dawson Road,
Glasgow G4 9SS, Scotland, UK
e-mail: timothymartincollins@gmail.com

R. Goto
e-mail: goto.reiko1801@gmail.com

M. Agnoletti and F. Emanueli (eds.), *Biocultural Diversity in Europe*,
Environmental History 5, DOI 10.1007/978-3-319-26315-1_20

20.1 Introduction

At around 1000 ha in size, the Black Wood of Rannoch in Highland Perthshire is one of the largest remnants of the ancient Caledonian pine forests that once extended across much of Scotland after the last glaciation. Designated as both a site of special scientific interest (SSSI) and a special area of conservation (SAC), it is of undoubtedly high biodiversity value and a focus for ecological restoration of a network of native pinewood habitat across the region (FCS 2009). While typically referred to as pine forests, and primarily comprising Scots pine (*Pinus sylvestris*), the Caledonian forests are diverse ecosystems also containing birch (*Betula* spp.), rowan (*Sorbus aucuparia*) and juniper (*Juniperus communis*) and provide habitats for iconic and threatened wildlife such as capercaillie (*Tetrao urogallus*), Scottish wildcat (*Felis silvestris silvestris*) and pine marten (*Martes martes*). Although less frequently acknowledged, the Caledonian forests are also culturally important: as a tangible embodiment of biocultural heritage, as a location and inspiration for aesthetic and spiritual experiences and artistic expression, and as places which members of local rural communities and those further afield identify with and feel a sense of attachment and belonging. In 2014, following a public vote, the Scots pine was declared the national tree of Scotland by the Scottish Government. Despite this, the social and cultural relationship to the Caledonian pinewood ecosystem is limited. For example, it is neither an image nor a concept that has much traction in archives, museums and galleries or parks and botanic gardens in the cities of Scotland. In the Black Wood, the values and benefits that have the most influence on decision-making about how the site is managed are those associated with biodiversity. This focus has been extremely effective in protecting the forest from destructive economic interests over the last 40 years. But it also shapes and constrains the ways in which its cultural benefits are understood, discussed and realised. It is argued that the protectionist policy in the Black Wood, which allows the forest to 'follow nature's way', is also the best way to maintain its cultural value. In this view, as reflected in our title, the high conservation status of a forest like the Black Wood is also an indicator that the forest has high cultural value. Is it that simple? What does cultural value actually mean? Is the only important cultural benefit of the Caledonian pinewoods the aesthetic experience of being in close contact with nature? Who knows about this opportunity and has access to it? What about the fact that the Black Wood is anything but an untouched wilderness—the product of interaction with humans within a historical context of conflict and change over several centuries? What about the diverse meanings and associations of members of local communities who visit or work in the forest? How important are representations of the forest in art, literature, advertising, media, museums and botanic gardens? How might the notion of biocultural diversity help our understanding?

Such questions formed the basis of a year-long creative enquiry into the ecological and cultural values associated with the Black Wood between July 2013 and July 2014. The project involved a series of residencies, workshops and discussions led by environmental artists, Tim Collins and Reiko Goto, in collaboration with David Edwards, a social scientist from Forest Research (the research agency of the Forestry Commission) and a diverse range of other partners, including Forestry Commission Scotland (FCS), who own and manage the Black Wood, the Landscape Research Group, the Perth and Kinross Countryside Trust (a local NGO concerned with access and planning), local artists and residents in Kinloch Rannoch, and academics from the arts and humanities. Over time, the effort would focus upon the development of a 'Future Forest' workshop held over two days in Kinloch Rannoch in November 2013. The event created a space for participants to reflect on their own experiences of the forest and imagine alternative futures that protect its ecological value, while exploring a more robust cultural relationship. The next sections focus on three types of cultural value or benefit that were identified during the project as being realised to different degrees through engagement with the Black Wood: 'aesthetic/spiritual', 'biocultural heritage' and 'community-held' values.[1] The discussion explores how these categories relate to recent literature on cultural ecosystem services (Church et al. 2014), how they might be delivered through different future management scenarios and how a process of realising the cultural benefits of Caledonian forests might be facilitated both locally and nationally. The conclusions return to the question in the title, and consider how the notion of biocultural diversity might contribute to the debate. Our account raises as many questions as it answers, but we feel this is appropriate during this early stage in what we hope will become a national-level conversation about the ecological and cultural values of the Caledonian forests of Scotland.

20.2 Engaging with the Black Wood (and with the People Who Care About It)

20.2.1 Aesthetic and Spiritual Experience

> Anyone who has ever stepped into one of the ancient pinewoods of Scotland will understand their allure: they are, quite simply, magical landscapes, resonating with almost primal echoes. But the pinewoods are also landscapes of contradiction—for alongside this aura of apparent resilience is a tender fragility. To journey through the woods is to immerse yourself in this blurring of time and meaning. Their stillness is inspirational: the outside world just falls away and you are left alone, awed—and humbled—by their enduring magnificence (Vanessa Collingridge in Bain 2013).

This emotive account of the experience of visiting a Caledonian forest, taken from the foreword to a recently published book, sits easily within the wider public

[1]Church et al. (2014: 24–5) define cultural values as "the collective norms and expectations that influence how ecosystems accrue meaning and significance for people", and cultural benefits as 'the dimensions of human well-being that can be associated with and that derive from… interactions between people and the natural environment' (cf. Kenter et al. 2014).

Fig. 20.1 'Gunnar's Tree' in the Black Wood of Rannoch, with community members, November 2013 (© Collins & Goto Studio, 2013)

imagination of places like the Black Wood: we like to think this is what we would experience during a visit to an ancient pine forest, even if we might never do so. How accessible is an experience of this kind to members of the public? Is this description really valid? How many people have had the opportunity to find out for themselves? The Black Wood borders the southern shore of Loch Rannoch (see Fig. 20.1). Whether one arrives by train, car, foot or bicycle, visitors can struggle to find their way into the forest. There is one sign, easily missed as it is set back and parallel to the Loch road. Another can be found half a mile down a rough track. To get into the forest one follows any one of four trails that move in a southerly direction. Moving gently uphill, the forest is alternately open and closed with a mix of birch and pine, and some rowan, across a range of age classes from saplings to mature trees. The most memorable trees are the huge 200–300-year-old pines with sprawling limbs, colloquially known as 'granny pines'. One is immediately struck by the relationship of the forest to its curious topography—a mix of small glacial moraine deposits or hillocks with a seemingly endless repetition of smaller hummocks of thick blaeberry, cowberry, bracken and heather, formed over large rocks and tree stumps, creating an unusual 'lumpy' forest floor that adds texture to the rolling mound-and-hollow topography. But it is the granny pines that capture one's attention: they are far from the traditional foresters' ideal of a tall straight trunk. Why are they here and why so many of them? Moving through the forest along the western-most trails the walker will notice changes to the underlying moraine topography, with alternating wetter and drier areas. Walking south, the forest opens up to reveal a bog clearly visible through the trees. Those who explore that area will discover the remnants of an old homestead site on higher, drier ground. Moving further east along the trail, the observer will realise that the ground vegetation changes, with wetland grasses replacing the robust blaeberry and cowberry understory in reaction to the increasingly wet ground underfoot. Further along the (raised and dry) trail, there are two spots where small open streams are first heard, then seen. These wet/dry transitions do two things. They provide a gradation of microhabitats that support a range of species. But they also provide an aesthetic complexity, which rewards the eye and ear, the nose and the kinesthetic (bodily) senses of those who walk attentively through this forest. At the centre of the forest, an area known locally as the 'potato patch' is often attributed to war-related food production in the first part of the twentieth century, but historical maps reveal that it was actually cleared by 1906, apparently for some other purpose. The potato patch is notable today for its broad even-aged stand of pines that looks like a plantation, straight and tall with little understory diversity. It provides an aesthetic counterpoint to the rest of the forest. What we are trying to establish here is that the Black Wood has a powerful aesthetic presence. We argue that it 'returns ones gaze': it is woodland of sufficient complexity that it cannot be seen in a day, and indeed evolves in one's eye and mind as it is visited over seasons and years. People look at forests in different ways: the appreciation of botany and birds or insects and fungi is just one framework for looking carefully, and considering the life of a place. One argument worth considering is that, what science tells us through data and detail needs to be complemented through aesthetic attention. In the layers of organisms, divergent reproduction cycles and ever-changing seasonal conditions lie a complex aesthetic experience that repays attention over time.

It may be tempting to go further and suggest that the dominance of a scientific discourse of conservation has blinded us to the appreciation of cultural value, by objectifying nature, and de-sensitising us from a holistic aesthetic engagement. But this is evidently not the case—at least not in such a simplistic way. Those responsible for the Black Wood were among the most passionate advocates of the aesthetic and spiritual values associated with the experience of visiting the wood. They saw their role as (at times, misunderstood) guardians of the natural features that give the wood these values. Local managers indicated that they personally valued the wood as much (if not more) for its cultural importance as its environmental importance. This appreciation was brought home to us clearly by the FCS manager who accompanied us during one of the first walk-and-talk visits. He spoke of the granny pines as "the fantastic matriarchs of the forest, something to respect, to be treated honourably", and the need to find the Black Wood "emotionally, intellectually and spiritually", "to make room for imagination", "to come feel the forest" and see the trees "as things of great beauty" (Collins et al. 2014). Perhaps, the answer to the question posed in the title is 'yes'—high biodiversity value means high cultural value. Perhaps, the forest needs only be managed for biodiversity, because in doing so its cultural value is realised and protected by default. Importantly, the cultural value referred to here is a kind of aesthetic experience that is compatible with, and realised through, the current policy of biodiversity conservation. Is something lost as a result of recognising only this particular kind of experience?

20.2.2 Understanding Biocultural Heritage

With access to the Forestry Commission Library and key staff members, the authors began to think in terms of the three-hundred year life cycle of a Scots pine tree and its changing relationship to the use of the land. Over three centuries the forest would have been picked through consistently. But what was it that kept the forest structure sufficiently open, especially during the eighteenth century, to produce the 'granny' form, rather than a tightly spaced forest where trees must grow straight and tall to reach the light? In a specific response to this question, Peterken and Stace (1987) conducted a resurvey of historic experimental plots, and identified a generation of trees dating from 1780 to 1835 with a predominantly straight form characteristic of close-grown trees. This, however, was preceded by an older generation dating from 1650 to 1700, which appears to have originated in an open condition, and would have allowed trees with the characteristic low branching to develop. The effect would have been accentuated by subsequent selective felling of any well-formed old generation trees (Peterken and Stace 1987). The eighteenth century would have been a difficult time to live in the Rannoch Valley, which was seen as a local stronghold of support for the Jacobite risings against the ruling government. Estate land was forfeited to the crown repeatedly and a barrack was built in the valley to suppress resistance by force. Agents brought into manage the forfeited estate were particularly worried about keeping tenants' goats out of the Black Wood. Tenant livestock would not be excluded from the forest by fencing until the later part of that century (Lindsay 1974). Speaking of Caledonian forests, in general, the historian Smout (2000) described their use by tenants for shelter and feed for goats, cattle, sheep and horses. The combination of grazing pressure and wood extraction would have a particular impact on regeneration and growth in the Black Wood, shaping both biodiversity and the form and perception of the forest overstory, its understory and its related ground flora. Partly as a means to repress the Jacobites, the Highland Clearances across northern Scotland resulted in forced evictions of Gaelic-speaking tenant farming communities in the Rannoch area. By the middle of the nineteenth century the clearances were just about complete. With vast herds of sheep replacing people, the forest was again struggling to regenerate itself. Steven and Carlisle (1959) tell us that at the end of the nineteenth century another fence was erected to create a deer forest for the hunting estate. Young trees that might have had a chance with free range sheep would have no chance with an enclosed population of red deer! In the twentieth century the fences were down again, deer were free to roam, but there were more sheep on the land than at any time in the past. Thus generation after generation of creatures would have eaten the young trees. Cattle, goats, sheep and deer would have changed the shape of saplings and older trees as well. In addition, there is evidence that capercaillie (*Tetrao urogallus*) and black grouse (*Tetrao tetrix*) can damage leading shoots, while strong winds, wet snow and ice storms will also have caused the loss of branches and whole trees over the years. These are the conditions that shaped a semi-open canopy where the few trees that could get away were able to grow into the light both horizontally as well as vertically. Through the years, the straight trees would have been targeted for timber harvesting. Overall, the historical narrative can be seen as a fluctuation between periods of exploitation and protection, as outlined in Box 1. It is interesting to think that the aesthetic form, considered to be so attractive in the Black Wood today, actually embodies, among other factors, the contested land use history of this region of Scotland. And yet, its representation is typically underpinned by assumptions of untouched 'wilderness' which shape our aesthetic experience and the meanings and associations we attribute to the forest. Does it matter that so few people are able (or encouraged) to interpret the physical form of places like the Black Wood in terms of their conflicted social and political history?

Box 1. The Black Wood of Rannoch—history shapes the forest form

1439	Rannoch estate given to the Robertsons of Struan for apprehending the murderers of King James 1st
1689–1745	Estate is forfeited in 1689, 1715, and again in 1745. Exploitation of 960 trees per year
1745	The forest was much feared by local people as a haunt for 'broken men', outlaws from the failed Jacobite rebellion. A garrison was established and Jacobite homes were burned
1750	The forest was judged to be in bad shape. Yet forfeited estates initiate felling at 1200 trees/year
1757	Sawmill burnt down by an evicted tenant. New sawmill built 1758
1781	The forest is completely enclosed to protect it from domestic animals
1784	The estate is returned to the Robertsons (until 1857)
Late C18	Swine put in forest to break up soil for regeneration
1803–15	Napoleonic Wars and significant felling occurs. Canals dug to float timber to market
Early C19	Opened again to farm stock; sheep farming in full swing
Mid C19	Highland clearances. Radical increases of the number of sheep. Human population of Rannoch is less than half the original number
1895	Enclosed as a deer forest. Roads constructed along canals
1889–94	1000 trees felled for West Highland Railway
1918	Opened for general grazing
1939–45	8000 trees cut for the 2nd World War effort
1947	Protected again from deer
1957–67	5000 trees cut by Forestry Commission
1974	Fully protected as Forest Nature Reserve, later becoming an SSSI

In addition to the tangible biocultural heritage embodied in the forest structure and granny pines, there are several unscheduled ancient monuments in the Black Wood that are little understood or appreciated. Within the local Perth Museum and Art Gallery there is an understanding of the scientific import of the site amongst the curators but a paucity of records in their entomology or botany collection.

The Jacobite history of conflict in Rannoch, coupled with the histories of the clearances and the practical management of the forest and landscape, would have constrained ‘cultural interests’ at the time that these collections were being developed. That said there are important archival records of the Black Wood, which document the repeated forfeiture from its owners, the Robertsons of Struan (Millar 1909).[2] There are histories of the brutal clearances and various Gaelic language translations of both histories and tales of the region found in the Inverness Gaelic Society archives as well as in the Scottish Studies Library at the University of Edinburgh. Together, these resources and artefacts remind us of the largely lost intangible cultural heritage associated with the Black Wood—the traditional forest and pastoral management practices that were historically embedded in the local society and economy, and the knowledge and world views associated with them. Once the Black Wood was part of an organic cultural landscape, but now it is best defined as ‘relict’ (UNESCO 2008). The Gaelic place names throughout the landscape, translated as part of this project, hint at the heritage lost in particular during the Highland Clearances (see Fig. 20.1; Collins et al. 2014; Murray 2014). Does it matter that so little attention is given to these features, and to their significance in understanding the current forest form and its changing place in the local culture and economy? As the southern-most large Caledonian pine forest, the Black Wood survived in modern times (where others did not) partly due to isolation and restricted access. There is one road in and out of Rannoch. It was spared from harvesting during World War I, although there was a significant harvest in the final years of the World War II. The Forestry Commission acquired the forest in 1947, with Lord Robinson, the FC Chairman, visiting the wood in that year and declaring that “this piece of old Caledonian Pine Forest should, if possible, be preserved” (Peterken and Backmeroff 1988). However, the actuality did not meet the ideal. Between 1947 and 1975 the Forestry Commission felled about 500 m^3 (from 5000 trees) to promote regeneration and clear up dead, dying and windblown trees. During the 1960s they planted exotic conifers on wetter sites unsuitable for local varieties of Scots pine (FCS 2009). By 1973, the Forestry Commission Conservator Gunnar Godwin was newly in charge of the Black Wood. With an awareness of this history and what he saw as potential threats from within his own organisation he began working closely with the Nature Conservancy Council to establish an agreement for the long-term management and conservation of the Black Wood in 1975. This was followed by a series of conservation-based remedial actions in the 1980s. The current management plan identifies long-term plans to extend the Black Wood significantly further, establishing a stronger functional link to neighbouring forests (FCS 2009). The Black Wood restoration is a success story that needs to be carefully described and widely communicated. It is interventions like these over the last 50 years that have contributed to the good condition of the Black Wood as we find it today. Recently, in recognition of Gunnar Godwin’s achievements, a plaque was erected in the forest at the base of an old pine tree (see Fig. 20.2). Does it

[2]See: http://archive.org/stream/selectionofscott00millrich/selectionofscott00millrich_djvu.txt.

Fig. 20.2 Initial mapping of translated Gaelic place names in Rannoch, showing the Black Wood to the south of the loch (© Ordnance Survey License number 100021242. Translation by Beathag Mhoireasdan, Collins & Goto Studio, 2014)

matter that so few people are aware of the historical tensions between conservation and production, and the precarious existence of the Black Wood at specific moments in its twentieth century history? These historical narratives are currently not how the Black Wood is portrayed in visitor interpretation and other representations, which tend to focus on biodiversity and naturalness. Yet, the long history of human intervention in the Black Wood is a far cry from an idealised notion of wilderness that lies behind public perceptions of Caledonian forest and underpins aesthetic appreciation. The granny pines have undoubtedly been shaped by human intervention, and if future management pursues a 'do nothing' route then they could eventually die out leaving them absent from the forest landscape. With such a limited biocultural record of the Caledonian forest in botanic gardens and museums, and without publications, memories, songs and stories, the Caledonian forest as a contemporary, living memory will also disappear. Debates within environmental aesthetics explore how aesthetic appreciation is shaped by knowledge and experience of a place (Eaton 2004). If it is not a wilderness, but a cultural landscape, shaped by humans over centuries, then how might this change the aesthetic and spiritual experience of visiting the forest? Perhaps, for some, the experience is enhanced by the idea that nature is regaining control over a forest with a history of human exploitation (Kirchhoff and Vicenzotti 2014).

20.2.3 Listening to Community-Held Values

In the Black Wood, FCS management practices meet the standards of the Land Reform (Scotland) Act of 2003, in which there are no fences or gates impeding movement. The current Black Wood Management Plan limits changes and 'formal recreational development' within the forest (FCS 2009). The plan affirms guided tours on request, and there is a threshold reserve sign and interpretation board describing timber harvest canals from the 1800s. As understood through dialogue with FCS and SNH, there would be little encouragement of or invitation to access; there would be no maps, or interpretative publications that might make way-finding easier, even where there were 'adopted core paths' or 'asserted rights of way' that were established over centuries during the transhumance. There would be no changes to public awareness that might increase public use of the Black Wood. One was left with the impression that the official invitation into the Black Wood, as it exists today, only addresses a limited group of interests.

The restrictions on awareness and access are attributed to the status of the site as a Special Area of Conservation regulated by SNH. The authors were told that any proposed changes that might affect the public awareness of the Black Wood or increase the number of people accessing it would need to provide scientific evidence that no possible harm would come to that sensitive ecosystem. The claim was that there are peripheral restoration areas that are more amenable than the 'jewel-like' core areas. Yet, the ecological importance of the Black Wood might be comparable to ancient pinewood sites at Glen Tanar and Glen Affric further north where responsible public access is outlined in management plans and fully supported. The culture of scientific conservation strategically embraced by Gunnar Godwin to protect the Black Wood from economic interests had over time (and without malice or intent) allowed the discourse and practice of conservation science to exclude nearly all other social and cultural interests.

Working closely with partners, the authors sought to establish an artistic and cultural discourse that might complement the dominant ethos. A first attempt at holding a collective 'walk and talk' in the Black Wood, with people invited to represent a range of interests, revealed overt tensions, but also a sense of unacknowledged common ground. Plans for the 'Future Forest' workshop evolved from this initial encounter. Thirty participants attended the two-day workshop, which was held in November 2013 along with a public discussion and visits to the forest. Invited participants included arts practitioners, humanities scholars, foresters and ecologists, government agency and NGO representatives, and local residents. The goal was to use the Black Wood as a setting to examine the ideas, knowledge, values and experiences that enable and constrain public access to, and awareness of, forests with significant ecological and cultural import. The workshop presentations began with local stakeholder perspectives, then ecological perspectives, a broad set of cultural perspectives, and followed by intensive 'Future Forest' break-out groups. The participants were introduced to a range of ideas, opinions and proposals. Break-out groups were asked to proceed from initial scoping of what mattered to them about the Black Wood, through more detailed consideration of problems and opportunities, and visions of the future forest that could be shared by most participants. Each group was given maps of the management zones in the forest and surrounding catchment, and a mapped overview of current strategic management objectives. These helped participants locate aesthetic and cultural interest and access opportunities within the forest. The event was recorded by video, and conversations were later transcribed and analysed as part of a residency at Forest Research.

The most important outcome from the workshop was that a constructive dialogue had now begun between diverse interests. There was agreement on a number of precepts. In particular, nothing should be done to change the generative capacity and biodiversity, structure and character of the Black Wood. Overall, the workshop and related events uncovered community-held values and perspectives that saw the Black Wood as a place of biocultural importance, and a desire to realise these values through closer engagement with the forest, and with those who protect and manage it. A number of tangible proposals emerged:

a. *Deep mapping*: A multi-layered, biocultural map, compiled using GPS and linked to texts and images, that celebrates the Black Wood. It would draw on workshops and events in the forest with local residents and experts to explore what is known, what is suspected, what is invisible to the untrained eye and what is not known but should be.
b. *Forest planning*: An inclusive planning process with FCS and other agencies to establish productive working agreements regarding awareness, access and branding. It would build on the deep mapping exercise, and involve a programme of rigorous discussion around agreed issues with speakers brought into provide critical and independent insight.
c. *Forest Way initiative*: A proposal to link specific areas, forests and communities through a landscape trail defined by arts, culture and Gaelic themes, including the life and poetry of Duncan Ban MacIntyre (1724–1812) who was born at neighbouring Loch Tulla. The project would require strong community support coupled with academic and professional expertise.
d. *Arts, humanities and ecology residencies*: An interdisciplinary residency programme to help establish new social and cultural relationships with the Black Wood. It could develop the ecological evidence base required for forest managers to move beyond the 'precautionary principle' while ensuring the Black Wood continues to be recognised and celebrated for its ecological value.

To conclude, many locals (and visitors) benefit from a sense of identity with, attachment to, and ownership of the Black Wood, and when the project began some residents saw its protectionist policy as a barrier towards fully realising these benefits. Despite understandable anxieties within FCS and Scottish Natural Heritage (SNH), it turned out that nobody involved wanted any harm to come to the wood, or substantially increase visit numbers, or introduce intrusive signage or trails. The ideas of local residents were largely compatible with a successful policy of biodiversity conservation.

20.3 Analysis and Discussion

To recap, we have argued that the values and benefits that justify decisions about the Black Wood are almost exclusively those associated with biodiversity conservation and ecological restoration. Over the last 40 years, this focus has proved very effective in the face of destructive economic interests. However, it could have at least two negative effects. First, it could continue to downplay the significance given to the full range of cultural values and benefits, and constrain their recognition and expression to those that are realised by a policy of conservation, i.e. a science-informed (cognitive) aesthetic experience limited to a few knowledgeable individuals. This 'cultural problem' is not restricted to those who make the decisions; it is reflected across society as a whole, which over centuries has gradually lost its former cultural and historical connections with the rural environment. During the Future Forest workshop, the depth of the responses from participants representing the arts and humanities proved to be very useful.

Second, it could be used as a 'precautionary principle' to justify exclusion of visitors ostensibly to protect the ecological integrity of the forest. As argued above, the risk that additional visitors, if properly managed, would pose to biodiversity value is unclear (see Marzano and Dandy 2012). Is there any reason why levels of public access cannot be similar to other comparable pine forests with relatively high visit levels? Exclusion of visitors might also be seen incidentally by decision-makers as a means to protect the 'wilderness' experience of the few people who already know and value the forest. But surely, to some extent, cultural value (unlike biodiversity value) is proportional to the number of people who are made aware of it, and can access it, appreciate it and respect it? An attitude towards the Black Wood as a potential cultural resource for the nation might take a more proactive approach to access, interpretation and engagement, while still successfully maintaining its protectionist functions (the legitimacy of which no one involved in the project has questioned). If the management of Caledonian pine forests, and protection of the Black Wood, was explicitly seen to be justifiable through a more equal balance of both biodiversity and cultural benefits, then it would no longer be appropriate for objectives to be defined primarily by a narrow group of expert interests. Indeed Mason et al. acknowledge that "Native pinewoods are an emblem of the natural and cultural heritage of Scotland" and that "discussions about their management have tended to take place within a relatively narrow community of landowners, foresters and conservationists" (Mason et al. 2004). To appreciate the full range of cultural values and benefits, and to incorporate them into decision-making, a debate is needed that goes beyond those currently involved. This project has prompted such a debate in relation to the Black Wood. What might be the outcome of this debate into the future, both locally and nationally? First, it could explore further the multiple meanings of 'cultural value' and create a shared language for discussing it. Second, it could explore potential future scenarios and their impacts on different types of value. Third, as well as allowing us to recognise existing ecological and cultural value, it could help create new values and meanings through an iterative and inclusive process. These points are outlined in turn below.

20.3.1 What Do We Mean by 'Cultural Value'?

Current debates around the cultural value of the environment are typically framed in terms of cultural ecosystem services (CES) (Millennium Ecosystem Assessment 2005). Church et al. (2014) provide some of the clearest thinking on how to define and assess these services in ways that support land management and policy-making. They understand CES as "the environmental spaces within which people interact with the natural environment and the cultural practices that define these interactions and spaces". Taken together, these environmental spaces and practices "shape and reflect a wider set of cultural (collective or shared) values about ecosystems". In turn, "values, spaces and practices interact in complex and non-linear ways to give rise to a range of cultural benefits to human well-being". Furthermore, this interaction leads to the production of 'cultural goods', such as "organised opportunities for recreation and tourism, food and drink of local provenance, local festivals etc" (Church et al. 2014). Although not explicitly stated by Church et al. we would suggest that both the production of such cultural goods, and use of the goods themselves, would also lead to cultural benefits.

Building on Chan et al (2011) they propose three kinds of cultural benefits:

- Identities—e.g. belonging, sense of place, rootedness, spirituality.
- Experiences—e.g. tranquillity, inspiration, escape, discovery.
- Capabilities—e.g. knowledge, health, dexterity, judgement.

With respect to 'identities', Church et al. highlight the "cultural meanings [associated with ecosystems] through which people understand themselves and their relationship to the world around them". An example is the idea of belonging: "ecosystems play a role in the process of place identification through which ideas of affiliation and attachment develop" (Church et al. 2014). Regarding 'experiences', Church et al. mean "benefits that are produced, mentally or physically, through immediate contact with ecosystems", for example feelings of calm or aesthetic pleasure. Importantly, they extend their definition of direct experience to include "dis-embodied and distant" forms of contact "such as the benefits associated with consuming nature through a television programme" (Church et al. 2014). To this, we would add experience of cultural artefacts in museums, galleries, etc., and other 'cultural goods' located outside the forest. Finally, regarding 'capabilities', Church et al. highlight "the role ecological phenomena play in shaping individual and social capabilities to understand and to take action", e.g. through acquisition of knowledge (e.g. making sense of biodiversity), skills, wisdom, judgement and insight, and acquiring employment (Church et al. 2014). In addition, we would interpret 'social capabilities' to include the sense of trust and cohesion within a community, and the capacity to act together to realise a common goal, which can develop when given opportunities to engage actively in the local environment and in the decisions that affect it. The previous section explored three broad categories of cultural benefits that emerged from discussions during the project: aesthetic/spiritual, cultural heritage, and communally held. Broadly speaking, these can be mapped onto the categories identified by Church et al. One caveat is that we would list 'spirituality' under 'experiences' rather than 'identities'. With this revision, 'experiences' maps onto our category of 'aesthetic/spiritual' and 'identities' maps onto 'communally-held'. 'Capabilities' maps onto 'biocultural heritage' in its intangible manifestations as the knowledge, skills and practices associated with traditional forest management, which are now largely lost or dissociated from a meaningful relationship with the contemporary cultural landscape. These capabilities are what UNESCO refers to as intangible cultural heritage (ICH). Meanwhile, the main contemporary expression of 'capabilities' relating to the Black Wood is evident in the discourses, practices and products of professional ecologists and foresters, and (to a lesser extent) artists, humanities scholars and professionals in the cultural sector. The typology and conceptual framework of Church et al. provides a nuanced language to support decisions about how to realise cultural benefits. Thus, opportunities exist in the Black Wood to realise 'experiences' and 'identities', but arguably they are restricted to a relatively narrow range of people. Perhaps, the most notable improvement could be made by creating further opportunities for 'capabilities'. Thus, from an ecological perspective, there might be greater involvement of volunteers and community representatives in contemporary conservation and restoration practice, helping to develop skills and create jobs in the local community, or in some locations a revival of traditional forms of woodland management and associated skills could be encouraged. From a cultural perspective, capabilities could also be enhanced through active engagement with representatives from the arts and humanities, locally and nationally, to encourage production and use of 'cultural goods' associated with the Black Wood. There is of course overlap between the categories, and, as Church et al. point out, one benefit can be understood differently through the lens of others. Inviting people to gain direct aesthetic 'experience' of the forest might motivate them to develop 'capabilities', i.e. skills and knowledge of the forest, which in turn might help strengthen 'identities' with the forest and its communities of interest. Ironically perhaps, this could also generate wider support for biodiversity conservation. A project seeking to realise the full range of biocultural values and benefits might try to build synergies of this kind.

20.3.2 What Might Be Done Instead?

In principle, how might the Black Wood and other Caledonian forests be managed differently in ways that enhance cultural benefits while still meeting conservation objectives? First, there are options that would not make direct changes to the physical structure of the forest (although there could be unintended impacts of an increase in visit numbers), e.g. improvements to public access, interpretation and representations of the forest through education, the media and the arts. Second, there are interventions that would seek to change the biophysical structure of the forest. Regarding the latter point, Mason et al. suggest that "Managers need to understand more about the spiritual values the forests provide and how these may be affected by management" (Mason et al. 2004). They call for better understanding, but the implication is that such understandings could inform changes to silvicultural practice to enhance 'spiritual values'.

Bringing 'interpretive' and biophysical interventions together into different combinations, a number of potential future scenarios for management of areas of Caledonian forest over the next 50 years can be outlined. A brainstorming exercise by the authors came up with six options:

1. **Untouched wilderness**—often talked about amongst conservationists, it implies a management style where utility is set aside and nature takes its own path. Human access is discouraged. Within the discourse of re-wilding it suggests robust natural expansion once human interest is retracted followed by a reintroduction of large mammals.
2. **Sacred and/or cultural ecology**—a reconsideration of landscape, forest and its experience within specific Celtic traditions, spirituality, song and literary frameworks: an interpretative approach that is of the forest, but without material impact in the forest (cf. Maddrell 2015).
3. **Native forest conservation**—preservation of the core and ecological restoration area to an expanding Black Wood perimeter (i.e. the approach closest to current practice). Following Peterken and Stace (1987) the forest evolves with no active attempt to maintain the specific form of the granny pines, which could eventually be lost from the landscape as a result.
4. **Social and ecological restoration**—an approach to native forest conservation which actively encourages interaction with relevant communities of place and interest in mutually beneficial ways, by engendering the formation of capabilities, identities and experiences, which in turn generate further support for biodiversity conservation.
5. **Revived cultural landscape**—introduction of historic forms of management to revive a relict 'organically-evolved' cultural landscape (as defined by UNESCO 2008), possibly based on transhumance and/or silvi-pastoralism, creating living forms that reflect processes and periods in social and environmental history.
6. **Community forest economy**—local community management and/or ownership for diverse social and economic objectives, including income generation from local visitor spending, timber production (only where appropriate) and grants to deliver public goods (e.g. education), in ways that find a balance with the objective of protecting the ecological integrity of the forest.

The scenarios have been listed in order of increasing levels of intervention. It is worth highlighting that there is no option that resembles a 'theme park', or involves intrusive interpretation or recreational infrastructure, since this was not seen by any participant as a desirable outcome. Also of note, loss of the current form of the granny pines is a possibility under all of these scenarios, not just under Scenario 3, which is probably closest to current practice. Rather than preserving their current form, the alternative scenarios might seek to produce a 'culturally meaningful' aesthetic form that also meets ecological (and in some cases economic) needs. The future scenarios outlined above impact differently on the three main types of cultural benefit adapted from Church et al. (2014) as well as on the biodiversity benefits of a Caledonian forest. Using the authors' judgement based on experiences with the forest and the local community, an indicative assessment was made as a means to illustrate possible levels of benefits provided by each option and the factors that influence them (see Table 20.1). Scores on a scale from 1–5 (1 = low; 5 = high) were assigned to each scenario to indicate the approximate level of cultural benefit that would be realised by any one person (e.g. a visitor to the forest) for each of the three categories from Church et al. (columns 'd', 'e' and 'f'). These scores were then averaged to derive an overall score for cultural benefit per person ('g'), which was then weighted by a score for the number of people (or 'beneficiaries') who would have access to these benefits ('c') to derive a score for total cultural benefit for each scenario ('h'). (Note that we have effectively given an equal weighting to the three types of cultural benefit.) A separate estimate was also made of the total biodiversity benefit judged to be delivered by each scenario ('i').

The scores in Table 20.1 suggest the following: (a) cultural benefits differ by scenario, (b) no single scenario is likely to maximise all types of benefit, and (c) high biodiversity benefit does not necessarily equate to high cultural benefit. If we are trying to provide a wide range of cultural benefits, for a wide range of people, then Scenarios 4 and 6 would score higher than Scenario 3 (the one closest to 'business as usual'). Broadening the analysis to include biodiversity value, arguably a high biodiversity score could be given to Scenarios 1 and 4. This suggests that overall Scenario 4 could offer the greatest aggregated public benefit, but only if we apply an equal weighting to total cultural benefit and biodiversity benefit—an assumption that is far from certain. These estimates are only indicative, and depend upon several assumptions: for example, the high biodiversity score given to Scenario 4 assumes there would be a synergistic ('win-win') relationship between community engagement and support for biodiversity conservation.

20.3.3 The Way Forward: An 'Authentic' Process

Table 20.1 is intended to offer an impressionistic analysis, and not everyone would agree with the scores that we have assigned. Also, we have not considered the costs of each option: community engagement can be very time consuming for everyone concerned, and cannot always be sustained. However, we suggest it would be neither necessary nor appropriate to seek an accurate, definitive, quantitative (or monetized) cost benefit analysis of these options. As it stands, Table 20.1 already illustrates the points highlighted above. It could be useful to conduct a similar exercise with representatives of different stakeholder groups helping to develop storylines, and select, assess and weight indicators, and aggregate them to derive an overall value for each scenario. But arguably its main contribution would be as a heuristic device to facilitate social learning and deliberations around planning rather than as a means to derive hard evidence for the best option to be selected for implementation. We make this argument for two reasons. First, using such an exercise, it is difficult to

Table 20.1 Indicative assessment of possible impacts of six management scenarios on the cultural and biodiversity benefits of a Caledonian forest (1 = low, 5 = high)

Scenario description			Cultural benefits per beneficiary (scale 1–5)				Total benefits (scale 1–5)	
a	b	c	d	e	f	g	h	i
Scenario name	Level of intervention	Relative number of beneficiaries	Identities	Experiences	Capabilities	Average cultural benefit[a]	Total cultural benefit[b]	Biodiversity benefit
1. Untouched wilderness	None	1	1	5	1	2.3	1.5	5
2. Sacred and cultural ecology	Insignificant	3	4	5	3	4.0	3.5	3
3. Native forest conservation	Significant but minimal	2	3	4	2	3.0	2.5	4
4. Social and ecological restoration	Significant	5	5	4	5	4.7	4.8	5
5. Revived cultural landscape	Substantial	2	4	3	4	3.7	2.7	3
6. Community forest economy	Substantial	5	5	3	5	4.3	4.6	2

[a]The score for average cultural benefit (column ‘g’) is the arithmetic mean of the scores in columns ‘d’, ‘e’ and ‘f’

[b]The score for total cultural benefit (column ‘h’) is the geometric mean of the average cultural benefit per beneficiary (column ‘g’) and the number of beneficiaries (column ‘c’). It is derived by multiplying ‘g’ by ‘c’, giving a score on a scale from 1–25, which is then converted to a score on a scale from 1–5 by calculating the square root. Thus: $h = \sqrt{(c \times g)}$

define, let alone quantify and aggregate, the cultural benefits of forests without important dimensions of value being lost or distorted (e.g. Chan et al. 2011). Second, scenario analyses that follow the technical-rational logic in Table 20.1 struggle to take account of the fact that values and benefits are not necessarily fixed and given, but are also created, in ways that are hard to predict, through the iterative process of participatory assessment, planning and implementation. In short, the process leads to changes in thinking and behaviour (Nutley et al. 2007). It is through this process that the synergistic relationship between community engagement and support for biodiversity conservation in Scenario 4 could be realised. While the authors were reflecting on the scenarios outlined in Table 20.1, Tim Collins prompted a debate on social media by posing the question: "What would an authentic cultural ecology look like?" One answer to this question is "One that resulted from an authentic process". In other words, the journey is as important as the destination: the answer is not to pick an idealised option off the shelf and try to implement it regardless of experience gained along the way. The role of scenarios would be to offer loosely held visions of the future that (a) inspire and guide an inclusive, iterative process of deliberation, decision-making and action; (b) combine aspects of ecological and cultural value, which are currently considered separately; (c) are unique to the Black Wood and the wider cultural landscape in which it is located; and (d) may challenge established discourse and its effects on management decisions in positive ways. This 'authentic' process might continue in the Black Wood through the deep mapping idea and other proposals identified in the Future Forest workshop.

20.4 Conclusion: Towards Biocultural Diversity?

Returning to the question in the title, in the Black Wood the answer is arguably 'yes'—high conservation status does indicate high cultural ecosystem value. But only up to a point, because managing for biodiversity conservation alone evidently shapes and constrains the realisation of a wider range of cultural values and benefits. We conclude that the way forward is to start a debate which leads to greater realisation of the diversity of cultural benefits associated with the Caledonian forest (where 'realisation' is defined as the simultaneous recognition of existing values and benefits and creation of new ones). How does this goal relate to the notion of biocultural diversity now enshrined in the 2014 Florence Declaration on the Links between Biological and Cultural Diversity (UNESCO 2014)? According to the Declaration, the European landscape is predominantly a biocultural multifunctional landscape. It goes on to state that "The involvement of local communities, and recognition of and respect for their cultural heritage, traditional knowledge, innovations and practices can assist in more effective management and governance of multifunctional (bio)cultural landscapes, and contribute to their resilience and adaptability". While the emphasis in this statement is on 'organic' cultural landscapes (as defined by UNESCO 2008) with unbroken traditions, knowledge, skills and practices (i.e. ICH), the passage reflects our own argument about the overall benefits of 'social and ecological restoration' (Scenario 4) whereby local community engagement is synergistic with realisation of both cultural value and biodiversity value. It might be worth noting that the interests arguing for conservation, aesthetic value and public access have shared historic origins in the Romantic movement, and at times have been close allies in their opposition to economic exploitation (Smout 2000). New integrated concepts such as biocultural diversity (but also 'living heritage', 'socio-ecological systems', 'cultural ecology', etc.) could help reunite these disparate interests and invite us to think imaginatively of future visions of the Caledonian pine forests of Scotland that might transcend the artificial categories of ecology and society.

Acknowledgements Primary funding for this project was provided by the 'Imagining Natural Scotland' programme of Creative Scotland. Match funding and support was provided by the Landscape Research Group and Forestry Commission Scotland, with additional in kind support from the Perth and Kinross Countryside Trust, the Perth and Kinross Museum and Art Gallery and Forest Research. In addition, we would like to thank everyone who kindly gave their time, expertise and enthusiasm to the project by participating in discussions, workshops, events and residencies, in particular: Anne Benson, Rob Coope, Peter Fullarton, Andy Peace, Chris Quine, Mike Smith, Bid Strachan, Paul Tabbush, Charlie Taylor and Richard Thompson.

References

Bain C (2013) The ancient pinewoods of Scotland: a traveller's guide. Sandstone Press, Ross-Shire, Scotland

Chan KMA, Goldstein J, Satterfield T, Hannahs N, Kikiloi K, Naidoo R, Vadeboncoeur N, Woodside U (2011) Cultural services and non-use values. In: Kareiva P, Tallis H, Ricketts TH, Daily GC, Polasky S (eds) Natural capital: theory and practice of mapping ecosystem services. University Press, Oxford, UK, pp 206–228

Church A, Fish R, Haines-Young R, Mourato S, Tratalos J, Stapleton L, Willis C, Coates P, Gibbons S, Leyshon C, Potschin M, Ravenscroft N, Sanchis-Guarner R, Winter M, Kenter J (2014) UK national ecosystem assessment follow-on. Work package report 5: cultural ecosystem services and indicators. UNEP-WCMC, Cambridge, UK

Collins T, Goto R, Edwards D (2014) Future forest: the Black Wood, Rannoch, Scotland. Landscape Research Group and Forest Research, UK

Eaton MM (2004) Fact and fiction in the aesthetic appreciation of nature, Chap. 9. In: Carlson A, Berleant A (eds) The aesthetics of natural environments. Broadview Press, Peterborough, UK

FCS (2009) Black wood of Rannoch management plan, 2009–2019. Tay Forest District, Forestry Commission Scotland, Dunkeld, Scotland

Kenter JO, Reed MS, Everard M, Irvine KN, O'Brien E, Molloy C, Bryce R, Brady E, Christie M, Church A, Collins T, Cooper N, Davies A, Edwards D, Evely A, Fazey I, Goto R, Hockley N, Jobstvogt N, Orchard-Webb J, Ravenscroft N, Ryan M, Watson V (2014) Shared, plural and cultural values: a handbook for decision-makers. UK national ecosystem assessment follow-on phase. UNEP-WCMC, Cambridge, UK

Kirchhoff T, Vicenzotti V (2014) A historical and systematic survey of European perceptions of wilderness. Environ Values 23:443–464

Lindsay JM (1974) The use of woodland in Argyllshire and Perthshire between 1650 and 1850, vol 2. Unpublished PhD Thesis, University of Edinburgh, Scotland

Maddrell A (2015) Contextualising sacred sites, pathways and shared heritage: local history, ecumenism and national identity. In: Maddrell A, della Dora V, Walton H (eds) Christian pilgrimage, landscape and heritage. Routledge, London, UK

Marzano M, Dandy N (2012) Recreation use of forests and disturbance of wildlife: a literature review. Forestry Commission Research Report, Forestry Commission, Edinburgh, Scotland

Mason WL, Hampson A, Edwards C (2004) Managing the pinewoods of Scotland. Forestry Commission Scotland, Edinburgh, Scotland

Millennium Ecosystem Assessment (2005) Current state and trends assessment. Island Press, Washington DC

Murray J (2014) Reading the Gaelic landscape: Leughadh Aghaidh na Tire. Whittles Publishing, Cunbeath, Caithness, Scotland

Nutley SM, Walter I, Davies HTO (2007) Using evidence: how research can inform public services. The Policy Press, Bristol, UK

Peterken GF, Backmeroff C (1988) Long-term monitoring in unmanaged woodland nature reserves. Nature Conservancy Council, Peterborough, UK

Peterken GF, Stace H (1987) Structure and development of the Black Wood of Rannoch. Scott For 41(1):29–44

Smout TC (2000) Nature contested: environmental history in Scotland and Northern England since 1600. Edinburgh University Press, Edinburgh, Scotland

Steven HM, Carlisle A (1959) The native pinewoods of Scotland. Oliver and Boyd Ltd, Edinburgh, Scotland

UNESCO (2008) Operational guidelines for the implementation of the world heritage convention. UNESCO, World Heritage Centre, Paris

UNESCO (2014) Florence declaration on the links between biological and cultural diversity. UNESCO/CBD, Florence (Italy), 11 Apr 2014

Chapter 21
Linking Biocultural Diversity and Sacred Sites: Evidence and Recommendations in the European Framework

Fabrizio Frascaroli and Bas Verschuuren

Abstract There is growing recognition that sacred natural sites (SNS) form hotspots of biocultural diversity and significantly contribute to conservation in traditional non-western societies. Using empirical evidence from SNS in Central Italy, we illustrate how a similar link between spiritual, cultural, and biological values can be fundamental also in relatively secular and modernized European contexts. We show that SNS are key to sustaining traditional practices and local identities, and represent important instances of biodiversity-rich cultural landscapes. Based on other case studies from across Europe, we suggest that these conclusions can be relevant also at a broader European scale. Greater awareness from planners and policy-makers, however, is needed to safeguard and emphasize the role of European sacred sites as refugia for biocultural diversity. We review policy guidelines on SNS previously developed by International Union for the Conservation for Nature (IUCN) and United Nations Educational, Scientific and Cultural Organization (UNESCO), and aimed at protected area managers and planners. We assess the applicability of these guidelines in European contexts, and complement them with findings and insight from Central Italy. We provide recommendations for guidelines that are suited to SNS related to mainstream faiths in Europe.

F. Frascaroli (✉)
Religion and Nature Group, Department of Religion, University of Florida, Gainesville, FL, USA
e-mail: fabrizio.frascaroli@ieu.uzh.ch

F. Frascaroli
Institute of Evolutionary Biology and Environmental Studies, University of Zurich, Zurich, Switzerland

B. Verschuuren
Sociology of Development and Change Group, Sub-department of Sociology and Anthropology of Development, Wageningen University, Wageningen, The Netherlands
e-mail: bas.verschuuren@wur.nl

M. Agnoletti and F. Emanueli (eds.), *Biocultural Diversity in Europe*, Environmental History 5, DOI 10.1007/978-3-319-26315-1_21

Keywords Biocultural landscape · Biocultural refugia · Biodiversity management and conservation · Cultural heritage · Policy guidelines · Sacred natural sites · Spiritual values · Traditional ecological knowledge · Traditional livelihoods · Biocultural diversity

21.1 Introduction

The paradigms of biocultural diversity and biocultural conservation place due emphasis on the religious and spiritual dimensions that characterize the relations between people and their environment (Maffi and Woodley 2010). The world's staggering varieties of spiritual beliefs, rituals, and celebrations constitute a primary instance of diversity across human cultures. The number of religions recorded in an area, often in combination with linguistic and ethnic diversity, has been repeatedly proposed as a robust proxy for cultural diversity at large (Loh and Harmon 2005; Maffi 2005). Religious and spiritual manifestations, such as myths, invocations, rituals, and taboos, also carry the teachings and norms of conduct that different societies have developed in synergy with their environment (Berkes 1999; Colding and Folke 2001). It is increasingly recognized that this living body of knowledge—which in conservation circles is frequently referred to as traditional (or indigenous) ecological knowledge (henceforth TEK; Berkes et al. 2000; Toledo 2002)—has played a central role in shaping and maintaining biodiversity-rich landscapes, as well as supporting sustainable livelihoods and resource management practices (Gadgil et al. 1993; Long et al. 2003; Tengö et al. 2007).

Sacred natural sites (SNS) are one of the clearest exemplifications of this tight interplay between spiritual beliefs, TEK, and biocultural diversity (Verschuuren and Wild 2012). SNS have been defined as 'areas of land or water having special spiritual significance to peoples and communities' (Wild and Mcleod 2008). As such, groves, forests, springs, rivers, caves, but also entire landscapes or nature patches surrounding particular human artifacts, like temples and shrines, can be SNS (Dudley et al. 2009). SNS have been recorded on every continent of Earth except for Antarctica (Bhagwat and Rutte 2006), and can be seen as universal manifestation of a deep bond connecting humans and nature (Hughes and Chandran 1998).

While this sacred bond between spatial features and people has been a source of inspiration and identity throughout human history (Eliade 1959; Tuan 1974), the biocultural importance of SNS has been consistently highlighted only in relatively recent times (Verschuuren 2010; Verschuuren et al. 2010; Pungetti et al. 2012). Over the last 15 years in particular, ecologists have shown that sacred areas in Asia and Africa tend to harbor especially valuable biodiversity, as reviewed by Dudley et al. (2010). Others have explored the spiritual significance of particular ecosystems, such as wetlands (Papayannis and Pritchard 2010) and mountains (Bernbaum 2006, 2010).

The importance of SNS as life-enhancing human activities (Zent and López-Zent 2007), hotspots of biocultural diversity (Verschuuren et al. 2010; Pungetti et al. 2012), and sources of human well-being (Guri and Verschuuren 2008; Delgado et al. 2010) has received recognition in policy frameworks internationally. The United Nations Educational, Scientific and Cultural Organization (UNESCO), the Convention for Biological Diversity, and the International Union for the Conservation for Nature (IUCN) alike have dedicated significant attention to the phenomenon (Lee and Schaaf 2003; Secretariat of the Convention on Biological Diversity 2004; Schaaf and Lee 2006; Wild and Mcleod 2008). These policy implications, however, have mostly concerned the SNS of indigenous populations, and barely touched on the relation between biodiversity and 'mainstream' faiths in the Western world. So far, and despite a few notable exceptions (Mallarach and Papayannis 2007; Papayannis and Mallarach 2009; Mallarach et al. 2012; Frascaroli 2013), the role of spiritual traditions in the relationship between biological and cultural diversity remains largely underexplored in Western modernized contexts. As a consequence, it is practically ignored at the level of European policy-making.

In this chapter, we pursue two objectives. First, we attempt to demonstrate the importance of SNS as biocultural hotspots in a Western European context. Second, we review the IUCN–UNESCO's *Best Practice Guidelines. SNS: Guidelines for Protected Area Managers* (Wild and Mcleod 2008), in light of our findings about European SNS. Based on this review, we make recommendations for the development of guidance that would be suitable to religious and holy sites in Europe. Our evidence and examples are mostly drawn from an ongoing socio-ecological investigation of SNS in Central Italy.

21.2 Sacred Natural Sites and Biocultural Linkages in Europe

21.2.1 Central Italy as a Case Study

The nexus between SNS and biocultural diversity in Central Italy has been addressed by a comprehensive research project, started in 2010 (Frascaroli 2013; Frascaroli et al. 2014). This specific regional focus discourages too broad generalizations of the results, but at the same time has permitted to establish a systematic investigation using quantitative as well as qualitative methods. We summarize the available conclusions and offer preliminary results from parts of the project that are currently underway.

21.2.1.1 Methods

The research project is sited in the six administrative regions that constitute Central Italy: Tuscany, Marche, Umbria, Lazio, Abruzzi, and Molise. Hills and mountains (the Apennine and pre-Apennine ranges) cover nearly 97 % of the total land and overlaps with one of the main biodiversity hotspots in Europe and the Mediterranean (Myers et al. 2000; Olson and Dinerstein 2002). The religious heritage covers a number of settlements related to seminal figures such as St Benedict of Norcia and St Francis of Assisi (who both had their birthplace in this area) and their respective orders. A number of smaller hermitages, shrines, and sanctuaries also dot the landscape.

The project design consists of three distinct phases: (1) identification, categorization, and mapping of SNS in the area; (2) floristic assessment and comparison of a sample of thirty representative SNS and equally many control non-sacred sites; and (3) interviews and participant observations at the same sample SNS.

During the first phase, in 2010, baseline information on sacred sites in Central Italy was derived from literature including the Ministry of Cultural Heritage's national census of Christian shrines in Italy (CSC 2003). Details on the environmental setting, historic development, and religious heritage of each site were inventoried and categorized into a database. Descriptive statistics were calculated to identify significant patterns in the distribution of sacred sites across different land covers. A list of SNS was compiled, including all the sacred sites located in 'natural' settings—forests, grasslands, cliffs, riverbanks, etc.—away from human infrastructures and populated areas (Frascaroli 2013).

Floristic assessments at a sample of thirty sites were carried out in Summer 2011 and 2012. These sites were selected among the overall pool of SNS so as to represent a balanced geographical distribution, different types of habitat, a gradient of religious importance, and different degrees of formal protection (some were included in official protected areas (PAs), while an equal number were not). Each SNS in the sample was paired with a non-sacred control site located nearby and with comparable environmental conditions (i.e., altitude, aspect, habitat type). The floristic assessments measured diversity and composition of herb, shrub, and tree layers, forest structure, and occurrence of endemic, threatened, and useful plants at all sites. Pairwise tests of statistical significance were performed to compare diversity measures at SNS and control sites. The impact of other anthropogenic variables (such as the presence of an official PA or the religious importance of the site) on floristic composition at all sites was also analyzed through analysis of variance (Frascaroli et al. 2014, in press).

Social science data have been collected regularly since 2010 at the same SNS on which the floristic assessment had focused. Observations were also carried out at several additional sites, in the context of a distinct sub-project in 2012. Religious rituals and ceremonies at the sites were attended as a form of participant observation. Further, semi-structured face-to-face interviews were conducted at each site with key informants, and additional informants were identified through snowball sampling (Bernard 2006). The interviews covered a range of subjects, including oral history, beliefs, and ritual practices related to the sites. Also, they collected information on site management and governance, and the relation of the sites with traditional livelihoods and economies. Data collection for this part of the research has continued in 2015.

21.2.1.2 Overview of Main Findings

The results obtained thus far highlight both the biological and cultural importance of SNS in Central Italy, and offer insight into the linkages between the two. Of all 539 sacred sites inventoried, more than half are located in natural or semi-natural settings, such as forests, tree stands, mountain grasslands, and agricultural lands. Moreover, natural features such as grottos, caves, and single trees were found to be explicit objects of worship at a large number of SNS.

These data suggest a much stronger prominence of nature-related spiritualities in Central Italy than originally hypothesized. Also, they suggest that the relation to the environment varies considerably not only between different strands of Christianity (Mallarach 2012b), but also within Catholicism itself. Indeed, variants closer to local folk beliefs and specific families like the Camaldolese are associated with significantly higher numbers of SNS than other groups. Nonetheless, the importance of nature-related worships seems to have been on the wane overall since the fourteenth century (Frascaroli 2013; see also Byrne 2010).

Biological values Floristic analyses of SNS in the area showed that SNS are significantly connected with old-growth forest and monumental trees (Frascaroli et al. 2014). Generally, the portions of old-growth forest in question are small patches of one to few hectares, although in some cases they can have a much larger extent, like the beech forests surrounding the shrines of Vallepietra and Canneto. Also, there are a few but noticeable cases where SNS are characterized by types of forest cover (mostly thermophilous forests dominated by *Quercus ilex*) that do not occur anywhere else at a landscape scale.

Significantly, higher species richness was recorded at SNS than at control sites. This appears to be largely related to the more heterogeneous habitat composition measured at SNS. SNS were also found to significantly contribute to diversity at the landscape scale. Indeed the number of species found uniquely at SNS was significantly higher than the number of species found only at control sites. Moreover, of the 97 plants found only at SNS, four are endemic and more than one-third are typical of forest and open-range habitats, such as Eastern white oak woods and *Festuco-brometalia* dry grasslands, considered priorities by the EU's Habitat Directive (EC 2013; Frascaroli et al., in press). SNS also harbor a greater number of plants of ethnobotanical importance. This difference is significant for species that have or used to have animal-related applications such as traditional veterinary practices (Frascaroli et al. 2014).

Finally, the floristic analyses revealed that being included in official PAs had a negative influence on the plant richness of SNS. Whereas SNS are significantly more diverse than control sides outside of PAs, this difference levels out within PAs (Frascaroli et al., in press).

Cultural values While social science data are still being collected, some themes have already emerged. There are indications that a spectrum of cultural values is consistently associated with SNS, and that these sites are important anchors of traditional practices and cultural distinctiveness.

The SNS investigated are often characterized by syncretism where a dominant religion, Roman Catholicism, has merged with previous indigenous and folks beliefs, resulting in *sui generis* variants (see also Verschuuren et al. 2010; Byrne 2010). Archeological evidence indicates that approximately 10 % of SNS in the area were also sites of worship in pre-Christian times (Frascaroli 2013). The presence of pre-Christian worships, however, is not a pre-condition for the occurrence of local folk variants, which appear to be a much broader phenomenon.

Claiming that each of these variants is a unique entity would be exaggerated, as numerous characteristics are shared across sites. For example, foundation stories tend to be quite similar, and revolve around two main themes: the original presence of a hermit and his miracles, or the accidental discovery of the site by a herder in search of a lost animal. Some ritual practices are also common, like the collection of miraculous spring water for therapeutic purposes, and the devotion to specific natural features (mostly rocks or trees) that are ritually rubbed. There are a few instances, however, where local customs display a much more distinctive character. Thus, some of the festivities annually held at SNS (named *festas*) remain manifestations of unique intangible heritage. Examples in this sense are the 'marriages of the trees' that are still celebrated in various places from Lazio to Basilicata (Fig. 21.1). Others are the pilgrimages and rituals at the Shrines of Madonna del

Fig. 21.1 Sacred natural sites in Central Italy are often locations of ritual practices and celebrations focusing on natural elements. In this 'marriage of the trees,' two live oaks are symbolically united in marriage in front of the Convent of St. Angel on Mount Fogliano, Vetralla (Lazio) (*Photos* K. Marsh)

Fig. 21.2 The confraternity Santissima Trinità of Subiaco (*left*) undertaking the annual pilgrimage to the sanctuary of Vallepietra (*right*) (*Photos* T. Fjeldsted)

Canneto and Santissima Trinità (Fig. 21.2). Here, Roman Catholicism, folk spiritualities, and Greek Orthodox influences fuse with the local passion for horseback riding and animal husbandry, giving rise to an incredible syncretism (see also Bernardini 2000).

To the organization of rituals and *festas* is connected another prominent trait of SNS, that is, the long-standing importance of local lay brotherhoods. Associations of this kind, which cooperate with but can be visibly independent from the local clergy, play a prominent role at several sites. For example, they can be charged with the maintenance of the shrines or organization of the main celebrations. Where ceremonies require intense planning and fundraising, the brotherhoods can be active during the whole year round and form a pivotal reference institution for local communities beyond the mere religious ambit. Similar instances are the confraternity in the town of Subiaco that supervises the complicated rituality around the *festa* of Santissima Trinità, or the 'brothers' of St. Angel in the small village of Balsorano (Marucci 1999).

The importance of animal husbandry is also a structural element in the symbolism and historical development of many SNS. The ancient relation between SNS and pastoralism in Central Italy is known (De Waal 2012), although not fully explicated. Plants used in traditional veterinary and animal feeding are especially common at SNS, which additionally confirms that deep connection (Frascaroli et al. 2014). Herders are still found to be amongst the most fervent devotees at many

SNS, and some are directly involved with care taking of the sites. Statements of herders recorded in Abruzzi and Lazio testify the pride in claiming that their profession is 'the closest to Christ.' Others commemorate days when animals grazing the mountain pastures were plentiful, and herders would find refuge within the shrine's walls during stormy nights.

Overall, SNS are emerging from this research as symbolic places connected to an important sense of existential and identity-making *continuity*. This is true at the personal level, as underlined by the many stories of individuals undertaking intimate pilgrimage to these sites to periodically mark special recurrences (return from emigration, anniversaries of healings or losses, etc.). However, continuity has also an inter-generational dimension. Thus, SNS are often described by informants as vehicles for transmitting a body of moral values and traditions from forefathers to descendants. And continuity, finally, can span over centuries if not millennia. SNS, in this case, are loci where historic identities, even from pre-Roman times, are grounded and conserved.

Biocultural linkages and SNS as *biocultural refugia* The research findings so far already provide some insight into the interplay between spiritual, biological, and cultural values at SNS in Central Italy.

A first indication is that the biological uniqueness of SNS is the result of anthropogenic practices, as well as original features. Indeed many of these SNS are found in outstanding geomorphological locations (Nolan and Nolan 1989; Frascaroli 2013), but their vegetation patterns also appear to be significantly influenced by specific human activities. This is visible in the selective conservation of individual trees (Schama 1995; Turner et al. 2009), and in the diversification of the microhabitat mosaic due to trampling, and in activities of traditional management, such as weeding, pruning, and pollarding. Indeed, it is known that moderate anthropogenic disturbances of this sort can have a positive influence on local species richness and habitat diversity (Naveh and Whittaker 1980; Selvi and Valleri 2012).

At a larger scale, the relation between SNS and biodiversity-rich cultural landscapes also involves the effects of pastoralism. Ongoing herbivore grazing, indeed, has maintained significant areas around several SNS as species-rich grasslands, or created distinct silvopastoral landscapes characterized by alternating pastures and stands of old-growth pollarded trees (cf. Oteros-Rozas 2013; Agnoletti 2014). The deep bond between religious symbolism and pastoralism at SNS may have contributed to the survival of transhumance and open-range herding in some of these areas, in spite of the general decline of such practices in the rest of the country.

The role of SNS in rural landscapes and livelihoods in Central Italy extends beyond pastoralism and acts (or used to act) as an important reference for farming at large. This view is supported by the temporal cadence of many *festas*, which often coincide with the dates of transhumance or key moments in the agricultural calendar. Similarly, informants at different sites revealed how the bell from the hilltop shrines marked the timings of the daily work in the fields, or warned against a coming storm, or was frantically rung by the local peasants to 'fight off' extreme weather events (cf. Gómez-Baggethun et al. 2012).

In this context, it is also relevant that there is a significant spatial association between SNS and portions of rural land collectively owned by local village communities (*usi civici*). These lands are largely managed for summer grazing and light extraction of forest products. The hypothesis that SNS may represent a particular variant of the common properties described by institutional economists (Ostrom 1990) had already been advanced (Rutte 2011). While the SNS in Central Italy are not conceived as common properties per se, they seem to act as symbolic centers that contribute to the sound functioning and management of collective areas. Admittedly, this distinction is based on a strict demarcation between sacred and non-sacred portions of land, which might have been perceived as scarcely meaningful or artificial in the traditional worldview of rural people up to a few decades ago.

In Central Italy as elsewhere, in conclusion, SNS appear as particular instances of biocultural landscapes, where both human practices and biodiversity are embedded in a system of symbolic and transcendental meanings. The presence of this religious framework has contributed not only to creating an aesthetic and environmental specificity, but also to strengthening socio-ecological resilience (Gómez-Baggethun et al. 2012) and maintaining relevant social memory throughout time (Barthel et al. 2013). In this sense, SNS can also be seen as *biocultural refugia*, that is, places that shelter not only defined species but also living knowledge and memory about biodiversity and ecosystems (Barthel et al. 2013).

21.2.2 From Regional to European Relevance

While the evidence presented strongly supports a view of SNS in Central Italy as biocultural refugia, the specific focus of the research makes it problematic to simply generalize the findings to the national or European level. Some additional indications on the biocultural importance of SNS across Europe, however, can be drawn based on case studies from the Delos Initiative.

The initiative, launched in 2005, was built around the identification and analysis of prominent SNS and landscapes located within PAs in the modernized world. Some 30 case studies, of which 22 in Europe, were investigated by local experts, and discussed at the three international workshops of the Delos Initiative (Mallarach and Papayannis 2007; Papayannis and Mallarach 2009; Mallarach et al. 2012). The conclusions reached through this process support that there is a significant overlap between biodiversity-rich areas and spiritual heritage, and that SNS and landscapes play a significant role in biological conservation even in Western and modernized contexts. Moreover, they confirm that nature maintains important spiritual and symbolic values also in those contexts, and that this offers potential for conservation approaches in modernized countries (Mallarach and Papayannis 2010).

European case studies of the Delos Initiative highlighted a range of cultural values that are comparable to those that we reviewed referring to Central Italy. For example, some of the sites revealed an important historical continuity between pre-Christian and Christian worships (Lyratzaki 2007; Wild 2012). Others keep hosting particular rituals, beliefs, and forms of veneration that are often linked to the sanctification of natural elements (Bosch and Varela 2007; De Waal 2012). Still in other instances, the location of SNS coincides with that of ethnic or linguistic minorities. In these cases, it was emphasized how they have historically contributed to the conservation of local and national identities, especially in the face of foreign occupations (Catanoiu 2007, 2009; Mallarach and Catanoiu 2009; Pesic et al. 2012).

21.3 Improving Management and Policy for European Sacred Natural Sites

Although additional studies are needed to assess regional variations and specificities in the relations between spiritual and biocultural heritages, our evidence underlines the significance of such linkages, and the important roles that sacred sites and landscapes have played in supporting biocultural diversity across Europe. These findings stand in stark contrast with the scarce attention accorded to spiritual and religious traditions in European land management and conservation approaches. While this lack of attention represents a significant hindrance to both theoretical and policy developments, it is not particularly surprising. In fact, it can be understood considering three possible causes.

The first cause is the biased and ideological self-perception of a fully modern, rational, and secularized "West," as opposed to a traditional, exotic, and superstitious "Rest" of the world (Latour 1993; Herzfeld 2001). Needless to say, according to this construct, talks of sacred forests, healing waters, or miraculous caves belong with the latter rather than the former (Tiedje 2007; Mallarach et al. 2012).

The second cause lies in the common assumption that Christianity (the dominant religion in Europe) is essentially anti-naturalistic, or at best scarcely interested in the world of nature (White 1967). The possibility that a monolithic understanding of Christianity may be a pure abstraction of a more diverse reality including a myriad of spiritual variants is hardly considered.

Whereas these issues stem from engrained perceptions, the third cause involves the modernistic regimes of land management that have been implemented across Europe since the nineteenth century. Changes in patterns of ownership and the role of expert scientific knowledge became distinctive tools in the pursuit of systematic control over natural and agricultural resources (Sponsel 2012). These developments were also key passages in the formation of nation states and the modernization of national economies (Scott 1998). In this process, the holdings of many religious communities were confiscated by the state (Mallarach and Papayannis 2009; Mallarach 2012a; Frascaroli 2013). Community rights over common properties and resources of local communities were also largely altered into private property regimes (Scott 1998). At the biological level, these changes acted as triggers for widespread and detrimental effects that were only partially compensated by the creation of protected areas and 'land-sparing' approaches to conservation (Antinori 2009; Mallarach and Papayannis 2009). The suppression of traditional land tenures also had destructive impacts on customary forms of resource management, TEK, and other cultural values of biodiversity (Mallarach 2012a). Overall, these developments created a permanent rift between many local and religious communities and secular authorities, and further contributed to the exclusion of the former ones from matters of land management.

Ethnographic and anthropological research can help to unpack and overcome the preconceptions that have stood against an unbiased understanding of SNS as biocultural refugia in Europe. Land management regimes, however, cannot be changed through research alone; they require adequate policies and management actions. In the next section we explore what guidance may be developed to achieve this purpose.

21.3.1 Assessing the IUCN–UNESCO Guidelines in the Context of European Sacred Natural Sites

The best practice guidelines of the World Commission on Protected Areas *SNS, Guidelines for Protected Area Managers* (henceforth 'IUCN-UNESCO Guidelines'), were devised by IUCN and UNESCO to enhance management and conservation of SNS in and around protected areas worldwide (Wild and Mcleod 2008; see Box 1). These guidelines were primarily focused on the SNS of indigenous people, as defined by Borrini-Feyerabend et al. (2004) and under the principle of self-determination of the United Nations Declaration on the Rights of indigenous peoples (UNDRIP 2008). In Europe, groups that self-identify or are officially recognized as indigenous are few, and most SNS are related to mainstream faiths. While the application of management guidelines would be beneficial also for SNS in Europe, that guidance will need to be adjusted to the specific spiritual and cultural contexts of European SNS.

In this last section, we present the results of an expert assessment that we conducted on the IUCN–UNESCO Guidelines. This effort responds to a call made in the IUCN–UNESCO Guidelines themselves, which state that more research should be undertaken to better understand the SNS of mainstream faiths and Western countries (Wild and McLeod 2008).

Box. 1 The IUCN–UNESCO Guidelines *Sacred Natural Sites: Guidelines for Protected Area Managers* have been developed by IUCN and UNESCO to promote cooperation between PA managers and site custodians, and thus enhance management and conservation of SNS around the world.

The Guidelines are clustered under six general principles. These principles were conceived as sufficiently broad and universal to allow their application across different cultural and religious settings. They are

Principle 1: Recognize SNS already located in protected areas
Principle 2: Integrate SNS located in protected areas into planning processes and management programs
Principle 3: Promote stakeholder consent, participation, inclusion, and collaboration
Principle 4: Encourage improved knowledge and understanding of SNS
Principle 5: Protect SNS while providing appropriate management access and use
Principle 6: Respect the rights of sacred natural site custodians within an appropriate framework of national policy

Thus far, the Guidelines have been translated into seven languages: English, Russian, Spanish, Estonian, French, Korean, and Japanese. The essential parts have also been translated into Italian, Persian, and Greek.

Sources: Wild and McLeod (2008); Verschuuren et al. (2015).

21.3.1.1 Methods

The IUCN–UNESCO Guidelines consist of 44 guidance points (see Appendix), grouped in six general principles (Box 1). We reviewed the 44 guidance points in light of our own findings on European SNS. We distinguished between guidance points that are 'applicable,' 'partly applicable,' and 'irrelevant or not applicable,' and noted any concerns with their application. We also identified gaps in thematic areas and concepts that are pertinent to European SNS and mainstream faiths, but not covered or treated only marginally in the current Guidelines. We develop our findings into recommendations for guidelines that are suitable to the SNS of mainstream faiths in Europe. This approach is similar to a survey conducted on the same topic in 2010 among participants in the third workshop of the Delos Initiative (Mallarach 2012b).

21.3.1.2 Results

Of the 44 guidance points, we concluded that 27 (61 %) can be directly applicable to SNS of mainstream faiths in European contexts, 11 (25 %) are at least partly applicable, while the remaining 6 (14 %) are irrelevant or non-applicable (see Table 21.1). The guidance points grouped under Principles 2, 3, and 5 appeared the least suitable to European and mainstream faith contexts. We found most difficulties

Table 21.1 Results of the expert assessment about the applicability of the IUCN–UNESCO Guidelines to SNS of mainstream faiths in Europe

Assessment outcome	Principles						Tot	Guidance points
	1	2	3	4	5	6		
Applicable	4 (80 %)	6 (55 %)	2 (40 %)	6 (75 %)	5 (50 %)	4 (80 %)	27 (61 %)	1.1, 1.2, 1.3, 1.5, 2.1, 2.6, 2.7, 2.8, 2.9, 2.10, 3.3, 3.5, 4.1, 4.2, 4.4, 4.5, 4.7, 4.8, 5.2, 5.3, 5.5, 5.5, 5.6, 5.7, 6.1, 6.2, 6.4, 6.5
Partly applicable	1 (20 %)	3 (27 %)	3 (60 %)	2 (25 %)	2 (20 %)	0 (0 %)	11 (25 %)	1.4, 2.2, 2.4, 2.5, 3.1, 3.2, 3.4, 4.3, 4.6, 5.1, 5.8
Irrelevant or non-applicable	0 (0 %)	2 (18 %)	0 (0 %)	0 (0 %)	3 (30 %)	1 (20 %)	6 (14 %)	2.3, 2.11, 5.4, 5.9, 5.10, 6.3

Number and proportion of applicable, partly applicable, and non-applicable guidance points in each General Principle are displayed to lie with the applicability of two terms, 'Indigenous' and 'secrecy.' Also, four concepts that hold particular significance with regard to SNS in European and mainstream faith contexts are not covered, or only partially developed in the IUCN–UNESCO Guidelines. They are (1) monasticism, (2) religious syncretism and layering, (3) pilgrimage, and (4) connectivity. A fifth concept, tourism, is mentioned in the IUCN–UNESCO Guidelines but is likely to have a substantial impact on European SNS, and therefore requires further elaboration into appropriate guidance. Our results are essentially in line with the 2011 Delos Initiative's survey regarding both the number of not directly applicable guidelines, and the absence of key concepts relevant to mainstream faiths and modernized contexts (Mallarach 2012b).

21.3.1.3 Discussion and Recommendations

Publicity in mapping and registration Several guidance points in the IUCN–UNESCO Guidelines (1.4, 2.2, 2.3, 2.4, 4.6) advise secrecy and confidentiality regarding the identification and public inventorying of SNS. This is done to secure that cultural protocol is followed and custodians remain in control over the information about their SNS. These indications, however, appear less relevant in relation to European SNS, as in many cases national policies have already led to disclosure and documentation of sites of worship. This means that the locations of most of these places are well known and possibly listed in heritage inventories or other public registers.

Because of legal protection and effective enforcement, SNS in European countries are under relatively little threat from destructive activities, excluding the need for secrecy and confidentiality as tools for their preservation. Nonetheless, care should be taken that SNS remain appropriately protected in those places where religious institutions, the state, and free market development pose threats to their preservation, and where effective law implementation fails. In such cases, mapping has often revealed itself to be an important tool for the conservation of SNS in Europe. In Lithuania, for example, special surveys recently identified and mapped over 2500 pre-Christian SNS that had originally fallen between the cracks of natural and cultural heritage protection (Vaitkevic 2010). With an increasing influence from the Christian Orthodox Church, some of those places were being forgotten or exposed to threats. Similar trends have been underway also in Estonia (Kaasik 2012; Valk 2012).

Furthermore, many local communities within Europe seem keen to celebrate their SNS and make them better known, rather than to conceal them. This is often motivated by religious reasons, and the genuine drive to make more people part of an important revelation. Social and economic considerations, however, including tourism development and the prestige that comes from well-attended celebrations, are also very common.

Managing tourism The IUCN–UNESCO Guidelines indicate tourism as offering 'the potential for economic benefits to indigenous and local communities' (point 5.4). European SNS appear to be consistently tied up with expectations about tourism development, and their potential to attract visitors and generate revenue (Shackley 2001). In many post-industrial economies, tourism is indeed perceived as one the most desirable forms of economic activity. Yet, tourism at SNS has revealed itself to be a double-edged sword. Excessive visitor pressures can have negative impacts on both the spiritual and biological values of SNS, which is visible at numerous monasteries in Central Italy (Frascaroli 2013), Spain (Mallarach and Papayannis 2010), and England (Wild 2012).

The reasons for inadequate visitor management can be diverse. Especially in countries where religious estates were once confiscated by the state—such as Italy, Spain, France, and the former communist countries—secular authorities may exclude monastic communities from the planning process, or have little consideration for the spiritual values of their sacred sites. In other instances, monastic communities themselves have embraced habits scarcely compatible with the original vocations of refrain and austerity, and give exclusive priority to generating revenue and increasing visitor access in the management of their sites (Mallarach and Papayannis 2009; Mallarach 2012a). Still in other cases, monastic communities may only be aware of the liturgical, artistic, and historical importance of their sites, and largely ignore their biological value (Frascaroli 2013).

Managing pilgrimage Together with tourism development, changes in the practice of pilgrimage can also lead to excessive visitor flows at many SNS. Since time immemorial, pilgrimage has represented a fundamental mode to experience and shape the environment. Some landscapes would not be what they are, without the networks of paths and the periodic disturbances that marching pilgrims have maintained over the centuries. This situation, though, has been dramatically modified in the last four decades, as even some of the most remote SNS in Europe have been connected to the road network, and walking or horseback riding has been replaced by private cars and coaches (cf. Mallarach and Papayannis 2010). While the wish of custodians to have as many people as possible partaking in the grace of a site is understandable, this can have severe repercussions on both the ecology and spiritual atmosphere of some SNS. Guidance for European SNS should expand on the current Guidelines (where pilgrimage is mentioned at point 5.2), and specifically address the challenges posed by current forms of pilgrimage in Europe.

Indigenous and mainstream faiths, syncretism, and layering The notion of 'Indigenous,' as presented in points 4.3 and 5.9 of the IUCN–UNESCO Guidelines, would require elaboration and complementation in the context of many European SNS. Besides the groups that are officially recognized as indigenous peoples, such as the Sami, much of the unique TEK and intangible heritage encountered at European SNS derives from pre-Christian or folk spiritualities. Although not formally defined as 'Indigenous,' these instances of syncretism and layering of religious traditions at SNS can pose similar management and ethical challenges, and practical guidance would be necessary on how to address them.

It should be recognized that mainstream religions are 'multi-scale' phenomena that vary at different levels of organization. Underneath the top layer of the highest clerical hierarchies, we find an intermediate level consisting of different strands of the same faith. In the case of Catholicism, examples of these strands include the different monastic and mendicant traditions (e.g., Franciscans, Benedictines, Carthusians, etc.), all with their distinctive features and forms of spirituality, which they maintain across regional and even national borders. A bottom layer, finally, consists of all place-bound and regional variants, characterized by the encounter of a mainstream faith with more specifically local beliefs and sensibilities.

Needless to say, situations of open or latent conflict between two or more of these layers are far from uncommon. Guidelines designed for SNS of mainstream faiths in Europe should acknowledge similar situations and help to sensitize managers about potential tensions. Whenever feasible, they should solicit the negotiation of agreements aimed to preserve local variants and syncretic spiritualities, without raising conflicts with the higher levels of religious organization. These local spiritual manifestations, indeed, often carry very high biocultural value, but tend to be opposed and normalized over time by religious hierarchies. Trying to mediate similar conflicts can be an important process, also in the perspective of renewing the dialog between religious authorities, conservationists, and other stakeholders at a broader scale.

In a similar way, and consistently with what the IUCN–UNESCO Guidelines already suggest for the SNS of indigenous peoples (4.3, 5.1, 5.6, 6.4), the TEK, cultural uses, and landscape preferences of local rural communities should be recorded and carefully considered by land managers and planners in and around SNS. Over the centuries, those communities have created the unique biocultural landscapes, which in many countries form the backbone of the present-day PA network. As illustrated, their agricultural and pastoral activities were often the part of a rich system of spiritual meanings and practices. Priority, therefore, should be given to guaranteeing survival, transformation, and full valorization of those activities, in ways that they are perceived meaningful by local communities. Other forms of management frequently favored by PA mangers, such as rewilding, should follow consensual process and appropriate zoning, especially in culturally sensitive areas with SNS.

Monasticism, tenure rights, and shared governance Monasticism is a fundamental way of organizing religious life typical of many mainstream faiths from Western Europe to East Asia. In Europe, monasticism is particularly associated with the Orthodox and Catholic traditions, and its impacts on European history and culture can hardly be overestimated (e.g., Salvatorelli 1929; Lawrence 1984). Despite many examples of how monastic practices have contributed to nature conservation (Romano 2010), the influence of monasticism on European nature is only seldom acknowledged (Frascaroli 2013). This also holds true for the substantial overlap between PAs and current or former monastic lands (Mallarach and Papayannis 2009).

The IUCN–UNESCO Guidelines do not expressly cover the particularities of monasticism with regard to SNS management. They do offer guidance points that affirm the rights of traditional custodians to govern and participate in the use and management of SNS (mostly 1.3, 1.4, 3.1, 3.2, 5.1, 5.4, 5.5, 6.4). These can form a starting point for guidance tailored to monastic lands. Such guidance, however, should also address the shifting values and habits of monastic communities, help to raise awareness of the biological values of monastic sites, and consider conflicts over tenure rights.

Tenure can be a primary source of contention in countries where religious estates were at some point expropriated by the State. Conversations with a number of informants in Central Italy indicated that even when confiscation programs were enacted over 100 years ago, they have left a profound drift between monastic communities and secular authorities. Property rights, in contrast, are often fundamental for the involvement of local communities in land management and planning (Agrawal 2005; Borrini-Feyerabend et al. 2007). This state of affairs can pose significant obstacles when attempting to incorporate monastic communities in the management of SNS, or soliciting greater sensibility toward conservation issues from their side. While a radical modification of land tenure may not be conceivable, ad hoc compromises and innovative solutions should be explored and experimented with. The current economic crisis, which has struck many PAs and public land management services especially in southern Europe, can represent an opportunity to return conservation of some SNS to local and religious communities.

Restoration and connectivity The IUCN–UNESCO Guidelines suggest revitalizing damaged or desecrated SNS, also as an important step in the restoration of wider areas (2.6, 5.8). These indications can be especially pertinent in European settings, where a number of SNS were abandoned for different reasons, and especially following state confiscations (Mallarach 2012a; Frascaroli 2013). Guidelines aimed at mainstream faiths and European SNS should more strongly emphasize the potential of abandoned SNS as nodes of biocultural diversity. Despite their current disrepair, in fact, many of these sites still harbor significant biodiversity and heritage values (Frascaroli 2013). As such, they can be starting points not only of ecological restoration (as the IUCN–UNESCO Guidelines already suggest), but also of social and cultural revitalization of conservation practices.

Similarly, the IUCN–UNESCO Guidelines underline the potential role of SNS for ecological connectivity (2.6, 2.8). These points should be expanded and contemplate other ways of networking, such as 'cultural heritage connectivity' (e.g., Mikusiński et al. 2013). Our review of SNS in Central Italy showed that these sites seldom exist in isolation. Rather, they tend to be nodes in existing webs of symbolic as well as physical relations. For example, SNS can be landmarks in a network of pilgrimage trails (Serenelli 2012), or settlements established along the itineraries of a charismatic founder (e.g., the Benedictine monasteries between Subiaco and Cassino). Strengthening, emphasizing, or reviving this network of relations can be important in the perspective of supporting the cultural identity of wider areas, or elaborating landscape scale management plans. Systematic and rigorous research would be pivotal to address this underexplored topic, and assist with the planning of cultural as well as ecological networks.

Stakeholder involvement and interconnectedness of values One of the key points in the IUCN–UNESCO Guidelines is that traditional custodians, as well as other possible stakeholders, should be recognized and actively included in the management and planning of SNS (1.3, 3.3). This can be a central tenet also in guidance aimed at the SNS of mainstream faiths in Europe. Our review, however, suggests that some common difficulties regarding the identification and involvement of relevant stakeholders in the European context should be considered.

Problems can be posed by the layering of religious traditions and institutions. For example, a number of shrines in Central Italy are tended by monastic communities, but have traditionally held their greatest significance to rural communities in the area. In some instances, these communities even retain the organization of some of the most intense pilgrimages and rituals celebrated at SNS in the area. Care should therefore be taken to include this 'local community layer' in participatory planning of SNS. Our observations suggest that this is very rarely the case, and groups of local believers are seldom acknowledged as stakeholders by PA managers and other administrators, even when they play an important role in the customary use of SNS.

Identifying and recognizing relevant stakeholders can also be a challenge with regard to abandoned SNS. In this case, representatives could be found among the last religious orders that inhabited the sites, or local communities that have maintained a special bond with them. As a last resort, management and governance of abandoned sites should be assigned to public institutions, such as heritage agencies, as already advised in the current Guidelines (point 6.1).

Evidently, the diverse values that different stakeholders carry can have implications in the planning process. Occasionally, this might lead to the emergence of conflicts over the meanings and values of particular sites, which can be further exacerbated by the erosion of traditional worldviews among custodians and worshippers. In such cases, adequate safeguarding of the biological value of SNS might no longer be guaranteed by local stakeholders alone. Nonetheless, care should be taken that the conservation measures implemented by land managers are not detrimental of local spiritual values and traditional uses of SNS.

In general, awareness of the interconnectedness of spiritual, biological, and cultural values of SNS is key and needs to be promoted through specific policy actions. In Central Italy, the trumping of one aspect over others often lies at the heart of poor and ineffective SNS management. A balance between those three sets of values should be indicated as a management priority, and achieved through forms of participatory planning, education, and mutual learning inclusive of traditional stakeholders as well as public administrators.

21.4 Conclusions

After its widespread secularization, incorporating spiritual values into biodiversity management has been a slow process across Europe. This has resulted in under-appreciating the contribution of spiritual and religious traditions to shaping and conserving local biocultural heritage, and in fact overlooking some of the deepest linkages between biological and cultural diversity.

Evidence from sacred sites in Central Italy supports the understanding that deep connections between spirituality and biocultural heritage are relevant also in a European context involving a mainstream faith. Empirical data demonstrate that SNS in Central Italy harbor important biological and cultural values, and support the conservation of habitats maintained through traditional livelihoods, such as pastoralism, which are otherwise quickly eroding across Europe (Fernández-Giménez and Fillat Estaque 2012). In all, SNS in Central Italy appear to act as biocultural refugia, that is, places that shelter significant biodiversity as well as practical knowledge on how to manage the environment (Barthel et al. 2013). These findings show similarities with those of case studies from the Delos Initiative, which in turn suggests that they might be at least partly generalizable to the wider European context.

Recognizing the importance of the spiritual, biological, and cultural values of SNS at the policy level, including their function as biocultural networks, would be essential to biocultural conservation in Europe. Guidance on how to achieve this is currently lacking. The IUCN–UNESCO *SNS, Guidelines for Protected Areas Managers,* offer a reference framework for conserving and valorizing SNS, but they need to be complemented with guidance specific to SNS of mainstream faiths in Europe. Based on research experience in Central Italy and an expert assessment of the IUCN–UNESCO Guidelines, we conclude that about two-thirds of the original guidance points could be retained in drafting guidelines applicable to European and mainstream faith contexts. At the same time, some gaps in the original guidelines should be filled. In particular, we suggest that policy guidelines aimed at Europe would need

- to move beyond the concept of 'Indigenous' in ways that can better fit modernized and syncretic settings;
- to reassess the need for confidentiality in SNS identification and mapping;
- to acknowledge the specificities of monastic SNS and historical conflicts regarding land tenure;
- to account for the syncretism and layering of religious traditions that characterize many SNS in Europe;
- to address the challenges posed by tourism and current forms of pilgrimage;
- to better integrate SNS in conservation planning as a means for linking biological, cultural, and spiritual values across biocultural networks.

The effectiveness of conservation efforts is often hampered by insufficient consideration for the traditions and heritage of local populations. This can lead to mistrust and resentment toward conservation planners, and European countries are no exception. Biocultural approaches, including more explicit recognition of SNS and their values, can contribute to ameliorating similar miscomprehensions. In the most positive cases, local sacred sites and beliefs can act as effective symbolic and social platforms for establishing partnerships between local communities and PA managers, even after initial mistrust (Frascaroli 2014). Producing and applying appropriate policy guidelines can help maximizing the full potential of SNS for conserving the precious biocultural heritage of Europe.

Acknowledgments We are thankful to the people of Central Italy for their hospitality and sharing the insights that are at the basis of this work. We also acknowledge Josep-Maria Mallarach, Marc Hall, and Vita de Waal for stimulating discussion. Fabrizio Frascaroli is funded by the Cogito Foundation.

Appendix: The Forty-Four Guidance Points of IUCN–UNESCO's *Sacred Natural Sites: Guidelines for Protected Area Managers*

Guideline 1.1	**Natural and cultural values:** Recognize that sacred natural sites (SNS) are of vital importance to the safeguarding of natural and cultural values for current and future generations
Guideline 1.2	**Ecosystem services and human well-being:** Recognize that sacred natural sites have great significance for the spiritual well-being of many people and that cultural and spiritual inspirations are part of the ecosystem services that nature provides
Guideline 1.3	**Recognition:** Initiate policies that formally recognize the existence of sacred natural sites within or near government or private protected areas and affirm the rights of traditional custodians to access and play an appropriate, ideally key, role in managing sacred natural sites now located within formal protected areas
Guideline 1.4	**Consultation:** Include the appropriate traditional cultural custodians, practitioners, and leaders in all discussions and seek their consent regarding the recognition and management of sacred natural sites within or near protected areas
Guideline 1.5	**Holistic models:** Recognize that sacred natural sites integrate social, cultural, environmental, and economic values into holistic management models that are part of the tangible and intangible heritage of humankind
Guideline 2.1	**Park planning:** Initiate planning processes to revise management plans to include the management of sacred natural sites located inside protected area boundaries
Guideline 2.2	**Identify sacred natural sites:** Where secrecy is not an issue and in close collaboration and respecting the rights of traditional custodians, identify the location, nature, use, and governance arrangements of sacred sites within and around protected areas as part of a participatory management planning process
Guideline 2.3	**Respect confidentiality:** Ensure that pressure is not exerted on custodians to reveal the location or other information about sacred natural sites and, whenever requested, establish mechanisms to safeguard confidential information shared with protected area agencies

Guideline 2.4	**Demarcate or conceal:** Where appropriate and to enhance protection, either clearly demarcate specific sacred natural sites, or alternatively, to respect the need for secrecy, locate sacred natural sites within larger strictly protected zones so exact locations remain confidential
Guideline 2.5	**Zoning:** Establish support, buffer, and transition zones around and near sacred sites, especially those that are vulnerable to adverse external impacts
Guideline 2.6	**Linkages and restoration:** Create ecological corridors between sacred natural sites and other suitable areas of similar ecology for connectivity, and in degraded landscapes consider restoring sacred natural sites as an important initial step to reviving a wider area
Guideline 2.7	**Ecosystem approach:** Adopt the ecosystem approach as the key strategy for the integrated management of land, water, and living resources that promote conservation and sustainable use in an equitable way and also include cultural and spiritual values
Guideline 2.8	**Landscape approach:** Take a landscape approach to sacred natural sites, recognizing their role in wider cultural landscapes, protected area systems, ecological corridors, and other land uses
Guideline 2.9	**Support development planning recognition:** Development planning authorities are the main planners of land use in areas outside many protected area systems. Seek their and other stakeholders' support for the recognition of sacred natural sites in the wider countryside
Guideline 2.10	**Protected area categories and governance:** Recognize that sacred natural sites exist in all of the IUCN protected area categories and governance types, and that those that fall outside formal protected area systems can be recognized and supported through different legal and traditional mechanisms according to the desires of their custodians, including as community conserved areas when appropriate
Guideline 2.11	**International dimension:** Recognize that some sacred natural sites, and the cultures that hold them sacred, cross international boundaries and that some may be within or may surround existing or potential transboundary peace parks
Guideline 3.1	**Prior consent:** Ascertain the free, prior and informed consent of appropriate custodians before including sacred natural sites within new formal protected areas and protected area systems and when developing management policies affecting sacred places
Guideline 3.2	**Voluntary participation:** Ensure that state or other stakeholder involvement in the management of sacred natural sites is with the consent and voluntary participation of appropriate custodians

Guideline 3.3	**Inclusion:** Make all efforts to ensure the full inclusion of all relevant custodians and key stakeholders, including marginalized parties, in decision making about sacred natural sites, and carefully define the processes for such decision making, including those related to higher level and national level policies
Guideline 3.4	**Legitimacy:** Recognize that different individuals and groups have different levels of legitimacy and authority in decision making about sacred natural sites
Guideline 3.5	**Conflict management:** Where relevant and appropriate, use conflict management, mediation, and resolution methods to promote mutual understanding between traditional custodians and more recent occupants, resource users, and managers
Guideline 4.1	**Multidisciplinary approach:** Promote a multidisciplinary and integrated approach to the management of sacred natural sites calling on, for example, local elders, religious and spiritual leaders, local communities, protected area managers, natural and social scientists, artists, nongovernmental organizations, and the private sector
Guideline 4.2	**Integrated research:** Develop an integrated biological and social research program that studies biodiversity values, assesses the contribution of sacred natural sites to biodiversity conservation, and understands the social dimension, especially how culturally rooted behavior has conserved biodiversity
Guideline 4.3	**Traditional knowledge:** Consistent with article 8(j) of the Convention on Biological Diversity (CBD), support the respect, preservation, maintenance and use of the traditional knowledge, innovations and practices of indigenous, and local communities specifically regarding sacred natural sites
Guideline 4.4	**Networking:** Facilitate the meeting of, and sharing of information between, traditional custodians of sacred natural sites, their supporters, protected area managers, and more recent occupants and users
Guideline 4.5	**Communication and public awareness:** Develop supportive communication, education, and public awareness programs and accommodate and integrate different ways of knowing, expression and appreciation in the development of policies, and educational materials regarding the protection and management of sacred natural sites
Guideline 4.6	**Inventories:** Subject to the free, prior, and informed consent of custodians, especially of vulnerable sites and consistent with the need for secrecy in specific cases, carry out regional, national, and international inventories of sacred natural sites and support the inclusion of relevant information in the UN World Database on Protected Areas. Develop mechanisms for safeguarding information intended for limited distribution

Guideline 4.7 **Cultural renewal:** Recognize the role of sacred natural sites in maintaining and revitalizing the tangible and intangible heritage of local cultures, their diverse cultural expressions, and the environmental ethics of indigenous, local, and mainstream spiritual traditions

Guideline 4.8 **Intercultural dialog:** Promote intercultural dialog through the medium of sacred natural sites in efforts to build mutual understanding, respect, tolerance, reconciliation, and peace

Guideline 5.1 **Access and use:** Develop appropriate policies and practices that respect traditional custodian access and use, where sacred natural sites fall within formal protected areas

Guideline 5.2 **Visitor pressures:** Understand and manage visitor pressures and develop appropriate policies, rules, codes of conduct, facilities, and practices for visitor access to sacred sites, making special provisions for pressures brought about by pilgrimages and other seasonal variations in usage

Guideline 5.3 **Dialog and respect:** Encourage ongoing dialog among the relevant spiritual traditions, community leaders, and recreational users to control inappropriate use of sacred natural sites through both protected area regulations and public education programs that promote respect for diverse cultural values

Guideline 5.4 **Tourism:** Well-managed, responsible tourism provides the potential for economic benefits to indigenous and local communities, but tourism activities must be culturally appropriate, respectful, and guided by the value systems of custodian communities. Wherever possible, support tourism enterprises that are owned and operated by indigenous and local communities, provided they have a proven record of environmental and cultural sensitivity

Guideline 5.5 **Decision-making control:** Strong efforts should be made to ensure that custodians of sacred natural sites retain decision-making control over tourist and other activities within such sites, and that checks and balances are instituted to reduce damaging economic and other pressures from protected area programs

Guideline 5.6 **Cultural use:** While ensuring that use is sustainable, do not impose unnecessary controls on the careful harvest or use of culturally significant animals and plants from within sacred natural sites. Base decisions on joint resources assessments and consensus decision-making

Guideline 5.7 **Protection:** Enhance the protection of sacred natural sites by identifying, researching, managing, and mitigating overuse, sources of pollution, natural disasters, and the effects of climate change and other socially derived threats, such as vandalism and

	theft. Develop disaster management plans for unpredictable natural and human caused events
Guideline 5.8	**Desecrations and re-sanctifying:** Safeguard against the unintended or deliberate desecration of sacred natural sites and promote the recovery, regeneration, and re-sanctifying of damaged sites where appropriate
Guideline 5.9	**Development pressures:** Apply integrated environmental and social impact assessment procedures for developments affecting sacred natural sites and in the case of the land of indigenous and local communities support the application of the Convention on Biological Diversity's Akwé: Kon Guidelines for minimizing the impacts of development actions
Guideline 5.10	**Financing:** Where appropriate, pay due attention to the suitable financing of sacred natural site management and protection, and develop mechanisms for generating and sharing revenue that take into account considerations of transparency, ethics, equity, and sustainability. Recognize that in many parts of the world poverty is a cause of the degradation of sacred natural sites
Guideline 6.1	**Institutional analysis:** Understand traditional management institutions and enable and strengthen the continued management of sacred natural sites by these institutions. Make appropriate arrangements for the adoption and management of sacred natural sites that have no current custodians, for example by heritage agencies
Guideline 6.2	**Legal protection:** Advocate for legal, policy, and management changes that reduce human and natural threats to sacred natural sites, especially those not protected within national protected areas and other land planning frameworks
Guideline 6.3	**Rights-based approach:** Root the management of sacred natural sites in a rights-based approach respecting basic human rights, rights to freedom of religion and worship, and to self-development, self-government, and self-determination as appropriate
Guideline 6.4	**Confirm custodians' rights:** Support the recognition, within the overall national protected area framework, of the rights of custodians to their autonomous control and management of their sacred sites and guard against the imposition of conflicting dominant values
Guideline 6.5	**Tenure:** Where sacred natural sites have been incorporated within government or private protected areas in ways that have affected the tenure rights of their custodians, explore options for the devolution of such rights and for their long-term tenure security

References

Agnoletti M (2014) Rural landscape, nature conservation and culture: some notes on research trends and management approaches from a (southern) European perspective. Landscape Urban Plan 126:66–73

Agrawal A (2005) Environmentality: technologies of government and the making of subjects. Duke University Press, Durham

Antinori A (2009) I sentieri del silenzio: guida agli eremi rupestri ed alle abbazie dell'Appennino Umbro-Marchigiano, 2nd edn. Società Editrice Ricerche, Folignano

Barthel S, Crumley C, Svedin U (2013) Bio-cultural refugia: safeguarding diversity of practices for food security and biodiversity. Glob Environ Chang 23:1142–1152

Berkes F (1999) Sacred ecology: traditional ecological knowledge and resource management. Taylor and Francis, London

Berkes F, Colding J, Folke C (2000) Rediscovery of traditional ecological knowledge as adaptive management. Ecol Appl 10:1251–1262

Bernard HR (2006) Research methods in anthropology: qualitative and quantitative approaches. Altamira Press, Oxford

Bernardini FF (ed) (2000) Nessuno vada nella terra senza luna: etnografia del pellegrinaggio al Santuario della Santissima Trinità. Provincia di Roma, Roma

Bernbaum E (2006) Sacred mountains: themes and teachings. Mt Res Dev 26:304–309

Bernbaum E (2010) Sacred mountains and global changes: impacts and responses. In: Verschuuren B, Wild R, McNeeley J, Oviedo G (eds) Sacred natural sites: conserving nature and culture. Earthscan, London, pp 33–41

Bhagwat S, Rutte C (2006) Sacred groves: potential for biodiversity management. Front Ecol Environ 4:519–524

Borrini-Feyerabend G, Kothari A, Oviedo G (2004) Indigenous and local communities and protected areas: towards equity and enhanced conservation. Guidance on policy and practice for co-managed protected areas and community conserved areas. IUCN, Gland and Cambridge

Borrini-Feyerabend G, Pimbert M, Farvar MT, Kothari A, Renard Y (2007) Sharing power: learning-by-doing in co-management of natural resources throughout the world. Earthscan, London

Bosch JF, Varela JG (2007) Doñana National and Natural Parks: Sanctuary de la Virgen del Rocío, Spain. In: Mallarach JM, Papayannis T (eds) Protected areas and spirituality: proceedings of the first workshop of the Delos initiative, Montserrat 2006. IUCN, Gland, pp 175–199

Byrne D (2010) The enchanted earth: numinous sacred sites. In: Verschuuren B, Wild R, McNeeley J, Oviedo G (eds) Sacred natural sites: conserving nature and culture. Earthscan, London, pp 53–61

Catanoiu S (2007) Vanatori Neamt Natural Park: Romanian Jerusalem, Neamt region, Moldavia, Romania. In: Mallarach JM, Papayannis T (eds) Protected areas and spirituality: proceedings of the first workshop of the Delos initiative, Montserrat 2006. IUCN, Gland, pp 289–309

Catanoiu S (2009) Buila-Vânturarita National Park, Valcea county, Romania. In: Papayannis T, Mallarach JM (eds) The sacred dimension of protected areas: proceedings of the second workshop of the Delos initiative, Ouranoupolis 2007. IUCN, Gland, pp 137–151

Colding J, Folke C (2001) Social taboos: 'invisible' systems of local resource management and biological conservation. Ecol Appl 11:584–600

CSC (2003) Censimento dei Santuari Cristiani in Italia. http://www.santuaricristiani.iccd.beniculturali.it/. Accessed 8 Jan 2013

De Waal V (2012) The cultural and spiritual sites of the Parco Nazionale della Majella, Italy. In: Mallarach JM, Papayannis T, Väisänen R (eds) The diversity of sacred lands in Europe: proceedings of the third workshop of the Delos initiative, Inari/Aanaar 2010. IUCN, Gland, pp 111–123

Delgado F, Escobar C, Verschuuren B, Hiemstra W (2010) Sacred natural sites, biodiversity and well-being: the role of sacred sites in endogenous development in the COMPAS network. In: Verschuuren B, Wild R, McNeeley J, Oviedo G (eds) Sacred natural sites: conserving nature and culture. Earthscan, London, pp 188–197
Dudley N, Higgins-Zogib L, Mansourian S (2009) The links between protected areas, faiths, and sacred natural sites. Conserv Biol 23:568–577
Dudley N, Bhagwat S, Higgins-Zogib L, Lassen B, Verschuuren B, Wild R (2010) Conservation of biodiversity in sacred natural sites in Asia and Africa: a review of the scientific literature. In: Verschuuren B, Wild R, McNeeley J, Oviedo G (eds) Sacred natural sites: conserving nature and culture. Earthscan, London, pp 19–32
EC (European Commission) (2013) Interpretation manual of European Union habitats—EUR28. http://ec.europa.eu/environment/nature/legislation/habitatsdirective/docs/Int_Manual_EU28.pdf. Accessed 12 Sept 2014
Eliade M (1959) The Sacred and the profane. Houghton Mifflin Harcourt, New York
Fernández-Giménez ME, Fillat Estaque F (2012) Pyrenean pastoralists' ecological knowledge: documentation and application to natural resource management and adaptation. Hum Ecol 40:287–300
Frascaroli F (2013) Catholicism and conservation: the potential of sacred natural sites for biodiversity management in Central Italy. Hum Ecol 41:587–601
Frascaroli F (2014) Note sulla dimensione ecologica del culto di San Franco: spunti per una rilettura contemporanea di una fede tradizionale. In: Fiordigigli I (ed) San Franco di Assergi: storia di eremitismo e santità alle pendici del Gran Sasso. Arkhé, L'Aquila, pp 257–263
Frascaroli F, Bhagwat S, Guarino R, Chiarucci A, Schmid B (in press) Shrines in Central Italy conserve plant diversity and large trees. AMBIO
Frascaroli F, Bhagwat S, Diemer M (2014) Healing animals, feeding souls: ethnobotanical values at sacred sites in Central Italy. Econ Bot. doi:10.1007/s12231-014-9290-7
Gadgil M, Berkes F, Folke C (1993) Indigenous knowledge for biodiversity conservation. Ambio 22:151–156
Gómez-Baggethun E, Reyes-García V, Olsson P, Montes C (2012) Traditional ecological knowledge and community resilience to environmental extremes: a case study in Doñana, SW Spain. Glob Environ Chang 22:640–650
Guri BY, Verschuuren B (2008) Community well-being in Ghana: an African perspective. In: Verschuuren B, Subramanian SM, Hiemstra W (eds) Community wellbeing in biocultural landscapes, are we living well?. Practical Action, Rugby, pp 78–100
Herzfeld M (2001) Anthropology: theoretical practice in culture and society. Wiley, Malden
Hughes JD, Chandran MDS (1998) Sacred groves around the earth: an overview. In: Ramakrishnan PS, Saxena KG, Chandrashekara UM (eds) Conserving the sacred for biodiversity management. Oxford and IBH, New Delhi, pp 69–86
Kaasik A (2012) Conserving sacred natural sites in Estonia. In: Mallarach JM, Papayannis T, Väisänen R (eds) The diversity of sacred lands in Europe: proceedings of the third workshop of the Delos initiative, Inari/Aanaar 2010. IUCN, Gland, pp 61–73
Latour B (1993) We have never been modern. University Press, Cambridge
Lawrence CH (1984) Medieval monasticism: forms of religious life in Western Europe in the middle ages. Longman, New York
Lee C, Schaaf T (eds) (2003) The importance of sacred natural sites for biodiversity conservation. UNESCO, Paris
Loh J, Harmon D (2005) A global index of biocultural diversity. Ecol Indic 5:231–241
Long J, Tecle A, Burnette B (2003) Cultural foundations for ecological restoration on the White Mountain Apache Reservation. Conserv Ecol 8:4
Lyratzaki I (2007) Meteora World Heritage Site, Thessaly, Greece. In: Mallarach JM, Papayannis T (eds) Protected areas and spirituality: proceedings of the first workshop of the Delos initiative, Montserrat 2006. IUCN, Gland, pp 251–261
Maffi L (2005) Linguistic, cultural, and biological diversity. Annl Rev Anthro 34:599–617

Maffi L, Woodley E (2010) Biocultural diversity conservation: a global sourcebook. Earthscan, London
Mallarach JM (2012a) Monastic communities and nature conservation: overview of positive trends and best practices in Europe and the Middle East. In: Mallarach JM, Papayannis T, Väisänen R (eds) The diversity of sacred lands in Europe: proceedings of the third workshop of the Delos initiative, Inari/Aanaar 2010. IUCN, Gland, pp 157–173
Mallarach JM (2012b) Applicability of the IUCN-UNESCO Guidelines for protected area managers on sacred natural sites: first assessment. In: Mallarach JM, Papayannis T, Väisänen R (eds) The diversity of sacred lands in Europe: proceedings of the third workshop of the Delos initiative, Inari/Aanaar 2010. IUCN, Gland, pp 260–270
Mallarach JM, Catanoiu S (2009) Rila Monastery Natural Park, Bulgaria. In: Papayannis T, Mallarach JM (eds) The sacred dimension of protected areas: proceedings of the second workshop of the Delos initiative, Ouranoupolis 2007. IUCN, Gland, pp 173–175
Mallarach JM, Papayannis T (eds) (2007) Protected areas and spirituality: proceedings of the first workshop of the Delos initiative, Montserrat 2006. IUCN, Gland
Mallarach JM, Papayannis T (2009) Reflections on the management of monastic lands and facilities. In: Papayannis T, Mallarach JM (eds) The sacred dimension of protected areas: proceedings of the second workshop of the Delos initiative, Ouranoupolis 2007. IUCN, Gland, pp 191–199
Mallarach JM, Papayannis T (2010) Sacred natural sites in technologically developed countries: reflections from the experience of the Delos initiative. In: Verschuuren B, Wild R, McNeeley J, Oviedo G (eds) Sacred natural sites: conserving nature and culture. Earthscan, London, pp 198–208
Mallarach JM, Papayannis T, Väisänen R (eds) (2012) The diversity of sacred lands in Europe: proceedings of the third workshop of the Delos initiative, Inari/Aanaar 2010. IUCN, Gland
Marucci G (1999) Fratelli in grotta: un rituale maschile di solidarietà. Edizioni Andromeda, Roma
Mikusiński G, Blicharska M, Antonson H, Henningsson M, Göransson G, Angelstam P, Seiler A (2013) Integrating ecological, social and cultural dimensions in the implementation of the landscape convention. Landscape Res 38:384–393
Myers N, Mittermeier NA, Mittermeier CG, da Fonseca GAB, Kent J (2000) Biodiversity hotspots for conservation priorities. Nature 403:853–858
Naveh Z, Whittaker RH (1980) Structural and floristic diversity of shrublands and woodlands in northern Israel and other Mediterranean areas. Vegetatio 41:171–190
Nolan ML, Nolan S (1989) Christian pilgrimage in modern Western Europe. University of North Carolina Press, Chapel Hill
Olson DM, Dinerstein E (2002) Priority ecoregions for global conservation. Ann Mo Bot Gard 89:199–224
Ostrom E (1990) Governing the commons: the evolution of institutions for collective actions. Cambridge University Press, Cambridge
Oteros-Rozas E, Ontillera-Sánchez R, Sanosa P, Gómez-Baggethun E, Reyes-García V, González J (2013) Traditional ecological knowledge among transhumant pastoralists in Mediterranean Spain. Ecol Soc 18:33
Papayannis T, Mallarach JM (eds) (2009) The sacred dimension of protected areas: proceedings of the second workshop of the Delos initiative, Ouranoupolis 2007. IUCN, Gland
Papayannis T, Pritchard D (2010) Wetland cultural and spiritual values, and the Ramsar convention. In: Verschuuren B, Wild R, McNeeley J, Oviedo G (eds) Sacred natural sites: conserving nature and culture. Earthscan, London, pp 180–187
Pesic N, Dingarac S, Pesic D (2012) Special nature reserve Milesevka and the Mileseva Monastery. In: Mallarach JM, Papayannis T, Väisänen R (eds) The diversity of sacred lands in Europe: proceedings of the third workshop of the Delos initiative, Inari/Aanaar 2010. IUCN, Gland, pp 189–201
Pungetti G, Oviedo G, Hooke D (eds) (2012) Sacred species and sites: advances in biocultural conservation. University Press, Cambridge
Romano R (ed) (2010) Codice Forestale Camaldolese: le radici della sostenibilità. INEA, Roma

Rutte C (2011) The sacred commons: conflicts and solutions of resource management in sacred natural sites. Biol Conserv 144:2387–2394
Salvatorelli L (1929) San Benedetto e l'Italia del suo tempo. Laterza, Bari
Schaaf T, Lee C (eds) (2006) Conserving cultural and biological diversity: the role of sacred natural sites and cultural landscapes. UNESCO, Paris
Schama S (1995) Landscape and memory. Harper and Collins, London
Scott JC (1998) Seeing like a state: how certain schemes to improve the human condition have failed. Yale University Press, New Haven
Secretariat of the Convention on Biological Diversity (2004) Akwé: Kon voluntary guidelines for the conduct of cultural, environmental and social impact assessment regarding developments proposed to take place on, or which are likely to impact on, sacred sites and on lands and waters traditionally occupied or used by indigenous and local communities. CBD Guidelines Series, Montreal
Selvi F, Valleri M (2012) Cork oak woodlands in the north Tyrrhenian area (Italy): distribution and plant species diversity of a relict forest ecosystem. Biodivers Conserv 21:3061–3078
Serenelli C (2012) Landscape linkages between cultural and spiritual values: the wetland of Colfiorito and the Lauretana pilgrimage route in the Plestian Plateaus. In: Mallarach JM, Papayannis T, Väisänen R (eds) The diversity of sacred lands in Europe: proceedings of the third workshop of the Delos initiative, Inari/Aanaar 2010. IUCN, Gland, pp 139–153
Shackley M (2001) Managing sacred sites: service provision and visitor experience. Continuum, London
Sponsel LE (2012) Spiritual ecology: a quiet revolution. ABC-CLIO, Santa Barbara
Tengö M, Johansson K, Rakotondrasoa F, Lundberg J, Andriamaherilala JA, Rakotoarisoa JA, Elmqvist T (2007) Taboos and forest governance: informal protection of hot spot dry forest in southern Madagascar. Ambio 36:683–691
Tiedje K (2007) The promise of the discourse of the sacred for conservation (and its limits). J Stud Relig Nat Cult 1(3):326–339
Toledo V (2002) Ethnoecology: a conceptual framework for the study of indigenous knowledge of nature. In: Stepp JR, Wyndham FS, Zarger RK (eds) Ethnobiology and biocultural diversity: proceedings of the 7th international congress of ethnobiology. International Society of Ethnobiology, Athens, pp 511–522
Tuan YF (1974) Topophilia: study of environmental perception, attitudes, and values. Prentice-Hall, Englewood Cliffs
Turner NJ, Ari Y, Berkes F, Davidson-Hunt I, Ertug ZF, Miller A (2009) Cultural management of living trees: an international perspective. J Ethnob 29:237–270
UNDRIP (2008) Declaration on the rights of indigenous peoples. United Nations, General Assembly, 61st session, agenda item 68, report of the Human Rights Council. http://www.un.org/esa/socdev/unpfii/documents/DRIPS_en.pdf. Accessed 28 Oct 2014
Vaitkevic V (2010) Use and reuse of ancient sacred places in Mikytai, Žemaitija National Park, NW Lithuania. In: Mallarach JM, Papayannis T, Väisänen R (eds) The diversity of sacred lands in Europe: proceedings of the third workshop of the Delos initiative, Inari/Aanaar 2010. IUCN, Gland, pp 75–83
Valk H (2012) Sacred natural places of Estonia: regional aspects. Folklore 42:45–66
Verschuuren B (2010) Arguments for developing biocultural conservation approaches for sacred natural sites. In: Verschuuren B, Wild R, McNeeley J, Oviedo G (eds) Sacred natural sites: conserving nature and culture. Earthscan, London, pp 62–71
Verschuuren B, Wild R (2012) Safeguarding sacred natural sites: sustaining nature and culture. In: Verschuuren B, Wild R (eds) Langscape. Terralingua, Salt Spring City, pp 12–19
Verschuuren B, Wild R, McNeeley J, Oviedo G (eds) (2010) Sacred natural sites: conserving nature and culture. Earthscan, London
Verschuuren B, Wild R, Verschoor G (in press, exp. 2015) Connecting policy and practice for the conservation of sacred natural sites. In: O'sarmiento F, Hitchner S (eds) Indigenous revival and sacred sites conservation. McGraw-Hill Press, New York
White LJ (1967) The historical roots of our ecologic crisis. Science 155:1203–1207

Wild R (2012) Holy island of Lindisfarne and the modern relevance of celtic 'nature saints'. In: Mallarach JM, Papayannis T, Väisänen R (eds) The diversity of sacred lands in Europe: proceedings of the third workshop of the Delos initiative, Inari/Aanaar 2010. IUCN, Gland, pp 125–137

Wild R, McLeod C (2008) Sacred natural sites: guidelines for protected areas managers. IUCN, Gland

Zent S, López-Zent E (2007) On biocultural diversity from a Venezuelan perspective: tracing the interrelationships among biodiversity, culture change, and legal reforms. In: McManis C (ed) Biodiversity and the law: intellectual property, biotechnology and traditional knowledge. Earthscan, London, pp 91–114

Chapter 22
Ingenious Agricultural Heritage in Cold Oases Connected to Collective Grazing Areas (Haut Atlas, Maroc)

Jean Gault and Seddik Saïdi

Abstract In the High-Atlas Mountains of Morocco, cold oases connected to rangelands and collective grazing areas have been created by Amazigh people despite extreme climate conditions. Throughout the centuries of isolation, they have maintained their cultural heritage (agricultural, social, artisanal and linguistic). These people and their culture are today recognised by the Moroccan Constitution as an integral part of national identity. Over many centuries, they have developed astute systems to manage natural resources, which are still in place today, to ensure food self sufficiency. Indeed, varieties of wheat, barley, corn, alfalfa, apricots, almonds, etc. are exceptional, as well as varieties of aromatic and medicinal plants. There is also one bee species, and associated biodiversity (fishes (salmonids), mammals, reptiles…). The rational use of rangelands for grazing is well established; fertile areas (agdals) are controlled by precise regulations dependent on rotational system (fallow time and grazing pressure limitation). Water management is controlled by a local hydraulic and legal culture: century old conducts: khettaras, and customary right and institutions. The authors are grateful to IFAD (International Fund for Agricultural Development) (through OXFAM (Oxford Committee for Famine Relief) Italia) to FAO (Food and Agriculture Organisation of the United Nations) programme GIAHS (Global Ingenious Agricultural Heritage Systems/Systèmes Ingénieux du Patrimoine Agricole Mondial, SIPAM). They describe an "ingenious agricultural heritage system" in Imilchil and Amellago:

http://www.fao.org/giahs/giahs-sites/httptypo3faoorgid31410/oases-system-in-atlas-mountains-morocco/fr/

J. Gault (✉)
Ingénieur général des ponts des eaux et des forêts, ministère français de l'agriculture, de l'alimentation, et de la forêt, CGAAER, 251 rue de Vaugirard, 75732 Paris, Cedex 15, France
e-mail: jean.gault@agriculture.gouv.fr

S. Saïdi
Chef de Département Amélioration et Conservation des Ressources Génétiques, Division Scientifique, INRA Maroc, BP 6356 Rabat Institut, 10101 Rabat, Morocco
e-mail: seddiksaidimya@gmail.com

M. Agnoletti and F. Emanueli (eds.), *Biocultural Diversity in Europe*, Environmental History 5, DOI 10.1007/978-3-319-26315-1_22

biodiversity and associated biodiversity, water and soil resources, cultural heritage, income generation for women, and tourism potentialities. Measures are proposed to dynamically conserve this culture and its associated biodiversity, in the present context of economic, cultural, technological globalisation. This paper (in itself, this partnership is already an achievement of cultural diversity, involving six languages, including Latin, and connected cultures and know-how. Some species names are in French, some varieties names are in Amazigh, all species names are in Latin) is a result of a fruitfull partnership between IFAD [financing this study to FAO through OXFAM Italia (Initially ,UCODEP, Unité et Coopération pour le Développement des Peuples)], GIAHS programme, focal institution INRA (Institut National de la Recherche Agronomique) Maroc, the French ministry of Agriculture (CGAAER) (CGAAER is a high-level advisory body to the Minister, one author is presently member of CGAAER, Conseil Général de l'Alimentation, de l'Agriculture et des Espaces Ruraux), local partners Adrar (this means in Amazigh: mountains) and ORMVAT (Office Régional de Mise en Valeur Agricole du Tafilalet, ministère marocain de l'Agriculture): in itself, this partnership is already an achievement of cultural diversity, involving six languages, including Latin, and connected cultures and know-how (Not to say: «Weltanschauung», vision of the world). They would also like to thank the Conseil Général du développement agricole (Morocco), for its support and advices.

Keywords Cold oases · Agdal · Khettara · Seguia · Ingenious agricultural heritage · Cultural values · Amazigh identity · Customary rights · Jmaa · Urf · Biocultural diversity

22.1 Location, Context

Location:
The Imilchil—Amellago site is located on the Eastern High Atlas of Morocco. It lies at 2000 m above sea level (Fig. 22.1).

Context:
It is submitted to extreme climatic conditions, as well as to scarce fertile soil resources. (Actually, arable soil is the limiting factor.)

Population:
The people of tribes Ait Marghad and Ait Hdidou (38,000 inhabitants) live on 309,000 ha, whose only 2 % are arable. These tribes are Amazigh (Berbers, pre-Arabic inhabitants). Roles are strictly dispatched between men and women; this in turn implies that to address specific issues, it is necessary to target specific groups: for instance (women) seed management, fire at home, harvesting or (men) farming, house building and cattle husbandry.

Fig. 22.1 Imilchil cold oasis in winter (*Photo* Dr. Saïdi)

The Imilchil–Amellago cold oasis system integrates three special components:

- limited availability of small fertile plots along wadi (rivers), subject to specific connected water management,
- arid rangelands accessible all year round,
- and pastures or fertile/wet rangelands subject to traditional old management and restricted access.

These indigenous people have overtime transformed the small cold oases and vast rangelands of Imilchil–Amellago into a model of sustainable and rational development. With the aid of knowledge, hard work and skills, they have been able to face various uncertainties of a harsh environment.

With strong community of solidarity and discipline, these people have developed a self-sufficient cold oasis system.

In addition to noteworthy agricultural diversity, lakes, mountains and plateaux, valleys, oases, caves and desert contribute to confer to this area an exceptional natural heritage. On the other hand, the cultural heritage includes expertise, architecture (Ksour (compact cities)), customary rights and associations to manage natural resources, grazing and water management (Altieri and Koohafkan 2008).

This geographical area provides food and other necessities to local populations, as much as water, wool, medicinal plants, fire wood, wood for construction, as well as others needs.

However, this system is threatened by globalisation and climate change at institutional, economic, cultural, physical and environmental levels.

22.2 Biological Diversity (Agricultural and Connected)

22.2.1 Rich Agricultural Biodiversity

Plants

Agricultural plant diversity is highly rich (Table 22.1).

- According to an inventory by INRA Maroc, there are 80 % local varieties among the 105 cultivated varieties, dispatched through 53 species.
- Applied biodiversity (Table 22.2): medicinal and aromatic herbs are carefully known and recovered by women,[1] as well as herbs used for natural dyes (Figs. 22.2 and 22.3).

Biodiversity is rich, because of a broad range of edaphoclimatic conditions: dry and high rangelands, rivers beds, etc. and very fragmented areas of cultivation.

In these cultivated areas, there are, for example, three varieties of durum, produced by three 'farmers diversity units':

- 'irks', a thin ear variety (4 rows);
- 'Abrioun': short, white ear, with black ribs; 6 rows; high quality for bred and semolina;
- 'Tialaline': 6 rows, good quality, good yields (45 quintals/ha),

22.2.1.1 Animal Husbandry (Fig. 22.4)

Animal husbandry is primarily pastoral (MADR/PNUD 1999); over grazing is causing heavy deforestation: in only two generations, entire cedar and red juniper forests have disappeared. Erosion is accelerating, because water or wind can move heavy quantities of soil. Consequently, there is a significant siltation in water dams, which in turn reduces the capacity of these dams to afford abundant or clean

[1]This local knowledge is being lost, because the younger generation tends to rely on modern technology.

Table 22.1 Agricultural plant diversity

Cultivation	Number of species	Number of varieties	
Cereals	7	14	*Triticum durum, T. aestivum, Hordeum vulgare, Zea maïs, Secale cereale*[a]
Legumes/pulses	7	25	*Vicia faba, Pisum sativum, Lens culinaris, Phaseolus vulg*[b]
Vegetable	11	20	*Solanum tuberosum, Daucus carota, Brassica campestris, Cucurbita maxima, C. pepo, Lycopersicon esculentum, Capsicum annuum, Solanum melongena, Cucumis melo, Citrullus vulg*
Fruit trees	13	31	*Juglans regia, Prunus dulcis amygdalus, Ficus carica, Malus pumila,*[c] *Prunus persica, Punica granatum, Vitis vinifera, Olea europea*[d]
Trees for wood	6	6	*Populus alba, P. nigra, Tamarix gallica, Nerium oleander, Salix* spp.
Condiments/spices	9	9	*Allium sativum, Cuminum cyminum, Trigonella foenum graecum, Mentha viridis*
Total	53	105	

[a]+2 minor cereal species: ("tafsoute"): Pennisitum typhoïdes, and ("anelli") Panicum milliaceum
[b]+(«ikiker») orobe Vicia ervilia
[c]Newly introduced in the cold oasis, bright extension
[d]Introduced during the 80s: Prunus armeniaca, Cydonia oblonga, Prunus domestica, Pyrus communis

drinking and irrigation water. This ultimately challenges national water policies.[2] Deforestation is irreversible with the ongoing climate change (Table 22.3).

22.2.1.2 Connected Biodiversity

Plants

There are 445 catalogued plant species, and 30 % are endemic.

Many of them are rare, or endangered.

According to the strict dispatching of roles in the local community, it incombs to women to collect fire wood and brush wood; but their excessive collection leads to further degradation of the vegetation, and in turn to more severe erosion. This ultimately challenges national policies for drinking and irrigation water.

Some species indicate this degradation:

[2]And, as consequence of a speed run between siltation and water demand, new water detention capacities are required.

Table 22.2 Applied biodiversity

Species	Common name
Thymus commutatus	Thyme
Teucrium polium	Germander
Artemisia mesatlantica	Wormwood
Artemisia nigrei	Wormwood
Mentha rotundifoila	Round leaf mint
Calamintha grandiflora	Calamint
Teucrium polium	Germander
Rosa damascena	Dades rose
Juniperus thurifera	Juniper
Ruta montana	Wild rue
Citrulus colocynthis	Coloquint
Salvia lavandulifolia	Sage
Lavandula brevidens	Wild lavender
Capparis spinosa	Capre
Peganum harmala	Harmel
Ononix natrix	Yellow bugrane
Nerium oleander	Oleander

Fig. 22.2 Juniper (*Photo* Dr. Saïdi)

Fig. 22.3 Dades rose

Fig. 22.4 Apis mellifera sahariensis

Table 22.3 Animal biodiversity

Animal husbandry	Number of races	endemic
Sheep	2	D'man, Rahali
Goats	3	Tacherguite , Rahali
Cattle	2	Local breed and crossed
Donkeys, horses	2	
Camels	2	
Dogs	2	Atlas dog
Poultry	2	1 local breed
Honey Bees	2	*Apis mellifera sahariensis*

90 % of meat production

in mid-mountain range areas: *Peganum harmala, Hammada scoparia, Hertia maroccana and Ononis natrix.*
on the plateaux: *P. harmala, Hamada scoparia, Reseda luteola and Reseda alba.*

Table 22.4 Endangered species

Class	Nbr. spEspèces	Endemic	In danger	Red list IUCN
Birds	115	5	20	2
Mammals	37	3	10	9
Reptiles	43	6	12	1

High mountainous areas are very diversified, as they are home to 57 species. (*Ormenis scariosa*, *Retama dasycarpa*, *Prunus prostrata*, *Stipa nitens*, *Buxus balearica*, *Adenocarpus anagyrofolius*, etc.).

In wet areas (where soil is rich and the grazing potential is high), one finds *Helianthemum* sp., *Aristida* sp., *Stipa parviflora*, *Artemisia herba-alba and Retama* sp.

22.2.1.3 Animals

As the terrain is difficult to access, it is home to a varied animal population, which lives in relative security:

Vertebrates species inventory (Southern High Atlas) (Table 22.4).

Among endemic vertebrates, may be mentioned are:

Birds: *Chlamydotis undulata*, *Picus vaillantii*, *Sylvia deserticola*, *Phoenicurus moussieri;*
Mammals: *Elephantulus rozeti*, *Atlantoxerus getulus*, *Gazella cuvieri*, *Ammotragus lervia;*
Reptiles: *Bufo brongersmai*, *Quedenfeltia moerens*, *Quedenfeltia trachyblepharus*, *Lacerta andreanszky*, *Chalcides montanus*, *Vipera monticola.*

Biodiversity remains under threat, and 12 vertebrates appear on the IUCN red list: among them are

Mammals: *Panthera pardus*, *Hyaena hyaena*, *Ammotragus lervia*, *Gazella dorcas*, *Gazella cuvieri.*

Conclusion Morocco is home to significant endemic biodiversity; this cold oasis and rangelands reflects this in every way, because of climatic, topographic, pedologic diversity of the environment.

22.3 Ingenious Systems and Connected Cultures

There are strong links between the above-mentioned biodiversity and the associated culture of Imilchil–Amellago; these can be summarised as "ingenious heritage systems", and as such this site is now registered by the Food and Agriculture

Organisation, on the list of "Globally Ingenious Agricultural Heritage Systems" (GIAHS) since 2011 (FAO 2008). (see http://www.fao.org/giahs/en/, also in French (SIPAM), Arabic and Spanish.)

22.3.1 Biodiversity Supports Culture

22.3.1.1 Food Security

Agricultural production in Imilchil–Amellago agro-system contributes to the balanced diet of local populations (FAO/PARplatform 2011).

- Meat: Sheep and goat grazing is the main activity in this area, and their forage is mainly provided by the rangelands. Some plant species are harvested by women in summer and stored for winter months. The meat is primarily mutton (and goat).
- Cereals (mainly durum wheat and barley) are used for bread and couscous production, which are elements of utmost importance in the local diet. Durum wheat varieties have specific uses. Irks is the variety that is intended for local breads, especially Tahtoucht (this bread can be stored for longer periods (long journeys on rangelands) and may soften when exposed to humidity). The Abrioune variety is of high quality, and is used for baking bread and for preparing semolina. Maize semolina is consumed with whey. It constitutes the daily meal for all the families during the period of cow lactation. Rye is prepared in soup for breakfast or dinner, especially in winter.
- Pulses are the main dietary source of protein for the local population.
- Potato is the second source of carbohydrates energy.
- Fresh consumption is most sought after, especially for broad beans, peas and haricots. Other vegetables are important, and fruits constitute an important part of the diet equilibrium, such as figs, apricots and almonds.
- Thanks to many melliferous species, the honey production is significant in the region. Beekeeping is an ancient and widespread activity, and provides an important resource for the population. In recent years, dry weather conditions have affected this activity. The prevalent melliferous species are *Thymus* spp., *H. maroccana*, *Zilla* spp., *Adenocarpus bacquei*, *Teucrium fruticans*, *O. natrix*, *Launaea arborescens* and *B. balearica*.

22.3.1.2 Water

Water percolation to groundwater is improved with plant biodiversity; the water filters through the soil to wadis and recharges the reservoirs.

It is vital to humans, their farming and wildlife.

22.3.1.3 Habitat

- Earth: rammed earth construction has been the prevalent building technique until the arrival of concrete. The unregulated and rapid development of concrete constructions also opposes the architectural traditions of the region and its cultural diversity.

Furthermore, concrete does not isolate from heat in summer, nor cold in winter (they can reach extreme values).

- Consequently, people use more fuel wood in concrete houses in winter, and needs for wood are rising sharply: today, these needs are covered mainly by polar wood and reed. It is necessary to encourage their planting: they also protect the dykes along wadi (water currents) against floods.

But however, in recent years, inhabitants are now convinced that despite their safety, buildings of concrete are not adapted to the climate of the region. Adobe construction allows to buffer the heat of summer and the bitter cold of winter without resorting to heating (Fig. 22.5).

- Timber wood: the demand for timber is such that illegal logging is increasing. As forestry used to provide quality timber wood for making ceilings of Adobe

Fig. 22.5 Traditional adobe building in High Atlas

houses, they have been significantly degraded. Juniper or Adghmam (*Juniperus* spp.) is a much rarer plant, and in some areas it is totally extincted. It is therefore necessary to regulate the use of these resources as the climatic conditions are not favourable for the regeneration of different local species that are under threat of extinction, or to promote substitutes like Eucalyptus from plantations in other regions, in order to provide timber for construction.

22.3.1.4 Wool for Winter Clothing in Mountainous Regions

The availability of wool as sheep co-product allows for woving winter clothes, mainly burnous and jilbab for men and, for women, tahandirt. The latter expresses through its colours and motifs, tribal identity. The wool is also used for the manufacture of blankets and carpets, and they are essential in every home (Fig. 22.6).

22.3.1.5 Medicinal Use

The know-how of rural women about medicinal plants is highly bright and diverse. Knowledge vary with women' age, plant richness of their immediate environment and probably also with the family income and distance to a health centre. Rural women are today developing a new trust for pharmaceutical products, and their expertise is going lost; up to now, it is so far kept by some senior persons. Older women have in fact more information than their juniors, for girls know only about the most common species.

Over forty species of plants are used by local women for their medicinal properties. The abundance, intensity of use and the area of distribution vary from

Fig. 22.6 Tahandirt patterns, according to tribes (*Photo* Dr. Saïdi)

one species to another and also from one ecosystem to another. Alili (Nerium oleander), Awjtam (Erinaceae), Flyou (*Mentha pulegium*), Izaghyoul (*Lavandula multifida*), Iziknou (*Artemisia herba-alba*) and Lharmal (*Perganum harmala*) are widely used and are found in all ecosystems.

22.3.1.6 Firewood

Nearly, sixty ligneous species are exploited in the area for domestic needs. Rural women hold unparalleled expertise about species of firewood. They distinguish between species by morphological characters such as the shape and size of the plant, foliage, flower colour, or smell, difficulty to cut or uproot. Women's preference is for slow combustion, no smoke, nice smell, lack of thorns and low humidification after the rains.

22.3.1.7 Immaterial, Cultural Heritage

- Rites: When mowing sheep, the owner of the animals slaughters mutton and prepares a meat meal he offers his guests and shearers. A similar practice marks the harvesting and threshing of wheat and barley: before starting work, the farmer cooks a meat meal (slaughtered muttons) for notables and workers. Also, symbolic meals are prepared during the repair of the irrigation system. These feasts used to be held to thank those who helped in the work and who have demonstrated sacrifice for the community (Fig. 22.7).
- Cultural heritage: multi-tribal festive meeting: Imilchil engagement moussem or *Agudoud n'Sidi Ahmad Oulmghani*.
  Each year a moussem stands after the grain harvest and earning almonds in mountains (usually the third week of September). This period also corresponds to the return of herds to mountains agdals (see further down). This is an appointment of great economic importance for social and cultural ethnic groups and families. It attracts more than 20,000 people for 3 days. This is a unique

Fig. 22.7 Brides at Imilchil engagement moussem

opportunity each year for pastoralists, poets, farmers, to meet and organise marriages of their sons and daughters who have reached the age where its name "engagement moussem". This is also an opportunity to exchange breeding animals, mules and other goods. Currently, the 'Engagement Moussem' brings together mountain communities across the High Atlas and is a place of exchange and important communication at local, national and international levels. Moussem is being proposed by Midelt prefecture to UNESCO office in Rabat, in partnership with Moroccan Culture ministry, to be registered as world cultural heritage.

- Tales: Ait Ihya village is well known for its famous poets nationwide. The population of Imilchil–Amellago is reputed throughout the Kingdom by its attachment to their identity. For it was able to keep traditions and ancestral customs, as it remained, until recently, an enclave in the heart of the High Atlas. Their Berber tales and riddles is a rich and diverse cultural heritage.
- Music festival: Since 10 years, a festival of traditional music has been joined to the moussem, to promote songs and dances of Ait Hdidou and Ait Marghad tribes, for they are unique to Morocco; the presidents of Imilchil and Bouzmou commons wish that an eventual registration of Engagement Moussem at Unesco world heritage integrates this festival.

22.3.2 Culture Cares for Biodiversity

The following part presents how human activities may care for biodiversity.

22.3.2.1 Seed Storage and Management

Seed is the strategic input in cold oasis agriculture. Each farmer operating a given area organises its own seed procurement from 1 year to the following, and local varieties grown under the same agricultural practices since decades, become specific to these places. This specific adaptation is used by farmers to distinguish between areas of crop species, and also serves to name the species. The seed is removed from the production immediately after harvest (by women) and placed in the storage area called "Lakhzin". This area bears a sacred character and is managed exclusively by the head of the family. Thanks to the cold and dry climate, seed conservation in this region does not pose any problem. Small seeds other than cereals (alfalfa, turnip, onion) are most often present in small quantities, and they are contained in earthen pots. In each village, there is often a seed holder of one or more cultures in the area.

22.3.2.2 Agriculture Systems

The adopted cultivation systems use techniques that are well adapted to the three interconnected components (oasis, open rangelands, restricted pastures). The practice of agro forestry (combination of tree and ground crops) and crop rotation (crop succession in time and space) contributes to improving soil fertility and conservation.

Animal husbandry (MADR/PNUD 1999), integrated to cropping, also uses the open arid rangelands all year round and the restricted pastures and rangelands. The practice of transhumance in exploiting these rangelands has the potential to ensure the continuous feed of livestock, while ensuring vegetation regeneration. Improving soil fertility contributes to soil conservation, and consequently to the maintenance and regeneration of the vegetation cover. Both contribute to the aquifer recharge, all which is the basis for the system sustainability, and essential to the preservation of natural resources and balance of basic natural cycles. Succession in time and space contributes to improving soil fertility and conservation.

The technique of "khettara" involves draining by gravity the aquifer through a tunnel. Every 50–100 m, depending on the nature of the ground, a channel is dug vertically to communicate outside and facilitate khettara maintenance.

22.3.2.3 Management of Soil and Water

- Fertile land is scarce (less than 2 % of the land area) and builds the limiting factor (Fig. 22.8). Cultivated fields are located along the ephemeral rivers (ouadi) and vulnerable to violent floods. Dykes are built against the current

Fig. 22.8 Terraces to expand cultivated land

streams to protect plots. In recent years, due to climate change, summer storms are more violent and a majority of these dykes are not strong enough anymore to contain floods. Farmers carry out planting poplar, reed and willows along river beds to protect the banks against erosion.

– Water Management: In the area, irrigation system dates back to centuries. Beyond its vital importance to the oasis populations, it is a cultural heritage of inestimable value.

  - Streams and springs are considered public goods of local communities;
  - Wells that are dug by individuals are either open to public use or reserved for the exclusive use of the dugger;
  - Seguia or irrigation canals belong to smaller groups, but with a stronger right to control communities;
  - Khettara: As the flow of water currents is variable along the seasons, scarcity must be managed during the dry seasons. In the southern part of Amellago, abstraction of groundwater sources by the method of khettara is a common technique. There are currently seven functional khettaras;
  - Jmaa manages the common water: The Jmaa is the elders committee, representing all the ethnic units constituting an ethnic group or tribe. Its power is recognised morally and applies without physical force. In irrigation, the role of Jmaa had and still has an unparalleled importance. Regarding flowing water taken from a thalweg through a diversion dam (uggug) and a supply channel (targa) or a source (taghbalut), the water authority remains today the case of the local community.

The system governing the distribution of irrigation water is of customary type, called urf, ensuring the individual and collective rights for the use of water. The Jmaa may punish crimes relating to any offence regarding agreements settled (since centuries) about irrigation water. This tradition is still perpetuated today; obviously, adaptations are necessary to the new political, economic, social and cultural context.

Indeed, the "modern" right is unfortunately not always compatible with the customary law and Moroccan justice does not take into account urf (customary rights), which may lead to inextricable situations. In this context, associations of agricultural water users are set up in "modern" right, and settle internal detailed rules of procedure about possible offences and sanctions.

22.3.2.4 Managing Animal Husbandry: Agdal

Animal husbandry relies on transhumance, and feeding happens outdoor, based mainly on pastoral resources. This system occupies the foothills and regards the rearing of sheep and goats Rahali. It has been adopted by former nomads who have abandoned their long hikes and reduced the amplitude of movement to paths from the plains to plateaux and surrounding mountains, taking advantage of the complementarity between the mountains, foothills and lowlands. This joint operation is the best way to buffer the effects of weather conditions and ensures revegetation. As a part of this system, rangelands generally have a collective traditional status. They are operated under customary law, urf.

The oldest typical form that characterises livestock keeping in the region is herd movements between the three main ecological units called "Agdal" (grazing whose opening is regulated). There are three types of agdals:

- The agdals that are open all year and are set at rest by the needs of the beneficiary population. They fall within villages (Agdal n'Ighram). Users of these courses are mainly sedentary people of the village and former nomads with less than 100 heads of small ruminants.
- The agdal closed to grazing between March and June (Example: Ait Brahim Izlan, Ait Iazza Izlan). These are the most important agdals in terms of area. Each fraction operates its agdal with transhumant (contractors) from lower elevations to the customary time. In case of drought, this date can be changed.
- Smaller agdal (10 ha) operated from July to March and are dedicated to cattle and horses. Only sedentary people from the Ksour (villages) have the right to these courses.

As per water disputes, modern right has the worst difficulties to integrate urf, regarding animal husbandry management; judges may simply ignore customary right and pronounce an inadequate sentence.

22.4 Challenges, Threats and Opportunities

With the opening experienced by Morocco from the late nineteenth century, the modern administration is completely or partially substituted for Jmaa at managing multiple fields of life. Cohabitation between two forms of institutions becomes the rule, and this is often difficult. The organisation of the territory is deeply transformed under the influence of new endogenous and exogenous factors. Of these,

- The establishment and forced settlement of the tribes, including nomads;
- The increase in quantitative and qualitative needs of the population due to the population explosion, and road building, or electricity connection (and consequently TV and radio);
- Exodus, national migration to large low-land cities, including international migration to Western Europe for example, and outward mental opening (monetisation of rural life, heavy presence of the media, school enrolment, increased urbanisation, cultural colonisation, etc.);
- Climate change is marked by a higher occurrence of longer droughts, and more violent weather events.

  This set of factors and pressure leads to a strong degradation of natural resources: water, soil, flora and fauna.

22.4.1 Results and Outcomes

- The vegetation is improperly cleared, some rangelands being re-cultivated, when they are not simply overgrazed; in second generations, deforestation has destroyed entire cedar and juniper forest areas;
- This overexploitation of natural resources, combined with an increase in climate accidents, leads to an intensification of erosion: water reservoirs are being silted much faster than expected, and government faces the urgency of building new dams, to secure water provision to cities (where demand increases);
- The traditional social structures (Jmaa) being weakened, they are less able to support the above-mentioned management systems. The khettaras for example are poorly maintained;
- The collective wedding (Moussem) is being somewhat out fashioned, as well as the experience of traditional midwives and traditional gynaecology;
- The local know-how relating to the conservation of biodiversity is being lost. Conversely, new varieties are introduced (cereals, vegetables, fruits (apples) and milk cows) and the local heritage lessens;
- In particular, the use of aromatic and medicinal plants is declining, and traditional medicine is used less and less;
- Local knowledge on adobe construction is also disappearing;
- The local oral cultural heritage is discredited [stories, songs and rituals (connected to harvesting, grain milling and to family celebrations)];
- Local folkloric groups are not included in major national festivals;
- The copyrights are not protected (Piracy of local cultural heritage: poetry, dress, etc.).

The Way Forward

'Ready to use' packages of technical solutions have long been put forward to solve the above-diagnosed problems; they have failed because these technocratic approaches were conceived outside the sociocultural context of the targetted populations; they were not "bought in". Various administrative systems are today conflicting often painfully; for modern and traditional local institutions are superimposed, they may assimilate each other or cohabit or simply reject each other. In this context, the protection of biological resources is relegated to second place: this raises questions about the future of new coming generations in areas as fragile as the mountains and oases.

> Thus, in order to preserve the rich flora and fauna in Atlas in general and in Imilchil Amellago specially, any initiative should primarily build on reconciliating institutions (traditional and modern) and should boost human resources. It is essential to promote social initiative and the emergence of project leaders: initiatives "bottom up" should be encouraged. Women's role has been very important for the whole performance and endurance of this system: women's initiative should be fostered, through adequate workshops and sensitisation.

In particular, the authors recommend to support partnerships being built with:

- Local NGOs, whether small or flawed they may be, in particular:

  a union of poets is created, its activities should be supported, so that they defend their copyrights and better participate to national cultural events;
  the existing union of mountain guides should also be supported to provide better service in response to touristic demand.

- International NGOs (like Oxfam) that have been involved many years in local development;
- Moroccan institutions [Office Régional de Mise en Valeur Agricole du Tafilalet, Haut Commissariat des Eaux et Forets et de Lutte contre la Désertification (HCEFLCD 1994), Agence du Sud, Agence des oasis…)];
- International agencies including UNESCO, UNDP. In particular, UNESCO should take into consideration the request of the prefect of Midelt, to register the engagement moussem as UNESCO World Heritage (Saidi 2013). For its part, FAO has classified the site Imilchil Amellago among the pilot sites GIAHS.

Parts of this site are also already ranked among the areas of UNESCO Biosphere and among wetlands in the Ramsar Convention.

The authors would also like to recommend to:

- Reboost the local museum Imilchil that presents heritage, and define its action within a sustainable process, including long-term funding; edit CD and booklets presenting song and tales, legends and dances;
- Better include the potentialities of this site among national priorities, which aim to promote small-scale agriculture (resources of plan Maroc Vert, Pillar II, are considerable): promote green tourism and better signal touristic roads and remarkable sites; broadcast leaflets on biodiversity;
- Promote food handicraft (example: packaging of honey), woving of traditional carpets, blankets and coats; helping families (women) to breed small livestock (poultry);
- Address energy demand and reduce pressure on local fire wood or ligneous products;
- Address overgrazing concerns, and identify solutions to safeguard the principles of present traditional animal husbandry, in reinforcing traditional law, and promoting in a participatory process new forage production;
- Leverage the law on labelling local products: this will certify their origin and quality and add value to some typical products (Almonds? Honey? Bread? Semolina? Clothes?), or services (hiking, rock climbing);
- Take better advantage of the ongoing regionalisation process, which is now enshrined in the new Constitution, and of the recognition of the Amazigh culture, which is now referred to as “component of the Moroccan identity”: cultural globalisation may be fought through better recognition of this identity and its components.

More generally, a dynamical conservation of this biological and cultural diversity cannot flourish under broad commercial and financial globalisation: the authors recommend to take advantage of waiver possibilities in international trade agreements, allowing regional trade areas, to facilitate the development of regional agriculture frames, and to promote food self sufficiency and to fight impoverishment (Saidi 2011).

Having regard to climate change (Altieri and Koohafkan 2008), the authors recommend to facilitate cross-cutting information exchange, so that the cold altitude oasis in Maghreb which are dispatched not only across the Moroccan Atlas, but also in Algeria mountain regions, exchange their success stories (or their failures): in this regard, the GIAHS programme may offer a useful workframe.

References

Altieri MA, Koohafkan P (2008) Enduring farms: climate change, smallholders and traditional farming communities, Published by third world network 131 Jalan Macalister 10400 Penang, Malaysia. Website www.twnside.org.sg

FAO (2008) Conservation and adaptive management of globally important agricultural heritage systems (GIAHS) PIMS 2050 terminal report

FAO/PARplatform (2011) Biodiversity for food and agriculture contributing to food security and sustainability in a changing world

HCEFLCD (1994) Le Parc National du Haut-Atlas Oriental, Plan d'aménagement et de gestion

MADR/PNUD (1999) PROJET DE CONSERVATION DE LA BIODIVERSITE PAR LA TRANSUMANCE DANS LE VERSANT SUD DU HAUT ATLAS MOR/99/G33/A/1G/99

Saidi S (2011) Oases systems in Morocco Mountains. International forum on: globally important agricultural heritage systems. Theme: dialogue among agricultural civilizations, Beijing, China, 9–12 June 2011

Saidi S (2013) Integration of globally important agricultural heritage systems in Morocco. International forum on globally important agricultural heritage systems, 29 May–1 June 2013 in Ishikawa prefecture, Japan

Chapter 23
The Integration of Modern and Traditional Agriculture in the Cultural Landscapes of Poland

Józef Hernik, Robert Dixon-Gough, Barbara Czesak and Maria Pazdan

Abstract Integration of modern and traditional agriculture in the cultural landscapes of Poland is related to the big events of recent decades, i.e. to socio-political transformation that took place in 1989 and EU accession in 2004 since these big events shaped the policies influencing the integration. The transformation officially restored the idea of private property and EU accession brought EU law into Polish legal system. Integration of modern and traditional agriculture in the cultural landscapes is strictly related to the policies. The most important policies in respect to agriculture and landscapes are planning, land use and taxation policies. Other important aspects are economic pressures and demographic changes. All these factors must be regarded in a perspective of a country that still has marks of past socialist times. The paper analyzes the policies and factors related to current condition of Polish cultural landscapes and attempts to interpret how they influenced the integration of modern and traditional agriculture in the cultural landscapes.

Keywords Rural cultural landscapes · Agriculture · Policy · Poland · Biocultural diversity

J. Hernik (✉) · R. Dixon-Gough · B. Czesak · M. Pazdan
University of Agriculture in Kraków, Department of Land Management and Landscape Architecture, ul. Balicka 253c, 30-149, Kraków, Poland
e-mail: rmhernik@cyf-kr.edu.pl

R. Dixon-Gough
e-mail: dixongough@tiscali.co.uk

B. Czesak
e-mail: b.czesak@ur.krakow.pl

M. Pazdan
e-mail: m.pazdan@ur.krakow.pl

M. Agnoletti and F. Emanueli (eds.), *Biocultural Diversity in Europe*,
Environmental History 5, DOI 10.1007/978-3-319-26315-1_23

23.1 Introduction

Communist rule in Poland was broken in 1989 and this transformation changed not only the political system from a centrally planned economy into market economy, but also indirectly changed the Polish cultural rural landscape through evolving land policies. Often neglected, the integration of modern and traditional agriculture in the cultural landscapes requires a proper and cohesive policy. This paper analyzes changes of land use policy during and after the transformation and shows the influence of those changes upon cultural landscapes. This paper provides a review of the way in which land use policies have evolved since transformation and provides description of the measures already taken by professionals and authorities to protect environmental friendly landscapes—often to the detriment of the traditional and cultural landscapes. It will also provide an evaluation of the land policies introduced to provide legal instruments and the scientifically based tools and solutions adopted to help improve the environment, human and natural quality of life, to strengthen the economy, but will seek to question whether these measures have helped protect or degrade the cultural landscapes. Cultural landscapes have been shaped over centuries by a combination of agricultural techniques, patterns of land ownership and, more recently, by evolving land policies. The introduction of market economy instead of a centrally planned economy following transformation in 1989, not only changed the Polish legal system and, as a consequence, the land policy—but also accelerated the speed of change. This rate of change was given further impetus through EU accession was imposed even more changes in land policy and the market economy of agriculture. These two events have imposed significant changes in the perception of rural land, which in turn was reflected by changes in land-related policies. The importance of agriculture in Poland was effectively diminished as a result of the import of subsidized food from Western Europe.

23.2 The General Changes in Land Use Policy in Rural Areas

Land use policies are important catalysts for shaping and altering cultural landscapes, particularly those policies that consider some land uses more desirable than others. The attractiveness of land use changes with time according to current economic pressures. During communism, agricultural land had the greatest appeal since the principal role of agriculture was to feed the nation and neighbouring states. Woś (1992) stated that during the period between 1946 and 1989, it was believed that every hectare of farmland should be capable of producing products that would go directly or indirectly into the food chain. The consequences of such a policy were that it was compulsory for designated agricultural land to be used exclusively for agricultural purposes. In Poland, as was also the case in most other countries in Europe—both East and West, there were problems in food supply following the end of WW2. This was dealt with in Poland by the act of 12th July 1966 on Agricultural Land Protection (Act 1966), which introduced controls and specifically prohibited changing agricultural land use (Czechowski et al. 1994).

This control was strengthened when production was assessed to be directly related to the area of land owned since food production was crucial in times of scarcity witnessed by the empty shelves of socialist shops.

Following transformation these policies changed. Until 1991 normative acts provided higher levels of protection for agricultural lands than for forests, which was the result of the necessary agricultural land use and a simplified procedure of changing forests into agricultural land. After 1991, significant changes were introduced to alter the legal protection of agricultural land and, as a consequence, forests presently receive a higher protection than agricultural land—largely for environmental purposes. Agricultural land can be easily afforested, but forest can only become agricultural land in justified situations. The legislator makes it clear that forests take precedence over agricultural land (Act 1991).

The main aim of the act on Agricultural Land Afforestation from 2001 was to increase the rate of forestation in Poland to bring it up to the average forestation rate of EU members. An important additional goal was to diminish the area of so-called 'vestigial' or 'marginal' agricultural land, i.e. that land with the lowest level of productive use—often in mountain regions and those where the cultural landscape is most evident.

A relatively high-level interest was generated by farmers in allocating agricultural land for afforestation through subsidies that were available for this purpose (Gawroński 2004). The Act of 2001 also determined the amount of subsidies paid to farmers who decided to afforest agricultural land on their own, but this Act was only in force for a very short time. Following Polish accession to the EU, a new Act of 28th November 2003 on Supporting the Development of Rural Areas from funds obtained from European Agricultural Guarantee Fund was introduced (Act of 2003).

Moreover, the EU regulations concerning the protection and shaping of land use are based on the extensification of agriculture. This policy reflects one of the prerequisites of the EU agriculture policy, which is that there is no need to maintain all agricultural areas. In direct contrast to all previous policies enacted in Poland, non-agricultural activities should be indentified in order to protection the best agricultural land. It is argued that by maintaining agricultural production on all designated agricultural lands there would be an overproduction of food, which could decrease the profitability of the production and, in turn, could result in the overall reduction in the area of prime agricultural production land. As a direct result of these policies, agriculture in marginal land—frequently coinciding with the important elements that make up the cultural landscape—is becoming fallow, a phenomenon that was absent in socialist times.

The growing area of abandoned lands and afforestation has inevitably changed the traditional cultural landscape of Polish villages. The land loses its appearance of a mosaic of individual land parcels. Extensive areas are now often covered with trees or bushes, which spread downward from the least fertile hilltops, and it is not easy to revert to agricultural use since permission and fees are required for a farmer to remove such growth from his/her own land (Act on Nature Conservation 2004).

A similar effect of degrading the cultural landscape is caused by urban sprawl and one of the many new non-agricultural land uses introduced in Polish villages has resulted in areas becoming built upon to satisfy the housing needs of former inhabitants of nearby urban areas, creating a landscape more akin to a city than a rural village (Hernik et al. 2013).

The great question in both cases identified above is whether it is possible to integrate modern land use policies within a cultural landscape without it becoming degraded.

Although both cases above address issues covering extensive areas of agricultural lands, the integration between current land use policies is often more subtle. In the case of Polish villages, changes may be small but, nevertheless, very significant. Buildings and the infrastructure are gradually being improved and the traditional style of building may be viewed as old fashioned and inefficient when compared to more modern structures. Despite these gradual changes, it is noticeable that in following years, during which there was little private ownership, the Poles are very focused on their property rights and are reluctant to adjust the use of their land to the global concepts for development.

In spite of this reluctance, there is one branch of Polish agriculture that is becoming more popular: organic farming, which already has a long history since Rudolf Steiner, who once educated a Polish duke Stanisław Karłowski, introduced the concept to Poland (Paull 2011a, b). The duke died during WW2 but the book written on the basis of the lectures given by Steiner during his visit at duke's domain survived and inspired Julian Osetek who founded the first organic farm in Poland (Duda-Krynicka and Jaskółecki 2010). This was the first and only such farm until the 1980s when organic farming became more popular and shortly after institutions certifying organic production were created (Duda-Krynicka and Jaskółecki 2010). The first Act on organic farming was introduced in 2001, with another two acts respecting EU law in 2004 and 2009. Nowadays organic farming has increasing numbers of followers and from 2002, it increased ten times during a decade (Kancelaria Senatu 2013). By 2013, there were ca. 20,000 organic farms (Table 23.1).

The development of organic farming is highly dependent on subsidies. Therefore, the Rural Development Program of the EU may have a positive influence on maintaining traditional land use, yet modern since organic farming is in fact

Table 23.1 Number of organic farms in Poland [data compiled from GUS (Polish Central Statistical Office)]

Year	Number of organic farms	Percentage of all farms
2005	4050	0.15
2007	8443	0.33
2010	17,173	0.75
2013	19,832	1.39

a modern concept, although it embraces the traditional forms of land management and the protection of traditional cultural rural landscapes.

23.3 Transformation in Spatial Planning in Rural Areas

After WW2, the planning system became one of the main instruments of a centrally planned economy (Gorzym-Wilkowski 2012). Plans approved by politicians and frequent expropriations (70 % of the land was private) defined the fact that planning was seen as a repressive measure. But it was during this period that some of the, still valid, rules of plan-making were created (Dutkowski et al. 2012). The Act on Spatial Planning from 1961 ruled that the spatial plan should implement the goals of central industrial plans. Because of the nature of this legislation and the protection it afforded, indirectly, to the cultural landscape the integration between modern farming policies and the cultural landscape was implicit.

This was followed by the Act on Spatial Development from 1994, which was the first regulation on spatial planning following the transformation and since planning was so closely associated with the centrally planned economy, the name of the first act did not include the word 'planning'. A new legal reality was introduced by this act whilst the next act on Spatial Development and Planning eliminated the compulsory nature of spatial plans. Instead of a spatial plan, each community should have a compulsory study of the conditions and directions of spatial development, which should form the basis for future spatial plans. The studies, unlike the traditional spatial plans, are not legal acts. Therefore, investments conducted with new tool, are produced in the absence of spatial plans, i.e. planning permits. Planning permits do not have to conform to the studies of conditions and directions of spatial development. Once again the relationship between modern farming policies and the cultural landscape was implicit.

One of the other important acts concerning land is the Act on the Protection of Agricultural Land and Forests from 1995. The amended act allowed for the automatic change of land use of arable lands (which until then required ministerial consent if over 0.5 ha). Currently, according to the new amendment, all changes of use of high-quality (I–III valuation class) agricultural land has to be approved by a Minister for Rural Development, and proper changes are to be made in the local spatial plan or included in the planning permit (Act 2013). This Act, was the beginning of the end to the implicit protection that had been granted to cultural landscapes by offering farmers a relatively easy means of applying for a change of use to their land.

An important factor that should be taken into consideration is that Polish spatial planning regulations did not necessarily regulate landscape protection. This was only achieved through the regulation of farming policies which had, up to this date, protected the cultural landscape through defining farming policies through land use regulations. Table 23.2 presents changes in attitude towards landscape in spatial planning regulations.

Table 23.2 Cultural landscapes and change of designation of agricultural land in spatial planning regulations before and after transformation

Timeline		Act	References to cultural landscape
1946	Socialism—Centrally planned economy	Decree on Planned Development of Poland from 2nd April 1946	• No requirements for changes of land use
			• Landscape not mentioned
		Act on Spatial Planning from 3rd January 1961	• Changes of land use possible after obtaining permission from authorities responsible for spatial planning. (If there is no construction planned)
			• Landscape not mentioned
1989		Act on Spatial Planning from 12 July 1984	• Proper and deep analysis of the area is necessary to change the land designation into non-agricultural
			• Spatial plans should regard protection of cultural heritage, landscapes and environmental protection
2004 2005 2014	Democracy—market economy	Act on Spatial Development from 1994	• Protection of agricultural land not mentioned. In order to change land designation certain requirements should be met
			• If land designation is to be changed, certain requirements of cultural heritage protection should be met
		Act on Spatial Planning and Development from 2003	• Spatial plans designate areas for non-agricultural use. The areas that are to be designated for non-agricultural purposes should be covered by the spatial plan. Spatial plan has to cover rules for environmental protection and cultural landscapes protection

Despite these changes, the Polish spatial planning systems still fails to take into consideration the European Landscape Convention, introduced in 2000, which was ratified in 2005 but for which there are no tools for its implementation. This alone would provide an adequate level of protection for the cultural landscapes that are gradually being eroded by farming and landscape policies that favour intensive farming and forested area over the traditional and cultural landscape. Effectively thought out and implemented spatial planning is one of the tools that provide support for the shaping of the landscape. Proper planning can efficiently protect landscapes and, for example, setting limits for the degree of change of land designation are a necessary addition that should be included in spatial plans. If such a measure was to be implemented, together with the European Landscape Convention, a far greater degree of integration could be achieved between modern and traditional agriculture in cultural landscapes.

23.4 The Current Fiscal Policies and Economic Pressures

The process whereby modern and traditional farming can be integrated is not helped by the current fiscal policies relating to land. There are presently three main taxes related with land in force in Poland, those of property, agricultural land and forestry land.

The first of these, property tax, applies to parcels of land smaller than 1 ha, and to buildings and apartments together with commercial lots. In certain villages close to major urban settlements, there is considerable encouragement given to landowners to change the zoning of their land from agriculture to residential (Dixon-Gough et al. 2013; Hernik et al. 2013). Although the level of taxes for property is higher than for agricultural land, the profits associated with re-zoning far exceed those of maintaining the current status quo.

The second group of taxation relating to agricultural land is riddled with uncertainty. The fundamental question is, 'who or what is a farmer and what exactly constitutes agricultural land'? According to the Polish Central Statistical Office, there are over 2 million farmers but according to the EU system responsible for agricultural subsidies (IACS) only 1.5 million farmers are registered. Everything depends on criteria and interpretation. Is farmer a person who produces food, or can a person be called a farmer if he or she owns agricultural land? In Poland, there are currently three different definitions for a farmer in three different acts. The respective acts are on the Agriculture, Social Insurance for Farmers and Services and Products Taxation. Those definitions are similar but not identical, which causes some interpretation problems as in certain situations, certain definitions should be applied, which are inconsistent. The definition of a farmer is crucial as it determines who can buy agricultural land, be insured in KRUS (Agricultural Social Insurance Fund), obtain EU subsidies, and take part in agricultural environmental programs. Therefore, a new single definition of a farmer is being discussed in a program of development and consolidation of public finances.

In addition to the problems of definition, this group of taxes is very controversial since agrarian taxes are calculated on the basis of the area of agricultural land and does not consider the income of a farmer. Currently, changes in the agrarian tax are being discussed.

At this juncture, it should also be considered that there has been a significant increase in the number of private farms since transformation for a number of significant reasons. The first was the attraction of low taxes and the ease by which any person could 'become' a farmer. The only prerequisite was to possess at least 1 ha of agricultural land. Before transformation, a farmer had to be suitably qualified having undergone higher education in farming or in a farming college). This led to an important concession, namely that a farmer could relatively easily build a house on agricultural land as building permits were easy to obtain. During socialism, only a person who had worked in a farm for at least for a year or who had been educated as a farmer, could inherit a farm. With time, the conditions to inherit a farm became less strict. Finally, the ruling of the Constitutional Tribunal from

Table 23.3 The number of private farms in Poland [table compiled from Polish Central Statistic Office]

Year	Number of private farms
1970	3,224,000
1980	2,897,000
1985	2,841,000
1989	2,143,000
1996	3,060,132
2000	2,854,374
2002	2,928,578
2005	2,728,909
2007	2,575,113
2010	2,273,284
2013	1,425,386

2001 introduced more freedom. This ruling, in turn, whilst having important tax implications, also introduced into rural and cultural landscapes a different type of person who was interested in farming purely as a means of acquiring land to build property in remote and often scenic locations—frequently to the detriment of the cultural landscape (Dixon-Gough et al. 2011; Hernik et al. 2013).

Table 23.3 shows that the number of farmers was bigger after the transformation. It may not be the only criterion.

Coupled with the problems of changing the use of agricultural land to property, it has to be accepted that working in urban areas is often much easier than working in a non-profitable agricultural environment. Small farm may be not be large enough to support families and there is a growing demographic trend that indicated an increasing average age of those who live in rural area, which are increasingly becoming depopulated by the ambitious young. Low profitability of agriculture is one of the factors that increase the number of abandoned fields. These fields could be used for something else but this would require a protection policy that could balance environmental protection and land development, or even to provide subsidies for farmers to maintain the landscape. Low profitability was also one of the reasons for the declining number of farms (Table 23.3).

Whilst there is a growing trend for rural depopulation, as discussed above, this trend is by no means universal. This outward migration from villages to cities has been observed for a period of more than 50 years. As the cities grew, the suburbs expanded—a trend that is evident through research by the Polish Central Statistical Office but, most importantly, the population of villages has started to increase since 2000 (Table 23.4) (GUS 2013). Although this is related to the fiscal and economic factors noted above, it should also be considered as a relatively modern trend and one that has witnessed the transformation of agricultural villages to suburbs of nearby towns and cities.

Residential developments and the increasing number of service developments have cut into the flesh of agricultural land. This is especially true in villages surrounding big cities. In those villages close to urban areas, there is often a

Table 23.4 Population of Polish cities and villages since 1990. Table compiled from the Polish Central Statistic Office

	Cities		Villages	
	Population (thousands)	Percentage	Population (thousands)	Percentage
1990	23,546.0	61.84	14,527.0	38.16
1995	23,675.0	61.84	14,609.0	38.16
2000	23,670.3	61.88	14,583.7	38.12
2005	23,423.7	61.39	14,733.4	38.61
2010	23,429.1	60.81	15,100.8	39.19
2013	23,257.9	60.42	15,237.8	39.58

growing population as people migrate out from the urban areas to enjoy a better quality of life. In such cases, their needs often justify changes in the designation of land in the villages. As a result, there have been changes in the spatial structure of the villages. It is especially present in the near vicinity of big cities, which are becoming densely populated and land becoming intensively developed (Fig. 23.1.) (Hernik et al. 2012).

Building permits issued in the previous system considered only the economic development of the area and failed to consider the effects the investments might have on the environment. Now the authorities are more aware of the negative influence of some investments on the environment and the reasons for that is not simply the result of European directives but of a growing environmental awareness. The opponents of the eco-related restrictions consider that the restrictions will limit

Fig. 23.1 Urban development of a village. (Michałowice village, Małopolska, Poland). *Source* thanks to the courtesy of Michałowice community

the development of Polish economy, whilst ecological NGOs, on the other hand, struggle for more eco-investments but this is often at the result of increased expenditure.

Third group of taxation applies to forests and it is the lowest of all three categories. A tax from 1 ha of forest is eight times lower than a tax from 1 ha of arable land. There are certain types of forest that are not charged any taxes. These are protection to forest or forest younger than 40 years old. If we consider only taxes then forest are the most viable which could lead to forests being preferred type of landscape.

All these factors contribute towards the conflict rather than integration of modern and traditional farming in cultural landscapes and indicate that there needs to be a careful balancing act between the needs of the local population and the need to maintain and conserve the cultural landscape.

23.5 Nature Conservation

One of the important factors in integrating of modern and traditional farming in cultural landscapes is the extensive, protected areas in Poland, which cover 23 national parks, together with natural reserves, landscape parks and areas with protected landscapes. In 2004, a new form of nature protection was introduced, that of Nature 2000, which covers almost 26.8 % of Poland. Natura 2000 in Poland includes 145 Special Protection Areas (Bird Directive) and 845 Special Areas of Conservation (Habitat Directive) (GUS 2013). An extensive element of Natura 2000 areas in Poland is the agricultural land in cultural landscapes.

The protection of these areas does not exclude its use as agricultural land but the development of agriculture on the areas should be adjusted to provide the appropriate level of environmental protection. This imposes limits on actions that can threat the habitats, which are perceived by the inhabitants of the area as inhibiting the development. Natura 2000 is also mistakenly believed in some quarters to be a threat to traditional agriculture (Mickiewicz and Gotkiewicz 2010).

One of the solutions for symbiosis between protected areas and agriculture is organic farming (discussed above) which allows for the improved use of existing potential but does not break the rules of sustainable development (Bera 2014). Economy and societies of the areas of significant natural values may develop better (Zielińska 2007).

23.6 The Current Situation

The European Landscape Convention stresses the role of landscapes as it is an important part of the quality of life in the cities and rural communities. It is not important if the area is thought to be ordinary or uniquely beautiful, the landscape should be protected in both cases as the development of technology and better quality of life accelerate changes in landscapes (European Landscape Convention 2000).

Work on the implementation of European Landscape Convention has been undertaken for a number of years. Experts commenced public consultations on the protection, development and management of landscapes according to European Landscape Convention. Directions of legal changes, Polish landscape evaluation, education, social participation, cultural and natural landscape protection have been widely discussed. The European Landscape Convention obliges member countries to identify, analyze and determine pressures and changes in Landscapes. Regional and national systems of typology and the identification of landscape are created in member countries to meet the obligations. An interdisciplinary team is working on the task in Poland. Recently, a methodology and main rules for the identification of landscape has been proposed.

The first task is to identify the landscape by dividing the country into physiographic microregions, which are going to be characterized in three ways according to the natural landscape; plant landscape and cultural and historical diversity. The first two approaches are based on existing methodologies but the historical-cultural approach needs to be created from the beginning (Solon et al. 2014). At the next stage, a set of analytic and synthetic features for landscape description should be created. The final stage in landscape evaluation is a 'card of landscape valuation' and a choice of criteria that will set apart the most important landscapes. Measures taken are still only theoretical.

23.7 Conclusion

It seems that paradoxically, rural cultural landscapes were better off during socialist time since agriculture was a priority during that period. Furthermore, extensive agriculture production meant that there was no abandoned land.

Nowadays, production is unviable in many marginal areas and there is more and more abandoned land and forests as forests are more desirable land designation than agricultural land.

In modern times, however, many forms of landscape and nature protection have been developed, which could help to save traditional agricultural landscapes. These positive processes may be observed in spatial planning regulations, which are beginning to include policies on landscape protection, and although the European Landscape Convention is not fully implemented yet, extensive discussions on the future shape of its implementation should bring excellent results in the integration of modern and traditional agriculture in the cultural landscapes of Poland.

Managing landscapes is not an easy task. It may be done with legal regulations but the regulations are often too inflexible, which is disliked by the people. Unfortunately, law making has many potential flaws as each and every law has its opponents from the very beginning, which encourages loopholes.

With the present rural landscape, particularly in marginal areas—often those areas including cultural landscapes, where the evidence of encroaching forests and abandoned lands are quite visible, there is a certain hope for maintaining traditional rural landscapes through organic farming.

The Polish attitude toward landscape has travelled a long way from generally being unrecognized in socialist times, to serious attempts to implement European Landscape Convention.

The regulations and legal instruments concerning land policies have evolved in a relatively short period of time (Table 23.5). It is clear that they are far from perfect but this is also true in many western European States. The great advantage that Poland has in the design and implantation of spatial planning regulations is that it has the opportunity of learning and correcting the mistakes found in many other state jurisdictions to provide a system that is both suited to the needs of the unique geography of Poland yet complies with national regional and European guidelines and conventions.

What has been achieved to date, particularly in the regulation of agriculture, the conservation of cultural landscapes—both rural and urban, as built on the system prior to transformation yet made it ‘fit for purpose’ in a Poland that is looking both towards the west and also towards its new neighbours.

We do not have any program of comprehensive landscape protection in Poland. We protect only certain elements of landscape:

- Environmental protection regulations provide several forms of protection, which focus only on nature (e.g. national parks, landscape parks). We should not forget that this protection cannot limit the development in some areas.
- Regulations on the protection of monuments provide protection of unique landscapes with historical monuments by creating cultural parks
- Spatial planning is narrowed to small areas and is often a short-term spatial planning. In the cities where there are many investments, the lack of a local spatial plan causes some controversies.

Table 23.5 Influence of traditional and modern agriculture on rural cultural landscape

Traditional agriculture	Modern agriculture (after transformation)
• No land abandonment	• Abandoned land
• Clear boarders between forest and agricultural land	• No farmer education is necessary to have a farm
• Traditions in land use	• Organic farming
• Agriculture education was required to have a farm	• Designating agricultural land for non-agricultural purposes
• Difficult change of designation of agricultural land use	• More environmental protection of landscapes
	• Vanishing boarders between forest and agricultural land

- Lack of control over the changes in the landscapes of the cities (city's spread) and villages. Lack of institution holistically responsible for spatial development and landscape with all its elements.

In Poland much needs to be done to adjust the land policy to landscape protection (so that land policy could stop negatively influence landscape protection). Apart from creating new areas with protected landscape, we should also focus on creating spatial plan to normalize construction and investments regulations.

References

Bera M (2014) Rolnictwo ekologiczne jako czynnik rozwojowy gmin położonych na obszarach chronionych. Progr Econ Sci 1:121–129

Czechowski P, Korzycka-Iwanow M, Prutis S, Stelmachowski A (1994) Polskie prawo rolne na tle ustawodawstwa Unii Europejskiej. PWN, Warszawa

Dixon-Gough R, Hernik J, Dixon-Gough J (2011) Sustainability and the cultural landscape. In: Hepperle E, Dixon-Gough RW, Kalbro T, Mansberger R, Meyer-Cech K (eds) Core-themes of land use politics: sustainability and balance of interests. vdf Hochschulverlag AG an der ETH Zürich, Zürich, pp 85–110

Dixon-Gough R, Hernik J, Gawroński K, Taszakowski J (2013) Peri-urban problems around Krakow. In: Hepperle E, Dixon-Gough R, Maliene V, Mansberger R, Paulsson J, Pödör A (eds) Land management: potential, problems and stumbling blocks. vdf Hochschulverlag AG an der ETH Zürich, Zürich, pp 133–140

Duda –Krynicka M, Jaskolecki H (2010) Historia i perspektywy rozwoju rolnictwa ekologicznego w Polsce. Problemy Ekologii 14(2):85–91

Dutkowski M, Tölle A, Wdowicka M (2012) Historia planowania przestrzennego. http://www.arl-net.de/commin/poland-polska/11-historia-planowania-przestrzennego. Accessed 15 Sept 2014

Gawroński K (2004) Problematyka zalesiania gruntów marginalnych terenów górskich. Infrastruktura i Ekologia Terenów Wiejskich, nr 2(2004):153–166

Gorzym-Wilkowski WA (2012) Planowanie zagospodarowania przestrzennego województwa lubelskiego w okresie powojennym - praktyka wobec teorii i ustawodawstwa. Studia Regionalne i Lokalne 1(47):24–42

GUS (Polish Central Statistic Office) (2013) Ludność. Stan i struktura demograficzno - społeczna, Warszawa

Hernik J, Dixon-Gough R, Taszakowski J (2012) Open source geospatial technologies in the context of peri-urban problems around Kraków, with particular respect to the commune of Michałowice. In: Service-oriented mapping. Verlag Wien, Wien, pp 201–214

Hernik J, Gawroński K, Dixon-Gough R (2013) Social and economic conflicts between cultural landscapes and rural communities in the English and Polish systems. Land Use Policy 30:800–813

Kancelaria Senatu, Biuro Analiz i Dokumentacji (2013). Szanse i zagrożenia polskiego rolnictwa ekologicznego w perspektywie lat 2014–2020, Warszawa. http://www.senat.gov.pl/gfx/senat/pl/senatekspertyzy/2340/plik/oe-210_inter.pdf. Accessed 15 Sept 2014

Mickiewicz B, Gotkiewicz W (2010) Znaczenie i rola obszarów Natura 2000 w życiu gospodarczym gminy. Ochrona Środowiska i Zasobów Naturalnych 45:145–152

Paull J (2011a) Attending the first organic agriculture course: Rudolf Steiner's agriculture course at Koberwitz, 1924. Eur J Soc Sci 21(1):64–70

Paull J (2011b) Biodynamic agriculture: the journey from Koberwitz to the world, 1924–1938. J Organ Syst 6(1):27–41

Solon J, et al (2014) Przygotowanie opracowania pt „Identyfikacja i ocena krajobrazów – metodyka oraz główne założenia" Raport etapowy 01. http://ochronaprzyrody.gdos.gov.pl/iii-krajowa-konferencja-poswiecona-wdrazaniu-europejskiej-konwencji-krajobrazowej. Accessed 15 Sept 2014

Woś A (1992) Strategie rozwoju rolnictwa. PWN, Warszawa

Zielińska A (2007) Konkurencyjność obszarów przyrodniczo cennych. In: Bernat T (ed) Przedsiębiorstwo i państwo - wybrane problemy konkurencyjności. Wyd. PRINT GROUP Daniel Krzanowski, Szczecin, pp 162–172

Acts and Other Legal Documents

Dekret o planowym zagospodarowani przestrzennym kraju z dnia 2 kwietnia 1946 r. (Dz. U. 1946 nr 16 poz. 109)

European Landscape Convention, Florence 20 Oct 2000 (Dz. U. 2006 nr 14 poz. 98)

Uchwała Rady Ministrów z 12. VII. 1966 r. w sprawie ochrony użytków rolnych (MP 1966, nr 40, poz. 20

Ustawa z dnia 31 stycznia 1961 r. planowaniu przestrzennym, (Dz. U. 1961 nr 7, poz. 47)

Ustawa z dnia 29 czerwca 1963 o ograniczeniu podziału gospodarstw rolnych. (Dz. U. 1963 nr 28, poz. 168)

Ustawa z dnia 12 lipca 1984 r. o planowaniu przestrzennym (Dz. U. 1984 nr 35 poz. 185)

Ustawa z 28 września 1991 r. o lasach. 1991(Dz. U. 1991 nr 101, poz. 444, z późn. zm.)

Ustawa z dnia 7 lipca 1994 r. o zagospodarowaniu przestrzennym (Dz. U. 1994 nr 89 poz. 415)

Ustawa z dnia 8 czerwca 2001 r. o przeznaczaniu gruntów rolnych do zalesienia (Dz. U. 2001 nr 73, poz. 764)

Ustawa z dnia 27 marca 2003 r. o planowaniu i zagospodarowaniu przestrzennym (Dz. U. 2003 nr 80 poz. 717)

Ustawa z dnia 28 listopada 2003 r. o wspieraniu rozwoju obszarów wiejskich ze środków pochodzących z Sekcji Gwarancji Europejskiego Funduszu Orientacji i Gwarancji Rolnych (Dz. U. 2003 nr 229, poz. 2273)

Ustawa z dnia 16 kwietnia 2004 r. o ochronie przyrody. (Dz. U. 2004 nr 92 poz. 880)

Ustawa z dnia 8 marca 2013 o zmianie ustawy o ochronie gruntów rolnych i leśnych (Dz. U. 2013. Poz. 503)

Chapter 24
Preserving the Mediterranean Diet Through Holistic Strategies for the Conservation of Traditional Farming Systems

Luigi Ponti, Andrew Paul Gutierrez and Miguel A. Altieri

Abstract The Mediterranean diet is described by the UNESCO Cultural Heritage of Humanity website (http://www.unesco.org/culture/ich/en/RL/00884) as encompassing more than just food of the various cultures. These diets are embedded in bio-cultural landscapes that are at risk from global markets, industrial agriculture, invasive species and climate change, and yet little research aimed at conserving this Mediterranean agricultural heritage is being conducted. A focus on preserving traditional Mediterranean agricultural systems provides unique opportunities to link UNESCO-SCBD's Joint Programme on Biological and Cultural Diversity (http://www.cbd.int/lbcd/) and FAO's Globally Important Agricultural Heritage Systems initiative (GIAHS, http://www.fao.org/giahs/) with the goal of developing strategies and policy to preserve this heritage and the food production systems that are its basis for future generations. An important step in this direction is the development of holistic ecosystem-level assessments of the stability and resilience of traditional Mediterranean farming systems to evolving global change including climate change

L. Ponti (✉)
Laboratorio Gestione Sostenibile degli Agro-ecosistemi, Unità Tecnica Agri, ENEA Centro Ricerche Casaccia, Via Anguillarese 301, 00123 Rome, Italy
e-mail: luigi.ponti@enea.it
URL: http://www.casasglobal.org

L. Ponti · A.P. Gutierrez
Center for the Analysis of Sustainable Agricultural Systems, Kensington, CA 94707-1035, USA
e-mail: casas.global@berkeley.edu
URL: http://www.casasglobal.org

A.P. Gutierrez
College of Natural Resources, University of California, Berkeley, CA 94720-3114, USA

M.A. Altieri
Department of Environmental Science, Policy, and Management, University of California, Berkeley, 130 Mulford Hall #3114, Berkeley, CA 94720, USA
e-mail: agroeco3@berkeley.edu

M. Agnoletti and F. Emanueli (eds.), *Biocultural Diversity in Europe*, Environmental History 5, DOI 10.1007/978-3-319-26315-1_24

and shifting economic patterns and associated landscape transformations. A holistic approach is an important step to ensure ecologically sustainable development, conserve cultural identities, improve farming community livelihood, preserve agro-biodiversity and ensure the continued provision of vital ecosystem services for humanity.

Keywords Traditional mediterranean farming systems · Agricultural heritage systems · Physiologically based demographic models (PBDMs) · Global change · Resilience to climate change · Bio-cultural diversity

24.1 Introduction

Cradle to the development of some of the largest and most powerful civilizations in the world, the Mediterranean Basin has witnessed millennia of interactions between humans and ecosystems, particularly through agriculture (Blondel 2006), with associated evolving dietary patterns that resulted in what we now call Mediterranean diet. In 2010, the United Nations' Educational, Scientific and Cultural Organization (UNESCO) first recognized the Mediterranean diet as an Intangible Cultural Heritage of Humanity (http://www.unesco.org/culture/ich/en/RL/00884) that encompasses 'a set of skills, knowledge, rituals, symbols and traditions concerning crops, harvesting, fishing, animal husbandry, conservation, processing, cooking, and particularly the sharing and consumption of food'. Embedded as they are in bio-cultural landscapes (Hong et al. 2014), these dietary patterns are but one aspect of traditional Mediterranean food systems (Kuhnlein and Receveur 1996) that in turn are grounded on traditional farming systems such as olive, grape and cereals, with olive being of particular ecological, economic, social and cultural (e.g. landscape) relevance (Loumou and Giourga 2003). In contrast to the increasing homogeneity in global food supplies with implications for food security, Mediterranean dietary patterns bring about significant health benefits (Khoury et al. 2014), and are increasingly linked to a reduced ecological footprint (Capone et al. 2013) that can be used as a metric of how sustainable Mediterranean farming systems are. However, little research aimed at conserving Mediterranean agricultural heritage systems is being conducted despite ongoing threats to their persistence such as global markets, industrial agriculture, invasive species and climate change (see e.g. Rosenzweig and Tubiello 1997; Souissi et al. 2013; Alessandri et al. 2014; Ponti et al. 2015).

Traditional agricultural systems have evolved through centuries of humans–nature interactions and continue to provide local food security, harbor high levels of genetic and organismal biodiversity free of transgenic contamination and are living libraries of indigenous knowledge and management of systems that are the foundation for contemporary and future agricultural innovations and technologies (Koohafkan and Altieri 2011; Lansing and Kremer 2011; Nicholls and Altieri 2012). Worldwide, a panoply of ecologically based small farm agricultural systems feed at least 70 % of the world population (i.e. global food security) and account for over a half of the agricultural output used for domestic consumption (i.e. food sovereignty) in Africa, Asia and Latin America (Altieri 2008; ETC 2009; Rosset 2011). However, the world now faces an era of climate change of global dimension and impact (Koohafkan et al. 2012), particularly in the developing countries where local knowledge systems and agricultural practices and techniques remain the dominant coping mechanisms/responses to climate change (Morton 2007; Altieri and Koohafkan 2008). Climate change is a developing additional level of complexity that adds to risks posed by global markets, industrial agriculture and invasive species, and together these interacting factors will make the world's traditional agroecosystems unprecedentedly hard to manage.

There is little doubt that traditional farming systems will be impacted by climatic change despite adaptations to cope with extremes of current climate (Altieri and Koohafkan 2008; Koohafkan 2009; Beddington et al. 2012). The level of resiliency to extreme climatic events that have occurred over the last two decades has been closely linked to the level of on-farm biodiversity of small farms (Altieri et al. 2012; Altieri and Nicholls 2013). However, a major challenge is to assess the limits of this resiliency in the face of climatic change, and for this the tools of ecosystem modeling and analysis and geographic information system (GIS) will be useful. We note, however, that assessing the impacts of climate change in traditional faming systems, especially in small farm systems, is a far greater challenge than in conventional monocultures because of the lack of standard definitions and data, the greater complexity of the farming systems and their vulnerability to a range of climate-related and other stressors (Morton 2007; Koohafkan and De La Cruz 2011).

A focus on preserving traditional Mediterranean agricultural systems provides unique opportunities to link UNESCO-SCBD's Joint Programme on Biological and Cultural Diversity (http://www.cbd.int/lbcd/) and FAO's Globally Important Agricultural Heritage Systems initiative (GIAHS, http://www.fao.org/giahs/) with the goal of developing strategies and policy to preserve this heritage including the food production systems that are its basis for future generations (Fig. 24.1). This chapter illustrates how an important step in this direction would be the development of holistic ecosystem-level assessments of the stability and resilience of traditional Mediterranean farming systems to evolving global change including climate change and shifting economic patterns and associated landscape transformations. A holistic approach would help guide ecologically sustainable development, conserve cultural identities, improve farming community livelihood, preserve agro-biodiversity and ensure the continued provision of vital ecosystem services.

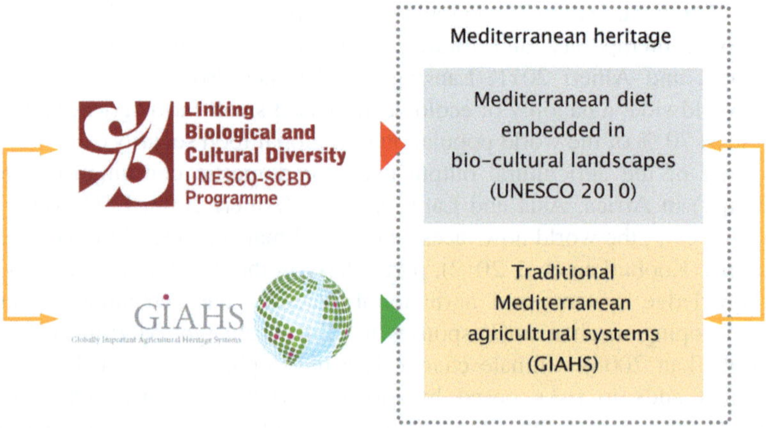

Fig. 24.1 Conceptual graph illustrating how the UNESCO-SCBD's Joint Programme on Biological and Cultural Diversity (http://www.cbd.int/lbcd/) and the FAO's Globally Important Agricultural Heritage Systems initiative (GIAHS, http://www.fao.org/giahs/) may be linked with the goal of developing strategies and policy to preserve the traditional Mediterranean agricultural heritage

24.2 Traditional Mediterranean Farming Systems

Because of isolation in a multitude of small and diverse river basins generated by a broken topography (Halstead 1987), most rural people of the Mediterranean Basin were forced until quite recently to rely for survival on wheat, olives, milk, cheese, wine, meat from domestic and wild animals, a wide range of domestic and wild fruits, as well as a variety of wild products that could be found in shrublands and woodlands (Blondel 2006). Many local land use systems emerged across the Basin, and the endless redesign of Mediterranean landscapes through traditional land use systems had an overall beneficial influence on biological diversity in the region (Blondel and Aronson 1995). Highest biological diversity probably never occurred in pristine oak woodlands but in systems that were moderately modified by man (see e.g. Gómez-Campo 1985; Blondel and Aronson 1995) such as traditional agro-silvo-pastoral systems (Fig. 24.2). Such pattern supports the diversity-disturbance hypothesis that intermediate levels of disturbance promote biological diversity (Huston 1994). Furthermore, many of today's agroforestry systems have often replaced natural forests (Scarascia-Mugnozza et al. 2000), with humans exerting a profound influence on forests virtually everywhere on the globe, and hence 'natural' defined as 'without human influence' is an almost entirely hypothetical notion, despite the fact that productive systems rely on natural biodiversity as their foundation and source of resilience (McNeely 2004).

Traditional farming systems like olive remain of great importance in many areas of the Mediterranean Basin even if they are often restricted to marginal areas, as they form a consistent landscape component within a variety of arable or pastoral

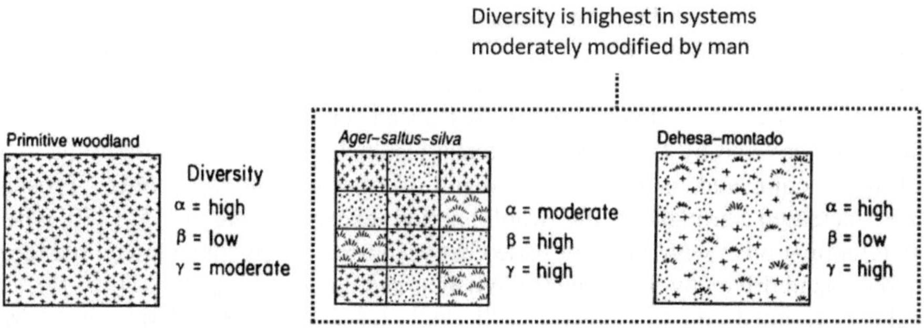

Fig. 24.2 Levels of the three components of diversity, namely α-diversity (within habitat), β-diversity (between habitat) and γ-diversity (regional), in primitive woodlands and in two ancient land use systems in the Mediterranean region: the *Sylva-saltus-ager* (woodland-pasture-field) system that was widespread during the Roman empire and is arguably the most influential and well-known of all ancient systems in the region, and the savannah-like *Dehesa-montado* system, typical of the Iberian peninsula and of many islands such as Sardinia (Italy). Cultivation, grazing, and harvesting of forest products are practiced according to a different spatial arrangement in the two systems, as the three activities are conducted in separate areas in the *Sylva-saltus-ager* system, whereas they are combined within a single area in the *Dehesa-montado* system. Highest overall level of diversity is found in these traditional agricultural systems that also show a high degree of resilience (modified from Blondel 2006). The text to the right of each system denotes levels of α, β and γ diversity

land uses (Eichhorn et al. 2006). This is especially the case for silvo-arable systems like the olive groves of central Italy (e.g. in Umbria, a relatively dry, hilly region) (Bertolotto et al. 1995) or the *dehesas* and *montados* typical of southwestern Spain and Portugal but also abundant in Sardinia, Italy (Joffre et al. 1988; Grove and Rackham 2003) (see also Fig. 24.2), where the trees themselves define the landscape even if they are not the dominant landscape component (Eichhorn et al. 2006). The pre-Roman practice of intercropping olives with cereals, vegetables and fodder crops still survives on several thousands of hectares in Italy, Greece and Spain, making olive a consistent element in the landscape throughout southern Europe (Lelle and Gold 1994; Papanastasis et al. 2009). Mediterranean agroforestry systems such as olive are relatively poorly studied in terms of their economic importance and role in preserving local and regional biodiversity, with agroforestry itself having mostly been considered a feature of the tropics (see Rodríguez et al. 2009). However, the role of temperate agroforestry systems in reconciling productivity with environment protection is gaining attention (see e.g. http://www.agforward.eu/), as it simultaneously addresses issues of considerable importance to the future of European agriculture (Eichhorn et al. 2006; Smith et al. 2013).

Olive inhabits a variety of different landscapes ranging from olive trees interspersed in natural forest vegetation to intensive monocultures, with many intermediate cases (Kizos and Koulouri 2010). Although olive has proven an environmentally sustainable farming system through millennia when growing as an agroforestry tree across the Mediterranean Basin, this traditional crop is not per se a guarantee for sustainability. An assessment of soil erosion rates in olive orchards performed by Vanwalleghem et al. (2011) in Andalusia, Spain over a 250-year period since 1752 using historical records and field measurements of relic tree mounds, showed that olive cropping that lasted only 153 to 291 years in areas previously covered by Mediterranean forest resulted in a 29–40 % loss of the total soil depth compared to the total soil depth of other current soil profiles. Erosion rates increased throughout the whole period except during 1935–1970 when intercropping olive with wheat or barley between rows (likely due to increased local food demand driven by the Spanish Civil War and the autarkic policy of the Franco regime) resulted in the lowest soil loss rates over the 250-year period (Vanwalleghem et al. 2011). Highest soil loss rates occurred from the year 2000 onward, because herbicides were adopted to control weeds that in combination with superficial harrowing during summer, decreased tillage erosion but reduced surface cover and roughness greatly enhancing water erosion (Vanwalleghem et al. 2011; see also García-Orenes et al. 2012).

Olive is the most emblematic plant species across Mediterranean cultures and landscapes (Loumou and Giourga 2003; Terral et al. 2004) and forms the basis of the Mediterranean rain-fed agroforestry system (Pasternak 2001), one the oldest (Renfrew 1973; Zohary and Spiegel-Roy 1975; Yasuda 1997; Blondel 2006), most ecologically sustainable rain-fed agroecosystems worldwide (Lüttge 2010). As such, cultivated olive provides a model to design sustainable rain-fed systems for the semiarid Africa (Pasternak 2001), including regions such as the Sahel (Pasternak and Schlissel 2001). Olive culture historically played an important role in rural development and poverty alleviation in marginal areas across the Middle East and North Africa (Lybbert and Elabed 2013) and parts of Europe (Fleskens and de Graaff 2010). Economic and social viability and persistence of this well-documented cosmopolitan agroecosystem have important implications for preventing and ameliorating desertification that is a major environmental threat to the whole Mediterranean region (Mtaita et al. 2001; Geeson et al. 2002; Schröter et al. 2005; de Graaff et al. 2010, 2011) under global climate change.

Farming communities in the Mediterranean Basin have developed traditional olive agroecosystems that ingeniously mimic the local subtropical dry forest ecosystems adapted to shallow soils and unreliable rainfall typical of the Basin (FAO 2001; see also Ponti et al. 2014). These olive based agroecosystems have been in use for many centuries without depleting the resource base, and hence illustrate that agricultural sustainability may be achieved when natural habitat complexity is emulated (Blondel 2006) at multiple spatial scales (Benton et al. 2003). It illustrates how systems that successfully mimic nature should seek complementary species according to the so-called 'M5' golden rule (Making Mimics Means Managing Mixtures) (Dawson and Fry 1998). Research has shown that the prevalent coevolved natural secondary plant associations of an area can provide a model for designing multispecies crop mixtures (Ewel 1976; Altieri and Letourneau 1982; Ewel et al. 1982; Altieri et al. 1983). This is why the recent debate on sustainable versus ecological intensification goes beyond pure semantics, as ecological intensification involves a landscape approach to designing multifunctional agroecosystems that are both sustained by nature and sustainable in nature (Tittonell 2014).

The below ground side of this story revolves around the ubiquitous soil glycoprotein glomalin. Until studies in 1996 by the United States Department of Agriculture (USDA) (Wright et al. 1996), glomalin was assumed an unidentifiable constituent of soil organic matter (Comis 2002). It is now known that this very stable molecule is produced by and abundantly coats the hyphae of arbuscular mycorrhizal (AM) fungi, and is highly correlated with increased stability of soil aggregates, as it binds soils particles together (Rillig 2004). Researchers speculate that abundant glomalin makes soil favorable to the symbiotic relationship of AM fungi and almost all vascular plants (Wright and Upadhyaya 1998). Long-established olive tree plantations common throughout Mediterranean landscapes feature a variety of root-associated AM species (Calvente et al. 2004) as well as a stable glomalin carbon pool that protects soil aggregates and enhances carbon sequestration (Ramachandran Nair et al. 2010; Emran et al. 2012; Pardini and Gispert 2013). This is especially important on shallow marginal soils in the Mediterranean Basin such as in terraced olive groves that require very low levels of agricultural management and where abandonment and the associated periodical wildfire occurrence trigger soil degradation processes (Gispert et al. 2013). However, the turnover times of the gomalin pool in the soil is on the order of decades, and hence the beneficial effects of glomalin on soil aggregate stability are slow to recover after soil degradation such as erosion by water (see Rillig 2004). This long turnover also means that management changes that are detrimental to the glomalin pool (e.g. tillage, see Wright et al. 1999) will affect soil aggregate stability for decades (Rillig 2004) and increase the loss rate of agricultural soil.

Abandonment of traditional farming systems as driven across the Mediterranean by shifting economic patterns increasingly leads to loss of biodiversity and ecosystem services. For example, a field experiment conducted in a long-term observatory located in Sardinia that is representative of local Mediterranean agroforestry systems, Bagella et al. (2014b) concluded that a complex landscape made of a variety of contrasting agro-silvo-pastoral land uses ranging from marginal cork oak forests to intensively managed grape monocultures, and including animal grazing in secondary grasslands, is able to buffer the impact of most intensive land uses on above ground biodiversity via specific below ground biodiversity. This study was one of the first to attempt an assessment of how maintaining the typical Mediterranean landscape diversity including traditional farming systems is important for conserving biodiversity and ecosystem function at the landscape level (Bagella et al. 2014b). These systems are the backbone of the regional food system, but only if actively managed can they promote sustainable use and conservation of bio-cultural diversity (Bugalho et al. 2011) including the Mediterranean diet. An appropriate eco-social context is therefore needed that provides human activity with the capacity to enhance plant biodiversity via traditional farming systems (Bagella et al. 2014a). For example, novel economic incentives that pay for ecosystem services are required to complement and replace a number of agri-environmental payment schemes that have been implemented in the recent past with checkered success (Pretty 2008). A related question is whether carbon offset credits such as those provided under the so-called REDD (i.e. reducing emissions from deforestation and forest degradation, see http://www.un-redd.org/) mechanism would be appropriate for supporting the mitigation potential of traditional farming systems, as it addresses but one of a panoply of benefits (i.e. ecosystem services) that these systems provide (Altieri and Nicholls 2013).

Mediterranean agricultural heritage systems including olive are at risk from global markets, industrial agriculture, invasive species and climate change, and yet currently little research aimed at conserving these systems is being conducted. A recent analysis using the Köppen–Geiger climate classification provided a first robust assessment of future northward and eastward expansions of the Mediterranean climate in the Euro-Mediterranean region and western North America, and posit a significant displacement of the southern margins of these climate (e.g. in southern Europe) by an arid type climate during the twentyfirst century (Alessandri et al. 2014). This evidence points to the importance of developing holistic ecosystem-level assessments of the stability and resilience of traditional Mediterranean farming systems to climate change (Ponti et al. 2014).

24.3 Holistic Assessment of Olive Under Climate Change

In this section, we explore the use of physiologically based demographic models (PBDMs) in a GIS context for holistic ecosystem-level assessments of the stability and resilience (Gutierrez 1996) of traditional Mediterranean farming systems to climate change. The Mediterranean Basin is a climate change hotspot of global relevance (Giorgi 2006; Diffenbaugh et al. 2007; Diffenbaugh and Giorgi 2012) where warming of about +2 °C will likely occur between 2030 and 2060 (Giannakopoulos et al. 2009) with unknown bio-economic impact on major crop systems including olive. Climate change will alter the interaction of olive with its herbivorous pests, and understanding these shifts is crucial to assessing the ecological and economic impacts. In contrast with mainstream assessments of climate change impact on agricultural and other ecosystems that have omitted trophic interactions (van der Putten et al. 2010), PBDMs in a the context of a GIS have the capacity to estimate the impacts of climate warming on the interactions of olive and its major obligate pest, the olive fly *Bactrocera oleae* (Gutierrez et al. 2009; see Daane and Johnson 2010 for a comprehensive review on olive fly).

How to analyze the tripartite ecological, economic and social effects of climate change has been vexing and largely unexplored, and olive is one of the first agroecosystems where these different factors have been analyzed holistically, albeit with different level of precision, using process-based modeling tools such as PBDMs (Ponti et al. 2014). The underlying assumption of the models is that all organisms in all trophic levels, including the economic one, are consumers that have similar resource acquisition (inputs) and allocation (outputs) priorities (Gutierrez et al. 1994; Gutierrez 1996; Regev et al. 1998). Based on analogies, the dynamics of say olive and olive fly can be captured using the same resource acquisition and birth–death rates sub-models imbedded in an age-mass structured population model (i.e. the PBDM). Resource acquisition (i.e. the supply, *S*) is a search process driven by organism demand (*D*), while allocation occurs in priority order to egestion, conversion costs, respiration and reproduction, growth and reserves. The ratio *S/D* is greater or equal to zero but always lower than unity due to imperfect consumer search, and in the model scales maximal growth rates of the species in a time-place varying manner. At high resource levels, *S/D* tends to unity. The model for olive is a canopy model with subunit populations of leaves, stem, root and healthy and attacked fruit.

The olive system model simulates the age-mass structured population dynamics of plant subunits and of olive fly numbers (Gutierrez et al. 2009). The olive model predicts flowering phenology controlled by vernalization, the age-structured dynamics of growth and yield, and fruit mortality due to temperature and fly attack. Olive fly biology is closely linked to olive fruit phenology, age and abundance. The effects of temperature on vital rates of olive fruit and the fly are captured by normalized concave scalar functions that approximate the net of *S* corrected for metabolic costs across temperature (Gutierrez et al. 2009). Several weather data sources can be used to drive weather-driven PBDMs (see Ponti et al. 2013), including satellite remote sensing (e.g. Neteler 2010) and state-of-the-art regional climate change projections (e.g. Artale et al. 2010; Dell'Aquila et al. 2012), while GRASS GIS (see http://grass.osgeo.org) (Neteler et al. 2012) is used to perform geospatial analysis and produce maps.

The bio-economic analysis of olive was driven by daily weather scenarios that include a fine scale representation of topography and the influence of the Mediterranean Sea on regional climate (see Ponti et al. 2014). Climate warming will affect olive yield and fly infestation levels differently across the Basin (not shown, see Ponti et al. 2014), and hence local profitability and system stability, resulting in regional economic winners and losers (Fig. 24.3a). Instead of the commonly used production and damage control function approach, the economic impact of climate warming was assessed using a process-based PBDM as the production function (Gutierrez et al. 2009). Profitability of family olive farms in many marginal areas of Europe and elsewhere in the Basin will decrease, leading in some cases to abandonment. These farms are critical to conserving soil, maintaining biodiversity and reducing fire risk. The analysis showed that understanding the interactions of olive and olive fly is critical to estimating the ecological and bio-economic effects of climate change on olive culture across the Mediterranean Basin, and provides a template for assessing climate impacts in other traditional Mediterranean farming systems such as grape, with extant and potential new invasive pests. The study suggests that climate warming could have far greater impact on less heat and drought tolerant crop systems such as grape and wheat (Ponti et al. 2014).

An additional step in this analysis would be to investigate to what extent areas where olive is grown that are projected to be at risk from climate change (see Fig. 24.3a) are also located in vulnerable environments such as those prone to soil erosion. Global datasets that may be used to this end include those resulting from

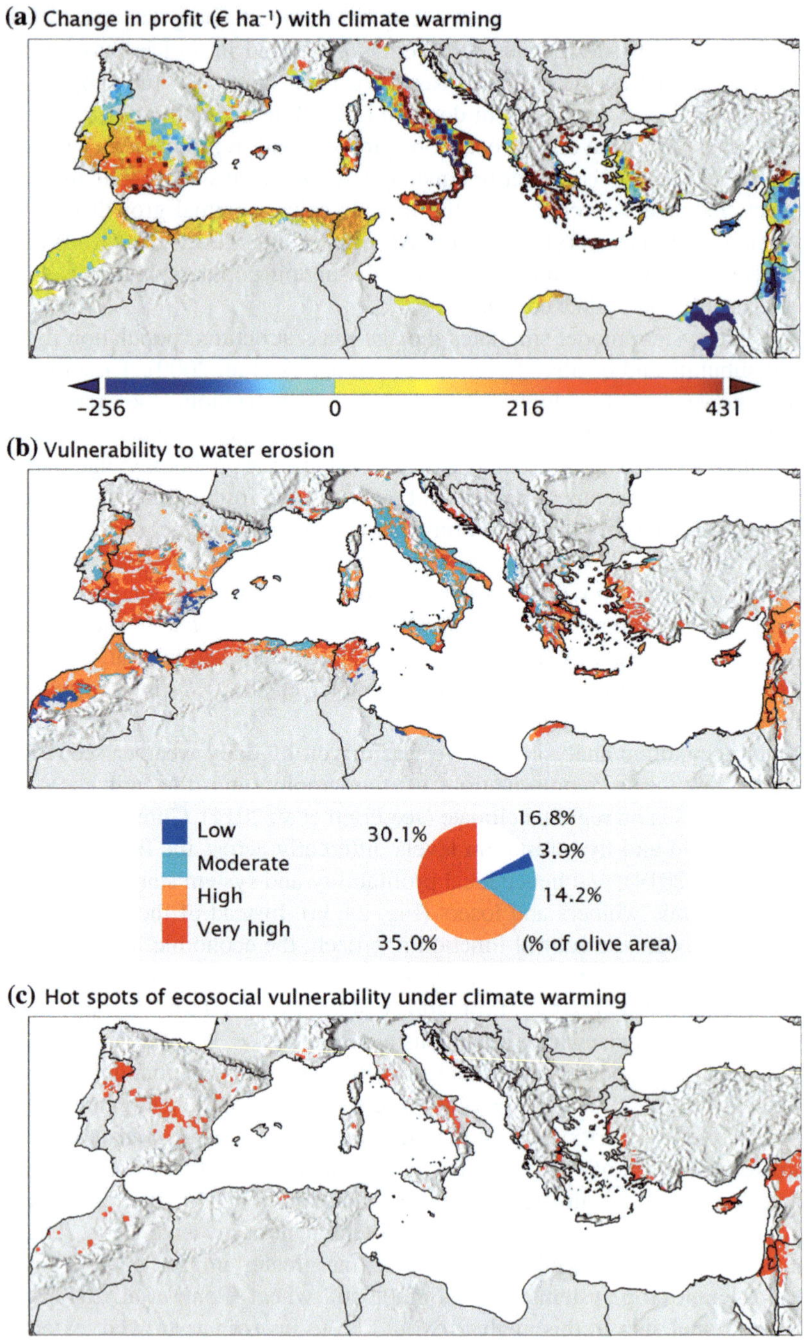
(a) Change in profit (€ ha-1) with climate warming
-256
0
216
431
(b) Vulnerability to water erosion
Low
Moderate
High
Very high
30.1%
16.8%
3.9%
14.2%
35.0%
(% of olive area)
(c) Hot spots of ecosocial vulnerability under climate warming

◀ **Fig. 24.3** Understanding eco-social resilience to climate warming in olive across the Mediterranean Basin. **a** Bio-economic multitrophic impact of climate warming on olive and olive fly as measured by projected change in profit (€ ha^{-1}) under an A1B climate scenario with +1.8 °C warming (from Ponti et al. 2014). To improve data visualization, statistical outliers were identified using the boxplot function in R (see www.r-project.org/) and mapped with *darker blue/red shades* indicated by the *triangles* at the extremes of the color legend. The full data interval is [−914.7, 3000.0]. **b** Soil vulnerability to water erosion (data from Reich et al. 2001) within the observed distribution of olive (data from Ponti et al. 2014). **c** Hot spots of eco-social vulnerability across the Mediterranean Basin are defined as areas with projected negative change in profit (*blue areas* in **a**) and high to very high vulnerability to water erosion (respectively *orange* and *red areas* in **b**)

the assessments by Oldeman et al. (1991), Batjes (1996) and Reich et al. (2001). Oldeman et al. (1991) performed the first Global Assessment of Soil Degradation (GLASOD) that although based on untested expert judgment, has been the most influential appraisal of land quality in terms environmental policy (Sonneveld and Dent 2009). Batjes (1996) used a water erosion model similar to the universal soil loss equation (Wischmeier and Smith 1978), and evaluated model output against GLASOD, showing fair geographic agreement. Reich et al. (2001) combined global soil and climatic data via GIS analysis, and assigned each soil unit a vulnerability class using soil moisture regime as a proxy for erosivity, based on how soils are known to behave under certain prevailing climatic conditions (see http://1.usa.gov/1AnepoQ). We used the more recent map of water erosion by Reich et al. (2001) as it attempts a quantitative assessment of global erosion extent and its causative parameters, including a human population density layer as a proxy for land management intensity. Human population is a particularly important risk factor in biodiversity hot spots such as the Mediterranean Basin (Cincotta et al. 2000). This dataset was made available on request from Paul F. Reich (Soil Science Division, USDA Natural Resources Conservation Services), and was imported, processed and mapped using GRASS GIS (GRASS Development Team 2014). We mapped this water erosion dataset (including four erosion classes: low, moderate, high and very high) for the observed distribution of olive in the Mediterranean Basin as estimated by Ponti et al. (2014), and the resulting map shows that about 30 % of the land where olive is currently grown is highly vulnerable to water erosion (Fig. 24.3b), 30 % is very vulnerable, 14 % is moderately vulnerable and only about 4 % has low vulnerability, with no vulnerability class assigned to about 17 % of the land in the dataset used (e.g. the Nile river area is assigned to the 'dry' class, with no particular water vulnerability). Based on projected bio-economic impact of climate warming (Fig. 24.3a) and soil vulnerability to water erosion (Fig. 24.3b), we mapped hot spots of eco-social vulnerability across the Mediterranean Basin as areas having projected negative change in profit and high to very high vulnerability to water erosion (Fig. 24.3c). These hot spots of eco-social vulnerability cover 13.6 % of the total olive area (i.e. the area mapped with blue to red color shades in Fig. 24.3a).

24.4 Conclusions

An important step for developing strategies and policy to preserve the heritage of traditional Mediterranean agricultural systems is the development of holistic ecosystem-level assessments of their stability and resilience to evolving global change including climate change and shifting economic patterns and associated landscape transformations. Using olive as an prominent case study, we illustrated a physiologically based weather-driven geospatial approach that is holistic in nature, and has the capacity to lay the groundwork for ecosystem analysis of system sustainability. Knowledge gathered via holistic ecosystem-level assessments such as the one illustrated here can be used to support interdisciplinary field methodologies designed to help farmers prepare for climatic variability (see e.g. Altieri et al. 2011; Rogé et al. 2014). Sound knowledge of the ecological processes that underpin the sustainability of traditional Mediterranean farming systems is an important prerequisite, but resilience to global change including climate change requires more than applying a set of practices: the challenge in both intensive and marginal agriculture and in small to large scale farms, is to reinstate social organization and collective strategies in farmer communities that make full use of holistic knowledge about food systems (Altieri et al. 2014; MacMillan and Benton 2014; Martínez-Torres and Rosset 2014).

Preserving the Mediterranean diet through holistic strategies for the conservation of traditional farming systems holds potential to help meet multiple concurrent goals in Mediterranean biodiversity, climate change and eco-social hot spots by ensuring ecologically sustainable development, conserving cultural identities, improving farming community livelihood, preserving agro-biodiversity and supporting food systems that are more healthy and secure, as well as continuously providing vital ecosystem services. Linking the UNESCO-SCBD's Joint Programme on Biological and Cultural Diversity to the FAO's Globally Important Agricultural Heritage Systems initiative represents a unique opportunity for developing strategies and policy to preserve the Mediterranean heritage.

Acknowledgments We thank Dr. Guido Bongi (Emeritus Director of Research, Italian National Research Council) for useful discussions on olive and climate change in the Mediterranean Basin including long-term effects on soil fertility mediated by glomalin.

References

Alessandri A, De Felice M, Zeng N et al (2014) Robust assessment of the expansion and retreat of Mediterranean climate in the 21st century. Sci Rep 4:7211. doi:10.1038/srep07211

Altieri MA (2008) Small farms as a planetary ecological asset: five key reasons why we should support the revitalisation of small farms in the global south. Third World Network, Penang, Malaysia

Altieri MA, Koohafkan P (2008) Enduring farms: climate change, smallholders and traditional farming communities. Third World Network, Penang, Malaysia

Altieri MA, Letourneau DK (1982) Vegetation management and biological control in agroecosystems. Crop Prot 1:405–430

Altieri MA, Nicholls CI (2013) The adaptation and mitigation potential of traditional agriculture in a changing climate. Clim Change. doi:10.1007/s10584-013-0909-y

Altieri MA, Letourneau DK, Davis JR (1983) Developing sustainable agroecosystems. Bioscience 33:45–49

Altieri MA, Lana MA, Bittencourt HV et al (2011) Enhancing crop productivity via weed suppression in organic no-till cropping systems in Santa Catarina, Brazil. J Sustain Agric 35:855–869. doi:10.1080/10440046.2011.588998

Altieri MA, Funes-Monzote FR, Petersen P (2012) Agroecologically efficient agricultural systems for smallholder farmers: contributions to food sovereignty. Agron Sustain Dev 32:1–13

Altieri MA, Koohafkan P, Nicholls C (2014) Strengthening resilience of modern farming systems: a key prerequisite for sustainable agricultural production in an era of climate change. Third World Network Brief Paper 70. http://www.twn.my/title2/briefing_papers/No70.pdf

Artale V, Calmanti S, Carillo A et al (2010) An atmosphere-ocean regional climate model for the Mediterranean area: assessment of a present climate simulation. Clim Dyn 35:721–740. doi:10.1007/s00382-009-0691-8

Bagella S, Caria MC, Farris E, et al (2014a) Traditional land uses enhanced plant biodiversity in a Mediterranean agro-silvo-pastoral system. Plant Biosyst. doi:10.1080/11263504.2014.943319

Bagella S, Filigheddu R, Caria MC et al (2014b) Contrasting land uses in Mediterranean agro-silvo-pastoral systems generated patchy diversity patterns of vascular plants and below-ground microorganisms. C R Biol 337:717–724. doi:10.1016/j.crvi.2014.09.005

Batjes NH (1996) Global assessment of land vulnerability to water erosion on a 1/2° by 1/2° grid. Land Degrad Dev 7:353–365. doi:10.1002/(SICI)1099-145X(199612)7:4<353:AID-LDR239>3.0.CO;2-N

Beddington JR, Asaduzzaman M, Clark ME et al (2012) What next for agriculture after Durban? Science 335:289–290

Benton TG, Vickery JA, Wilson JD (2003) Farmland biodiversity: is habitat heterogeneity the key? Trends Ecol Evol 18:182–188. doi:10.1016/S0169-5347(03)00011-9

Bertolotto C, Pisanelli A, Cannata F (1995) Pratiche agroforestali nella regione Umbria. Monti E Boschi 46:5–11

Blondel J (2006) The "design" of Mediterranean landscapes: a millennial story of humans and ecological systems during the historic period. Hum Ecol 34:713–729

Blondel J, Aronson J (1995) Biodiversity and ecosystem function in the mediterranean basin: human and non-human determinants. In: Richardson DDM, Davis DGW (eds) Mediterr.-Type Ecosyst. Springer, Berlin, pp 43–119

Bugalho MN, Caldeira MC, Pereira JS et al (2011) Mediterranean cork oak savannas require human use to sustain biodiversity and ecosystem services. Front Ecol Environ 9:278–286

Calvente R, Cano C, Ferrol N et al (2004) Analysing natural diversity of arbuscular mycorrhizal fungi in olive tree (*Olea europaea* L.) plantations and assessment of the effectiveness of native fungal isolates as inoculants for commercial cultivars of olive plantlets. Appl Soil Ecol 26:11–19

Capone R, Iannetta M, El Bilali H et al (2013) A preliminary assessment of the environmental sustainability of the current Italian dietary pattern: water footprint related to food consumption. J Food Nutr Res 1:59–67

Cincotta RP, Wisnewski J, Engelman R (2000) Human population in the biodiversity hotspots. Nature 404:990–992

Comis D (2002) Glomalin: hiding place for a third of the world's stored soil carbon. Agric Res 50:4–7

Daane KM, Johnson MW (2010) Olive fruit fly: managing an ancient pest in modern times. Annu Rev Entomol 55:151–169. doi:10.1146/annurev.ento.54.110807.090553

Dawson T, Fry R (1998) Agriculture in nature's image. Trends Ecol Evol 13:50–51

de Graaff J, Duarte F, Fleskens L, De Figueiredo T (2010) The future of olive groves on sloping land and ex-ante assessment of cross compliance for erosion control. Land Use Policy 27:33–41

De Graaff J, Kessler A, Duarte F (2011) Financial consequences of cross-compliance and flat-rate-per-ha subsidies: the case of olive farmers on sloping land. Land Use Policy 28:388–394

Dell'Aquila A, Calmanti S, Ruti P et al (2012) Effects of seasonal cycle fluctuations in an A1B scenario over the Euro-Mediterranean region. Clim Res 52:135–157

Diffenbaugh N, Giorgi F (2012) Climate change hotspots in the CMIP5 global climate model ensemble. Clim Change 114:813–822

Diffenbaugh NS, Pal JS, Giorgi F, Gao X (2007) Heat stress intensification in the Mediterranean climate change hotspot. Geophys Res Lett 34:L11706. doi:10.1029/2007GL030000

Eichhorn MP, Paris P, Herzog F et al (2006) Silvoarable systems in europe—past, present and future prospects. Agrofor Syst 67:29–50. doi:10.1007/s10457-005-1111-7

Emran M, Gispert M, Pardini G (2012) Patterns of soil organic carbon, glomalin and structural stability in abandoned Mediterranean terraced lands. Eur J Soil Sci 63:637–649. doi:10.1111/j.1365-2389.2012.01493.x

ETC action group on E (2009) Who will feed us? Questions for the food and climate crises. ETC Commun 102. http://www.etcgroup.org/content/who–will–feed–us

Ewel JJ (1976) Litter fall and leaf decomposition in a tropical forest succession in eastern Guatemala. J Ecol 64:293–308

Ewel J, Benedict F, Berish C et al (1982) Leaf area, light transmission, roots and leaf damage in nine tropical plant communities. Agro-Ecosyst 7:305–326

FAO F, AO of the UN (2001) Global ecological zoning for the global forest resources assessment 2000: final report. Working Paper 56. FAO, Rome, Italy

Fleskens L, de Graaff J (2010) Conserving natural resources in olive orchards on sloping land: alternative goal programming approaches towards effective design of cross-compliance and agri-environmental measures. Agric Syst 103:521–534. doi:10.1016/j.agsy.2010.05.005

García-Orenes F, Roldán A, Mataix-Solera J et al (2012) Soil structural stability and erosion rates influenced by agricultural management practices in a semi-arid Mediterranean agro-ecosystem. Soil Use Manag 28:571–579. doi:10.1111/j.1475-2743.2012.00451.x

Geeson N, Brandt CJ, Thornes JB (2002) Mediterranean desertification: a mosaic of processes and responses. Wiley, Chichester, England

Giannakopoulos C, Le Sager P, Bindi M et al (2009) Climatic changes and associated impacts in the Mediterranean resulting from a 2 °C global warming. Glob Planet Change 68:209–224

Giorgi F (2006) Climate change hot-spots. Geophys Res Lett 33:L08707. doi:10.1029/2006GL025734

Gispert M, Emran M, Pardini G et al (2013) The impact of land management and abandonment on soil enzymatic activity, glomalin content and aggregate stability. Geoderma 202–203:51–61. doi:10.1016/j.geoderma.2013.03.012

Gómez-Campo C (1985) Plant conservation in the Mediterranean area. W. Junk Publishers, Dordrecht

GRASS Development Team (2014) Geographic Resources Analysis Support System (GRASS) Software, Version 6.4. Open Source Geospatial Foundation. http://grass.osgeo.org

Grove AT, Rackham O (2003) The Nature of Mediterranean Europe: an ecological history. Yale University Press, New Haven

Gutierrez AP (1996) Applied population ecology: a supply-demand approach. Wiley, New York

Gutierrez AP, Mills NJ, Schreiber SJ, Ellis CK (1994) A physiologically based tritrophic perspective on bottom-up-top-down regulation of populations. Ecology 75:2227–2242

Gutierrez AP, Ponti L, Cossu QA (2009) Effects of climate warming on olive and olive fly (*Bactrocera oleae* (Gmelin)) in California and Italy. Clim Change 95:195–217. doi:10.1007/s10584-008-9528-4

Halstead P (1987) Traditional and ancient rural economy in mediterranean europe: plus ça change? J Hell Stud 107:77–87. doi:10.2307/630071

Hong S-K, Bogaert J, Min Q (eds) (2014) Biocultural landscapes: diversity, functions and values. Springer, Dordrecht

Huston MA (1994) Biological diversity: the coexistence of species on changing landscapes. Cambridge University Press, Cambridge

Joffre R, Vacher J, de los Llanos C, Long G (1988) The dehesa: an agrosilvopastoral system of the Mediterranean region with special reference to the Sierra Morena area of Spain. Agrofor Syst 6:71–96. doi:10.1007/BF02344747

Khoury CK, Bjorkman AD, Dempewolf H et al (2014) Increasing homogeneity in global food supplies and the implications for food security. Proc Natl Acad Sci 111:4001–4006. doi:10.1073/pnas.1313490111

Kizos T, Koulouri M (2010) Same land cover, same land use at the large scale, different landscapes at the small scale: landscape change in olive plantations on Lesvos island, Greece. Landsc Res 35:449–467. doi:10.1080/01426390802048297

Koohafkan P (2009) Conservation and adaptive management of globally important agricultural heritage systems (GIAHS). Resour Sci 31:4–9

Koohafkan P, Altieri MA (2011) Globally important agricultural heritage systems: a legacy for the future. Food and Agriculture Organization of the United Nations, Rome

Koohafkan P, De La Cruz MJ (2011) Conservation and adaptive management of globally important agricultural heritage systems (GIAHS). J Resour Ecol 2:22–28

Koohafkan P, Altieri MA, Gimenez EH (2012) Green agriculture: foundations for biodiverse, resilient and productive agricultural systems. Int J Agric Sustain 10:61–75

Kuhnlein HV, Receveur O (1996) Dietary change and traditional food systems of indigenous peoples. Annu Rev Nutr 16:417–442

Lansing JS, Kremer JN (2011) Rice, fish, and the planet. Proc Natl Acad Sci USA 108:19841–19842. doi:10.1073/pnas.1117707109

Lelle MA, Gold MA (1994) Agroforestry systems for temperate climates: lessons from Roman Italy. For Conserv Hist 38:118–126

Loumou A, Giourga C (2003) Olive groves: "The life and identity of the Mediterranean". Agric Hum Values 20:87–95

Lüttge U (2010) Runoff-rainwater for sustainable desert farming. In: Ramawat KG (ed) Desert plants biolology and biotechnology. Springer, Berlin, pp 461–477

Lybbert TJ, Elabed G (2013) An elixir for development? Olive oil policies and poverty alleviation in the Middle East and North Africa. Dev Policy Rev 31:485–506

MacMillan T, Benton TG (2014) Agriculture: engage farmers in research. Nature 509:25–27. doi:10.1038/509025a

Martínez-Torres ME, Rosset PM (2014) *Diálogo de saberes* in La Vía Campesina: food sovereignty and agroecology. J Peasant Stud 41:979–997. doi:10.1080/03066150.2013.872632

McNeely JA (2004) Nature versus nurture: managing relationships between forests, agroforestry and wild biodiversity. Agrofor Syst 61–62:155–165. doi:10.1023/B:AGFO.0000028996.92553.ea

Morton JF (2007) The impact of climate change on smallholder and subsistence agriculture. Proc Natl Acad Sci USA 104:19680–19685. doi:10.1073/pnas.0701855104

Mtaita TA, Manqwiro BK, Mphuru AN (2001) The role of horticulture plants in combating desertification. In: Pasternak D, Schlissel A (eds) Combating desertification plants. Kluwer Academic/Plenum Publishers, New York, pp 33–43

Neteler M (2010) Estimating daily land surface temperatures in mountainous environments by reconstructed MODIS LST data. Remote Sens 2:333–351

Neteler M, Bowman MH, Landa M, Metz M (2012) GRASS GIS: a multi-purpose Open Source GIS. Environ Model Softw 31:124–130

Nicholls CI, Altieri MA (2012) Ecologically based food production systems for the XXI century. Agroecología 6:29–37

Oldeman LR, Hakkeling RTA, Sombroek WG (1991) World map of the status of human-induced soil degradation: an explanatory note, Second revised edition edn. International soil reference and information centre, United Nations Environment Programme, Wageningen and Nairobi

Papanastasis VP, Mantzanas K, Dini-Papanastasi O, Ispikoudis I (2009) Traditional agroforestry systems and their evolution in Greece. In: Rigueiro-Rodróguez A, McAdam J,

Mosquera-Losada MR (eds) Agroforestry in Europe: current status and future prospects. Springer, Netherlands, pp 89–109

Pardini G, Gispert MA (2013) Soil quality assessment through a multi-approach analysis in soils of abandoned terraced land in NE Spain. Cuad Investig Geográfica 38:7–30

Pasternak D (2001) Combating poverty with plants. In: Pasternak D, Schlissel A (eds) Combating desertification plants. Kluwer Academic/Plenum Publishers, New York, USA, pp 17–30

Pasternak D, Schlissel A (2001) Combating desertification with plants. Kluwer Academic/Plenum Publishers, New York

Ponti L, Gutierrez AP, Basso B et al (2013) Olive agroecosystems in the Mediterranean Basin: multitrophic analysis of climate effects with process-based representation of soil water balance. Procedia Environ Sci 19:122–131. doi:10.1016/j.proenv.2013.06.014

Ponti L, Gutierrez AP, Ruti PM, Dell'Aquila A (2014) Fine scale ecological and economic assessment of climate change on olive in the Mediterranean Basin reveals winners and losers. Proc Natl Acad Sci 111:5598–5603. doi:10.1073/pnas.1314437111

Ponti L, Gutierrez AP, Altieri MA (2015) Holistic approach in invasive species research: the case of the tomato leaf miner in the Mediterranean Basin. Agroecol Sustain Food Syst. doi:10.1080/21683565.2014.990074

Pretty J (2008) Agricultural sustainability: concepts, principles and evidence. Philos Trans R Soc B Biol Sci 363:447–465. doi:10.1098/rstb.2007.2163

Ramachandran Nair PK, Nair VD, Mohan Kumar B, Showalter JM (2010) Carbon sequestration in agroforestry systems (chapter five). In: Sparks DL (ed) Advances in agronomy. Academic Press, Waltham, pp 237–307

Regev U, Gutierrez AP, Schreiber SJ, Zilberman D (1998) Biological and economic foundations of renewable resource exploitation. Ecol Econ 26:227–242

Reich P, Eswaran H, Beinroth F (2001) Global dimensions of vulnerability to wind and water erosion. In: Stott DE, Mohtar RH, Steinhardt GC (eds) Sustaining the Global Farm. Selected papers from the 10th International Soil Conservation Organization Meeting, 24–29 May 1999, West Lafayette IN. International Soil Conservation Organization, United States Department of Agriculture Agricultural Research Service National Soil Erosion Research Laboratory, and Purdue University, West Lafayette, IN, USA, Purdue University and the USDA-ARS National Soil Erosion Research Laboratory, pp 838–846

Renfrew JM (1973) Palaeoethnobotany: the prehistoric food plants of the Near East and Europe. Methuen, London

Rillig MC (2004) Arbuscular mycorrhizae, glomalin, and soil aggregation. Can J Soil Sci 84:355–363

Rodríguez AR, McAdam J, Mosquera-Losada MR (2009) Agroforestry in Europe: current status and future prospects. Springer Science & Business Media

Rogé P, Friedman AR, Astier M, Altieri MA (2014) Farmer strategies for dealing with climatic variability: a case study from the Mixteca Alta region of Oaxaca, Mexico. Agroecol Sustain Food Syst 38:786–811. doi:10.1080/21683565.2014.900842

Rosenzweig C, Tubiello FN (1997) Impacts of global climate change on Mediterranean agrigulture: current methodologies and future directions. Mitig Adapt Strateg Glob Change 1:219–232

Rosset P (2011) Food sovereignty and alternative paradigms to confront land grabbing and the food and climate crises. Development 54:21–30

Scarascia-Mugnozza G, Oswald H, Piussi P, Radoglou K (2000) Forests of the Mediterranean region: gaps in knowledge and research needs. For Ecol Manag 132:97–109

Schröter D, Cramer W, Leemans R et al (2005) Ecosystem service supply and vulnerability to global change in Europe. Science 310:1333–1337. doi:10.1126/science.1115233

Smith J, Pearce BD, Wolfe MS (2013) Reconciling productivity with protection of the environment: Is temperate agroforestry the answer? Renew Agric Food Syst 28:80–92. doi:10.1017/S1742170511000585

Sonneveld BGJS, Dent DL (2009) How good is GLASOD? J Environ Manage 90:274–283. doi:10.1016/j.jenvman.2007.09.008

Souissi I, Temani N, Belhouchette H (2013) Vulnerability of mediterranean agricultural systems to climate: from regional to field scale analysis. In: Pielke RA (ed) Climate vulnerability: understanding and addressing threats to essential resources. Academic Press, Amsterdam, pp 89–103

Terral J-F, Alonso N, Chatti N et al (2004) Historical biogeography of olive domestication (*Olea europaea* L.) as revealed by geometrical morphometry applied to biological and archaeological material. J Biogeogr 31:63–77

Tittonell P (2014) Ecological intensification of agriculture—sustainable by nature. Curr Opin Environ Sustain 8:53–61. doi:10.1016/j.cosust.2014.08.006

Van der Putten WH, Macel M, Visser ME (2010) Predicting species distribution and abundance responses to climate change: why it is essential to include biotic interactions across trophic levels. Philos Trans R Soc B Biol Sci 365:2025–2034. doi:10.1098/rstb.2010.0037

Vanwalleghem T, Amate JI, de Molina MG et al (2011) Quantifying the effect of historical soil management on soil erosion rates in Mediterranean olive orchards. Agric Ecosyst Environ 142:341–351. doi:10.1016/j.agee.2011.06.003

Wischmeier WH, Smith DD (1978) Predicting rainfall erosion losses: a guide to conservation planning. Agriculture Handbook No 537, U.S. Department of Agriculture, Washington DC 62

Wright SF, Upadhyaya A (1998) A survey of soils for aggregate stability and glomalin, a glycoprotein produced by hyphae of arbuscular mycorrhizal fungi. Plant Soil 198:97–107. doi:10.1023/A:1004347701584

Wright SF, Franke-Snyder M, Morton JB, Upadhyaya A (1996) Time-course study and partial characterization of a protein on hyphae of arbuscular mycorrhizal fungi during active colonization of roots. Plant Soil 181:193–203. doi:10.1007/BF00012053

Wright SF, Starr JL, Paltineanu IC (1999) Changes in aggregate stability and concentration of glomalin during tillage management transition. Soil Sci Soc Am J 63:1825–1829. doi:10.2136/sssaj1999.6361825x

Yasuda Y (1997) The raise and fall of olive cultivation in Northwestern Syria: palaeoecological study of Tell Mastuma. Jpn Rev 8:251–273

Zohary D, Spiegel-Roy P (1975) Beginnings of fruit growing in the Old World. Science 187:319–327. doi:10.1126/science.187.4174.319

Chapter 25
Conserving Biological and Cultural Diversity Along the Latin Arc: The Role of Protected Areas

Emma Salizzoni

Abstract The Euro-Mediterranean coastal area is an 'artificial' landscape par excellence, having been moulded by thousand-year-old anthropic activities. Here, biodiversity values are largely determined by human actions, so that biological and cultural diversity are inextricably linked. A landscape approach to planning and management of these areas, one that focuses both on natural and cultural aspects of conservation, is therefore necessary. This statement, which could seem quite obvious, has been only recently recognized by the main international policies for Euro-Mediterranean coastal areas, where landscape has until recently played a minor role compared to naturalistic and socio-economic aspects. An example of landscape-oriented strategies aimed at conserving both natural and cultural diversity is represented by the policies developed by some Regional Parks—all classified as 'Protected Landscapes' (category V Protected Areas, according to the International Union for Conservation of Nature classification system)—situated along the coast of the Latin Arc countries. This paper presents the management experiences developed inside these areas, highlighting their innovative approach to conservation and envisaging the more general role that Protected Areas, and Protected Landscapes in particular, could play in the wider context of the Euro-Mediterranean coastal landscape.

Keywords Protected areas · Protected landscapes · Euro-Mediterranean coastal areas · Landscape approach · Biocultural diversity

E. Salizzoni (✉)
Interuniveristy Department of Regional and Urban Studies and Planning, Politecnico and Università di Torino, Viale Mattioli 39, 10125 Turin, Italy
e-mail: emma.salizzoni@polito.it

M. Agnoletti and F. Emanueli (eds.), *Biocultural Diversity in Europe*, Environmental History 5, DOI 10.1007/978-3-319-26315-1_25

25.1 The Euro-Mediterranean Coastal Landscape: Features, Challenges and International Responses

The Euro-Mediterranean region, and especially its coastal area (with specific reference to the countries that make up the so-called 'Latin Arc',[1] Spain, France and Italy), can be regarded as an 'artificial' landscape par excellence, a 'man-made world' (Houston 1964: p. 105), having been moulded by thousand-year-old anthropic activities. Among the anthropic activities that historically have mostly influenced the structure of Euro-Mediterranean coastal landscapes, a main role has been played by agriculture. Also other activities have contributed to 'build' the Latin Arc landscape—such as salt production and fishing (Braudel 1992; Matvejevic 1991; Miossec 2004)—but we have to remember that 'despite the proximity of the sea (...) the coastal zone was assessed, in the first instance, for its potential as an agricultural resource area, just as the inland districts were' (Delano Smith 1979: p. 326). Already in the Neholitic Age large woodland areas were felled to make way for agriculture (Vogiatzakis et al. 2005) and since then people have kept transforming the Euro-Mediterranean coastal landscape, fighting against environmental and climate conditions not suitable for agricultural activity (Ribeiro 1972; Hackens 1997). Some 'correctives' (Bethemont 2000) were used to adapt to and to modify these conditions: terracing along the slopes and draining in the plains to extend the cultivable area, irrigation to widen the range of species to be cultivated, and crops selection (choosing those species, such as olive, vines, wheat, that are able to survive in the Euro-Mediterranean context). Ways of living closely tuned to natural processes were elaborated (Isnard 1973), interacting with biological diversity through adaptive processes for thousands of years[2]: the Euro-Mediterranean coastal landscape is based on a fragile balance between man and nature (Hackens 1997). Consequently, along these areas biological and cultural diversity (meaning traditional management practices) are inextricably linked, both with relation to agrobiodiversity and wild biodiversity.

The main, current challenges in these areas are strictly related to two socio-economic processes that have taken place since the second postwar period and that started a real 'revolution' with regard to the landscape structure. The first one was a progressive 'littoralisation', namely the concentration of people and activities (tourism above all) along the seashore. Already by the 1970s, along the coasts of Spain, France and Italy, the coastal population density had reached

[1]For a closer examination of the concept of 'Latin Arc' see Daviet (1994), Voiron Canicio (1994). In the paper we shall refer to a wider notion of 'Latin Arc' (Salizzoni 2012: pp. 12–13), which includes the Italian Adriatic and Ionian coast.

[2]'Through the centuries, rural communities have managed their environment and farmed the land in their own natural way, creating a rich diversity of landscapes, choral representation of historical identity of the territory and cultural human heritage. We now tend to recognize in that model of development the precursor of "sustainability"' (Guarino et al. 2015). 'Many (...) traditional cultures have developed livelihood practices that inevitably alter the landscape, but do so with care so as to ensure natural resource security into the future (...)' (Pretty et al. 2008: p. 5).

decidedly higher values compared to the national average (Benoit and Comeau 2005). The second socio-economic process was complementary to the first and was connected to the resizing of agricultural and grazing activity, which has caused a massive rural exodus from inland areas. The joint action of these processes has led to a socio-economic and territorial fracture between coastal and inland areas, which is typical of the current Euro-Mediterranean context: in most cases coastal and inner areas are worlds apart and hardly communicate.[3] A series of phenomena, which is highly critical from an environmental point of view, is a corollary of this fracture. Among these, we can cite the unrestrained process of coastal 'artificialisation' [a real 'urban tsunami' (Forman 2010) which has invaded the Mediterranean coast since the second postwar period[4]], the pollution and artificialisation of wetlands, an excess of human pressure in beach areas, the intensification of agricultural activity along the coastal plains and the uncontrolled and spontaneous renaturalisation of abandoned inland rural areas.

All these phenomena have led to a continuous loss in biodiversity. With specific reference to rural landscapes (on which this paper is going to focus, see Chap. 3), the development of intensive agriculture along the coast (monocultures) and the processes of renaturalisation in abandoned inland rural areas have jointly caused an important reduction of species. Cultural diversity has been affected as well, above all in terms of loss of traditional agricultural practices, which has had, in turn, an enormous impact on biodiversity loss: '(…) just as varieties of domestic plants and animals depend on the continuation of traditional farming systems, so are many wildlife species equally reliant on such forms of land management' (Phillips and Stolton 2008: p. 10). In these areas, the best way to conserve biological diversity seems to be conserving cultural diversity, namely traditional management practices. A landscape approach to the planning and management of these areas—that focuses both on natural and cultural aspects of conservation [landscape is a bridge between nature and culture (Gambino and Peano 2015)]—is therefore necessary.

This statement, that could seem quite obvious, has been only recently recognised by the main international policies for Euro-Mediterranean coastal areas, where landscape was usually confined to secondary roles, especially if compared to naturalistic and socio-economic aspects of such policies. Actually, the first and the most renowned policies for the Mediterranean environmental context, the Mediterranean Action Plan (MAP) and the Convention for the Protection of the

[3]Today, there is an evident '(…) spatial dichotomy between strong, heavily populated coastal areas, characterised by high intensity of land use and consumption, and inevitably weaker, thinly populated inland areas with lower housing density and a less dynamic economy' (UNEP, MAP, PAP/RAC 2001: p. V). On the contrary, until the first half of the Twentieth Century, coastal and inner areas have been strictly related, particularly due to the integration between fishing and agricultural activity and to the transhumance practice.

[4]It is estimated that more than 70 % of the coast of Spain and Italy, and 60 % of France, has now been artificialised (Benoit and Comeau 2005; EEA 2006), due to tourist and residential pressures.

Mediterranean Sea Against Pollution issued by United Nations Environment Programme (UNEP) in the 1970s (1975–1976), did not mention landscape at all. The focus of the MAP and of the Convention was on the pollution problems of the Mediterranean Sea. Twenty years later (1995), perhaps also thanks to the issuing, in 1993, of the 'Charte du paysage Mediterranéen',[5] the new MAP and the Convention addressed the topic of landscape, even though not in an extensive and systematic way. Eventually, in 2006, in the wake of a process started in 2003, an Expert Meeting on Landscape Management in the Mediterranean was organised by the MAP Priority Actions Programme/Regional Activity Centre (PAP/RAC). Participants acknowledged the value and the risks affecting Mediterranean coastal landscapes.[6] They stated that 'coastal landscapes of the Mediterranean have never been studied or elaborated in the MAP projects per se (…)' and that actions should be implemented in order to better know, manage and plan the Mediterranean landscape. Two years later (2008), the Protocol on Integrated Coastal Zone Management (UNEP, MAP, PAP/RAC) was issued. This document contains several references to the Mediterranean landscape features and problems[7] and attests a desirable and progressive, but still ongoing, opening up to the landscape dimension of the main international policies for Euro-Mediterranean coastal areas.

[5]A pioneer document, signed by the Presidents of the Regions of Andalucia, Toscana and Languedoc Roussillon, that was an important step towards the drawing up of the European Landscape Convention (Nègre 2001): 'In March 1994 (…) the Standing Conference of Local and Regional Authorities of Europe (…) called on its succeeding body, the CLRAE, to draw up, on the basis of the Mediterranean Landscape Charter (…) a framework convention on the management and protection of the natural and cultural landscape of Europe as a whole' (CoE 2000).

[6]'The diversity of Mediterranean landscapes contributes to local and regional identity, reflecting the past and present relationships between the man and his natural and built environment. Very rich cultural landscapes have been developed through many millennia when different human populations, cultures, and religions flourished around the Mediterranean and developed coastal landscapes as a result of land transformations in order to produce food, build living habitats, art, etc. Nowadays, however, increasing threats to cultural identity, heritage and landscape diversity of the region due to external (e.g. globalisation) and internal factors (e.g. rapid urbanisation of coastal areas with consequent impacts on traditional socio-economic structures) can be witnessed constantly. As a result, natural and cultural (man-made) landscapes have deteriorated significantly in several coastal areas' (UNEP, MAP, PAP/RAC 2006: p. 1).

[7]Contrary to the EU Recommendation 412/2002 on ICZM, which does not mention landscape. In the ICZM Protocol there is even an article (art. 11, 'Coastal landscapes') that specifically addresses the need for protection of the coastal landscape: 'The Parties, recognising the specific aesthetic, natural and cultural value of coastal landscapes, irrespective of their classification as protected areas, shall adopt measures to ensure the protection of coastal landscapes through legislation, planning and management (…)' (UNEP, MAP, PAP/RAC 2008: art. 11).

25.2 A Special Observatory of Policies for Conserving Biological and Cultural Diversity Along the Euro-Mediterranean Coastal Areas: Protected Areas Classified as Protected Landscapes

25.2.1 The Protected Landscape Category

An interesting example of landscape-oriented strategies aimed at conserving both natural and cultural diversity of the Euro-Mediterranean coastal landscape is offered by the policies developed by some Regional Parks. These Parks are situated along the coast of Spain, France and Italy and classified as 'Protected Landscapes', that is category V Protected Areas (PAs), according to the Protected Area classification system of the International Union for Conservation of Nature, IUCN.[8]

What are 'Protected Landscapes'? We should start saying that Protected Landscapes are those Protected Areas whose natural conditions, compared to other IUCN PA categories, resemble most those of areas which are *not* protected (Dudley 2008). They are lived-in, working landscapes, structured over time by an interaction between natural and anthropic factors, the conservation of which is essential to the survival of the self-same biodiversity values. As Jessica Brown (member of the IUCN Protected Landscapes Task Force) remembers, 'these landscapes are rich in biological diversity and other natural values not in spite of, but *rather because of* the presence of people' (Brown et al. 2005: p. 3).

Therefore, the so-called 'Protected Landscape Approach' pursues the primary objective of all Protected Areas, namely biodiversity conservation,[9] but acknowledges that, in highly anthropic landscapes, biodiversity can be preserved only if the interaction between man and nature, which is at the basis of self-same biodiversity values, is preserved too (Amend et al. 2008; Dudley and Stolton 2012). As a consequence, the management focus of these PAs is not on nature per se (Phillips 2002), but on those human processes which have influenced nature and moulded the landscape,[10] looking simultaneously at the naturalistic dimension and at the cultural and socio-economic dimension. It is not by coincidence that Protected

[8]IUCN recognises six categories of Protected Areas, defined by their management category: Strict Nature Reserve (Ia) and Wilderness Area (Ib), National Park (II), Natural Monument (III), Habitat/Species Management Area (IV), Protected Landscape/Seascape (V) and Protected Area with Sustainable Use of Natural Resources (VI).

[9]'(…) there is a strong global consensus within the conservation community that the principal role of protected areas is nature conservation' (Watson et al. 2014: p. 68). A protected area is: 'A clearly defined geographical space, recognised, dedicated and managed, through legal or other effective means, to achieve the long-term conservation of nature with associated ecosystem services and cultural values' (Dudley 2008: p. 8). In the context of this definition 'nature always refers to biodiversity, at genetic, species and ecosystem level' (Dudley 2008: p. 8).

[10]The main objective of the category, as defined by the IUCN guidelines, is 'to protect and sustain important landscapes/seascapes and the associated nature conservation and other values created by interactions with humans through traditional management practices' (Dudley 2008: pp. 20–21).

Landscapes are indicated as the 'conservation model for the twenty-first century' (Beresford and Phillips 2000). More than other PAs categories, Protected Landscapes take up the challenge launched by the so-called 'new conservation paradigms' (Phillips 2003) of widening the management objective of PAs, looking both at nature conservation and at socio-economic development.

25.2.2 Planning and Management Strategies for Conserving Rural Landscapes

In the previous paragraph we talked about the *model* of Protected Landscapes, namely the theoretical definition of the category and its management objectives.

Hereafter we are going to present the management experiences actually developed inside some of these areas situated along the coast of the Latin Arc countries, highlighting the strategies, actions and tools aimed at conserving both natural and cultural diversity. As mentioned, the focus of this analysis will be on rural, and specifically agricultural, landscapes. Rural landscapes actually best exemplify the relationship existing between biodiversity and cultural diversity, also in relation to the critical phenomena discussed above (agriculture intensification and abandonment of inner areas, see paragraph 1). As it has been noted, 'in the North Mediterranean Countries (…) the main threat to biodiversity (…) is the gradual disappearance of open rural environments and traditional agricultural practices' (Benoit and Comeau 2005).

25.2.2.1 Processes

Despite the different geographic and socio-economic contexts that characterise the three cases of Protected Landscapes (Regional Parks) here considered (the *Parque Natural de la Albufera de Valencia*, PNAV, 21,000 ha, Spain; the *Parc Naturel Régional de la Narbonnaise en Méditerranée*, PNRNM, 70,000 ha, France and the *Parco Naturale Regionale del Conero*, PNRC, 6000 ha, Italy), starting from the second postwar period processes of abandonment of the inner agricultural areas and intensification of agricultural activity in the coastal plains have concerned the three Parks in a very similar way.

They are particularly evident in the French Park (PNRNM). Here the inland, mountain rural landscapes, mainly vineyards (extensive cultivation), have been progressively abandoned during the last 30 years and renaturalisation processes are now spreading in these areas. The '*garrigue fermée*' (a very thick, shrubby vegetation typical of the Euro-Mediterranean region) is gaining on the land of abandoned fields (see Fig. 25.1) and biodiversity and agrobiodiversity is importantly decreasing (PNRNM 2010). In the meantime, along the coastal plains, intensive cultivation of vines has been growing, creating a landscape made up of large fields

Fig. 25.1 Renaturalisation processes in the inner rural areas of the *Parc Naturel Régional de la Narbonnaise en Méditerranée* (*source* Salizzoni 2010)

that lack in those landscape elements that are not only important from a scenic point of view, but also from an ecological one (e.g. hedges, rows, walls).[11]

These kinds of processes are less intensive in the Italian Park (PNRC), where extensive agriculture persists on a large part of the Protected Area, but even here inner areas are increasingly affected by rural exodus, while coastal plains host intensive cultivations (PNRC 2010).

With regard to the Spanish Park (PNAV), during the past 50 years the ricefields in the inner areas underwent important abandonment processes (the so-called '*crisis del arrozal*') and were gradually attacked by urbanisation (see Fig. 25.2) or transformed in *huerta*s. *Huertas* are a very intensive agricultural model (fruits and vegetables), much more profitable and less demanding than rice farming. The reduction of ricefields is a serious threat to biodiversity in these areas, with particular reference to avifauna (PNAV 2004).

[11]Also traditional practices changed from the *vigne en gobelet* cultivation method (low vines, able to survive in the winding inner, mountain areas) to the *vigne palissée* cultivation method (high vines, situated in the plains, that facilitate the use of mechanic tools and pesticides).

Fig. 25.2 The ricefields of the *Parque Natural de la Albufera de Valencia*, a precious habitat for avifauna, have been increasingly affected, during the past 50 years, by urbanisation processes and by the extension of *huertas* (*source* Salizzoni 2010)

25.2.2.2 Policies

In the three Parks, the social and economic changes concerning agricultural activities, mainly consisting of a loss of traditional management practices, have caused an overall biodiversity reduction. In order to face such changes, the three Parks put in action strategies mainly aimed at maintaining traditional management practices. Table 25.1 hereafter synthetically outlines the main tools used by each Park Authority to implement these strategies.

Looking at the strategies and actions implemented by the three Parks as a whole, it is worth mentioning four main general aspects. The first one is directly related to the above-mentioned 'landscape approach'; the others are co-related aspects.

Integrated Policies: All the three Parks put into practice, besides 'material' actions addressing the spatial dimension of the landscape (such as the restoration of the network of dry stone walls dividing fields in the inner rural areas of the French Park), first and foremost 'immaterial' actions addressing the cultural and socio-economic dimension of the landscape. Among these, we can cite the education activities implemented by the French Park in order to promote traditional agriculture management practices among students of the local *licée agricole*. The making available of volumetric incentives to traditional agricultural undertakings

Table 25.1 Tools for conserving biodiversity and cultural diversity in agricultural landscapes

Parque Natural de la Albufera de Valencia (Spain)	Parc Naturel Régional de la Narbonnaise en Méditerranée (France)	Parco Naturale Regionale del Conero (Italy)
Intensification processes		
Rules for countering the conversion of ricefields in *huertas* (land use control)[a]	Projects for improving the eco-compatibility of intensive viticulture (reduction of pollution and of waste production)[b]	Rules for regulating the introduction of new crops[c]
	Awareness campaigns among farmers about environmental impacts of intensive agriculture	
Abandonment processes		
Rules for safeguarding rice farming[d]	Projects for restoring the dry stone walls providing divisions in the rural inner areas	Rules for maintaining traditional farming[e]
Economic incentives addressed to rice farmers	Education initiatives on traditional agriculture management practices, directed to students (*licée agricole*)	Volumetric incentives addressed to traditional farming[f]
Protected Designations of Origin (PDO) for rice	Projects for tourist pathways among vineyards and promotional campaigns of accommodation facilities (agritourism) in the inner areas	Branded Park products (*Marchio Agricolo del Parco*)
Projects for improving eco-compatibility of rice farming (reduction of pollution and of waste production)[g]	Promotional campaigns of wine estates in the inner areas	Projects for tourist pathways among vineyards (*Strada del Rosso Conero*) and incentives for *agriturismi* on the basis of a verification of farming effectiveness (*Piano Agricolo Aziendale*[h])

[a]The conversion from ricefields to *huertas* is forbidden inside the Park (PNAV 2004: art. 93.2)
[b]The 'Contrat d'etangs' project (2005–2009) envisaged actions concerning the promotion of biologic agriculture and the collection of no more usable pesticides
[c]The introduction of new crops must be done considering soil pedological features and the specific vocation to agriculture activity of each area (PNRC 2010: art. 25)
[d]Traditional rice farming is protected all over the Park because of its ecological, social, economic and cultural value (PNAV 2004: art. 8.2)
[e]A particular attention must be given to the preservation of traditional farming and of endangered species (PNRC 2010: art. 96)
[f]Traditional firms can restore buildings to be used as agricultural laboratories and/or points of sale (PNRC 2010: art. 22)
[g]The LIFE projects Biocompost (2001–2004) and Eco-rice (2005–2007) promoted alternative methods with respect to the practice of *paja de arroz* (organic wastes of rice farming) fires, which have a deep impact on landscape (pollution of air and water)
[h]The *Piano Agricolo Aziendale* (Agricultural Farm Plan) must clearly show the economic efficiency of farms (PNRC 2010: art. 26.1)

by the Italian Park is another example of 'immaterial' action, aimed at keeping alive traditional management practices. Consistently with a 'landscape approach' to conservation, and more generally with the above cited 'new paradigms' of conservation, the three Parks make a clear effort to focus on the interaction between nature and humans and so on the social, economic and cultural roots of biodiversity. Consequently they develop not only *spatial planning* policies, but also *cultural and socio-economic programming* policies. This integration is an important requirement for strategies intended to face the loss of biodiversity and cultural diversity.

Active Conservation: The three Parks face the problems of agricultural intensification and abandonment of rural inner areas through both a normative approach and a proactive approach. In addition to mandatory rules (e.g. aimed at countering the conversion of ricefields in *huertas* in the Spanish Park), other non-binding measures are enacted by the Park Authorities, such as projects or marketing initiatives. The French Park strategy, in particular, strongly relies on this kind of measure, since the *Charte du Parc* (i.e. the management plan of the Regional French Parks) is not a regulatory tool with direct legal value (it does not entail rules), but a voluntary agreement between different territorial subjects.[12] This 'active conservation' is in line with the concept of 'innovative conservation' (Gambino 1997), meaning a conservation that acknowledges the dynamic nature of landscape[13] and, instead of 'freezing' it, tries to guide landscape changes towards a sustainable development perspective.[14] 'Innovative conservation' acts first and foremost through projects or other dynamic actions, more than through rules (passive conservation).

Innovating Traditions: If an 'innovative conservation' approach acknowledges the dynamic nature of landscape, it therefore accepts and favours changes if these are useful to conserving biodiversity. This is all the more true with relation to the cultural and socio-economic dimension of the landscape. To this regard, we can point out that the three Parks in the case studies do not simply try to maintain as they are the traditional agricultural practices which are acknowledged to be at the basis of biodiversity values, but they also intend to innovate them. This is particularly clear in the Spanish Park, where the Park Authority wishes to maintain traditional rice farming, but tries also to improve its eco-compatibility, envisaging new methods and approaches for reducing organic waste production. In the Italian and in the French Parks, traditional agriculture is considered from a new perspective, namely as a potential driving force for specific types of tourism such as rural tourism, which could represent an important socio-economic support for inner areas and for the traditional practices themselves. If we want to 'keep the farmers on

[12]For a closer examination of the matter, see FPNRF (2008).

[13]Landscape is a dynamic entity par excellence: 'it inevitably changes and evolves over time, in response to natural processes and to the changing needs and activities of people' (Phillips 2005).

[14]'Protective measures (…) should not be designed to stop time (…). They may guide changes in sites in order to pass on their specific, material and immaterial features to future generations' (CoE, Recommendation of the Committee of Ministers to member states on the guidelines for the implementation of the European Landscape Convention, 2008).

the land' (UNEP, MAP, PAP-RAC 2006) in the inner areas, it is necessary to find new ways to keep alive traditions in the current socio-economic context, favouring also their evolution.[15]

Agreements versus Rules: Who decides what has to be conserved and how? This is a *vexata quaestio* with regard to conservation processes. In the above-mentioned cases, the Park Authority decides about conservation policies, considering biodiversity conservation as a priority ['biological diversity is the basis of human welfare' (IUCN 2009)], and traditional management practices as the main tool to guarantee it. However, in order to make effective the envisaged strategies for conserving both biodiversity and cultural diversity in rural landscapes, Park Authorities seem to know that this vision must be widely shared among farmers, who are the leading actors of these landscapes: strategies can be actually implemented only thanks to the people who mould the landscape day after day. Nevertheless, farmers generally see intensive agriculture along the coast as a much more profitable and less demanding activity than extensive agriculture in the inner areas, so that the conflict between Park Authorities and farmers is usually at its height in these areas (and more generally in highly anthropic landscapes to be protected). An agreement must be patiently defined in each situation, looking for a balance between local interests (farmers) and interests which transcend the local dimension (Park Authorities). This is especially true in category V Protected Areas.[16] In this perspective, it is important to build a common vision which aligns farmers' actions with the Park management objectives. This can be done not only through rules, which are not always easy to implement or desirable or effective in such situations, but also through other tools, such as the awareness campaigns carried on among farmers by the French Park, in order to sensitise them about environmental impacts of intensive agriculture, or the incentives provided to keep traditional farming in the Spanish and Italian Parks. In this regard, also the above-mentioned strategies aimed at innovating tradition are important in order to meet as much as possible the farmers' needs and at the same time to pursue conservation objectives.

---

[15]'Category V protected areas can seek to maintain current practices, restore historical management systems or, perhaps most commonly, maintain key landscape values whilst accommodating contemporary development and change (…)' (Dudley 2008: p. 22).

[16]'Since people are considered to be the stewards of the landscape or seascape in category V protected areas, it is important to understand (…) the extent to which decision making can be left to local inhabitants and how far a wider public interest should prevail when there is conflict between local and national needs' (Dudley 2008: p. 22).

25.3 Which Role for Protected Areas in the Wider Territorial Context?

The above-mentioned strategies and actions implemented by the Parks draw an interesting framework of attempts to conserve biological and cultural diversity in Euro-Mediterranean coastal and rural landscapes. Although these strategies and actions are not always, or not still completely effective (abandonment and intensification processes are still ongoing in a certain measure in the three Parks),[17] they represent experiences worthy of being highlighted because of two main reasons: (i) unlike many bodies responsible for planning and managing the 'ordinary landscapes', Park Authorities, as Protected Areas, put biodiversity and cultural diversity conservation at the heart of their planning and management policies; (ii) moreover, the three Parks, in line with the new paradigms of conservation and with the features of the IUCN category of Protected Landscapes, try to put in action an innovative approach to conservation, namely *integrated policies* implemented through *active conservation strategies* that try also to *innovate tradition* and to operate *with and throughout farmers*. This approach is perhaps the best way to face the continuous loss of biodiversity and cultural diversity in highly anthropic-influenced landscapes such as the coastal, Euro-Mediterranean landscapes. Therefore, the three cases studies analysed suggest that Protected Areas and Protected Landscapes in particular are important actors for the conservation of biological and cultural diversity along the Latin Arc.

However, until these efforts are confined to Protected Areas, they will be only partially useful for the Euro-Mediterranean landscape. The major challenges concerning the conservation of biological and cultural diversity are often situated outside Protected Areas. PAs risk representing virtuous 'islands' where biodiversity and cultural diversity are maintained, as opposed to the rest of the landscape—which represents the main part of the landscape[18]—where biodiversity and cultural diversity are constantly decreasing.[19]

Which role should Protected Areas, and Protected Landscapes in particular, play in the wider context of the Euro-Mediterranean landscape? They should of course act as models of a harmonious interaction between humans and nature (Gambino 1997), but they should also care about the dissemination of these models ('*Benefits*

[17]The effectiveness of PAs is a crucial topic indeed. There are studies that point out the current, significant shortfall in PAs effectiveness (for an overview on the matter, see Leverington et al. 2010). However, PAs can still be considered as extraordinary experimental laboratories of territorial and landscape policies, above all in those countries where there is an adequate financing and a good governance system (Watson et al. 2014).

[18]The current set of the European coastal Protected Areas is quite poor and highly fragmented and covers a reduced portion of the coastal areas, especially in the Mediterranean region (Gambino et al. 2008).

[19]On the contrary, we know that the concept of biodiversity is much more linked to the concept of "network", instead of that of "island".

beyond boundaries'[20]). That could be achieved not only by improving communication regarding Protected Areas policies and tools,[21] but also envisaging coordinated actions between areas situated inside and outside PAs, to be put into practice at least within a transition zone.[22]

Actually, the approach to conservation we have underlined above (see Sect. 25.2.2.2) makes Protected Areas policies (at least the policies of those most anthropic-influenced PAs, as Protected Landscapes) exportable also in the 'ordinary' landscape context. The paradigm of an innovative and dynamic conservation, which allows and favours changes in the landscapes if they go in the direction of conserving biodiversity, breaks up the dangerous distinction between protected landscapes, where nothing is permitted, and landscapes which are not protected, where everything is accepted, and challenges not protected landscapes to pursue, as much as possible, the same objective as PAs: that is, a transformation consistent with biodiversity and cultural diversity.

References

Amend T, Brown J, Kothari A, Phillips A, Stolton S (eds) (2008) Protected landscapes and agrobiodiversity values, volume 1 in the values of protected landscapes and seascapes series. IUCN & GTZ, Heidelberg

Benoit G, Comeau A (eds) (2005) A sustainable future for the Mediterranean. The Blue Plan's Environment and Developmnet Outlook, Earthscan, London

Beresford M, Phillips A (2000) Protected landscapes: a conservation model for the 21st century. The George Wright Forum 17(1)

Bethemont J (2000) Géographie de la Méditerranée. A. Colin, Paris

Braudel F (1992) Il Mediterraneo. Lo spazio, la storia, gli uomini, le tradizioni. Bompiani, Milano

Brown J, Mitchell N, Beresford M (eds) (2005) The protected landscape approach. linking nature, culture and community. IUCN, Gland

CoE (2000) Explanatory report on the European landscape convention

Daviet S (1994) L'Arc latin, histoire et problematiques d'un concept. Méditerranée 79:3–6

Delano Smith C (1979) Western Mediterranean Europe. A historical geography of Italy, Spain and Southern France since the neolithic. Academic Press, London

Dudley N (ed) (2008) Guidelines for applying protected area management categories. IUCN, Gland

Dudley N, Stolton S (eds) (2012) Protected landscapes and wild biodiversity, volume 3 in the values of protected landscapes and seascapes series. IUCN, Gland

EEA (2006) The changing faces of Europe's coastal areas. Report 6

---

[20]This key message was launched on the occasion of the III IUCN World Park Congress held in Durban (2003).

[21]*'Les Parcs disposent d'outils, de moyens. Ils développent des idées efficaces, originales, pertinentes, inventives. Il faudrait seulement que ces idées soient plus partagées, plus connues pour être réutilisées, recomposées, réinterprétées et donner lieu à de nouvelles actions'* (Kempft 2006).

[22]In this regard, the French Park is an interesting example, since it carries on coordinated actions with institutional bodies and privates situated outside the Park boundaries on topics such as the hydrographic resources management.

Forman RTT (2010) Coastal regions: spatial patterns, flows and a people nature solutions from the lens of landscape ecology. In: Mininni M (ed) La costa obliqua. Un atlante per la Puglia. Donzelli Editore, Roma

FPNRF (2008) Argumentaire. 50 questions - réponses sur les Parcs naturels régionaux

Gambino R (1997) Conservare Innovare. Paesaggio, Ambiente, Territorio. UTET, Torino

Gambino R, Peano A (eds) (2015) Nature policies and landscape policies. Towards an alliance. Springer, Dordrecht

Gambino R, Talamo D, Thomasset F (eds) (2008) Parchi d'Europa. Verso una politica europea per le aree protette. Edizioni ETS, Pisa

Guarino R, Menegoni P, Pignatti S, Tulumello S (2015) A territorial contradiction. In: Gambino R, Peano A (eds) Nature policies and landscape policies. Towards an alliance. Springer, Dordrecht

Hackens T (1997) Le paysage culturel Méditerranéen: lecture synthetique et leçon d'humanisme. In: *Il paesaggio mediterraneo: segno della storia, messaggio della civiltà*, Atti del convegno, Capri, 13–15 ottobre 1995, Centro Internazionale di studi sul paesaggio mediterraneo. Napoli: Ed. Crea

Houston JM (1964) The Western Mediterranean World. An introduction to its regional landscapes. Longsmam, London

Isnard H (1973) Pays et paysages méditerranéens. Presses universitaires de France, Vendome

IUCN, Shaping a Sustainable Future, Programme 2009–2012

Kempft M (2006) Urbanisme et paysage: inventaire des methodes et outils. FPNRF

Leverington F, Costa KL, Pavese H, Lisle A, Hockings M (2010) A global analysis of protected area management effectiveness. Environ Manage 46:685–698

Matvejevic P (1991) Breviario mediterraneo. Garzanti, Milano

Miossec A (2004) Les littoraux entre nature et amenagement. A. Colin, Paris

Nègre F (2001) La Charte du paysage méditerranéen: une expérience de coopération entre des régions méditerranéennes. In: Consiglio d'Europa, CoE, Landscape Heritage, Spatial Planning and Sustainable Development, Proceedings, Lisbona, 26–27 November 2001

Phillips A (2002) Management guidelines for IUCN category V protected areas, protected landscapes/seascapes. Cardiff University, Cambridge

Phillips A (2003) Turning ideas on their head. The new paradigm for protected areas. In: The George Wright forum, vol 20(2)

Phillips A (2005) Landscape as a meeting ground: category V protected landscapes/seascapes and world heritage cultural landscapes. In: Brown J, Mitchell N, Beresford M (a cura di) The protected landscape approach. linking nature, culture and community. IUCN, Gland

Phillips A, Stolton S (2008) Protected landscapes and biodiversity values: an overview. In: Amend T, Brown J, Kothari A, Phillips A, Stolton S (eds) Protected landscapes and agrobiodiversity values, volume 1 in the in the values of protected landscapes and seascapes series. IUCN & GTZ, Heidelberg

PNAV (2004) Plan Rector de Uso y Gestión del Parque Natural de la Albufera (PRUG)

PNRC (2010) Variante generale al Piano del Parco Naturale del Conero (PdP)

PNRNM (2010) Charte du Parc 2010-2022. Le projet de territoire adopté par tous

Pretty J, Adams B, Berkes F, Ferreira de Athayde S, Dudley N, Hunn E, Maffi L, Milton K, Rapport D, Robbins P, Samson C, Sterling E, Stolton S, Takeuchi K, Tsing A, Vintinner E, Pilgrim S (2008) How do biodiversity and culture intersect? In: Plenary paper for conference "sustaining cultural and biological diversity in a rapidly changing world: lessons for global policy". Organized by American Museum of Natural History's Center for Biodiversity and Conservation, IUCN, and Terralingua, 2–5 April 2008

Ribeiro O (1972) Il Mediterraneo. Ambiente e tradizione. Mursia, Milano

Salizzoni E (2012) Paesaggi Protetti. Laboratori di sperimentazione per il paesaggio costiero euro-mediterraneo. Firenze University Press, Firenze

UNEP, MAP, PAP/RAC (2001) White paper: coastal zone management in the Mediterranean. Split

UNEP, MAP, PAP/RAC (2006) Expert meeting on landscape management in the Mediterranean. Report, Dubrovnik

UNEP, MAP, PAP/RAC (2008) ICZM protocol in the Mediterranean. Madrid

Vogiatzakis IN, Griffiths GH, Cassar LF, Morse S (2005) Mediterranean coastal landscapes, management practices, typology and sustainability. Report commissioned by UNEP, MAP, PAP/RAC, Split

Voiron Canicio C (1994) A la recherche d'un arc méditerranéen. Méditerranée, 79

Watson JEM, Dudley N, Segan DB, Hockings M (2014) The performance and potential of protected areas. Nature 515

Chapter 26
Identification of Values of the Designed Landscapes: Two Case Studies from the Czech Republic

Markéta Šantrůčková and Martin Weber

Abstract Designed landscapes or big landscape park-integrated gardens, agricultural and woody landscapes have created an environment for human well-being. Several of these areas were created in the Czech lands during the nineteenth century by rich noblemen and they covered one or more cadastral units. The designed landscapes were intentionally managed to be beautiful, sustainable and productive and they had, and in many cases, continued to keep outstanding cultural and natural values. Political changes in the second half of the twentieth century caused deep changes in the ownership of landscape and management. The designed landscapes were either totally or partially disintegrated; economical profit came first from disregarding sustainability. Nevertheless, the designed landscapes still were areas with serious cultural and natural value and many of them became protected areas with more sustainable management than a common landscape. Today, the same driving forces as in other European countries influence the landscape in the Czech Republic: landscape abandonment in marginalized areas and intensification in core areas. Designed landscapes are found in both types. The paper will present new management approaches and problems for preserving both cultural and natural values of designed landscapes based on two case studies from the Czech Republic (Petrohrad and Žehušicko).

Keywords Designed landscapes · Management · Czech Republic · Culture · Nature · Biocultural diversity

M. Šantrůčková (✉) · M. Weber
Silva Tarouca Research Institute for Landscape and Ornamental Gardening,
Květnové nám. 391, 252 43 Průhonice, Czech Republic
e-mail: santruckova@vukoz.cz

M. Weber
e-mail: weber@vukoz.cz

M. Agnoletti and F. Emanueli (eds.), *Biocultural Diversity in Europe*,
Environmental History 5, DOI 10.1007/978-3-319-26315-1_26

26.1 Introduction

A landscape is a key concept for responsible protection of cultural and natural heritage because all natural processes and human activities take place in landscapes and have an impact on the landscape. Due to this, a growing interest in landscapes as an important perspective in sustainable development is currently observed. A landscape-based approach presents landscapes as a possible new paradigm for a developmental model which would harmoniously integrate social, economic and environmental factors in space and time (Agnoletti 2014).

Landscape is a repository of cultural values and is an original synthesis of the beauty of a place with its historical character and diverse habitats. Moreover, landscapes are documents of past civilizations. For these reasons, landscape heritage is a fundamental resource which must be safeguarded (Agnoletti 2014). Nowadays, a shift from a rural, agriculture- and manufacturing-based economy is related to commoditization of the countryside as a predominantly residential and recreational area. In general terms, the landscape is no longer essentially devoted to food production but has become an essential lifestyle component (Rogge et al. 2013). This fact implicates the rising interest in landscape (landscape changes, landscape history and landscape protection) not only by researchers from different scientific fields but also by people and consequently politics. On the other hand, this fact is accompanied by lot of problems in landscape management since the present appearance of the European landscape has been formed by long-term human activities in the landscape and the landscape is a mirror of society. Society's demands for new functions in rural landscapes are rapidly changing and diversifying. As a result, highly valued landscapes that developed during centuries vanish or are completely transformed within a limited number of years. The push factors are those connected with trends in agriculture, which may involve intensification or extensification (Vos and Meekes 1999; Agnoletti 2014; Sklenička et al. 2009). This leads to the following crucial question: how can a sustainable future for old cultural landscapes, based on sound economics and the commitment of the relevant actors, be achieved (Vos and Meekes 1999)?

Designed landscapes are a specific type of old cultural landscapes. The designed landscapes integrate parks and gardens, agricultural and woody landscapes and create an environment for human well-being. They were intentionally managed to be beautiful, sustainable and productive and had, and in many cases continue to keep, outstanding cultural and natural values. The main design principle was the use of the existing natural topography and cultural landscape (Kümmerling and Müller 2012). Designed landscapes were created at the end of the eighteenth century and during the nineteenth century by rich noblemen, which covered one or more cadastral units. Designed landscapes, as well as landscape parks, are characterized by design elements which combine landscape features with scattered trees, vistas, water bodies, flowerbeds and flowing lawns (Abendroth et al. 2012). Kümmerling and Müller (2012) stressed vistas as typical design elements that usually extended cuttings through forest patches within the park, which connect features of special interest (sculptures and pavilions) and open new views from different locations.

Studying of the cultural landscape could be divided into three main groups of research tasks. First, the study of landscape development, mainly land use/land cover changes; second, the identification of values; third, landscape management and protection. A number of papers have been published about landscape development (Timár et al. 2008; Skaloš et al. 2011; Musaoglu et al. 2005; Larsson and Nilsson 2005). This knowledge is basic not only for understanding the landscape history but also the present state of the landscape. The main problem of many of these papers is that they only cover a short time period due to lack of data. Ortophotos and aerial photographs are widely used; however, they are available only for the second half of the twentieth century (Mallinis et al. 2011; Nahuelhual et al. 2014; Zald 2009). It was possible to use old maps for certain, mainly European, regions dating back to the eighteenth and nineteenth centuries (Kanianska et al. 2014; Petrovszki and Mészáros 2010; Pătru-Stupariu et al. 2011; van Eetevelde and Antrop 2009). They are key sources of information about traditional landscape since it has been possible to observe an increase to the speed and magnitude of changes to many features since the eighteenth century, associated with the increase of population and the growth of urbanization (Antrop 2005; Gustavsson et al. 2007). Archive files and (old) grey literature form another important source of information which were produced and preserved especially for exceptional areas such as designed landscapes. These unpublished documents included, e.g. filed reports, management notes, plans, historic etchings and drawings; however, they are scattered in different archives, libraries and other institutions and written mostly in national languages. For these reasons, they are used less often (Nestor and Mann 1998; Nutt et al. 2013).

Before natural and cultural heritage could be treasured, it first had to be recognized (Lowenthal 2005). Information about landscape development plays a key role in landscape value identification. A complex method of identification and assessment of characteristic landscape appearance was presented by Slámová et al. (2013). The four-step assessment involves: (1) optometric parameters for evaluating visual landscape appearance include spatial parameters about dimensions of the valleys and spatial characteristics of visual dominants; (2) identification of landscape types by statistically significant combinations of land forms and land use categories; (3) assessment characteristic features (originality, age, scarcity, visual harmony, identity of local inhabitants, significant biotopes); (4) determination of a risk of landscape value degradation. Methods of identifying historical landscape structures on old maps and aerial photographs occur more frequently as does the assessment of their change in time (Sklenička et al. 2009; Black et al. 1998; Eetvelde and Antrop 2009). Capelo et al. (2011) presented a 17th criterion for definition and heritage valuation of landscape study cases which combined cultural and natural values (built heritage, natural biotic heritage, natural abiotic heritage, rarity of the heritage landscape type, antiquity, scientific potential, recreational potential, pedagogic potential, historic record, conservation statue, symbolic importance, coherence degree, conservation degree, aesthetical quality, monumentality, range and craft-related value).

Most of the designed landscapes are important areas for both protecting cultural and natural heritage, and they have became protected areas with more sustainable management than a common landscape while still suffer from common problems (Lowenthal 2005). They feature on the list of monuments of national cultural heritage which also fall under nature protection (Nutt et al. 2013, Rosická and Sýkorová 2011). They are credited for their ecosystem services (Kümmerling and Müller 2012), are rich in native species, habitat diversity and capable of sustaining endangered biodiversity (Kümmerling and Müller 2012; Lohmus and Liira 2013; Liira et al. 2012; Jonsell 2012). In the Czech Republic, the initial heritage conservation legislative instrument refers to the Act on State Heritage Preservation (ČNR 1987), nature conservation is mainly based on the Act on the Protection of Nature and the Landscape (ČNR 1992). Area development is coordinated and directed by spatial planning instruments made mandatory by the Building Act (Parlament 2006).

Many complete or fragmented historic cultural landscapes have been preserved in the Czech Republic which requires responsible and systematic care. This primarily involves areas protected under the Heritage Act. Traditional protection categories include Cultural or National Cultural monuments. Cultural monuments include buildings, often with the land on which they stand or with larger surroundings. Along with castles, this ensures that a number of historical gardens and parks are protected. Selected parts of a cultural landscape may also be declared a Conservation Area. A Conservation Area is a settlement formation area or parts of it with fewer cultural monuments, a historical environment or part of the landscape with significant cultural value. The term Landscape Conservation Area is used for protected parts of the landscape. Conservation Areas are declared by the Ministry of Culture of the Czech Republic after consultation with the appropriate regional office. Today, 24 Landscape Conservation Areas have been declared in the Czech Republic. The complete and participatory protection of areas of historical value including cultural landscapes contrasts with the segregated protection of individual sites used in the past, and is based on the European Landscape Convention (COE 2000), the Convention on Protecting Architectural Heritage (COE 1985) and especially the Convention on Protecting the World Cultural and Natural Heritage (UN 1973) including the related outputs of international expert teams (Mitchell et al. 2009).

The values defined in Petrohrad and Žehušicko focus on identification of cultural values and composition, as well as natural values. The aim is to create a complete and modern professional base for qualified professional decisions of state authorities (especially conservation) and local authorities, owners, administrators and users of different parts of the landscape, as well as a base for spatial planning and other plans for protecting and developing areas.

26.2 Methods

26.2.1 Model Areas

Two model areas were chosen for studying landscape values, e.g. cultural heritage and natural values (Fig. 26.1), one of them is in the marginalized region endangered by landscape abandonment, the second one refers to a core region endangered by intensification. The first model area is a landscape park in the small village of Petrohrad in western Bohemia. Petrohrad has only 700 inhabitants and is situated at the half-way between Prague and Karlovy Vary (Carlsbad). Almost the entire cadastral area was transformed to a landscape park (designed landscape/ferme ornée) between 1780s and 1860s. The older Baroque castle is a central part of the designed landscape which is made up of park meadows, forested hills, a game park, a pheasantry and ponds. Petrohrad is surrounded by a hilly landscape with an altitude of 350–450 m above sea level. Granite boulders are often on hills and slopes. Cambisoles prevail in the area, Fluvisoles and Gleysoles could only be found around streams. The most important stream is called Podvinecký. According to the division of the climatic regions framework (Quitt 1971), the area ranks as a semi warm region, characterized by moderate winters with average January temperatures of −2 to −3 °C. The summer months are moderate warm with average June temperatures of 17 °C. Average precipitation during the year fluctuates around 500 mm. The Petrohrad landscape park was chosen as an example of a

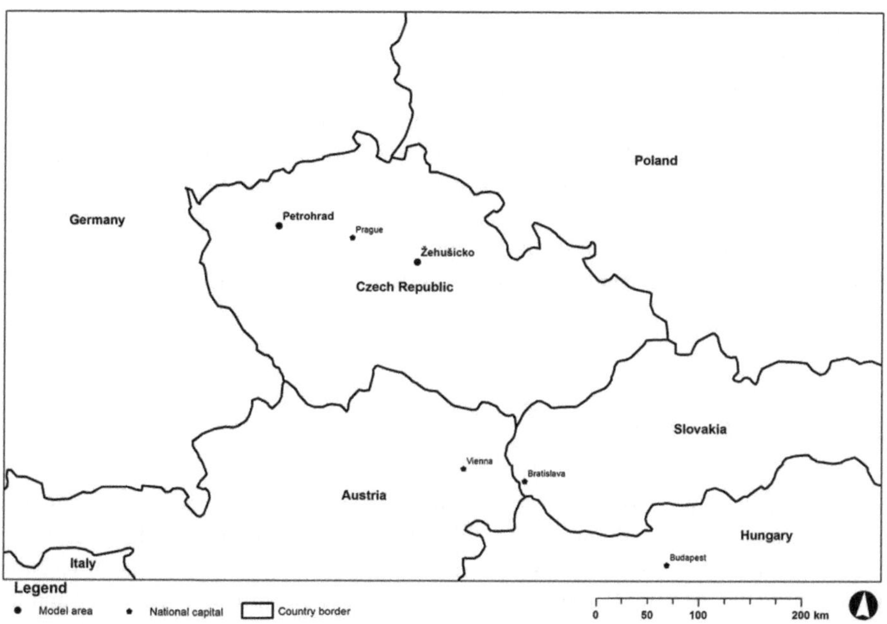

Fig. 26.1 The study areas

"small-scale" analysis (less than one cadastral unit) in a marginalized area threatened by landscape abandonment and loss of open space which endangered both cultural (patterns and appearance of the park) and natural (natural habitats) heritage protection. Ownership of the landscape is dispersed, a number of owners do not express an interest in their land or are unknown. Cultural heritage and natural heritage protection do not offer protection for the whole designed landscape.

The second study area is located in the intensively farmed landscape of Central Bohemia in the Czech Republic, to the east of Kutná Hora (Fig. 26.1). It covers a total area of 3091 ha. It is a typical rural landscape of the Middle Elbe on the eastern foothills of the Kaňk Hills. The area is situated in the basin of the Elbe tributaries Doubrava and Klejnárka. There are still traces of great landscape changes during the eighteenth and nineteenth centuries preserved in the surroundings of Nové Dvory, Kačina and Žehušice, which are the main reason for being declared a Landscape Conversation Area. The changes in the Nové Dvory area were carried out according to baroque-classicist principles. The landscape adaptation around Kačina castle and Žehušice was approached in a classicist and landscape way. The Žehušice area is morphologically part of the Czech Cretaceous Plateau. The terrain in the region predominantly consisted of plains and moderately undulating lands with altitudes of 200–230 m. The Kačina Ridge is morphologically clear and composed of Cretaceous sandstones and marlstone, with sandstone sheet in some places. Based on the ecological soil quality evaluation unit map, most of the farmland consists of alluvial and gleysolic soils. While, Cambisols prevail at higher altitudes in the Kačina Ridge. According to the division of the climatic regions framework (Quitt 1971), the area belongs to a drier warm region that is characterized by moderate and dry winters. The summer months are warm. The average temperature is 8.5–9 °C, annual precipitation amounts to 550–600 mm. Today the landscape is affected by intensive farming and housing construction. Žehušicko was chosen as an example of a "large-scale" analysis in the core area threatened by intensification. The landscape is cultivated by big farms with heavy mechanisation who rent agricultural land and are not interested in sustainability. Ownership of the landscape is dispersed, some of the owners do not express any interest in their land or are unknown. Cultural heritage does not offer protection for the whole designed landscape; on the other hand it also protects common agricultural land of large open fields.

26.2.2 Data Mining

A great deal of grey literature has been published on landscape gardening and model areas are no exceptions to this. The garden in Petrohrad is relatively unknown; hence, it is the least covered in (grey) literature. The most valuable information can be found in the old descriptions from the nineteenth century and the beginning of the twentieth century. The topographic description of the Petrohrad Estate by Ponfikl (1821) offers the oldest prove of landscaping activities there. The castle in Petrohrad was listed in the castle register by Heber (1844), including information on its landscape changes. Detailed documentation of the park's condition at the beginning of the twentieth century can be found in geographic account of the Podbořany administrative unit by Rott (1902). Information on the garden, upon the completion of all the landscape changes in the 1930s, can be found in the manuscript report of the former gardener Synek. The last account by Houda (1977) informs us of the garden's deterioration in the second half of the twentieth century. Only one thesis was written about the garden in Petrohrad, in which Sadilová (1983) made an inventory of the trees in the garden and also records the terrain situation. The History of Petrohrad was published in 2000 (Špilar 2000) and it gives, inter alia, a chronological account of the events and changes that took place in the garden.

The model area Žehušicko is divided into two core areas—Nové Dvory with the Kačina manor house and Žehušice. Two topographic descriptions were made for the whole area, an older description by Zavadil (1912) and a more current one by Novák (2001). Both the history and present day of this area are presented in books by Lipský et al. (2011, 2013). Local historians wrote highly informative books about local history, important persons and owners, and landscape parks [books by Ledr (1884) about Nové Dvory and by Novák (1932) about Žehušice]. Nové Dvory and especially the Kačina manor house are widely known, the Kačina landscape park was reconstructed four years ago and a detailed project was carried out (Šimek et al. 2005). The description of the Žehušice park is less current (Faustusová-Tomsová 1958).

When studying the history of landscape architecture and design, it is necessary to work with historical documents. They provide us with useful background information regarding the overall scheme, such as spatial orientation, road patterns, tree species, etc. They may be divided according to their typical characteristics—written sources, map sources and iconographic sources (Nutt et al. 2013). The major amount of archive data sources are those dating back to the eighteenth and nineteenth centuries. Map sources are largely popular and quite widely used as they can show the landscape and its layout graphically. The most important comparative map series are the first, second and third military survey maps, Topographical maps of the General Staff and maps of the stable cadastre. Many authors have dealt with these works and most of the maps are easily accessible and widely used when assessing the landscape.

Most of the sources (Table 26.1) for the Petrohrad landscape park can be found in the State Regional Archive Třeboň, department Jindřichův Hradec, in the Petrohrad Estate Fund. The most useful are the folders where budgets of works done in the park are stored, documents about the baroque garden, the design of the honour court, the works reports from the nineteenth century, and the instructions for the gardener with the account of his duties. The oldest plan of the Petrohrad Castle, park and game park comes from the first half of the eighteenth century and portrays the baroque design of the site. The following maps from the nineteenth century show the advancing changes and expansion of the park. The iconographic sources

Table 26.1 List of the archival and other sources for studying history and cultural values of the model areas

Data sources	Petrohrad	Žehušicko
State regional archives	State regional archive Třeboň/ Jindřichův Hradec, Estate Petrohrad	State regional archive Praha, Estate Nové Dvory
	State regional archive Třeboň/ Jindřichův Hradec, Czernins family archive	State regional archive Praha, Choteks family archive
	State regional archive Plzeň/ Nepomuk, map collection	State regional archive Praha, Estate Žehušice
		State regional archive Praha, Central administration of Choteks Estates
National archive	National archive, map collection	National archive, map collection
		National archive, patriotic-economic society
State district archives		State District Archive Kutná Hora
Prague city archives		Prague city archives, collection of handwritings and papers
Archive of the Czech National Bank		Archive of the Czech National Bank, credit department
Regional museums	Regional museum Rakovník/Jesenice	Czech Museum of Silver Kutná Hora
	Regional Museum Louny	
National Agriculture Museum	National Agriculture Museum, Museum of Czech Countryside Kačina	National Agriculture Museum, Museum of Czech Countryside Kačina
Museum of Czech Literature	Subcollection "visual art"	
National Heritage Institute	National Heritage Institute Ústí nad Labem	National Heritage Institute střední Čechy
	National Heritage Institute Praha, collection of photographs	National Heritage Institute Praha, collection of photographs
	National Heritage Institute Praha, drawing collection	
	National Heritage Institute, castle Jindřichův Hradec	
National gallery	Drawing collection	
Academy of Science, Institute of Art History	Collection of graphic arts	Map collection
Charles University Prague	Faculty of Science, map collection	

(continued)

Table 26.1 (continued)

Data sources	Petrohrad	Žehušicko
Old state maps series, medium scale-based maps	First, second, and third military maps	First, second, and third military maps
Old state maps series, large-scale-based maps	Stable cadastral maps	Stable cadastral maps
Ortophoto CR	Ortophoto from 1950s	Ortophoto from 1950s
Libraries	Grey literature before 1945	Grey literature before 1945
	Grey literature after 1945	Grey literature after 1945
Other		Municipality Nové Dvory

to Petrohrad are quite rich. There is a drawing from the eighteenth century, graphic sheets, a collection of watercolours from the nineteenth century.

The files for the Žehušicko study area (Table 26.1) are deposited in the State Regional Archives Prague and may be divided into institutional (invoices, land registers) and personal (letters). The oldest map of the Nové Dvory estate by Jan Glocksperger from 1734 is an exceptional source and has been used for a detailed investigation of the core area in the Baroque era. The detailed handwritten maps of the surrounding of Kačina manor house from the end of the eighteenth century and the beginning of the nineteenth century have been used to investigate these landscape designs. The second large group of the non-written sources is represented by old paintings and drawings. The painting of Nové Dvory in the Baroque period featuring the magnificent Baroque garden was studied in detail. This painting likely dates back to the first half of the eighteenth century. Several drawings and graphics from the nineteenth century show landscapes of the Classicistic period, especially the Kačina manor house and its surroundings and pavilions in the landscape. These drawings and paintings have been starkly criticised and compared to other sources to ascertain whether they depicted the real or only an ideal planned state.

26.2.3 Cultural and Natural Heritage Evaluation

Several methods of evaluating landscape values were proposed and they differ from each other in terms of the selected criteria and data sets used (Sklenička et al. 2009; Black et al. 1998; Eetvelde and Antrop 2009; Capelo et al 2011; Slámová et al. 2013). The cultural landscape values evaluation, which we proposed, is based on a historical approach and divided into a small-scale and large-scale analysis. Detailed archival research and knowledge of historical development form the basis of both analyses. Cultural historical values, composition, and natural values were defined using the ArcGIS software. The data gained from archives, grey literature, geographical materials and field research was digitalised in geographical database.

Table 26.2 Small-scale analysis: legend of the map of cultural and natural heritage as an example of identified values

Type	Legend items
Identifies patches of the composed landscape	Composed landscape determining core space
	Composed landscape developing core space
	Follow-up space of valuable landscape
	Borders of composed landscape areas
Legal protection of values	Borders of the cultural monument
	Borders of the natural monument
Axes	Existing visual connection
	Non-existing visual connection
	Riding route
Tree lines	Existing alley
	Non-existing alley
Vistas and follies	Existing view-point
	Non-existing view-point
	Existing building
	Non-existing building

The small-scale analysis was made for areas of one or less than one cadastral unit. We distinguished several important periods in a park designing based on historical studies, a time and scale—appropriate old map is found for each period. Maximum three types of designed areas/patches are distinguished in the studied park according to the intensity of the landscaping activities (composed landscape determining core space, composed landscape developing core space, follow-up space of valuable landscape). The composed landscape determining core space are areas of the most intensive landscaping activities or the oldest ones, appearance of these patches is significantly changed by man. The composed landscape developing core space are areas of less intensive landscaping activities, the patches have more of a natural look but have been recreated by man (too natural to be nature). The follow-up spaces of valuable landscape are natural areas with scattered designed points (follies) and lines (pathways). The shape of each patch in the studied period is drawn into the prepared old map. Based on these partial old maps, the best shape of each patch is drawn into a final map. The best shape refers to the shape that the patch assumed in a period when it was well-maintained. The final map not only contains these patches as keystones of the landscape park but other important elements of the composition too (axes, vistas, follies) (Table 26.2). Meanwhile, the cultural heritage analysis should be fully performed by the research team; the natural heritage evaluation could use materials of the state nature protection agency which were only revised.

The large-scale analysis was made for wide areas. Two partial maps were created (cultural and composition values map and natural values map), their intersection produced the final synthetic map of natural and cultural heritage values. The maps were produced with the help of the state Fundamental Base of Geographic Data

(ZABAGED), at scale 1:10,000. The results were interpreted by texts and tables, in the context of historical development and the present state. The first partial map shows the cultural historical values and compositional phenomena significant for Žehušicko model area. Apart from the current values and phenomena, this map also includes phenomena only partially preserved and vanished. Equally the map reveals distortions, which lower the landscape value and cut the historical connections and composition of the landscape. The map reveals familiar categories of values (for example historically and architectonically valuable buildings and building complexes) as well as newly defined categories. The specific newly defined categories include these phenomena: the core area of a historical cultural landscape and settlements, historical cultural landscape—designed landscape areas, historical cultural landscape—former land cover (forests, water areas, arable land and grasslands), historical cultural lines, composition and distortions (Table 26.3). The second partial map reveals natural habitats in the Žehušice area which are not only important for maintaining sustainable landscape development but also for their cultural historical value. To make a map of types of natural habitats which provides basic information about vegetation in the area, for nature conservation, digitalized output from mapping biotopes in the Czech Republic was used—this mapping was carried out to create a system of a protected area, important in the European context of Natura 2000. These materials were reviewed by the field research in accordance with biotope mapping methodology (Guth et al. 2002; Guth and Kučera 2005). The Czech Republic national interpretation manual for biotopes Natura 2000 (Chytrý et al. 2010) was used to classify biotopes. Zoologically significant localities were declared on passing critical judgment of the accessible data and reviewing it by the field research. The map includes natural values of the model area and also how well they are preserved (Table 26.3). The final map of natural and cultural heritage values was achieved as the intersection of the two partial maps mentioned above.

26.3 Results

26.3.1 Landscape Abandonment in Marginalized Areas in Small-Scale Analysis: Petrohrad Case Study

Petrohrad castle and estate was owned by Czernins' family from 1622 until 1945. The oldest garden was founded there around 1700 when the Baroque castle was built by the owner Herman Jakub Czernin and the famous Baroque architect Giovanni Battista Alliprandi. The garden with a small orchard surrounded the castle and an enclosed game park was situated next to it in the south. This Baroque garden was used for almost 100 years until the end of the eighteenth century. The landscape park was founded there by Johann Rudolph Czernin in 1780s and gradually built by J.R. Czernin and his son Eugen Karl Czernin until the 1860s. To begin with only the appearance of the existing Baroque garden was changed from formal to informal with a few scattered trees and small meadows. The landscape park later

Table 26.3 Large-scale analysis: legend of the map of cultural and natural heritage as an example of identified values

Type	Legend items
Landscape conservation area	Border of the landscape conservation area
	Core area of the landscape conservation area
Settlement	Built-up area
	Valuable built-up area
	Settlement, village
Valuable buildings	Summerhouse
	Extinct summerhouse
	Calvary
	Manor farm
	Extinct manor farm
	Gamekeeper's lodge
	Extinct gamekeeper's lodge
	Tavern
	Cemetery
	Other building
	Extinct other building
	Folly
	Extinct folly
	Church
	Cross
	Bridge
	Memorial
	Statue
	Spring
	Well, pump
	Fortress
	Extinct fortress
	Castle
	Bell tower
Composition—axes	Primary
	Extinct primary
	Secondary
	Extinct secondary
Composition—landscape dominants	Positive dominant
	Negative dominant
Composition—others	Vista
	Extinct vista
	View-point
	Local view horizon
	Place of historical event

(continued)

Table 26.3 (continued)

Type	Legend items
Historical cultural landscape—designed landscape areas	Game park
	Park
	Preserved "pluzina" (historical agricultural structure)
	Island
Historical cultural landscape—former land cover	Park and ornamental garden
	Water area
	Arable land
	Meadows and pastures
	Forest and scattered vegetation
	Rocks
Historical point vegetation	Group of trees
	Solitaire tree
Distortions	Distortions (points)
	Distortions (areas)
Historical cultural lines—historical vegetation lines	Four lines alley
	Extinct four lines alley
	Extinct four lines fruit alley
	Two lines alley, originally fruit
	Two lines alley
	Extinct two lines alley
	Two lines fruit alley, originally not fruit
	Two lines fruit alley
	Extinct two lines fruit alley
	Tree line alley
	Extinct tree line alley
	Extinct tree line fruit alley
	Scattered vegetation
Historical cultural lines—historical lines	Historical water stream
	Historical pathway
	Historical agricultural line
	Extinct agricultural line
	Historical wall
Zoological valuable habitats	Zoological valuable line
	Zoological valuable area
Formation groups of valuable natural habitats	Forests
	Secondary grasslands
	Wetlands and riparian vegetation
	Water streams and areas
	Shrubs

crossed over the walls of the Baroque garden which were demolished. The park in Petrohrad assumed a representative role and was a pleasant background to the manor house. Mainly open spaces were made use of, where meadow plants with groups of trees or single trees were grown. On the other hand, diverse and picturesque rock formations in the valley of the Podvinecký Stream, on Zámecký Hill and Kozí Hill and also in the game park were used. These formations complemented the scenery of important paths, or served as viewpoints. They were welcome as an enlivening element for walks in the garden meadows and helped to overlook them. The broader surroundings of Petrohrad are quite hilly, with boulders and deep valleys. The landscape surrounding the park was composed of fields, pastures, meadows, and forests and aimed to be picturesque too. The borders between the park and the agricultural landscape were fuzzy, important park ways went continuously to the landscape.

The garden itself is located on an undulating slope rising up from north to south. The horizon on the south is formed by a number of peaks beyond which the terrain falls down to forest bushes of the former Petrohrad game park. The northern horizon is not so dominant and is formed by a slope edge which rises up from Podvinecký Stream towards the north. A garden was situated between these two horizons which could be overlooked from the viewpoints in the south as well as in the north. The area around Petrohrad castle had several strengths which support the foundation of a picturesque park. One particular advantage was the Zámecký Hill featuring rocks, a chapel, a castle ruin and impressive view. The elevation between the lowest and highest points of the garden is also rather big, specifically 150 m. A weakness is that there is no other larger and richer water course near the castle which would allow for water elements to be introduced. The most changed places are in the immediate surroundings of the manor house and the Všech Svatých Hill, in other words, places changed by construction activities. Also, establishing meadows required a lot of efforts in Petrohrad, especially in the south, on the steep slopes of the Zámecký and Kozí Hills. The soil is very stony here, requiring for stones to be dug out and removed as well as surfaces relevelled before new plantations could be made.

The aforementioned appearance and development of the park was completely changed in the second half of the twentieth century. First of all the castle and park were nationalized in 1945 and along with a great many German speaking inhabitants, the Czernins were forced to move to Germany. The castle was transformed to a psychiatric hospital and the park was split among different owners. The surroundings of the castle is owned by the psychiatric hospital, the woods belong to the Czech Forests (Lesy ČR), while the pastures and meadows were given to the state farm. The farm is in liquidation for several years already. This means that nobody takes care of the landscape and the number of inhabitants is lower than it used to be. On the other hand, cultural and natural heritage protection was established in 1950s and the core area of the landscape park became a cultural monument. A natural monument was declared in the Zámecký Hill and was significantly enlarged on the park meadows four years ago. Despite this double protection, both the cultural and natural landscape values have slowly but continuously deteriorated.

Three main problems are associated with this marginalized area: (1) cultural heritage and natural heritage protection do not protect the whole designed landscape; (2) there are many owners of the landscape and some of them do not express an interest in their land; (3) landscape abandonment and loss of the open space has endangered both cultural (patterns and appearance of the park) and natural (natural habitats) heritage protection. The main result of the time analysis based on the old maps and ortophotos is that the shape of some patches of both types of the composed landscape (composed landscape determining the core space and composed landscape developing core space) is smaller, i.e. some parts of the composed landscape were lost for other purposes (e.g. football playground, small sewage treatment plant). Other patches have maintained their shape yet deteriorated due to a

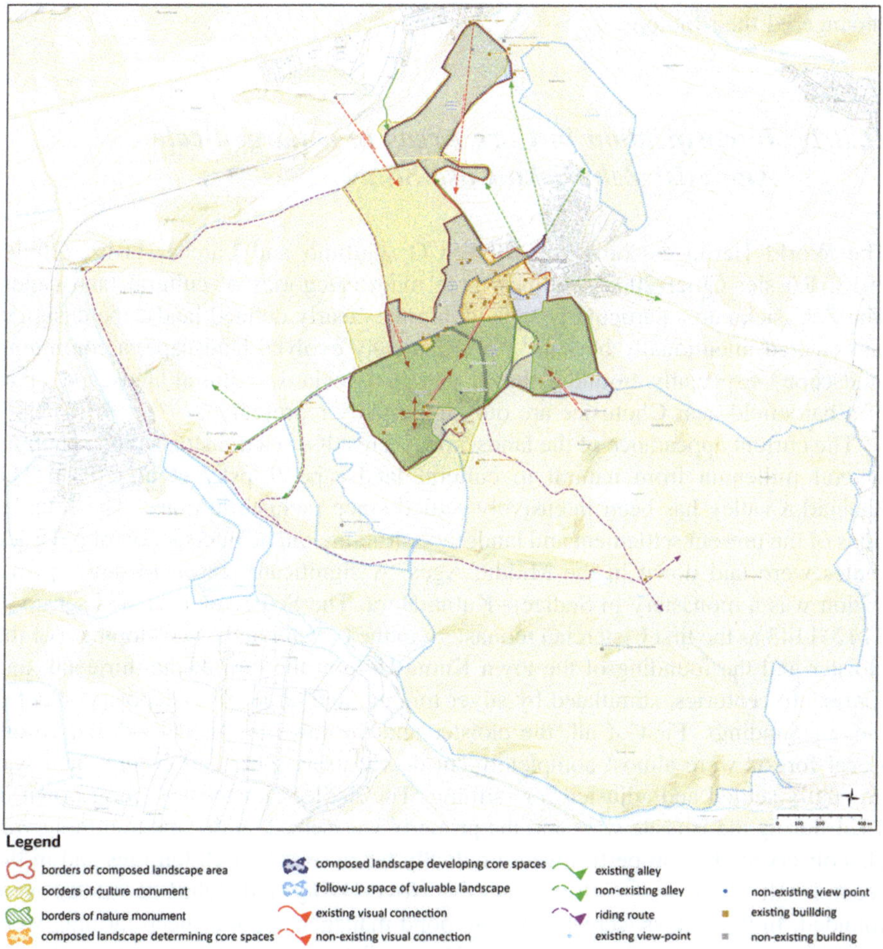

Fig. 26.2 The integrated map for heritage protection and management of the designed landscape in Petrohrad

lack of maintenance (Fig. 26.2). The main features of lack of maintenance in meadows are overgrowing by shrubs and young common or expansive trees, and decline in the old valuable trees. The main features in the woods are similar, especially the decline of old valuable trees and thickening of forest cover. As a result, a decline in both cultural and natural heritage has been observed. The overgrowing and decline of old trees have meant that the composition is less clear, vistas are overgrown and their main features are not able to maintain their function. The axes are not clear or do not keep their function as well. Many of the follies (garden pavilions, statues, small lakes) and tree lines have either completely disappeared or are seriously damaged. The parkways are also in a bad state and many of them could not be passed due to being overgrown or unrecognizable. The nature values have also suffered since the protected saproxylic beetles, which live in the park, prefer old sunny or half-sunny trees, meaning the decline of these trees has endangered their biotops.

26.3.2 Intensification in Core Areas in a Large-Scale Analysis: Žehušicko Case Study

The World Heritage Committee UNESCO (Gullino and Larcher 2013; Gfeller 2013; Rössler Chief 2007) outlines three main categories of cultural landscapes. The Žehušicko area particularly falls within the clearly defined landscape designed and created intentionally by man. An organically evolved landscape, a continuing landscape—especially around Rohozec and an associative cultural landscape—part of a battlefield near Chotusice are only marginally represented.

The current appearance of the landscape is a result of its transformation spanning several millennia from natural to cultural landscape (Lipský et al. 2011). The Klejnárka valley has been intensively settled since prehistoric times. The foundations of the present settlement and landscape structure on the intersection of old trade routes were laid down in the Middle Ages. A significant reason to develop this region was a monastery in Sedlec—Kutná Hora. The Sedlec cloister was set up in 1142–1143 as the first Cistercian monastery in the Czech Lands. Development of the cloister and the founding of the town Kutná Hora at the turn of the thirteenth and fourteenth centuries, stimulated by silver mining, and large silver deposits, affected the surroundings. First of all, the cloister and the town required food and wood. Local forests were almost completely cut down, impassable and swamp land was gradually settled and adjusted for farming. The Sedlec cloister was burnt down in 1421 during the Hussite wars and the property was seized. A decline in mining was also observed. The properties were gradually taken over by rich families and in the second half of the sixteenth century the Nové Dvory and Žehušice estates were founded. In those days, cereal growing, pond management and sheep farming were very important parts of management. During the sixteenth century, a number of ponds and pond systems were built. Economic growth was interrupted again by the

Thirty Years War during the first half of the seventeenth century. The end of the war was followed by the gradual development of aristocratic castles and manor farms.

Baroque renewal was one of the most important landscape designing period. After the post-war renewal, both owners of the estates (from 1661 Thun-Hohensteins in Žehušice, from 1679 Věžníks in Nové Dvory) wanted to build new residences, corresponding to the changing lifestyle. New castles were built in Žehušice in 1679 and in Nové Dvory in 1686. Woods, as pheasanteries and game parks, started to appear again in the newly designed landscape. At the turn of the seventeenth and eighteenth centuries, Bernard Věžník (1651–1714) radically changed the centre of Nové Dvory estate. The unique cultivated landscape emerged as a result, deliberately designed according to Baroque principles. This landscaping formed the estate for the next century and continues to be visible today. The author is not verifiably known. Based on indirect connections, it is thought that Jean Trehet may have been the author. Trehet was a student of Le Nôtre and he was a French garden architect who worked in Vienna. The designed landscape included the baroque castle, a cloister residence as well as parts of the surrounding landscape and town Nové Dvory, composed in accordance with baroque principles. On the seventeenth May 1742 a battle between the Prussian king Friedrich II and the Austrian prince Karl Lothringen, took place near Chotusice. It was one of the most important battles of the Austrian heritage war. Chotusice was burnt down. During the reconstruction of the town, a great deal of wood was cut down for building material. The Enlightenment signified an important era—a period of extensive landscape management changes and landscaping activities. The Enlightenment in the Nové Dvory area is connected with the Chotek family who bought the estate in 1764. Jan Rudolf Chotek (1787–1824) had built a new Empire manor house Kačina. It stood dominantly in the middle of open countryside and was surrounded by a landscape park. This designed ferme ornée landscape combines classicistic and natural landscaping elements. The Žehušice landscape area owned by the Thun-Hohenstein family underwent similar adaptations. The castle stands on the border of a large game park famous for white deer almost two centuries (Fig. 26.3).

Social changes in the middle of the nineteenth century, accompanied by the abolition of patrimonial administration and revolutionary farming changes were first reflected by the local landscape. The agriculture was the most important. The development of food industry and road and railway systems was encouraged. Land Reform from the 1920s divided up the large grounds, except for the so-called residual farmhouses, among smallholders and the landless. Despite efforts made to strengthen general farming profitability as well as other parts of the rural economy and despite local fertile grounds, the economic importance of the countryside gradually decreased. The start of the central economy based on directive administration in the second half of the twentieth century, that rejected previous development, radically affected this landscape: collectivisation, gradual transition to large-scale open fields and other drastic landscape changes which resulted in the disappearance of the majority of landscape structures. Large buildings for large-scale farming were built around villages. In the post-war period a military airport was established right next to Chotusice which was extended after 1989. The

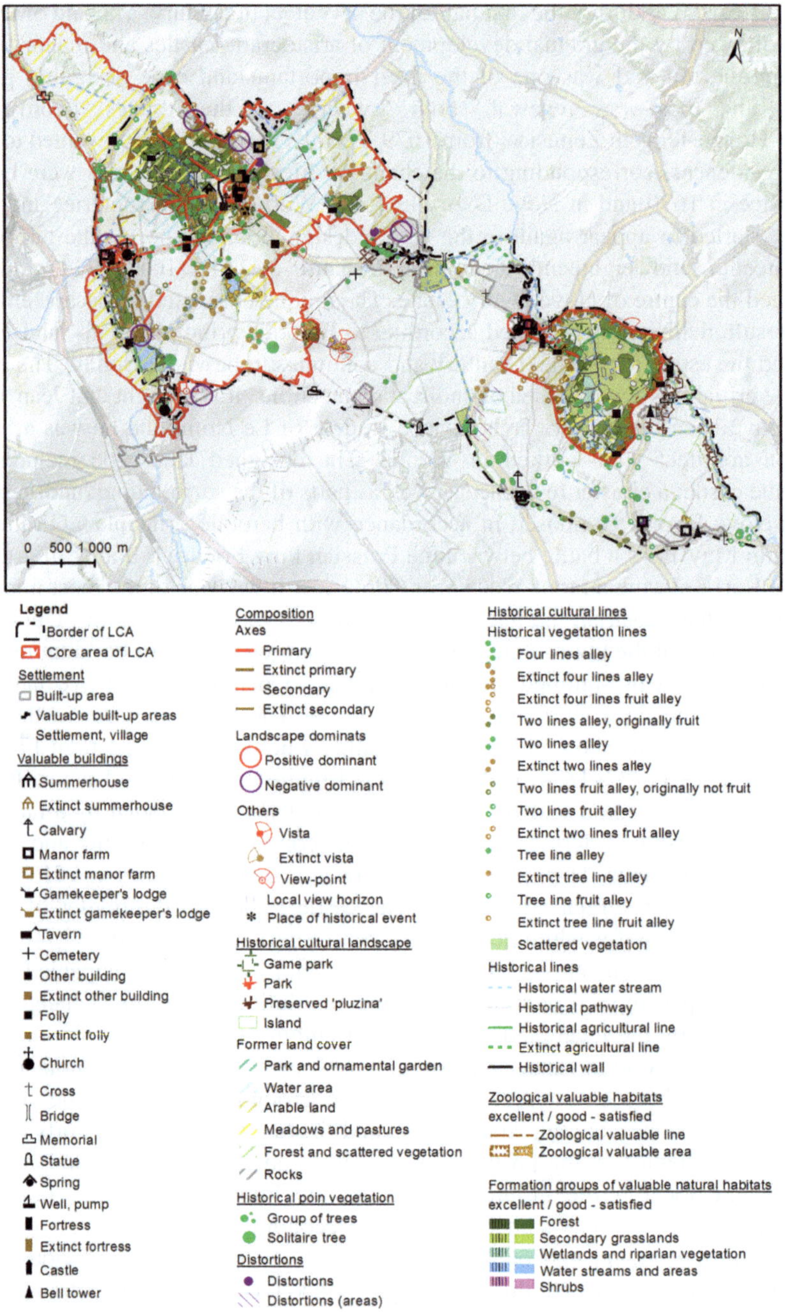

Fig. 26.3 Map of cultural and natural heritage values of the Žehušicko designed landscape

landscape has changed dramatically especially as a result of intensive agriculture and forestry. The land is mainly used as arable land and the most frequently grown crops are cereals, rape plants and corn. Remnants of old landscape structures have a very important role. Natural habitats are especially gathered in two localities, the Žehušice game enclosure and the surroundings of Kačina manor house.

Only 12 % of the total area consists of a natural habitat area. The vegetation of 49 % of this area has massively deteriorated and it is difficult to say if it still ranks as a natural habitat system of Natura 2000. This vegetation includes forests and meadows. Half-natural forests are mainly in the Žehušice game park as a result of deer rearing. The meadow and pasture area is very small. The natural habitat state is only perfect in 19 % of their total area. The model area has the least number of natural biotopes in the Czech Republic and these biotopes have been degraded to a relatively high extent (Boucníková and Kučera 2005). The fact that natural habitats are mostly found in Kačina and the Žehušice game park shows the importance of parks surrounding aristocratic manors. Establishing these places was determined by human activities needed to preserve the natural habitats of the original flora and vegetation. Due to deer rearing in Žehušice, the herbaceous level in the bottomland forests enclosure has deteriorated, however thanks to proper enclosure care, the level of trees with hundreds of years old oaks has been excellently preserved and the state of the mesic meadows is good. The game park is important not only from a cultural historical point of view but also in terms of protecting nature. The surrounding countryside lacks the habitats that are found here (Lipský et al. 2013). The positive impact of human settlements is supported by the nearness of most valuable meadow cover, in gardens, large orchards and on regularly cut, unfertilised areas. These areas are a potential regional source of a meadow cover gene pool (Losvik 2007).

26.4 Discussion

Cultural and designed landscapes are important parts of the world heritage, and create the local/regional identity. Culture landscapes are historical documents par excellence, simultaneously, they are live conditions for people and their activities and diverse habitats for plants and fauna (Agnoletti 2014). Cultural and designed landscapes are constantly changing and the key question is how to maintain a balance between modernisation and heritage protection (Vos and Meekes 1999). Generally speaking, the attitude of the population (especially in Europe) towards cultural and natural heritage has changed in the past 200 years, and monuments have started to awaken people's emotions. Keeping and preservation of heritage became a matter of public interest leading to legislative protection of cultural and natural heritage (Rosická and Sýkorová 2011). In the Czech Republic, cultural landscapes could be protected by the Act on State Heritage Preservation which covers the area of the cultural heritage protection (cultural heritage objects, landscape conservation areas) or by the Act on the Protection of Nature and the Landscape which covers the area of the nature protection (nature monuments, protected landscape areas). Economic activities and landscape changes could be moderate and limited there.

Identification of values is the first step to setting limits for use and economic activity in the landscape. Many approaches could be used, some of them focus on the historical agricultural landscape, other on parks and gardens. The most often approaches are based on digitalisation of former land cover (Skaloš et al. 2011; Pătru-Stupariu et al. 2011; van Eetevelde and Antrop 2009) and landscape structures (Sklenička et al. 2009; Capelo et al. 2011), using old maps and ortophotos. This method is suitable for large areas, for this reason, we used it in the large-scale analysis in Žehušicko. Digitalisation of old maps is very time consuming so only core areas were digitalised. One even more problematic task is which time period should be chosen for the analysis. First, a landscape is a vivid organism and deciding what time was the most valuable is tricky. Second, old maps are only available for a certain time period. In the end we decided on the third military maps for Žehušicko and stable cadastre for Novodvorsko. The time propriety was considered to be more important than the same old map for the whole model area. On the other hand, the small-scale analysis could use old maps from different time periods because it is not so time consuming. The analysis was focused on the patches and lines of the designed landscape and their shape and qualitative changes.

The designed landscapes are defined not only in terms of specific land cover and landscape structures but also in terms of composition (Abendroth et al. 2012; Kümmerling and Müller 2012). The first step is to identify the lines and points which mark the composition, than assessing their function and preservation (Tables 26.2 and 26.3). The composition analysis is quite similar in both small-scale and large-scale analysis. More differences could be observed according the way in which it is endangered; if by landscape abandonment or by intensification. The manner of endangerment is also important to preserve nature heritage. Both landscape abandonment and intensification are caused by different landscape management than it used to be when designed landscapes were created (Antrop 2005). The structure and composition of the designed landscapes are based on traditional landscape management, using man and animal forces, and relatively extensive agriculture, however all these conditions were changed during the twentieth century (Grešlová Kušková 2013). Landscape abandonment means overgrowing of the landscape, increasing the share of forests, increasing or decreasing the share of pastures and decreasing the share of meadows and arable land (Šantrůčková et al. 2013). The designed landscapes are endangered by a loss of open space and vistas, so axes became less clear. Buildings, follies, statues, small lakes and old trees are challenging their decline which has degraded the whole landscape. Overgrowing of the landscape is suitable for common forests species but rare species, requiring traditional management and a semi-open space, are endangered. On the other hand, intensification means large open fields cultivated by huge mechanisation, an increase of built-up areas (housing, transport and industry), decrease of meadows, pastured, and scattered vegetation (Šantrůčková et al. 2013). The designed landscapes endangered by intensification are simplified by a loss of traditional structure (e.g. water streams, hedgerows, other agricultural lines, pathways). Buildings, follies, statues, small lakes and old trees are challenging the lack of interest and decline; axes are cut by new structures in the landscape or lose their perspective. Biodiversity has declined due to loss of suitable habitats and pollution.

Possible solutions include implementing present tools for protecting general landscapes (e.g. landscape character assessment and protection); implementation of the European Landscape Convention; protection of designed landscapes in present development strategies and landscape planning documents. The designed landscapes should be vivid, they could play an important role in tourism in less touristy regions (Kozak 2013). Of course, all kinds of development should be sustainable for cultural and natural heritage.

26.5 Conclusion

Some general methodology—especially identifying and assessing cultural landscape values—can also be used universally for cultural landscapes outside protected areas. They can also be used for general heritage preservation and as a base for landscape planning and nature protection instruments. Identifying nature values is a very important step towards increased local public, state and local administration and other entities and landscape users' awareness of the importance of designed landscapes, as well as a guideline for adopting an approach to preserving these areas. This paper is a component contribution to proposing methods of identifying, interpreting and presenting cultural, historical and natural values and composition, including, detecting and evaluating key elements, patches and spatial relationship of landscape in GIS.

Acknowledgement This study was supported by the NAKI grant of the Ministry of Culture, Czech Republic, No. DF12P01OVV001: Protection and management of historic cultural landscapes through Landscape Conservation Areas.

References

Abendroth S, Kowarik I, Muller N, Lippe M (2012) The green colonial heritage: woody plants in parks of Bandung, Indonesia. Landscape and Urban Plan 106:12–22. doi:10.1016/j.landurbplan.2011.12.006

Agnoletti M (2014) Rural landscape, nature conservation and culture: some notes on research trends and management approaches from a (southern) European perspective. Landscape and Urban Plan 126:66–73. doi:10.1016/j.landurbplan.2014.02.012

Antrop M (2005) Why landscapes of the past are important for the future. Landscape and Urban Plan 70:21–34. doi:10.1016/j.landurbplan.2003.10.002

Black AE, Strand E, Wright RG, Scott JM, Morgan P, Watson C (1998) Land use history at multiple scales: implications for conservation planning. Landscape and Urban Plan 43(1–3):49–63. doi:10.1016/S0169-2046(98)00096-6

Boucníková E, Kučera T (2005) How natural and cultural aspects influence land cover changes in the Czech Republic? Ekológia Bratislava 24:69–82

Capelo S, Barata FT, Mascarenhas JM (2011) Why are cultural landscapes of various values? Thinking about heritage landscape evaluation and monitoring tools. J Landscape Ecol 4(1):5–17

Chytrý M, Kučera T, Kočí M, Grulich V, Lustyk P (2010) Katalog biotopů České republiky. Agentura ochrany přírody a krajiny ČR, Praha

ČNR (1987) Zákon České národní rady č. 20, o státní památkové péči. Sbírka zákonů Československé socialistické republiky 6:214–227

ČNR (1992) Zákon České národní rady č. 114/1992 Sb., o ochraně přírody a krajiny. Sbírka zákonů České a Slovenské federativní republiky 28:666–692

COE (1985) Convention for the protection of the architectural heritage of Europe, Granada, ETS no. 121. Council of Europe, Strasbourg

COE (2000) European landscape convention, Florence, CETS no. 176. Council of Europe, Strasbourg

Faustusová-Tomsová S (1958) Žehušický park. Botanický průvodce. Krajský dům osvěty v Pardubicích, Pardubice

Gfeller AE (2013) Negotiating the meaning of global heritage: 'cultural landscapes' in the UNESCO world heritage convention, 1972–92. J Glob Hist 8(3):483–503. doi:10.1017/S1740022813000387

Grešlová Kušková P (2013) A case study of the Czech agriculture since 1918 in a socio-metabolic perspective—from land reform through nationalisation to privatisation. Land Use Policy 30:592–603. doi:10.1016/j.landusepol.2012.05.009

Gullino P, Larcher F (2013) Integrity in UNESCO world heritage sites. A comparative study for rural landscapes. J Cult Heritage 14:389–395. doi:10.1016/j.culher.2012.10.005

Gustavsson E, Lennartsson T, Emanuelsson M (2007) Land use more than 200 years ago explains current grassland plant diversity in a Swedish agricultural landscape. Biol Conserv 138(1–2):47–59. doi:10.1016/j.biocon.2007.04.004

Guth J, Kučera T (2005) Natura 2000 habitat mapping in the Czech Republic: methods and general results. Ekológia Bratislava 24:39–51

Guth J et al (2002) Metodika mapování biotopů soustavy Natura 2000 a Smaragd. AOPK ČR, Praha

Heber FA (1844) Böhmens Burgen, Vesten und Bergschlösser, Zweiter Band. C. W. Medau und Comp., Prag

Houda J (1977) Zámecký park v Petrohradě. Kulturní měsíčník 3:13–14, 4:13–14, 6:11–13

Jonsell M (2012) Old park trees as habitat for saproxylic beetle species. Biodivers Conserv 21:619–642. doi:10.1007/s10531-011-0203-0

Kanianska R, Kizeková M, Nováček J, Zeman M (2014) Land-use and land-cover changes in rural areas during different political systems: a case study of Slovakia from 1782 to 2006. Land Use Policy 36:554–566. doi:10.1016/j.landusepol.2013.09.018

Kozak MW (2013) Innovation, tourism and destination development: Dolnośląskie case study. Eur Plan Stud. doi:10.1080/09654313.2013.784597

Kümmerling M, Müller N (2012) The relationship between landscape design style and the conservation value of parks: a case study of a historical park in Weimar, Germany. Landscape Urban Plan 107:111–117. doi:10.1016/j.landurbplan.2012.05.006

Larsson S, Nilsson C (2005) A remote sensing methodology to assess the costs of preparing abandoned farmland for energy crop cultivation in Northern Sweden. Biomass Bioenergy 28:1–6. doi:10.1016/j.biombioe.2004.05.003

Ledr J (1884) Děje panství a města Nových Dvorů. Karel Šolc, Kutná Hora

Liira J, Lõhmus K, Tuisk E (2012) Old manor parks as potential habitats for forest flora in agricultural landscapes of Estonia. Biol Conserv 146:144–154. doi:10.1016/j.biocon.2011.11.034

Lipský Z, Šantrůčková M, Weber M, Skaloš J, Novák P, Vávrová V, Kučera Z, Kukla P, Stroblová L (2011) Vývoj krajiny Novodvorska a Žehušicka ve středních Čechách. Karolinum, Praha

Lipský Z, Weber M, Stroblová L, Skaloš J, Šantrůčková M, Kučera Z, Dostálek J, Trantinová M (2013) Současnost a vize krajiny Novodvorska a Žehušicka ve středních Čechách. Karolinum, Praha

Lõhmus K, Liira J (2013) Old rural parks support higher biodiversity than forest remnants. Basic Appl Ecol 14:165–173. doi:10.1016/j.baae.2012.12.009

Losvik MH (2007) Regional species pools of hay meadows: a case study. Appl Veg Sci 10 (2):239–248. doi:10.1658/1402-2001(2007)10[239:RSPOHM]2.0.CO;2

Lowenthal D (2005) Natural and cultural heritage. Int J Heritage Stud 11(1):81–92. doi:10.1080/13527250500037088

Mallinis G, Emmanoloudis D, Giannakopoulos V, Maris F, Koutsias N (2011) Mapping and interpreting historical land cover/land use changes in a natura 2000 site using earth observational data: the case of Nestos delta, Greece. Appl Geogr 31:312–320. doi:10.1016/j.apgeog.2010.07.002

Mitchell N, Rössler M, Tricaud PM (2009) World Heritage Cultural Landscapes. A Handbook for Conservation and Management. World Heritage Centre UNESCO, Paris

Musaoglu N, Tanik A, Kocabas V (2005) Identification of land-cover changes through image processing and associated impacts on water reservoir conditions. Environ Manage 35(2):220–230. doi:10.1007/s00267-003-0270-4

Nahuelhual L, Carmona A, Aguayo M, Echeverria C (2014) Land use change and ecosystem services provision: a case study of recreation and ecotourism opportunities in Southern Chile. Landscape Ecol 29:329–344. doi:10.1007/s10980-013-9958-x

Nestor BA, Mann WA (1998) An archival restoration of the horticultural and design elements of Barnsley Gardens, Georgia. Landscape Urban Plan 42:107–122

Novák F (1932) Dějiny městyse Žehušic. Obec Žehušice, Žehušice

Novák P (2001) Paměť krajiny. Novodvorsko—Žehušicko. Martin Bartoš—Kuttna, Kutná Hora

Nutt N, Nurme S, Hiob M, Salmistu S, Kotval Z (2013) Restoring Manor Parks: exploring and specifying original design and character through the study of dendrologous plants in estonian historical Manor Parks. Baltic For 19(2):280–288

Parlament ČR (2006) Zákon č. 183, o územním plánování a stavebním řádu (stavební zákon). Sbírka zákonů České republiky 63:2226–2290

Pătru-Stupariu I, Stupariu MS, Cuculici R, Huzui A (2011) Understanding landscape change using historical maps. Case study Sinaia, Romania. J Maps 7(1):206–220. doi:10.4113/jom.2011.1151

Petrovszki J, Mészáros J (2010) The Great Hungarian Plain in the sheets of Habsburg military survays and some historical maps—a case study of the Körös/Criş drainage basin. Acta Geod Geoph Hung 45(1):56–63. doi:10.1556/AGeod.45.2010.1.9

Ponfikl JE (1821) Vollständige Topographie der Fideikommiss—Herrschaft Petersburg im Saazer Kreise in Böhmen. Gottlieb Haase, Prag

Quitt E (1971) Klimatické oblasti Československa. Studia. Geographica 16:1–73

Rogge E, Dessein J, Verhoeve A (2013) The organisation of complexity: a set of five components to organise the social interface of rural policy making. Land Use Policy 35:329–340. doi:10.1016/j.landusepol.2013.06.006

Rosicka Z, Sykorova J (2011) Planning cultural heritage protection is easier than managing an actual disaster. Acta Univ Agric et Silvic Mendel Brun 59(4):281–286

Rössler Chief M (2007) World heritage cultural landscapes: a UNESCO flagship programme 1992–2006. Landscape Res 31(4):333–353. doi:10.1080/01426390601004210

Rott W (1902) Der politische Bezirk Podersam. Bezirkslehrervereines Podersam, Podersam

Sadilová L (1983) Dřeviny zámeckého parku v Petrohradě. MSc thesis, Charles University in Prague

Šantrůčková M, Weber M, Lipský Z, Stroblová L (2013) Participative landscape planning in rural areas: a case study from Novodvorsko, Zehusicko, Czech Republic. Futures 51:3–18. doi:10.1016/j.futures.2013.04.005

Šimek P et al (2005) Regenerace zámeckého parku Kačina. Mendel University in Brno, Lednice na Moravě

Skaloš J, Weber M, Lipský Z, Trpáková I, Šantrůčková M, Uhlířová L, Kukla P (2011) Using old military survey maps and orthophotograph maps to analyse long-term land cover changes—case study (Czech Republic). Appl Geogr 31:426–438. doi:10.1016/j.apgeog.2010.10.004

Sklenička P, Molnarová K, Brabec E, Kumble P, Pittnerová B, Pixová K, Salek M (2009) Remnants of medieval field patterns in the Czech Republic: analysis of driving forces behind their disappearance with special attention to the role of hedgerows. Agric Ecosyst Environ 129:465–473. doi:10.1016/j.agee.2008.10.026

Slámová M, Jančura P, Daniš D (2013) Methods of historical landscape structures identification and implementation into landscape studies. Ekologia (Bratislava) 32(3):267–276. doi:10.2478/eko-2013-0023

Špilar V (2000) Z historie obcí Petrohrad, Černčice, Bílenec. Žaket, Petrohrad

Timár G, Székely B, Molnár G, Ferencz C, Kern A, Galambos C, Gercsák G, Zentai L (2008) Combination of historical maps and satellite images of the Banat region—re-appearance of an old wetland area. Glob Planet Change 62:29–38. doi:10.1016/j.gloplacha.2007.11.002

UN (1973) Convention concerning the protection of the world cultural and natural heritage. Paris, November 16, 1972. 1037 U.N.T.S. 151, 27 U.S.T. 37, 11 I.L.M. 1358

Van Eetvelde V, Antrop M (2009) Indicators for assessing changing landscape character of cultural landscapes in Flanders (Belgium). Land Use Policy 26(4):901–910. doi:10.1016/j.landusepol.2008.11.001

Vos W, Meekes H (1999) Trends in European cultural landscape development: perspectives for a sustainable future. Landscape Urban Plan 46(1–3):3–14. doi:10.1016/S0169-2046(99)00043-2

Zald HSJ (2009) Extent and spatial patterns of grass bald land cover change (1948–2000), Oregon Coast Range, USA. Plant Ecol 201:517–529. doi:10.1007/s11258-008-9511-1

Zavadil AJ (1912) Kutnohorsko slovem i obrazem. Nakl. Karla Šolce, Kutná Hora

Chapter 27
Convention on Biological Diversity and European Landscape Convention: An Alliance for Biocultural Diversity?

Bianca Maria Seardo

Abstract While the Convention on Biological Diversity (CBD) addresses the safety of species and ecosystems diversity, the European Landscape Convention (ELC) stresses the importance of preserving, managing and creating high-quality landscapes encompassing both natural and man-affected ones, in accordance to cultural values and taking into account people's perceptions of landscapes. This recent pan-European policy may be a strong support to widen the application of the CBD, but is this really happening? As a matter of fact, although almost all of the European Member States have ratified both the CBD and the ELC, it is not obvious at all that respective national policies proceed in an integrated way. The paper will focus on the potential interactions between the CBD and the ELC in sustaining biocultural diversity, then a brief overview, will show how the landscape conception promoted by the ELC is influencing sectorial biodiversity national policies.

Keywords Landscape policies · Biodiversity policies · Landscape planning · Landscape multifunctionality · Biocultural diversity

27.1 Preliminary Assumptions: "Landscape and Biodiversity", a Couple not to Be Taken for Granted

In her definition of biocultural diversity Luisa Maffi includes "the diversity of life in all of its manifestations: biological, cultural, and linguistic, which are interrelated (and possibly coevolved) within a complex socio-ecological adaptive system" (Maffi 2008).

B.M. Seardo (✉)
DIST, Dipartimento di Scienze, Progetto e Politiche del Territorio, Politecnico e Università di Torino, Viale Mattioli 39, 10125 Turin, Italy
e-mail: bianca.seardo@polito.it

M. Agnoletti and F. Emanueli (eds.), *Biocultural Diversity in Europe*,
Environmental History 5, DOI 10.1007/978-3-319-26315-1_27

Recently practice and research in territorial and landscape planning are increasingly called to commit to the task of safeguarding this complexity throughout the variety of ecosystems, biomes, landscapes (MEA 2005) as long as nature and culture are intended as strictly intertwined.

From a theoretical point of view, the concept is perhaps not so new as shown by the debate within other scientific fields at least since the past century—e.g. ethnobiology, anthropology, linguistics (Maffi 2008), historical ecology (as well documented by Cevasco 2007). But the broader reception of such a sophisticate approach to nature and culture within international conventions affecting spatial and landscape policies could probably give a broader impulse to put that principle on the ground, as in the case of the European Landscape Convention (ELC—CoE 2000).

The official policy definitions of landscape and biodiversity encompass both dimensions—natural and cultural—stressing the importance of their interrelations: the Convention on Biological Diversity (CBD, United Nations 1992), besides recognizing to biodiversity "ecological, genetic, social, economic, scientific, educational, cultural, recreational and aesthetic values" (Preamble), stresses the fundamental role of indigenous and local communities in conserving life on Earth by means of "knowledge, innovations and practices [...] embodying traditional lifestyles" (Art. 8j). The relevance of such an issue requires a specific implementation programme (COP 2000) to encourage the equitable sharing of the benefits arising from the utilization of such knowledge, innovations and practices and to enhance the role and involvement of indigenous and local communities. Finally the recent joint programme between UNESCO and the SCBD remarks "a holistic approach consistent with cultural and spiritual values, worldviews and knowledge systems and livelihoods that contribute to conservation and sustainable and equitable use of biodiversity" (General Principles for the Implementation of the Joint Programme, p. 1).

On the other hand, landscape—as recently defined by the ELC—is shaped as a holistic dimension linking biological and cultural diversity as long as its "character is the result of the action and interaction of natural and/or human factors" (Art. 1a). Moving beyond a strictly ecological approach, social perception plays now a primary role for the landscape to such an extent that its inclusion is recommended into "regional and town planning policies and cultural, environmental, agricultural, social and economic policies, as well as in any other policies with possible direct or indirect impact on landscape" (ELC, Art. 5d).

The study of people's perceptions towards the landscape basing on cultural diversity is currently relevant to natural resource and protected areas management. Resolution n.4.099 (IUCN 2008) "Recognition of the diversity of concepts and values of nature" recognizes nature conservation as a "human action rooted in diverse, evolving cultures and world views" thus setting the task of enhancing and promoting "nature conservation actions including and reflecting practices and traditions that are rooted in culture and embody the cultural values of the diversity of peoples of the world". But the ELC stresses the importance of broadening this attention to the whole territory. Thus spatial planning is asked also to take charge of people's values and aspirations towards everyday landscape (attempts are summarized in Cassatella 2014).

Nevertheless, scientific research should consider "landscape and biodiversity" as a couple not to be taken for granted.

At a theoretical level, "diversity" is not a desirable condition of a landscape a priori: ecological diversity does not coincide with visual diversity as well as human-perceived naturalness does not coincide with naturalness as defined by ecologists (Daniel 2001); conversely, homogeneity is the key criteria for identification of landscape character (Tudor 2014) and often an outstanding value to be preserved in the landscape, as stated by the Convention concerning the protection of the world cultural and natural heritage (UNESCO 1972).

At a planning level, biodiversity and landscape are key issues of environmental and landscape planning as long as they are assumed in the cultural ecosystem services framework (Cassatella and Seardo 2014). Attention is increasingly accorded to the importance of ecosystem services in providing a framework for taking into account more systematically the ecological impacts of alternative planning scenarios, pointing out trade-offs between environmental and landscape goals (Chan et al. 2006; Seardo 2012a). Moreover, spatial planning is commonly assumed as the place to achieve such integration of policies (Beately 1995; van Asschea and Djanibekov 2012).

At the policy level, "territory", "landscape", "biodiversity" and "protected areas" are often regulated by specific sectorial and distinct policies. But, whereas on the one hand it can be easy to converge on abstract principles and policy tasks, on the other hand divergences may arise when it comes to put them on the ground. The word biodiversity never appears within the ELC official text, nor the ELC is mentioned by the SCBD Joint Programme (Annex 1. Examples of Relevant Declarations and Guidelines).

This suggests that even though the ELC endorses a multifunctional concept of landscape, focusing on its biocultural roots, thus potentially supporting the biocultural diversity issues, at a policy level the assumption of such an approach must be deeply investigated. As a matter of fact, although almost all of the European Member States have ratified both the CBD and the ELC, it is not obvious at all that respective sectorial policies proceed in an integrated way.

Suggesting that the integration should be investigated at least at three different levels (theoretical, policy and planning), the paper focuses on the policy dimension intended as a precondition to a well-integrated planning, specifically investigating the potential interactions between the CBD and the ELC in sustaining biocultural diversity.

27.2 The Analysis Framework

Following the ratification of the CBD, States are committed to develop National Biodiversity Strategies and Action Plans (NBSAPs or NBSs), while the ratification of the ELC foresees the integration of its principles cross-cutting sectorial policies.

The research has focused on the EU Member States committed in the implementation of the CBD and the ELC and has concerned the examination of the NBSAPs. At the time of closing the research (April 2012, revision in March 2014), the state of the art was the following:

- 23 EU Member States out of 28 had ratified the ELC;
- all EU Member States had signed the CBD (168 Countries all around the world had signed it), but the development of National Strategies and Action Plans are at different stages of progress.

Since then some changes have occurred, for example Croatia has joined European Union on 1 July 2013 (ratification of the ELC in 2003).

Do NBSAPs take into account the multifunctional dimension of landscape of the ELC to sustain biocultural diversity?

A comparative discourse analysis of EU Member States NBSAPs has been carried out on the basis of two criteria identified according to two relevant articles of the ELC.

Article 1(a) suggests that landscape multidimensionality is the result of the action and interaction of natural and/or human factors shaping its historic, natural and aesthetic visible features. According to this, the analysis has investigated the presence of social and cultural values related to biodiversity and landscape cultural services (e.g. spiritual, inspiration, aesthetic, recreation services).

Second, holism indicates that each landscape element gains its significance, importance or existence not only from its intrinsic properties but also in accordance to its relationships with the context (Antrop 2006). The spatial implication of this idea is an extended attention from few outstanding elements to their broader contexts, thus including "natural, rural, urban and peri-urban areas (…) land, inland water and marine areas (…) landscapes that might be considered outstanding as well as everyday or degraded landscapes" (ELC, art.2). According to this, the analysis has put in light the attention to biocultural landscapes not listed by the CBD.

The official website of the CBD has been the source for the official documents; last access dates to March 2014. In case of no English versions available for this year, reference has been made back to the last English versions published.

27.3 Results and Discussion

From the survey on the coherence of the NBSs with articles 1(a) and 2 of the ELC, different landscape conceptions seem to emerge.

The assumption of the multidimensional conception of landscape varies depending on the State, with a general oscillation among three main approaches.

These approaches can coexist within the NBSs either shaping a balanced mix or rather typifying them.

Although far from an ultimate and proper evaluation, the research has tried to extrapolate the main aspects characterizing the three approaches summarized in Table 27.1 and discussed below.

Table 27.1 Main aspects characterizing the landscape conceptions within National Biodiversity Strategies and Action Plans of the EU Member States

		ELC landscape conception defined by	
		Article 2, horizontal approach	Article 1(a), vertical approach
Landscape conceptions within NBSAPs	Landscape as large-scale for biodiversity conservation	Landscape as a large-scale for ecosystem knowledge and monitoring	
		Connection among ecosystems (natural and semi-natural)	
		Basis for national ecological networks	
			Adequate scale to establish partnerships for ecosystem management
			Reinforce conservation around PAs
	Rural landscape	Prevalent attention to cultural landscapes. Countryside as wildlife habitat, rural landscapes as buffer zones for PAs	Cultural practices related to genetic, species, ecosystem and landscape diversity
			Relevance of natural and man-made landscape elements
			Scenic, perceptive, symbolic, cultural identity values of rural landscape features—multifunctionality of landscape features
	Multifunctional landscape	Addresses to deeply human-affected and human-shaped landscapes (usually beyond those recommended by the CBD) e.g. following a gradient from natural protected areas to urban environments	Highlighting of historical coevolution and present connections between biodiversity and human practices
			From ecosystem functions to landscape values
			Role of people's perception
			People's awareness and participation in biodiversity conservation
			Cross-sectoral policy integration

27.3.1 Landscape as Large-Scale for Biodiversity Conservation

ELC article 2 suggests a spatial or horizontal enlargement of the landscape approach, moving from isolated protected areas to biomes and from natural to urban areas, while article 1(a) explicates a vertical deepening of the landscape conception concerning the functional and symbolic interrelation between cultural and natural factors shaping the landscape.

Evidences from the research suggest that strong consistence with the horizontal approach but weak consistence with vertical approach lead to NBSs not completely consistent with the ELC and usually centred on landscape as a large-scale factor exclusively aimed to support biodiversity conservation. As shown in Table 27.1, the main aspect characterizing this approach is a great emphasis to spatial connections among ecosystems, even to support planning of nation-wide ecological networks. Attention to non-natural systems is present, but usually in the perspective of restoring them for strictly ecological purposes.

Such an ecological-oriented approach stresses the priority of reinforcing the management of protected areas especially in connection with the surrounding environments. Bulgarian primary efforts are put in employing the principles of conservation biology and landscape ecology to designate effective buffer zones and to connect and coordinate reserves at the broader landscape scale.

Protected areas are to be intended as biodiversity hotspot of a national ecological network to be expanded by identifying high-priority regions to be integrated into this system. Biological diversity is intended to be supported by integration into all aspects of land, water and biological resources management, especially stimulating habitat restoration of wetlands, forests, lands supporting intensive crop agriculture, pastures, riparian zones and industrial zones degraded or destroyed by past management. Simultaneously impulse to environmental education and the development of an ecotourism policy shall strengthen the main target.

In the case of Czech Republic, the Territorial System of Ecological Stability of the Landscape—(TSES) is a mutually interconnected set of natural and semi-natural, ecosystems aimed to maintain the national natural balance. Territorial systems of ecological stability are classified as local, regional and supraregional systems and are defined, modified and further specified within preparation of the land use planning documentation.

In the UK a more effective landscape-scale approach is pointed out as the way to effectively establish coherent and resilient ecological networks on land and at sea, shifting away from piecemeal conservation actions. To achieve this, landscape-scale approach to ecological restoration will be applied to national parks and areas of outstanding national beauty. This approach is seen also to help achieving outcomes in not protected areas such as towns and open countryside. The landscape-scale approach is also seen as a strategy to be applied to the highly fragmented woodlands where small and isolated patches of ancient woodlands are particularly vulnerable to climate change, thus enlarging and buffering ancient woods is identified as a priority. Particularly, support for local delivery in the implementation of a landscape-scale approach is considered essential to achieve the objectives of the strategy.

Forman (1995) defines “landscape” as a “mosaic formed by a group of ecosystems that are repeated in space with similar shape, in a mileage range, with identifiable boundaries (…). Specific level of biological organization of life”. Landscape ecologists link the concept of landscape to the idea of a specific “spatial scale” ideal for the investigation of ecosystems’ particular structures and processes not detectable at other levels of analysis: ecotones, connectivity between ecosystems, porosity of the landscape matrix, metastability strategies (Ingegnoli 1999).

The implementation of scientific instruments (e.g. gap analysis, ecological network model, etc.), in the development of place-based policies has to be considered a positive gain for NBSs; moreover, such approach generates scientific knowledge relevant to support planning in the identification of priority areas of intervention. A risk lies however, in the predominance of this approach in case of omitting to consider other cultural ecosystem services generally attached to nature-valuable areas or scarcely considering the wider set of biocultural dynamics affecting them.

27.3.2 Rural Landscape

Positive consistence with ELC article 2 and 1(a) can lead to rural-landscape-centred NBSs, focusing on nation-wide rural landscapes as wildlife habitat or to be managed as suitable buffer zones for Protected Areas. Unlike the above-mentioned approach, the present gives specific attention to cultural landscapes.

Austria focuses on the connections between landscape and genetic resources. Indeed farming highlights the variety of the national cultural landscapes mosaic, but the preservation of landscape needs the support of biodiversity policies, from species to the genetic level: as long as six autochthon sheep breeds are considered to be endangered, the Austrian NBS calls for breeding programmes designated to avoid genetic impoverishment also in the framework of broader cultural landscape conservation (especially in the Alpine regions). On the other hand, genetic resources in the Alps, such as grasslands plant associations, need to be identified, studied and supported by proper corresponding farming practices that also need protection. Austria NBS acknowledges that nature conservation and landscape protection have to be integrated also for their relevance to a wide range of legal entities (e.g. regional planning, agriculture, transportation). Moreover, biodiversity and landscape services have to be properly balanced when it comes to consider functional and aesthetic interactions between adjoining habitats (forest, forest edge, meadow) and to minimize the landscape impacts of energy lines, transmitter masts and windmills.

Irish NBS aims primarily to conserve and restore biodiversity and ecosystem services in the wider countryside, due to the consideration that much of the Irish biodiversity lies outside the protected areas. The so-called “Burren Farming for

Conservation" programme is pointed out as an emblematic agri-environmental initiative assisting the maintenance and recovery of ecosystems outside traditional protected areas with benefits for farmers and habitats both inside and outside the protected land. Landscape is the regarded spatial scale at which to tailor partnerships with and among farmers, but with great attention to local landscape features: for example, hedgerow and scrub are objects of a regulation for their removal and management (Irish NBS, target 9). Significantly, in the first version of its NBS, Ireland did not employ the term "landscape", although having ratified the ELC, and the biodiversity conservation policy had a very sectorial approach focusing on the role of the rural landscape in terms of wildlife suitability.

The English NBS points out the historical role of mankind in the diversification of plant and animal species through traditional practices of agricultural land use. The main issue is to encourage farms multifunctionality (rural landscape cover almost the 80 % of the nation) and to apply environmentally suitable land management techniques emphasizing the historic landscape characteristics.

In Estonia, one of the issues for biodiversity policies is landscape fragmentation due to the rapid re-privatization. The NBS is a means to develop a unified landscape policy aimed both at preserving natural habitats (for example by establishing protection zones around water courses and lakes) and maintaining the richness and diversity of the national landscape. Such objectives are pursued linking conservation to agricultural policies which can help to restore and create woodlands, wetlands, dunes, riverbeds, strips, dry stone walls, etc.

Usually, multidimensionality of the landscape is taken into account by according importance to cultural practices related to genetic, species, ecosystem and landscape diversity, to the diversity of landscape elements both natural and man-made as well as to scenic, perceptive, symbolic, cultural identity values of rural landscape features. Nonetheless sometimes this approach can lack of the same multifunctional perspective while treating other ecosystems.

27.3.3 Multifunctional Landscape

Finally, strong consistence with both ELC article 1(a) and 2 is the most favourable condition.

This approach satisfies ELC article 2 by intending landscape in a holistic manner, not only in its spatial dimension but also as a means to reveal and manage mutual interactions and dynamics among natural and cultural factors.

Namely, this approach is generally characterized by: (a) enlargement of attention from ecosystems functions to landscapes values, (b) highlighting of the connections between biodiversity and human (traditional) land use/management practices, also in a co-evolutionary and historical perspective; (c) addresses to deeply human-affected or human-shaped landscapes and (d) a strong integration among sectorial policies.

Addresses to deeply human-affected and human-shaped landscapes beyond the ecosystems recommended by the CBD, are given by France, Germany, Latvia, Lithuania and the United Kingdom which dedicate specific sections to the urban landscape. Latvia focuses on the monitoring of invasive species within urban ecosystems creating inventories of genetic resources stored in urban parks and botanical gardens, seeing these places as "oasis of biodiversity" (a dualistic view of the urban environment prevails, considering only open spaces).

In Germany, the *Eingriffsregelungen*, regulating the ecological compensation of new building interventions, includes not only environmental but also scenic measures for the "preservation of the aesthetic character of the landscape". On the other hand, Germany's attention is given also to open spaces within the urban landscapes which need to be preserved from soil consumption with a more "systemic" approach, incentivizing densification in specific areas. At the federal level, mapping areas not yet fragmented by major traffic arteries help establishing priority interventions for the safeguard and the restoring of ecological corridors.

Moreover, France devotes a specific action plan to the urban environment (Plan d'action urbanisme) focused on the reform of the planning law and of the financial instruments supporting a sustainable spatial development. The construction of Ecoquartiers and EcoCités and the elaboration of a specific action plan for the ecosystems multifunctionality in urban areas are other specific issues included in the NBS.

Portugal focuses on the importance of recreation services of bio- and geo-diversity landscapes safeguarded by the Regional Natural Monuments of Geological Interest.

Aesthetic and inspirational ecosystem services are put in light by the Finnish NBS, as long as 2009, since Finland has at its disposal National Urban Parks aimed to safeguard special situations of coexistence between natural and cultural values and to "protect and maintain the beauty of natural or cultural landscapes, biodiversity, historic features or other social and recreational values associated to the urban environment".

"Putting people at the heart of biodiversity policy" is a priority for the United Kingdom as people's awareness towards biodiversity is intended as the basis for any action. The prevision of a new green areas designation, empowering communities to protect local environments that are important to them, is significant.

The Conference of the Parties (COP, the governing body of the CBD which advances implementation of the Convention through the decisions taken at its periodic meeting) has established seven thematic programmes of work corresponding to some of the major biomes on the planet (agricultural, arid and sub-humid lands, forests, inland waters, marine and coastal ecosystems, mountain ecosystems, island ecosystems, protected areas). Each NBS takes into account such biomes in order to address specific conservation measures. Besides these, the comparative analysis of NBSs puts in light the need for attention towards other ecosystems and other landscapes, encompassing above all cultural landscapes (see Fig. 27.1).

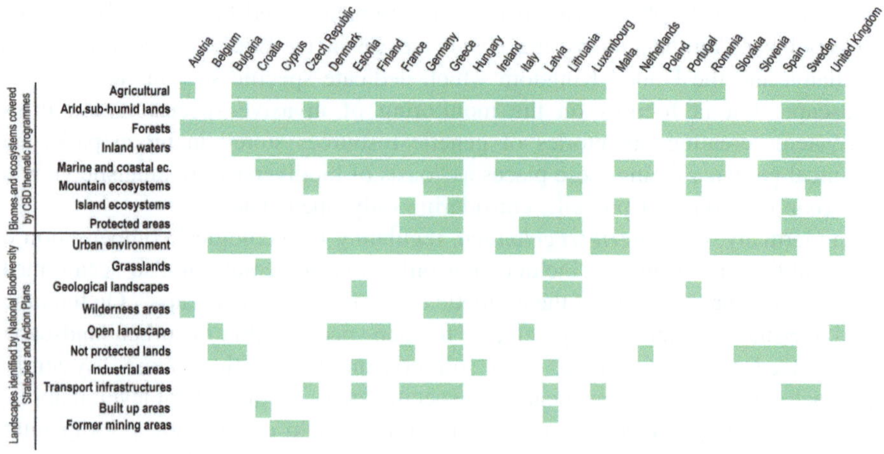

Fig. 27.1 A panorama of biomes, ecosystems and landscapes from the National Biodiversity Strategies and Action Plans of the EU member States (last update: March 2014): not all of them are covered by specific CBD thematic programmes, thus reflecting diverse national needs to spread the attention to landscapes

27.4 Conclusions

Policy integration is foreseen both by the CBD (taking part to the joint programme between UNESCO and the SCBD) and the ELC (Art. 5.d).

As a matter of fact, in the field of sustainability studies, environmental policy integration is commonly understood as balancing economic, social and environmental interests and policies in a way that trade-offs among them are minimized and synergies (or win-win-win opportunities) maximized (adapted from Berger et al. 2009). Moreover, policy integration on landscape issues influences spatial planning and having the power of jeopardizing its effectiveness (Seardo 2012b).

As shown in the discussion, coherence of biodiversity policies to ELC has been investigated, revealing a variety of approaches not always completely fitting to ELC conception of landscape. Namely, a sectorial approach to biodiversity—focused mainly on its intrinsic value–seems to be still present in NBSs. The major shift regards the assumption of a strictly ecological approach to landscape as a scale for planning and intervention thus not enhancing the interrelations with human practices and cultural ecosystem services and people's aspirations. Indeed, NBSs should establish on a strong nature conservation basis, provided that social and cultural dimensions are not excluded.

Not to be forgotten is the temporal shift among the ELC and the NBSs, a few of which dating back before the opening to ratification of the ELC and not yet updated.

At a policy level, addresses should be given for further policy harmonization in order to increase the relevance of the multidimensional conception of landscape in sustaining biocultural diversity within NBSs, namely: launching joint initiatives between the two conventions or stressing the role of landscape within the present ones such as the UNESCO-SCBD Joint programme; references to the ELC and the multidimensionality of landscape should be strengthened within the CBD COP guidance on NBSAPs (addressing both to new on-going NBSs and the updating ones).

The outputs of the research regard exclusively the NBSs and cannot be transferred to a general judgment on the complex policy framework of each State on the biodiversity issues. Thus the research should not be considered fully satisfactory until other aspects will be investigated: the presence of specific national policies addressed to landscape, the implementation of biodiversity–landscape programmes or the existence of integrated funded projects (Waldron et al. 2013 have drawn a picture of the current national expenditure), but also relevant immaterial aspects should be investigated such as the possible fear of watering down the conservation efforts by introducing into biodiversity policies a broader set of cultural values attached to biodiversity.

All of these aspects are not matter of this contribution, but should be collected in order to integrate and improve the present results.

References

Antrop M (2006) From holistic landscape synthesis to transdisciplinary landscape management. In: Tress B, Tres G, Fry G, Opdam P (eds) From landscape research to landscape planning. Springer, Dordrecht, pp 27–50

Beatley T (1995) Planning and sustainability: the elements of a new (improved?) paradigm. J Plan Lit 9:383–395

Berger G, Steurer R (2009) Environmental policy integration and sustainable development: conceptual remarks and governance examples. European Sustainable Development Network—ESDN Quarterly Reports, June 2009

Cassatella C (2014) Linking public participation and landscape planning: the experience of the landscape observatory of AMI (Italy) at a sub-regional scale, people and landscape III. European Landscape Convention in Practice. Lednice na Morave. Czech Republic. 29–30 May 2014

Cassatella C, Seardo BM (2014) In search for multifunctionality: the contribution of scenic landscape assessment. In: Rega C (ed) Landscape planning and rural development. Key issues and options towards integration. Springer, Berlin, pp 41–60

Cevasco R (2007) Memoria verde. Nuovi spazi per la geografia, Edizioni Diabasis, (RE) Italy

Chan KMA, Shaw RM, Underwood EC, Daily GC (2006) Conservation planning for ecosystem services. PLoS Biol 4(11) e379:2138–2152

Conference of the Parties to the Convention on Biological Diversity (COP) (2000) Report of the fifth meeting of the conference of the parties to the convention on biological diversity, Decision V/16, Article 8(j) and related provisions, Fifth meeting, Nairobi, 15–26 May

Council of Europe (2000) European landscape convention. Treaty Series n. 176, Florence

Daniel TC (2001) Whiter scenic beauty? Visual landscape quality assessment in the 21st century. Landscape Urban Plan 54:267–281

Forman RTT (1995) Land mosaics. Cambridge University Press, New York

Ingegnoli V (1999) Ecologia del paesaggio. In: Massa R, Ingnegnoli V (eds) Biodiversità, estinzione e conservazione: fondamenti di conservazione biologica. UTET, Torino, pp 189–210
IUCN (2008) Resolution n. 4.099. Recognition of the diversity of concepts and values of nature, In: The World Conservation Congress, 4th World Congress, Barcelona, 5–14 Oct 2008
Maffi L (2008) Biocultural diversity and sustainability. In: Jules P, Andy B, Julia G, David O, Max P, Hugh W (eds) The Sage handbook of environment and society. Sage Publications Ltd, London, pp 267–277
Millennium Ecosystem Assessment (MEA) (2005) Ecosystems and human well-being: synthesis. Island Press, Washington DC
Seardo BM (2012a) The implementation of a multifunctional-landscape project: the emblematic case of Hanover-Kronsberg. Urbanistica 148:76–79
Seardo BM (2012b) Conditions favouring landscape multifunctionality in German landscape planning system. Urbanistica 148:79–82
Tudor C (2014) An approach to landscape character assessment, Natural England
UNESCO (1972) Convention concerning the protection of the world cultural and natural heritage. In: Adopted by the general conference at its seventeenth session, Paris, 16 Nov 1972
United Nations (1992) Convention on biological diversity
van Asschea A, Djanibekov N (2012) Spatial planning as policy integration: the need for an evolutionary perspective. Land Use Policy 29:179–186
Waldron A, Mooers AO, Miller DC, Nibbelink N, Redding D, Kuhn TS, Roberts JT, Gittleman JL (2013) Targeting global conservation funding to limit immediate biodiversity declines. PNAS 110(29):12144–12148

National Biodiversity and Action Plans Cited in This Paper

Cabinet of Germany (2007) National strategy on biological diversity
Council of Ministers of the Republic of Bulgaria (1999) National Biodiversity Conservation Plan, Sofia
DEFRA, Department for Environment, Food and Rural Affairs (2011) Biodiversity 2020: a strategy for England's wildlife and ecosystem services, London
Department of Arts, Heritage and the Gaeltacht (?) Actions for biodiversity 2011–2016. Ireland's national biodiversity plan
Estonian Ministry of the Environment, United Nations Environmental Program (UNEP), Environmental Protection Institute of the Estonian Agricultural University (1999) Estonian Biodiversity Strategy and Action Plan, Tallin-Tartu
Finnish Government (2012) Government resolution on the strategy for the conservation and sustainable use of biodiversity in Finland for the years 2012–2020, 'Saving Nature for People'
Latvia (2003) National Programme on Biological Diversity
Ministère de l'Écologie, du Développement durable, des Transports et du Logement (2004) Strategie nationale pour la biodiversité – National biodiversity strategy 2011–2020
Ministrstvo za okolje in prostor Republike Slovenije, Ministry of the Environment and Spatial Planning (2002) Biodiversity conservation strategy of Slovenia
Ministry of the Environment of the Czech Republic (2005) National biodiversity strategy of the Czech Republic
Österreich, Umweltbundesamt (2005) Weiterentwickelte österreichische Strategie zur Umsetzung des Übereinkommens über die biologische Vielfalt
Poland (2005) The national strategy for the conservation and sustainable use of biodiversity
Portugal Ministerio di ambiente e do ordenamento do territori (2001) Estratdgia Nacional de Conservação da Natureza e da Biodiversidade
Republic of Lituania, Environmental Protection Ministry (1996) Biodiversity conservation. Strategy and action plan

Chapter 28
The Nexus Between Landscape Elements and Traditional Practices for Cultural Landscape Management

Mateja Šmid Hribar and Mimi Urbanc

Abstract Landscape diversity consisted of heterogeneous landscape elements is largely dependent on human activities such as traditional practices and knowledge related to land use which could be recognized as a contribution to cultural diversity. Losing traditional practices may result in impoverishing of landscape and biological diversity. We present cases illustrating connections between certain landscape elements and traditional practices typical for the cultural landscape of Ljubljansko barje (Ljubljana Marshes) in Slovenia. The study was carried out on the selected case study sites during 2012–2013 using study visits and interviews with locals. The aim of identifying these connections was to foster synergies between management of cultural landscape, traditional practices and modern way of living. However, against expectations, the study revealed that in the Ljubljansko barje area not many such practices and knowledge remain. The most useful practices that help to sustain extensive meadows and tall-herb communities are horse breeding and late mowing, and the local knowledge concerning agricultural and building land safe against floods. Moreover, we found important the fact that the first 'victims' of modern farming are particularly those landscape elements that are the result of a considerably lower level of technological development.

Keywords Landscape element · Traditional practices · Diversity · Cultural landscape · Slovenia · Biocultural diversity

M. Šmid Hribar (✉) · M. Urbanc
Anton Melik Geographical Institute, Scientific Research Centre of the Slovenian Academy of Sciences and Arts, Gosposka ulica 13 SI-1000, Ljubljana, Slovenia
e-mail: mateja.smid@zrc-sazu.si

M. Urbanc
e-mail: mimi.urbanc@zrc-sazu.si

M. Agnoletti and F. Emanueli (eds.), *Biocultural Diversity in Europe*,
Environmental History 5, DOI 10.1007/978-3-319-26315-1_28

28.1 Introduction

A cultural landscape is a combination of human activities in the natural environment and is created across time and space. The basic characteristic of the landscape is its dynamics, since it changes constantly despite its apparent permanence and solidity. Landscape does not comprise only tangible (material) elements, but also the intangible (mental, imagined) part. As such it is recognized as ecological as well as socio-economic system at the same time. Cultural landscapes are significantly characterized by this duality. On the one hand, they are most profoundly influenced by their elements and their functions, and by changes. While on the other, they reflect social structure, cultural tradition, economic activities and political models which play a key role in shaping the landscape. The cultural landscape teaches us about the relationships between man and nature, and reflects their interdependence. In the past, people continuously adapted to the natural conditions of an area, and their landscape management was more sustainable than today, albeit due to technical constraints and technological limitations. Various landscape elements were generated due to this adapted land use and enhanced landscape diversity increased, which was reflected in a mosaic landscape. The variety of human activity had a direct impact on both landscape diversity and biodiversity. Landscapes with distinctive internal diversity allow for a wide range of services and products that in both material (food, crops, building materials) and immaterial aspects (recreational, aesthetic, symbolic, spiritual and educational) enrich the life of their inhabitants. The latter is often taken for granted; nevertheless, the decline of so-called cultural ecosystem services coincides with the decline in landscape diversity as a result of increasingly intensive and mechanized land use. In this regard, during the recent decades contemporary landscape management has seen the need to attribute greater importance on those practices and knowledge that evolved on the basis of experience and observations. In the literature, they are referred to as indigenous knowledge, traditional ecological knowledge (Johannes 1989; Berkes et al. 1995), traditional practices (The Convention on … 1992; Agenda 21 1992), knowledge and practices concerning nature and the universe (Intangible heritage 2014), and similar. Losing traditional practices may result in impoverishing of landscape and, furthermore, biological diversity.

This paper aims to present the practices that we identified in the Ljubljansko barje Landscape Park area, Slovenia (135 km^2, close proximity to the capital city, Ljubljana), using study visits and interviews with locals during 2012–2013, as well as draw attention to their role in this rural landscape. We were interested in the practices and knowledge that had an impact on the formation of the present landscape, and the practices and knowledge that are nowadays still known among the local inhabitants and experts, including in terms of their impact on contemporary processes in the landscape. The impacts of policies cannot be avoided completely in such treatment—in this particular case those of agricultural policies. Furthermore, we questioned how these practices and knowledge, both 'traditional' and contemporary, were linked with the existing landscape elements. The aim of identifying these connections was to foster synergies between management of cultural landscape, local knowledge and modern way of living.

28.2 Background and Role of Traditional Practices and Knowledge

In 1992, by adopting the Convention on Biological Diversity and Agenda 21, traditional practices and knowledge relating to the conservation of biodiversity and landscape diversity became part of professional discourse. The Convention on Biological Diversity (1992) spotlights the role of knowledge, innovation and practices of indigenous and local communities with their traditional ways of life in the conservation of biodiversity (article 8j). Agenda 21 (1992) mentions the traditional, local practices and the traditional knowledge of indigenous inhabitants in various contexts, essentially providing the understanding that these practices and knowledge developed on the basis of long-standing relationships between the locals and their surroundings, and the formal recognition of their contribution to more sustainable management of land use and land resources and their positive impact on the environment. A working group on Traditional Ecological Knowledge was established at IUCN which introduced a collection of essays on traditional ecological knowledge (Johannes 1989). Berkes et al. (1995, 282) defined 'traditional ecological knowledge' as "*cumulative body of knowledge, practice and belief, evolving by adaptive processes and handed down through generations by cultural transmission, about the relationship of living beings (including humans) with one another and with their environment.*" Traditional ecological knowledge was understood as a complementary way of knowing abstract scientific ecology that is accumulated in the form of resource use practices and preserved particularly in nonindustrial or less technologically advanced societies. According to the authors, adaptive management is essential for the conservation of biodiversity and ecological systems. Antrop (2005) stresses the importance of traditional practices and knowledge in landscape management. According to Antrop, traditional knowledge related to land management and land use is of particular assistance in this regard, since it reflects ancestral knowledge about natural landscape conditions and ancestral relationship towards the landscape. The studying of these practices may reveal the beneficial elements that these individual practices might offer, and help to find the ways to include these elements in modern cultural landscapes.

FAO[1] understands traditional practices as an important part of conservation and provision of self-sufficiency in food, hence connecting food, agriculture and ecosystems (FAO and Traditional … 2009). As such, they are a fundamental contribution to the world's natural and cultural heritage, as well as to biodiversity in rural areas. In the extensive publication, *Gardens of Biodiversity* Batello et al.

[1]Food and Agriculture Organization of the United Nations.

(2010) stressed the importance of traditional practices that had proven effective in the past, particularly on the case of a detailed study of local agro-ecological and social conditions in the Southern Caucasus. This involves various practices of food production and processing, where gardens, land cultivation, animals raising, pasture grazing, use of wild plants and wildlife occupy an important position. Such practices generally require few external energy inputs (in the form of mechanization, chemical fertilizers, etc.) and contribute to food provision and resource management, as well as the conservation of local genetic resources and traditional landscapes. In this sense, the knowledge and application of such practices in contemporary landscape management contributes to their more sustainable development. An important position within traditional practices is occupied by man-made adaptations to the specific natural situation. These adaptations were mostly based on observations and experience, which were handed down from generation to generation as inherited knowledge, while nowadays they are recognized—as part of the living heritage—as economic and environmental knowledge. The domains of the Convention for the Safeguarding of the Intangible Cultural Heritage (2008) include knowledge and practices concerning nature and the universe, which include 'knowledge, know-how, skills, practices and representations developed by communities by interacting with the natural environment' (Intangible Heritage 2014). In Slovenia, according to the Rules on the Registry of Types of Heritage and Protection Guidelines (2010, Article 7), the knowledge about nature and the environment is defined as "traditions of knowledge about flora and fauna, weather, water, territory and space," while economic knowledge and skills are defined as "knowledge and skills of handicraft, craft-trade, trade and transport and knowledge and work practices of foraging, hunting and agriculture (e.g. arable farming, livestock farming, viticulture, apiculture), forestry, mining and industry." This knowledge and the practices resulting from it connect the tangible and the intangible heritage.

Recently, agroforestry has reclaimed its position among traditional practices. In the past, it was wide spread and had an important effect on the development of agro-silvo-pastoral landscapes. In agroforestry practices, which include traditional and modern land use systems, the inclusion of trees with crops and/or animal production systems in agricultural settings plays a crucial role (FAO 2013). Despite many advantages and benefits of agroforestry (enhancing ecosystems by storing carbon, preventing deforestation, increasing biodiversity, protecting water resources and reducing erosion), its practices are not included enough in contemporary rural landscape management, nor in agricultural policy. Indeed, the efficiency of practical application will be subject to successful inter-sectoral participation (particularly between agricultural, forest and environmental sectors), and strongly influenced by socio-economic relations.

The Paris Declaration on the 'Satoyama Initiative' signed in 2010 deals with 'socio-ecological production landscapes' which "'*are dynamic mosaics of habitats and land uses that have been shaped over the years by the interactions between people and nature in ways that maintain biodiversity and provide humans with goods and services needed for their well-being*" (Paris declaration on … 2010, article 1). Within this declaration seascapes are also included. As part of the Satoyama Initiative an internet-based portal was established, which among other information provides well-presented case studies of socio-ecological production landscapes around the globe (IPSI 2014). The Satoyama Initiative introduces the three-fold approach consisted of (1) consolidation of wisdom on securing diverse ecosystem services and values, (2) integration of traditional knowledge and modern science and (3) exploring new forms of co-management systems (new commons) (Takeuchi 2014).

Diacon-Bolli et al. (2012), who studied calcareous grasslands in agricultural areas of Switzerland, found that former practices created a heterogeneous landscape with a patch mosaic of various land uses, many transitional elements and various succession stages, which contributed to the overall biodiversity. Agnoletti (2013) draws attention to two types of biodiversity. The first one concerns natural habitats and wild flora and fauna, while the second one is the result of the long-standing relationship between man and nature, where, indeed, traditional practices play a key role. Over the past decade in Europe, the interest has grown considerably in traditional ecological practices that are likely to make a significant contribution to management and environmental policy. Such studies, however, are rare and mostly relate to transhumant grazing (Hoberg et al. 2012; Oteros-Rozas et al. 2013). An important step towards the recognition of interdependence of both types of diversity (i.e. biodiversity and cultural diversity) was the implementation of the UNESCO-SCBD programme (Joint Programme between UNESCO and the Secretariat of the Convention on Biological Diversity on the links between biological and cultural diversity) and the Florence Declaration on the Links between Biological and Cultural Diversity, adopted in April 2014, under which the concept of *biocultural diversity* was recognized and exposed.

The intangible sphere, in the form of various cultural practices affected by thinking, values and ideas, is manifested in the material landscape. Any landscape is composed not only of what lies before our eyes but what lies within our heads (Whyte 2002). The landscape is the external world as conveyed by human experience (Whyte 2002). The abandonment of traditional practices and knowledge in land use and management of natural resources, as a result of the response to subjective and objective factors, leads to landscape changes that are often manifested as the loss of landscape diversity and the decoupling of socio-natural interaction. Rotherham (2013) stresses that we need to recognize the 'Eco-Cultural' nature of landscapes in order to establish social and economic links between nature, landscape and ecology. This will require a political shift in thinking and planning and a paradigm shift in conservation and environmentalism.

28.3 Practices and Knowledge in the Ljubljansko Barje Landscape Park

The practices and knowledge concerning management of land use and land resources were studied in Ljubljansko barje (the Ljubljana Marshes) in the period 2012–2013. Our assumption was that the intangible sphere becomes tangible, i.e. the material landscape, through various practices. The existing cultural landscape is either preserved or modified due to changed natural conditions, new knowledge and technologies. Through interviews among local inhabitants and experts, we assumed that we would find such practices and old knowledge concerning land management and land use that reflect the relationship of their ancestors to the natural conditions of the landscape. Based on the interviews, we studied whether the practices had an impact on landscape structure and, if so, which ones and how strongly they were connected. The knowledge about such relationships will be helpful in future landscape management, as they might reveal the beneficial elements that these individual practices generate, and help to find ways of including these practices in modern cultural landscapes. We were particularly interested in man-made adaptations to the natural conditions of a given landscape. These were mostly based on observation and experience handed down from generation to generation as inherited knowledge.

28.3.1 Ljubljansko Barje Landscape Park

Ljubljansko barje is situated in the southern part of the Ljubljana Basin, near Slovenia's capital city, Ljubljana (Fig. 28.1). This is one of the Slovenian landscapes that has been most heavily changed. The greatest landscape changes occurred at the end of the eighteenth and the beginning of the nineteenth century, when extensive drainage was carried out in order to obtain new farmland. The plan was to make the Ljubljana Marsh, the breadbasket of the Habsburg Monarchy. In addition, the newly acquired areas were planned to be settled because up until then only the sides of isolated hills (e.g. the village of Bevke) and the marsh edges were inhabited. The drainage works were finished in 1829, after which colonization of the area followed. However, the marsh has never been fully drained because floods continue to be common in this area (the last major flood occurred in September 2010). During the 1820s, a road was built across the drained marsh, followed by a railroad in 1857. A major change to the landscape was also caused by intense peat extraction, which lowered the surface by several metres in many areas and further increased the risk of floods. The most important driving forces that shape the landscape today include intensive farming, nature protection and urbanization due to the proximity of Ljubljana, which continues to spread onto the marsh despite its great flood risk. In 2008, a large part of this area was declared as the Ljubljansko barje Landscape Park, and in 2011 the remnants of two pile dwellings were added

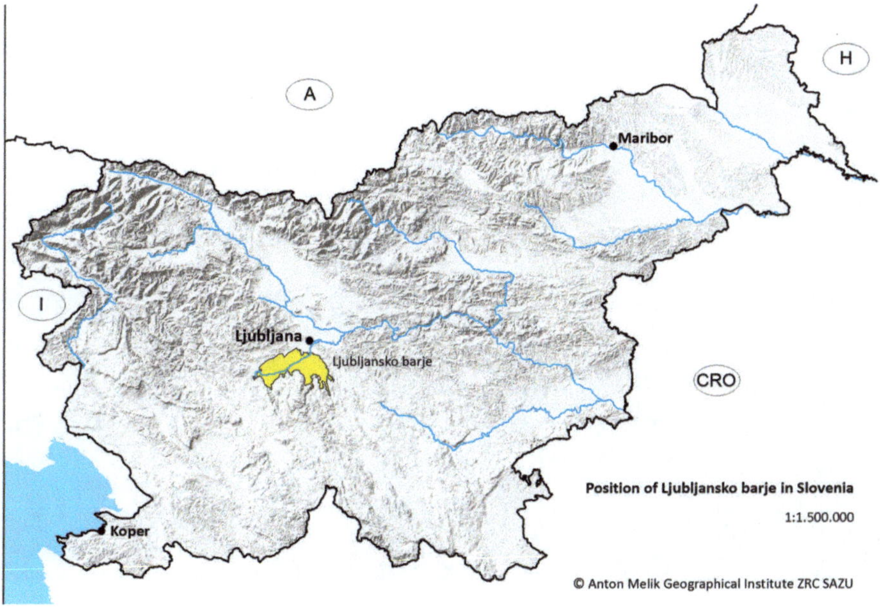

Fig. 28.1 Location of Ljubljansko barje in Slovenia

to the UNESCO Cultural Heritage List as part of the serial nomination "Prehistoric Pile Dwellings around the Alps."

28.3.2 Practices and Knowledge on Management of Land Use and Land Resources

In studying these practices, we did not focus only on traditional practices, but we also wanted to know about modern practices and knowledge in relation to spatial management, hence those which establish and affect landscape elements. The practices and knowledge were identified by interviewing experts and local inhabitants in selected case study sites (two older settlement areas—Iška Loka and Bevke, and one new settlement area in drained marshes—Črna vas). We identified 23 practices, of which four are former practices and are no longer used. Past practices had a significant impact on the appearance of the landscape nowadays: *initial drainage*, *colonizatio*n and *peat harvesting*. The initial drainage works introduced the construction of many canals and river engineering works within the landscape in grid-like patterns. The once extensive raised bog with wet meadows and pastures has been replaced by a mosaic landscape composed of meadows, cultivated fields, numerous channels and hedges along them, tall-herb communities, shrubs and thick bush, as shown in Fig. 28.2.

Fig. 28.2 Mosaic landscape of the Ljubljansko barje Landscape Park (*Source* Viktor Šmid)

Fig. 28.3 Hedges in Črna vas (*Source* Mateja Šmid Hribar)

Once drained, the land was settled, thus leading to the establishment of built-up land, additional canals and hedges along plot boundaries, roads and paths, ornamental and vegetable gardens, orchards, meadows, pastures and arable land. Concerning the appearance of the landscape today, the planting of hedges along the canals was particularly important (Fig. 28.3). Valuable trees were planted, particularly alder (to a smaller degree ash and oak trees), which were used as building materials (piling). In this way, distinctive hedges were introduced along ditches, a symbol of the present Ljubljansko barje landscape, which are, however, disappearing as they obstruct mowing. For many farmers in Črna vas, hedges were the only source of firewood, as there were no forests on the dried and newly acquired land. Shortly after the land was settled, peat harvesting began, which lasted only a few decades, but left an indelible mark on the landscape. This practice led to the irreversible disappearance of raised bog from the landscape, while the few peatlands that remained are small and fragmented, usually seen as thick bush and groves. The concept of harvesting grass seed should be mentioned among the former practices that might be of interest in the future. This refers to the collection of seed for restoring grasslands. With regard to the seeds for medium-intensive meadows, the growing of fodder crops appropriate for cattle is important, and with extensive meadows it is beneficial for biodiversity and genetic diversity. The essence of the practice is to leave the appropriate extensive meadow grow until the seed production stage, followed by sowing the harvested seeds in destroyed areas, thus restoring the meadows. Seed harvesting used to be one of the means of survival, while today the economic profitability of the practice is questionable, but it certainly contributes to the conservation of biodiversity.

Similar to former practices, the existing practices and knowledge differ in the level of complexity and impacts on landscape elements (Fig. 28.4). They are mostly related to livestock farming and arable farming, land drainage and the choice of appropriate building land and agricultural land in older settlement areas (Iška Loka and Bevke).

The Ljubljansko barje cultural landscape is maintained by agriculture. *Cattle raising* is nowadays Ljubljansko barje's most intensive practice, i.e. importantly contributing to arable land (particularly corn fields), intensive meadows, medium-intensive meadows, pastures and canals. *Pasture grazing* is increasingly pursued; however, it mostly involves intensive grazing, and the excessively high grazing pressure can lead to overgrazing and land degradation, as pointed out by Miličić et al. (2011). Cattle production is strongly linked with the measures relating to the EU Common Agricultural Policy, which allows for indirect and direct payments made to farmers for cultivation of land. While direct payments are not subject to the type of crop or breeding, the eligibility for indirect payments is limited to specific cases and specific conditions. The local inhabitants believe that direct payments are necessary for the existence of farms, as otherwise farming would not be economically viable. This is important, because the abandonment of agriculture in Ljubljansko barje would lead to overgrowth by non-native plant species, shrubs and thick bush, thus leading to a loss of landscape diversity.

To avoid overgrowth, *grass mowing i*s particularly relevant, which establish various types of meadows and tall-herb communities. For conservation of biodiversity, *late mowing* is important; however, it has been revealed that in the past this was not the result of a special know-how, but rather of limitations on technology and human resources. A mosaic of differently mowed meadows was created, providing the habitat to various animals, particularly endangered species of birds, such

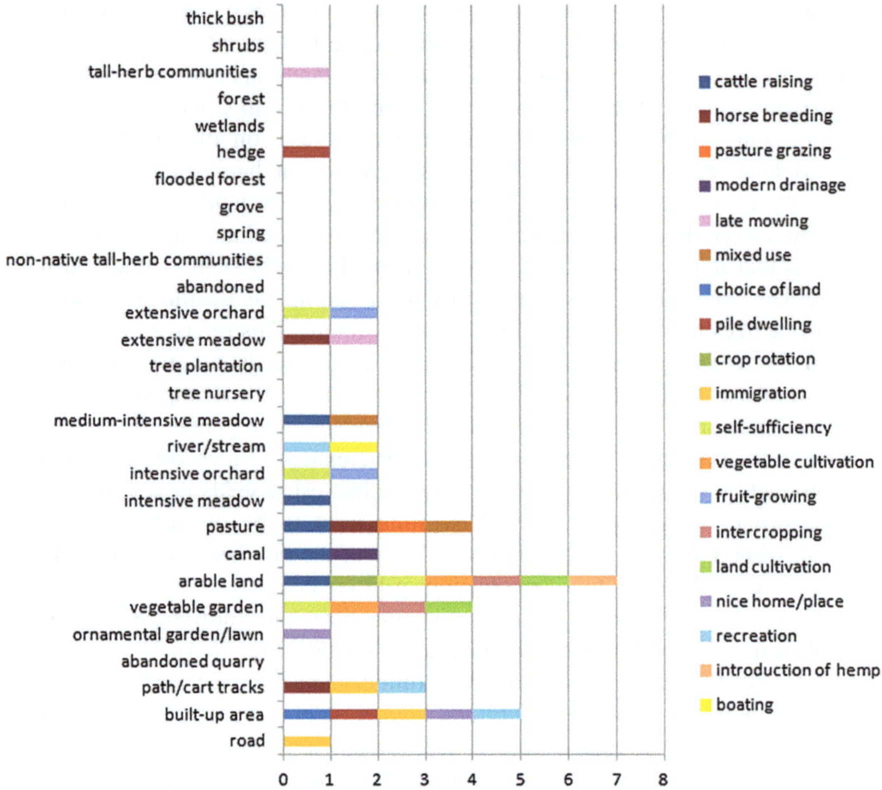

Fig. 28.4 Identified cultural practices and their integration with landscape elements in Ljubljansko barje

as the corncrake (*Crex crex*). Late mowing is studied at DOPPS—BirdLife Slovenia (Jančar 2013) where they find that to preserve extensive meadows and tall-herb communities, the selected areas have to be mown alternately—while some areas in the first year are mown after 1st August, others are mown several times; the regime switches in the second year, so that each meadow is ultimately revitalized. However, in the age of technological development, late mowing depends on agricultural policy and indirect payments made for cultivation of the land that would otherwise not be cultivated to grow something that is not needed. Late-mown extensive meadows (after 1 August) and even tall-herb communities are almost redundant for modern farmers who no longer use litter. However, in Slovenia such payments are relatively modest and farmers are constantly weighing up whether to apply for direct payments or employ other more profitable options, such as earlier mowing for horse fodder. One of the key challengers regarding the preservation of extensive meadows and tall-grown communities will be to find a sustainable practice that will include greater quantities of unused grasses or less valuable hay.

In the sense of sustainable landscape management, the following practices and knowledge are of interest in Ljubljansko barje: *horse breeding*, *self-sufficiency farming*, *intercropping*, *fruit growing*, *mixed use* and the *choice of agricultural land and building* land. *Horse breeding*, i.e. the raising of horses for food and partially for tourism (horse-riding or horse-drawn carriage rides through Ljubljansko barje) is important in Ljubljansko barje, because horses, contrary to cattle, are not susceptible to common horsetail (*Equisetum arvense*) and therefore do not need specially cultivated fodder, thus contributing to the preservation of extensive meadows. *Self-sufficiency farming,* which provides fewer, but more diverse crops (e.g. cereals, potatoes, root crops, fruit and similar, but not energy self-sufficiency) used to be a self-evident practice; nowadays, it is being reintroduced and results in the increase of vegetable gardens and intensive orchards. It contributes to more diverse and less intensive food production. It affects the abundance of crops in fields and vegetable gardens, and as such it is closely linked with *vegetable cultivation*, *intercropping* and *fruit growing*. Traditionally, *vegetable cultivation* had a strong presence and was an important source of income due to the proximity of the capital city. Ljubljansko barje's beans were a sought-after product. Besides, nowadays courgettes, cucumbers and cabbages are also grown. In the future, the practice could expand, since food does not have to be additionally processed for transport due to the proximity of the market (Ljubljana). However, younger generations show no particular interest in this regard. *Intercropping* is the next traditionally established practice, which is, however, rarely used today. This refers to a special way of growing specific types of vegetables, which is also adapted to floods: bean vines climbing the stalks of corn, while sometimes squash covers the ground under. Despite the fact that due to winds and frequent frost, the area is not favourable for *fruit growing*, in Bevke and Črna vas many intensive orchards flourish next to houses. The former extensive orchards, which used to be an important landscape elements, are disappearing (Iška Loka) or they are overgrown in thick bush and groves (Bevke). In Bevke, we observed *mixed use* land, which is first mown, and from July on turned into pastures. In older settlement areas, the knowledge about *the choice of agricultural land and building land adapted to natural conditions* is very important, and still present. This knowledge is taken into consideration the most in Iška Loka where the rule was to restrict any new house to the lowest quality land available, while still being safe from flooding. The built-up area continuity where the alluvial fan meets the marshes helped to both preserve the fertile land on the alluvial fan, as well as to protect the built environment against floods. Only cornfields, which are not affected by floods, except in early spring, are located in Barje's floodplains. In Bevke, the knowledge was still present, but it has been widely disregarded and much of the most fertile land along isolated hills has been built on. In Črna vas, where there are no higher elevation gravel alluvia or isolated hills, residents face flood-related problems on a yearly basis. Regarding newer settlements, we found no knowledge about adaptation to mitigate floods. This is partially the result of unfavourable natural conditions, and partially due to the quickly changing micro area, making it difficult for the population to develop sufficient knowledge.

Recreation is among the recently introduced practices that has already greatly affected the development of the landscape. It is uncontrollable and could lead to a conflict regarding the use of paths and cart tracks which are presently maintained by farmers. In Bevke in Črna vas, tourism is becoming an important driver of development, and it has the potential to grow into one or several practices. Nowadays, from the case study sites there are nice views of the Mali plac wetlands in Bevke, while tourism is developed in Črna vas.

Field visits to the case study sites suggested that arable land associated with as much as seven practices assured the greatest chance of existence (Fig. 28.4). Built-up areas are associated with five identified practices; with four vegetable gardens and pastures; with three paths; with two extensive and medium-intensive meadows; extensive and intensive orchards, canals and a river or a stream. Roads, ornamental gardens/lawns, intensive meadows, hedges and tall-herb communities are related to one practice only. Extensive meadows, tall-herb communities and hedges are endangered the most; indeed, the practice of pile dwellings has been modernized, and nowadays there is no need for alder piles, which were, in fact, the reason that the hedges were planted in the first place. Recently, some hedges were cut, and during our site surveys we did not see any new ones planted.

28.4 Discussion and Conclusions

The search for practices and old knowledge on landscape management and management of natural resources helping to preserve the individual landscape elements revealed, against expectations, that in the Ljubljansko barje area not many such practices and knowledge remain. The most useful practices for sustaining extensive meadows and tall-herb communities are horse breeding and late mowing, and the knowledge concerning how to sure up agricultural and building land safe against floods, which was not always taken into consideration. It should be stressed that due to up-to-date machinery, which allows for much quicker mowing, and the use of mineral fertilizers that yield more cuts per season, late mowing is exclusively a factor of financial measures, rather than of land use knowledge and adjustments to natural conditions. Understandably, farmers consider land management as a business generating profit, not only from the perspective of ensuring landscape diversity.

Practices such as self-sufficiency farming, vegetable cultivation, intercropping and fruit growing are associated with the vegetable garden, field and orchard, while some of them, e.g. intercropping, are disappearing. These aforementioned practices contribute to the preservation of biodiversity, but primarily in terms of species, rather than on the landscape level. Nevertheless, these practices are important because they connect people with nature and contribute to food self-sufficiency, even though this takes place within the small scope provided by vegetable gardens. From the viewpoint of identified practices concerning the vegetable garden and arable land, we agree with Van der Stege et al. (2012, 279) who emphasize that vegetable gardens ‘create highly resilient small agroecosystems that are nested in larger agroecosystems’.

Seed harvesting is a practice that has seen a growing interest in Europe. The reason for the lack of this and other practices and knowledge that could importantly contribute to landscape diversity may be that Ljubljansko barje is one of the most anthropogenically impacted rural landscapes in Slovenia, strongly influenced by the proximity of the Ljubljana capital city. Landscape diversity in Ljubljansko barje, which due to its location on a plain is beneficial to agriculture, is endangered mostly by intensification and urban development. In the case of two communities in Patagonia, Eyssartier et al. (2013) showed that the community with less access to market economy had greater richness in practices of cultivation and gathering of species, and they collected more seeds than the community that had stronger ties with urban centres. The other reason is probably deagrarization. Land is the source of survival only to a small proportion of people; most people commute to Ljubljana daily, while land accumulates in the hands of the farmers whose farming practices are increasingly intensive. Land is transferred from the hands of small farmers, to be rented, rather than owned, by larger farmers, who follow the logic of economy and make use of contemporary agro-technical achievements. The first ‘victims’ of modern farming are particularly those landscape elements that are the result of a completely different way of life and a considerably lower level of technological development. Hedges are such example, enhancing the mosaic nature of the landscape and its diversity, and are also an element of identity in Črna vas. In the sense of modern mechanization, however, they are seen as an obstacle.

A change or abandonment of a certain use or practice has a consequential impact on the change of landscape elements, and, hence, landscape diversity. Consequently landscape is becoming more and more monofunctional, providing less common goods and ecosystem services for the public but more benefits for the owners. Due to intensified farming enabled though modern technology, the traditional land use practices and subsequently the mosaic nature of the landscape have been disappearing. The meadows are turning into cultivated fields, the channels are either dredged, which causes even more intense draining, or abandoned, which results in overgrowth.

Site protection, i.e. the establishment of a landscape park, is beneficial for the preservation of a specific level of landscape diversity, allowing for acquisition of further indirect payments for land management under specific conditions. However, it should be noted that nature protection is not identical to landscape protection, as also pointed out by Agnoletti (2013) who observed that the two concepts can, indeed, be mutually exclusive. Nature protection can give an advantage to the protection of processes, where certain landscape elements associated with specific uses are lost, e.g. hedges, forest pastures and similar. In this scheme, these elements are protected only when they also provide a habitat for endangered species, otherwise they are left to natural succession. However, the overgrowing of hedges into thick bush is not positive in terms of the landscape, even though thick bush have a large value for the ecosystem. With the loss of hedges we lose identity, practices and knowledge concerning a specific land use, aesthetic value, as well as some other services, such as wood for construction. Hence, we propose, with regard to protection of specific landscape elements, that the agricultural policy starts to include and subsidize the practices that will, with due consideration of the natural conditions of the landscape, help create more sustainable landscape elements, which will also provide certain economic benefits to the owners, and offer cultural ecosystem services to others. With modern technologies and traditional knowledge on benefits and ecosystem services of individual landscape elements, selected elements could be maintained in a modern way with sustainable modern technology. This would significantly contribute to preserve traditional landscape elements and as such preserve landscape- and biodiversity. If it is more or less clear that farmers will sustain and establish specific landscape elements almost exclusively on the basis of subsidies, then the question remains whom to include, besides the farmers, in the preservation of landscape diversity in the future, and in what way. Based on various initiatives and papers (e.g. Satoyama initiative; Takeuchi 2014), possibilities lie in the new approaches to the management of the common good, a sort of 'new commons' or new forms of co-management adapted to the needs of modern society. Finally, more studies are needed to further identify and preserve traditional practices and knowledge, which will play a crucial role in the sustainable management of ordinary rural landscapes across Slovenia and Europe.

Acknowledgement This study was conducted as part of PhD studies of Mateja Šmid Hribar, under mentorship of Mimi Urbanc, PhD, and Valentina Brečko Grubar, PhD. We also gratefully acknowledge the help of local people and experts who, by sharing their knowledge, made this study possible.

References

Agenda 21 (1992) United Nations, Rio de Janeiro. http://sustainabledevelopment.un.org/content/documents/Agenda21.pdf Accessed 3 Oct 2014

Agnoletti M (2013) Valorising the European rural landscape: the case of the Italian national register of historical rural landscapes. In: Rotherham ID (ed) Cultural severance and the environment. Springer, London

Antrop M (2005) Why landscapes of the past are important for the future. Landscape Urban Plan 70:21–34

Batello C, Avanzato D, Akparov Z, Kartvelishvili T, Melikyan A (2010) Gardens of biodiversity. Conservation of genetic resources and their use in traditional food production systems by small farmers of the Southern Caucasus. http://www.fao.org/docrep/014/i1687e/i1687e.pdf Accessed 10 Oct 2014

Berkes F, Folke C, Gadgil M (1995) Traditional ecological knowledge, biodiversity, resilience and sustainability. In: Perrings CA, Mäler KG, Folke C, Jansson BO, Holling CS (eds) Biodiversity conservation. Kluwer Academic Publishers, Dordrecht

Convention for the Safeguarding of the Intangible Cultural Heritage (2003) UNESCO, Paris. http://unesdoc.unesco.org/images/0013/001325/132540e.pdf Accessed 14 Nov 2012

Diacon-Bolli J, Dalang T, Holderegger R, Bürgi M (2012) Heterogeneity fosters biodiversity: linking history and ecology of dry calcareous grasslands. Basic Appl Ecol 13:641–653

Eyssartier C, Ladio AH, Lozada M (2013) Traditional horticultural and gathering practices in two semi-rural populations of Northwestern Patagonia. J Arid Environ 97:18–25

FAO and Traditional Knowledge: The Linkages with Sustainability, Food Security and Climate Change Impacts (2009) Food and Agriculture Organization of the United Nations, Rome. http://www.fao.org/docrep/011/i0841e/i0841e00.htm. Accessed 14 Oct 2014

FAO (2013) Advancing Agroforestry on the Policy Agenda: A guide for decision-makers, by Buttoud G, in collaboration with Ajayi O, Detlefsen G, Place F, Torquebiau E. Agroforestry Working Paper no. 1. Food and Agriculture Organization of the United Nations. FAO, Rome. http://www.fao.org/docrep/017/i3182e/i3182e00.pdf. Accessed 14 Oct 2014

Florence Declaration on the Links Between Biological and Cultural Diversity (2014) http://landscapeunifi.it/images/pdf/UNESCO-CBD_JP_Florence_Declaration.pdf

Hoberg J, Oteros-Rozas E, Plieninger T (2012) Traditional ecological knowledge in Europe: status quo and missing links of empirical evidence. SULiNews 2 https://www.iucn.org/about/union/commissions/sustainable_use_and_livelihoods_specialist_group/sulinews/issue_2/sn2_tek/. Accessed 21 Oct 2014

Intangible Heritage (2014) http://www.unesco.org/culture/ich/?pg=56. Accessed 14 Oct 2014

International Partnership for the Satoyama Initiative (IPSI) (2014) http://satoyama-initiative.org/ Accessed 25 October 2014

Jančar T (2013) Mowing in Ljubljansko barje. (personal correspondance)

Johannes RE (ed) (1989) Traditional ecological knowledge: a collection of essays. International World Conservation Union (IUCN), Gland. http://books.google.si/books?id=SZM0oBN5IF4C&pg=PP4&lpg=PP4&dq=iucn+johannes+1989+traditional+ecological+knowledge&source=bl&ots=paUaU5w7Hy&sig=_vmNLr2QAwzfXfjEjOEYgxCcGtI&hl=sl&sa=X&ei=U7pGVJuqCsHe7-Aba7ICwCQ&ved=0CCkQ6AEwAQ#v=onepage&q=iucn%20johannes%201989%20traditional%20ecological%20knowledge&f=false. Accessed 14 Oct 2014

Miličić V, Perpar A, Kramarič F, Udovč A (2011) Analiza stanja kmetijstva na območju Krajinskega parka Ljubljansko barje. Končno poročilo. Univerza v Ljubljani, Biotheniška Fakulteta, Ljubljana. http://www.ljubljanskobarje.si/uploads/datoteke/KPLB_KP_JUL2011_13MB.pdf. Accessed 10 July 2014

Oteros-Rozas E, Ontillera-Sánchez R, Sanosa P, Gómez-Baggethun E, Reyes-García V, González JA (2013) Traditional ecological knowledge among transhumant pastoralists in Mediterranean Spain. Ecol Soc 18(3):33. doi:10.5751/ES-05597-180333

Paris declaration on the Satoyama Initiative (2010). http://satoyama-initiative.org/wp/wp-content/uploads/2011/09/Paris_Declaration_EN_april2010_revised03_low.pdf. Accessed 20 Oct 2014

Rotherham ID (2013) Cultural landscapes and problems associated with the loss of tradition and custom: an introduction and overiew. In: Rotherham ID (ed) Cultural severance and the environment. Springer, London

Rules on the Registry of Types of Heritage and Protection Guidelines. (2010) Uradni list RS 102/2010, Ljubljana

Takeuchi K (2014) Revitalizing Socio-ecological production landscapes and seascapes: toward creating resilient and sustainable societies. The Satoyama initiative regional workshop in Florence, 27–29 May 2014 (Keynote speech). https://satoyama-initiative.org/wp/wp-content/uploads/2014/08/Keynote-1-Takeuch.pdf. Accessed 20 Oct 2014

The Convention on Biological Diversity (1992) United Nations, Rio de Janeiro

Van der Stege C, Vogl-lukasser B, Vogl CR (2012) The role of homegardens in strengthening social–ecological resilience: case studies from Cuba and Austria. In: Plieninger T, Bieling C (eds) Resilience and the cultural landscape: understanding and managing change in human-shaped environments. Cambridge University Press, Cambridge

Whyte ID (2002) Landscape and history since 1500. Reaktion Books

MIX
Papier aus verantwortungsvollen Quellen
Paper from responsible sources
FSC® C105338

If you have any concerns about our products,
you can contact us on
ProductSafety@springernature.com

In case Publisher is established outside the EU,
the EU authorized representative is:
Springer Nature Customer Service Center GmbH
Europaplatz 3, 69115 Heidelberg, Germany

Printed by Libri Plureos GmbH
in Hamburg, Germany